KU-298-039

Handbook of Engineering Management

Handbook of Engineering Management

Edited by

Dennis Lock

Heinemann Newnes

Heinemann Newnes
An imprint of Heinemann Professional Publishing Ltd
Halley Court, Jordan Hill, Oxford OX2 8EJ

OXFORD LONDON MELBOURNE AUCKLAND SINGAPORE
IBADAN NAIROBI GABORONE KINGSTON

First published 1989

British Library Cataloguing in Publication Data
Lock, Dennis, *1929–*
Handbook of engineering management.
1. Engineering. Projects. Management
I. Title
620′.0068

Photoset by Wilmaset, Birkenhead, Wirral
Printed and bound in Great Britain by
Hartnolls Ltd, Bodmin, Cornwall

ISBN 0 434 91170 4

Contents

Foreword

I am pleased to have been invited to contribute a foreword to this impressive handbook of engineering management.

Formal training as an engineer provides a good background for management, because engineering necessarily involves the exercise of judgement in the absence of complete data: a situation frequently encountered in management problems. But it is difficult for the professional engineer to acquire other skills, including that of management, while practising engineering—itself a very demanding occupation.

A number of attractive MBA courses exist but tend to concentrate on management as a subject in itself rather than, as this handbook does, as an extension of the engineer's own skills. There is the further important point that engineering management is difficult for non-engineers, because of the way in which the engineering discipline pervades many of the problems to be tackled.

Because of this, the practitioner slant of the book is to be highly commended, and will prove of immediate help to many engineer-managers. Subjects such as CAD/CAM, Project Control, Simultaneous Engineering, Intellectual Property Rights, Contracts, Costing, Quality Assurance and many others are dealt with in a pragmatic and sensible way which will be of immediate use to engineers faced with the transition to engineering management.

They will have cause to be grateful to the authors of the book.

Francis Tombs
Chairman, Rolls Royce plc

Preface

The management of engineering activities requires a unique combination of engineering, management and business abilities in order effectively to design, develop, manufacture and market the firm's products and processes. These requirements are not a new phenomenon. The ability to develop and use engineering innovations has been a hallmark of industrialization in all countries.

Despite its apparent long history, it is only in recent years that engineering management – or the management of technology, as it is sometimes known – has started to attain the status of a recognized discipline. The impediment to such recognition has been mainly a combination of factors existing in firms and universities.

In firms, engineering is often thought to be solely the domain of the research and engineering personnel of the organization. Yet successful use of engineering requires strategic decisions about engineering by personnel in other functional areas such as marketing, sales, finance and so on. The successful engineering manager bridges these two cultures and links engineering and management disciplines to address the planning, development and implementation of engineering capabilities and shape and attain the strategic and operational objectives of an organization.

A similar cultural situation too frequently exists in universities. As a result, there is often institutional and personal resistance to the idea of collaboration between business and engineering schools. Hence traditional MBA and management diplomas rarely address the educational needs of engineering development and manufacturing managers.

The impediments in firms and universities mean that education and research efforts in engineering management are fragmented and uncoordinated. The field receives little research funding and the number of faculty involved is relatively small.

Today's climate of rapid technological change and the current intensively competitive global environment demand renewed emphasis on effective engineering management and a re-evaluation of traditional attitudes and approaches. An important initiative in this field was taken early in 1982 by the Institution of Mechanical Engineers when it launched the Post-graduate Diploma in Engineering Management. This is described in Chapter 2.

As well as providing the means of examining the competence of engineers in management skills, the syllabus for the Diploma was designed to cover particular aspects of engineering management in a way that would appeal to post-graduate educational establishments and thereby encourage them to provide courses specifically created to meet the requirements of the constituent parts of the Diploma.

Progress was initially slow in terms of both the number of engineers registering for the Diploma and, more particularly, the number of relevant courses available. However, momentum is now building up on both scores. Of special note is that universities, polytechnics and distance learning organizations now offer courses on specific subjects of the Diploma and a number of universities are either already offering or planning courses for the whole Diploma.

A related initiative was launched in the USA in May 1986 under the auspices of the National Research Council in Washington, DC, entitled 'Management of Technology: the Hidden Competitive Advantage'. This has proved to be a major stimulus for changes in thinking and action on the management of technology in the USA. Several new initiatives in education have appeared, many of which are along similar lines to the Diploma.

By 1988 a further landmark had been reached in the UK: it was recognized that the Diploma had evolved to a stage where it was applicable to engineers of all disciplines. A decision in principle was made for it to be progressively co-sponsored by the major professional engineering institutions. It has therefore become a post-graduate qualification in engineering management accepted and supported by the whole engineering profession.

123–125 Harley Street
London W1N 1HE

Norman White
Chairman, IMechE Committee on DEM
Chairman, Inter Institutional Working
Party on Engineering Management

Editor's Preface

This handbook has been written principally for practising engineering managers and for those engineers and students who want to reach management positions. In this context the term 'engineering manager' includes the managers of engineering and drawing offices, chief engineers, managers of specialist engineering groups, engineering project managers, project engineers, senior engineers and engineering supervisors.

Other potential readers are those managers who, although not engineers themselves, have engineering departments within their span of control. They may be asked to approve capital expenditure for engineering equipment and facilities, authorize recruitment and appraise or appreciate the performance of their companies' engineering departments. The ability of non-technical managers to make such judgements will be enhanced if they have a greater insight into the needs and norms of sound engineering management practice.

In terms of scale, engineering management ranges from the demanding challenge of directing vast technical resources across large organizations or huge engineering projects down to the humdrum, day–to–day administration of the most routine documents. In functional terms, individual engineers have to learn to interact increasingly with other company departments and to become more aware of external influences on the business as they progress up the management ladder. A practical appreciation of these other functions and of how they interface with engineering organizations therefore becomes a necessary part of the engineering manager's fund of knowledge.

In order to deal effectively with this very broad area of management, authorship of our handbook has been entrusted to a team of contributors, each member of which is expert in his or her subject. We set out with the objective of achieving a book which is essentially practical, and which will serve as a comprehensive reference source on the engineer's bookshelf.

The preparation of this book has been a considerable project in itself, conducted with the invaluable cooperation and assistance of many individuals and organizations. Source references and acknowledgments have generally been given by the contributors in their chapters, but I must now list at least some of those who have given me material, information, suggestions, advice and encouragement. In addition to the handbook contributors themselves, these include:

The British Standards Institution

The Chartered Institute of Management Accountants
The Chartered Institute of Patent Agents
The Design Council
The Imtec Group plc
The Institute of Marketing
The Institute of Purchasing and Supply
Professor K. G. Lockyer
Edmund Young (whose initial suggestion led to the handbook).
During the formative stages of this work, valuable discussions were held with officers of the Institution of Mechanical Engineers to help to ensure that the scope and purpose of the handbook would meet the needs of today's engineering manager. The late Lionel Plumb was enthusiastic and prominent in these discussions, and the publishers and I wish to record our special debt to him.

Dennis Lock
1989

Contributors

Trevor C. Ashton is employed by GEC Measurements Limited, Stafford. He has worked on standardization for over twenty-five years, including preparation of the full range of company standards for design, manufacture, purchasing and quality assurance in the light electronics and electromechanical field. In his early career he served an apprenticeship with the former English Electric Company and spent ten years in the mechanical design of protective relays and household meters. Mr Ashton is an active member of the British Standards Society, and has chaired both its Midland Region and Electronics Group as well as serving on the management committee. He lectures for the British Standards Society and is a principal contributor to the booklet PD 3542 *Operation of a Company Standards Department* published by the British Standards Institution.

Iain C. Baillie is an international lawyer who has practised in the USA and Europe in intellectual property, licensing and allied fields for over thirty years. He has a BSc. degree from Glasgow University and a JD from Fordham University, New York. He is a member of the New York Bar, a registered US patent attorney, a UK chartered patent agent and European patent attorney. Mr Baillie is presently resident in London as European Partner of an international law firm, Ladas & Parry and is also Senior Partner of a firm of UK chartered patent agents. He has written and lectured extensively on all aspects of intellectual property, international licensing and the management of information and reputation.

Gordon Bell is senior partner in Gordon Bell & Partners, an organization teaching communications skills for industry and commerce. The company runs training courses in the UK and in many other parts of the world. Its students range from the heads of large multinational groups to people at all levels who need to know about presentation skills. Mr Bell's early training as a scientist and his wide experience in the entertainment business – writing and appearing in films, TV and the theatre – provided the bases from which he developed his teaching philosophy. His recent book is *The Secrets of Successful Speaking and Business Presentation* (Heinemann).

Elaine Bimpson has spent over twenty years in information and library work for a major British oil company, and she now runs a library for its mining subsidiary in which most aspects of the business are covered. She has a BA (Hons) degree in geography, and holds the Certificate of the Institute of Information Scientists. Her interest in the earth sciences is maintained through membership of the Geologists' Association and she is a Fellow of the Geological Society of London.

Darek Celinski, now operating at home and overseas as a freelance training consultant, has over thirty years' experience of training and development in industry. Previously he was with Coventry Management Training Centre, which he joined in 1972. In addition to running personnel and general management courses he specialized in the training of trainers (training managers, training officers, training advisors and similar people from commercial and industrial organizations). Before that, he was training manager at Herbert-Ingersoll. His earlier appointments in training were with the Aviation Division of Smiths Industries and with the South Wales Division of the British Steel Corporation. Mr Celinski is a Fellow of the Institute of Training and Development.

Peter M. Coughlin, after graduating in mining engineering from the Royal School of Mines, spent eight years working for a large mining company on the Zambian Copperbelt. In 1960 he joined the engineering division of Selection Trust in London, (which has since become Seltrust Engineering Limited, a consulting engineering organization within the British Petroleum Group). Mr Coughlin managed several major overseas projects and spent a number of years as manager of the project engineering department. He was appointed Administration Director in 1977, this post included responsibility for personnel management. He recently left Seltrust Engineering and now works for a national charity organization. Mr Coughlin is a Chartered Mining Engineer and a Fellow of the Institution of Mining and Technology.

D. W. Diamond is Director of Procurement at the Home Office. Previously he was Senior Procurement Manager for Crown Agents Services Limited. Before that he was Manager of Procurement and Contracts for many years with Fluor (Gt Britain) Limited, following a similar appointment with Stone and Webster Engineering Limited. Transferring from an engineering/project management role to procurement in the middle of his career, Don Diamond has worked in West and East Africa, the Middle East and other countries in Europe. He is a Fellow of the British Institute of Management, and a Member of both the Institute of Purchasing and Supply and the Institute of Engineering Designers.

W. Jack, after serving throughout the Second World War with the Lovat Scouts and the Queen's Own Cameron Highlanders, was engaged in engineering administration from 1947 until 1970. In 1962 he was appointed a Director of Hopkinsons Limited. He fulfilled a three year contract in charge of administration with Roan Consolidated Mines, Zambia, as General Manager of their Central Services Division in Ndola. Mr Jack became Managing Director of Coles Cranes Ltd in 1974 and was subsequently Group Managing Director of Acrow plc from 1976 until resigning from the group in 1979. Among his current directorships he is Chairman of Brisch, Birn & Partners Ltd, and a Director of Thomas Broadbent & Sons Ltd. Mr Jack is a Fellow of the Chartered Institute of Management Accountants and of the Chartered Institute of Secretaries and Administrators.

Dan Johnston, OBE, BSc. (Tech), FCSD, trained as a designer and technologist at the Institute of Science and Technology, Manchester University. He then worked as a designer, manager and director in the textile industry. After war service in the army, Mr Johnston returned to the textile industry and then joined the staff of The Design Council in preparation for the opening of The Design Centre. He worked as an industrial officer, specializing in a variety of industries before becoming Head of Industrial Design. Since retiring from The Design Council he has worked as a design consultant, concentrating on design management. For many years he has also been active in the affairs of the Chartered Society of Designers, of which he was President in 1973–4. Much of Dan Johnston's career has been devoted to design development and design promotion. His concern that design should be properly integrated into management and commerce led to authorship of the book *Design Protection* published by The Design Council. For his services to design, Mr Johnston was awarded the OBE in 1974.

Andrew Kell is a senior consultant with Ingersoll Engineers. He graduated from Sunderland Polytechnic with a degree in mechanical engineering and then worked as a product design engineer for three years, winning a Design Council prize for innovative solutions to mining transport problems. He has since been involved with the development and implementation of computer-aided design applications in design, engineering, analysis, drafting, printed circuit boards and electronics. In recent years he has undertaken a number of projects in the manufacturing and service industries involving CAD strategy, feasibility studies, system specification and selection, financial appraisal and implementation. Mr Kell is an Associate Member of the Institution of Mechanical Engineers.

Peter Lebus, is Managing Director of C. E. Planning, a leading firm of office planning, design and project management consultants. He has worked in this field for over twenty years and pioneered many of the techniques now in common use. He holds a BA degree and his memberships include the British Institute of Management and the Institute of Administrative Management. He is chairman of the Facilities Management Group within the Institute of Administration Management. One of his responsibilities is the biennial Office of the Year Award Scheme, which is designed to promote good working environments in their widest sense. Mr Lebus has lectured widely and contributed to several books on this subject.

Thomas J. Lindem is Vice President of Technology with the Ingersoll Milling Machine Company, Rockford, Illinois where he is responsible for the technical planning and application of new product technology and advanced metalcutting systems. Before his present position he was responsible for all engineering activities of Ingersoll's Special Machine Group. He has held a number of other management positions during his long service with Ingersoll, including Director of Engineering at Herbert-Ingersoll in England, where he organized and managed the engineering function during the late 1960s. Since graduating from the Illinois Institute of Technology in 1954, Mr Lindem has had many years of engineering experience. He has been responsible for a number of industry technical advancements and he is a recognized authority in the machine tool field in the United States and in Europe.

Dennis Lock qualified in applied physics and began his career with the General Electric Company as an electronics engineer. His subsequent management experience has been long and exceptionally wide, in industries ranging from subminiature electronics to mining engineering and heavy machine tool engineering. He is a Fellow of the Institute of Management Services, and his other memberships include the British Institute of Management and the Institution of Industrial Managers. Mr Lock has carried out lecturing and consultancy assignments in the UK and overseas, and he has written or edited many management books. He is now a full time writer, specializing in management subjects.

Roland Metcalfe is a senior lecturer in engineering design at Trent Polytechnic, Nottingham. He has been developing design in engineering courses for over twenty years and pioneered the academic use of computer aided design in undergraduate and post-graduate programmes. He is currently course tutor for the MSc. in computer-aided engineering and Chairman of the design group. Mr Metcalfe is Chairman of the Information Technology group of the Institution of Mechanical Engineers, which promotes learned society events and provides the Institution's response to relevant UK government reports and initiatives.

S. J. Morrison graduated in electrical engineering at Glasgow in 1940. After serving a post-graduate apprenticeship with the British Thomson-Houston Company he specialized in quality, his engineering career being spent with Associated Electrical Industries, Standard Telephones and Cables and, finally, with British Ropes. In 1963 he began teaching operational research at the University of Hull, where he was appointed Director and Head of Department in 1968. Mr Morrison retired in 1982 and became an honorary research associate at Hull, his particular interest being quality assurance. He has been a Senior Fellow of the University of Hull since 1986. He is a Fellow of the Royal Statistical Society, a Fellow of the Institute of Quality Assurance, a Senior Member of the American Society for Quality Control, Member of both the Institution of Mechanical Engineers and the Institution of Electrical Engineers, and a Fellow of the British Institute of Management.

Patrick O'Connor received his engineering training at the Royal Air Force Technical College and served for sixteen years in the RAF Engineer Branch, including tours on aircraft maintenance with the Reliability and Maintainability Branch in the Ministry of Defence (Air). He joined British Aerospace (Dynamics) in 1975 and was appointed reliability manager in 1980. Mr O'Connor has written many papers and articles, and he lectures on reliability. He is the author of *Practical Reliability Engineering*, Wiley, 1985 and he is U.K. editor of the journal *Quality and Reliability Engineering International*. In 1984 he won the Allen Chop award, presented by the American Society for Quality Control for his contribution to reliability technology.

Paul Plowman is Finance Director of Glynwed's Tubes and Fittings Division, which consists of their copper tube, steel tube and international plastics interests. He graduated with a BSc. degree in chemistry from Birmingham University, and has had experience of management accounting in a variety of large industrial groups including GKN and Massey Ferguson. Mr Plowman is a Fellow and Member of council of the Chartered Institute of Management Accountants and he is the author of its publication *Accounting in Mechanical Engineering*.

Michael Tayles, MBA, ACMA, BA, is a senior lecturer specializing in cost and management accounting at Huddersfield Polytechnic. He has industrial experience of accounting in the food processing, light engineering and service industries. His teaching experience covers management, professional accountancy and degree courses. Mr Tayles writes articles for accounting journals and he is an external examiner in accounting for non-accountants.

David Warby is chief executive of Jordan Engineering (Bristol) Limited. After reading mechanical engineering at King's College, London he trained with Metropolitan Vickers Electrical Company Limited, where his first appointment was on the installation and commissioning of turbo alternators and associated plant. He joined AEI-John Thompson at Berkeley nuclear power station at the start of the civil nuclear programme and became deputy resident engineer, subsequently occupying a similar post with the Nuclear Power Group for Oldbury power station. In 1966 he joined his present company and was responsible for manufacturing followed by design and construction before taking up his current duties. During his years as a director and shareholder he has helped the company to become an established leader in its field with a thirtyfold growth. Mr Warby is a Fellow of the Institution of Mechanical Engineers, and has a keen interest in management and corporate matters. This has led to the publication of a number of papers on project management, safety, contract administration and quality assurance.

Stephen H. Wearne, BSc (Eng), PhD, DIC, CEng, FBIM is a consultant and works with the Project Management Group at UMIST on problems of projects and contract management and leads Institution short courses and in-company training. He was originally a mechanical apprentice and 'sandwich' student. After training in water power engineering he worked on the design, economic planning and coordination of projects in Spain, Scotland and South America. In 1957 he joined turnkey contractors, on construction and design and then contract and project management of projects in the UK and Japan. In 1964 he moved into research and teaching on engineering management, first at the University of Manchester Institute of Science and Technology (UMIST), and from 1973 to 1984 as Professor of Technological Management at the University of Bradford. His research has included studies of design and project organizations, engineering and construction contracts, project control, plant commissioning teams, and the managerial tasks of engineering in their careers. Professor Wearne was first chairman of the UK Engineering Project Management Forum, initiated in 1985 by the National Economic Development Office and the Institutions of Civil, Mechanical, Electrical and Chemical Engineers.

Norman A. White, PhD, MSc, CEng, FIMechE, FBIM graduated in engineering at the University of Manchester Institute of Science and Technology and University of London, and also in industrial economics at the London School of Economics and from the Advanced Management Programme at Harvard Business School. He was awarded a Ph.D by the University of London in 1973. After initial training in the civil engineering and aircraft industries, Dr White's business career began with the Royal Dutch/Shell Group, where he reached senior executive status with board appointments in several Shell oil and mining companies, finally becoming Chief Executive of New Enterprises Division (metals, coal mining, energy and other projects on a global basis). In 1972 Dr White began his second career, which is a synergistic blend of counselling, academic appointments and a number of independent company directorships. He is Principal Executive of his own company, Norman White Associates, and is Chairman of several public and private companies in the UK. His distinguished professional appointments and memberships are many: too numerous to list here. In addition to articles for professional journals, Dr White has written two books, and he has achieved international recognition for his counselling and lecturing services. He is visiting professor at several universities in the UK and USA. Dr White is a Freeman of the City of London.

Dennis Whitmore is Director of Research and Development at the Harrow College of Higher Education and is also a management consultant. Previously he was Senior Production Manager with Mullard and a Senior Consultant with Philips. Dr Whitmore was educated at London, Brunel and Surrey Universities and is a Fellow of the British Institute of Management. His numerous books include *Work Study and Related Management Services*, Heinemann, 1976; *Work Measurement*, Heinemann, 1987; and *Management of Motivation and Remuneration*, Business Books, 1977.

Edmund Young is senior lecturer in engineering management at the South Australian Institute of Technology. A graduate of four disciplines, civil engineering, industrial management, economics and educational administration, he first worked for six years as a civil engineer on design and investigations with the Australian Government. His next six years were spent with a large Australian company, Concrete Industries (Monier), as field and project engineer, project manager and finally as executive engineer responsible for management training, working in several Australian states and in South-East Asia before taking up full-time academic teaching in management. In 1976 he was a visiting lecturer in engineering management at the University of Newcastle-upon-Tyne and visiting lecturer in production management at the University of Bradford in England. He is currently Chairman of the Engineering Management Branch, South Australian Division, Institution of Engineers, Australia and a member of that institution's National Committee on Engineering Management. Mr Young is a Fellow of both the Institution of Engineers, Australia and the Australian Institute of Management and he is a Member of the British Institute of Management and of the American Society for Engineering Management. He is on the editorial panel of *Engineering Management International* and in 1988 was a visiting lecturer to the Department of Engineering Management at the University of Missouri-Rolla, USA.

Illustrations

PART ONE

Introduction

1

Engineering management*

Norman A. White

Sound and innovative engineering is essential to the success of an engineering enterprise but this alone will not ensure business success. For the creation of wealth, engineering developments must be harnessed to provide improved or new products to timescales and costs commensurate with competitive market needs. Second, in those engineering enterprises expected to have the best growth prospects, i.e. those engaged in high technology or subject to rapid technological change, it has been demonstrated that the engineer has a unique contribution to make to the management of the broader functions (such as marketing, commercial or top management) as well as in the more traditional role of the management of technical functions. This unique contribution is not always self-evident and needs to be carefully identified in each instance.

Engineering is a technical and logical discipline; management is concerned with technical, non-technical and social factors and, particularly at the higher levels, largely concerned with the evaluation of risks and uncertainty in both social and technical fields. In order to make their unique and full contribution to the wealth creating activities of their enterprises, therefore, engineering managers must be capable of blending their abilities to apply engineering principles with additional management knowledge and skills, particularly the ability to 'navigate areas of ignorance' and unpredictability.

Definition of engineering management

As with so many other terms in the management area, there are various interpretations of the title 'engineering management'. To many people, engineers and non-engineers alike, engineering management is viewed as being synonymous with 'workshop management' or 'production manage-

*Some of the material in this and the following chapter has been reproduced from the *Proceedings of the Institution of Mechanical Engineers*, Part B, 1983, by permission of the Council of the Institution.

ment'. This narrow interpretation of 'engineering management' has caused many engineers to propose that the prefix 'engineering' should be dropped altogether. This is a dangerous proposition, because it suggests that there are no special skills required for the management of engineering enterprises compared with non-engineering organizations.

The definition of engineering management with which I personally find the greatest affinity is the following:

> The good engineering manager is distinguished from other good managers by the fact that he simultaneously uses an ability to apply engineering principles and a skill in organizing and directing resources, people and projects. He is thus uniquely qualified equally for two types of job:

1 The management of technical functions (such as research, design, production or operations) in almost any engineering enterprise.
2 The management of broader functions (such as marketing, commercial, project management or top management) in the high-technology enterprise or the engineering enterprise subject to rapid technological change.

The key word in this definition is 'uniquely': in other words, only the engineering manager can effectively manage both types of job. Even in engineering enterprises which are dealing with well established technology and not subject to rapid change, and where the engineering manager may not offer significant advantage over other professionals in the management of broader functions (such as marketing or top management), he will be uniquely qualified to manage the technical functions. And in this type of role, as a member of a 'management team', he will have both responsibility and opportunity to contribute to the wealth creating plans and performance of the enterprise.

This definition has been justified by another writer (Babcock, 1978) with the following observations:

1 Since high-technology enterprises* make a business out of doing things that have not been done before, extensive planning is essential to ensure things are done right the first time. And since the critical factors in new systems are often (certainly not always) technological, the engineer is especially qualified to recognize them and manage their resolution.
2 The organizational structure must ensure that the functions essential to the success of the enterprise have sufficient visibility and clout to be effective. In the non-technical enterprise, marketing and finance tend to have the highest profile. In the high-technology organization, research,

*In this chapter, reference to high-technology enterprises will apply equally to engineering enterprises which are subject to rapid technological change.

engineering design, and quality assurance are often accorded equal stature.

3 The key staff in a high-technology enterprise are largely technical specialists, and their selection, training and evaluation require technical insight.
4 Leading and motivating high-technology effort requires that the manager understands the nature of the technical specialist and of the engineering process.
5 Although the manager of a high-technology project or enterprise of any complexity cannot expect to understand more than one or two technical specialities as well as the specialists supporting him, he must have the broad technical background to absorb the critical factors of a new technology from his specialists quickly enough to use this understanding in decision making.
6 Effective control has many dimensions. Most of them have a significant technical component when applied to a high-technology project or enterprise, and require technical judgement if they are to be successful.

It follows from this discussion that the term 'engineering management' can be correctly applied to any managerial activity in which there is a requirement for a unique engineering contribution to be made to the business decision-making process. This contribution is required in many kinds of enterprises as well as the more obvious engineering companies. To mention just three examples: banks and financial institutions making loans to or investing in engineering projects; insurance companies providing insurance cover for engineering activities and public bodies considering investment in industry.

The management of technical activities

The management of technical activities essentially relates to organizing and controlling the work of a team of specialists. It embraces both the first level of supervision in a technical activity as well as the more senior levels of management.

Perhaps the most difficult thing for an engineering manager to learn is how to live continually with areas of uncertainty, if not of outright ignorance. Some years ago, Lord Ashby, who was then Master of Clare College at Cambridge University, made an observation which is especially applicable to engineers in management. He pointed out that it is the daily experience of a manager to make decisions in areas outside his expertise on what a scholar would consider to be insufficient evidence. He must do so because his time constraints allow him no choice. Time is a dimension which quite properly plays only a subordinate role in a scholar's life. The manager, however, cannot disengage himself from the calendar.

Lord Ashby went on to say that 'the secret of good management lies not in the manager's vast and exact knowledge, but in his skill at "navigating areas of ignorance" '. Clearly this implies that the engineering manager must learn to combine effectively his engineering skills and knowledge with the different skills and knowledge of others so as to make his unique contribution in day-to-day decision making.

The nature of this 'unique' contribution of the engineering manager will vary significantly from one situation to another. Identifying and analysing possibilities and then formulating the contribution represents a challenge to be grasped enthusiastically and pursued vigorously. In my experience it invariably includes three interrelated and interdependent areas of uncertainty:

1 The probability of success or failure of the engineering development, i.e. the technological risk factor.
2 The most likely level of costs to be incurred, i.e. the financial uncertainty.
3 A sense of timing, i.e. to know when to stop a development, when to accelerate it or when to switch emphasis to match changing market requirements.

In addition to identifying the nature of the unique contribution it is also essential for the engineering manager to present it to his colleagues in the management team in such a way that it will be understood and therefore accepted by them. If this whole process is to be carried out successfully it must be recognized and tackled as an intellectual challenge fully equal to the resolution of any engineering problem.

The management of broader functions

The definition of engineering management given earlier defines 'broader functions' as marketing, commercial or top management. It asserts that there is a requirement for a unique engineering contribution in those functions in respect of all engineering enterprises except possibly those dealing with well-established technology and not subject to rapid change. In the latter enterprises the engineering manager may not have a significant advantage over other professionals in management.

Too often when engineers move into senior management or board positions they either become so preoccupied with the new facets of their work that they cease to make a contribution as engineers, or they do not adequately recognize the changed nature of the engineering contribution required from them at this level of management. The proper contribution of senior engineering managers is paramount to the success of any engineering company particularly if that company is engaged in high technology or rapidly changing technology. The development of broader management skills in engineers is discussed in Chapter 2.

Engineering management in manufacturing industry

To develop general thoughts in more detail, the sector of mechanical engineering which is concerned with the manufacture and marketing of engineering products will now be considered. The discussion is, however, equally applicable to other fields of engineering.

The challenges facing engineering management

We are moving through a period of rapid technological change in many industries. A missed opportunity could mean the demise of the firm or even the industry. It is essential not to overlook the fact that it is the application and consequences of these changes which are wealth creating – not the technological developments themselves. For the creation of wealth, technological change must be harnessed to provide improved or new products within timescales and costs commensurate with competitive markets needs. This is the central theme of engineering management in manufacturing industry. It involves many tasks, as well as the manufacturing process, including product innovation and the management of the complete product realization process. A conscious recognition and understanding of these activities is vital if change is to be exploited expeditiously, effectively and efficiently for the economic wellbeing of those both directly and indirectly involved.

At the risk of oversimplifying a complex subject, I suggest there are at least six main challenges faced by engineering managers concerned with manufactured engineering products, as outlined in Figure 1.1.

1. Getting the right people
2. Aiming at the right targets
3. Continuing action throughout the product realization process
4. Creating the climate so as to encourage innovation
5. Developing awareness of the social and other consequences of new technology
6. Increasing competition of an international nature

Figure 1.1 *Challenges facing engineering management in manufacturing industry*

In too many British engineering manufacturing companies these challenges, especially product innovation, development and quality assurance have been accorded low priority. For instance, in British Leyland poor product policy was as much to blame as bad industrial relations, and it was a major factor in the bankruptcy of other great British companies in recent years. In these catastrophes, senior engineering managers must shoulder much of the responsibility for failing to make their influence felt. Rarely has failure been due to the shortcomings of engineers acting in their purely technical capacity if they have received timely and appropriate engineering leadership from senior management.

For success to be achieved there must be a positive corporate attitude to these challenges which results from leadership by senior management in the company. This leadership is unlikely to be timely or entirely appropriate in engineering enterprises unless it is stimulated and continuously monitored by senior engineering managers.

Of the six challenges listed, priority must be given to getting the right people at all levels. This consists primarily in recruiting people with the right basic personal qualities who also have the appropriate university and other types of training; systematic professional and managerial development, about which I will say more later, is also of great importance. Personnel considerations are more important in engineering than for most other functions – the premium for excellence is high. Identifying opportunities for dramatic improvements in the benefit-to-cost relationship requires sound abstract as well as pragmatic reasoning capabilities in engineers. We must design products that work effectively and efficiently, continue to work well for long periods and can be sold at a profit. The products must be safe, the engineering must be timely.

Product development

Aiming at the right targets is choosing products for development that will meet a customer's need, offer substantial benefits compared with the investment and risk involved and fit the company's goals and capabilities. This choice relates even more to being in or developing into a prosperous and growing line of business than to having outstanding research and development (R and D), design, capital or labour inputs. Undoubtedly the key to success is choosing the best products to develop at the right time; these will not necessarily be those offering the most interesting engineering challenge. The engineering manager must therefore be prepared to recommend a course of action which will satisfy a growth market opportunity in business profitability terms. This is becoming increasingly difficult in an era of rapid social and technological change, increasing international competition and reduced availability of capital. Picking a product which is too 'far out' can be costly; it may not work or may not sell. At the other end of the scale, picking too modest a product improvement may well lose an opportunity; this may be an even more serious error, although the damage is less visible and safer for reputations!

Aiming at the right targets for engineering products is probably our most difficult challenge.

For many years the Stanford Research Institute in the USA has been studying the reasons why certain companies have outstanding growth records and others merely drift along at the national rate of economic growth. Their major conclusions cover six points which are particularly applicable to manufacturing industry:

1 Such companies systematically seek out, find and reach for growth products and growth markets.
2 They characteristically have organized programmes to identify and promote new business opportunities.
3 They are consistently self-critical about the adequacy of their present operations and products and therefore equally consistently demonstrate superior competitive abilities in their present lines of business.
4 Their top management slots are staffed by uniquely courageous, adventurous, high-spirited executives who bubble with dissatisfaction and are driven by an energetic zeal to lead rather than to follow.
5 The companies almost invariably have established formal systems for discovering opportunities and off-setting extreme risks by 'planning for the unseeable' within the context of clearly defined and growth-inspired statements of 'company goals'.
6 The chief executives consciously and continuously by word and by deed, establish an organizational environment of ruthless self-examination and effervescent high adventure.

Top management must see that the entire company becomes saturated with the idea of useful innovation and the merits of self-criticism. It must develop and transmit some guiding philosophy about the vital necessity of the innovation process – for a really effective business enterprise.

Management of innovation

A key consideration in selecting the right products to gain the desired business results is recognizing the linkage between 'technology push' and 'market pull' in achieving commercial success (see Pearson, 1982). This is illustrated diagrammatically in Figure 1.2. The 'technology push market pull' dichotomy is not an 'either/or' condition: success requires the best combination, taking into account the likely risks and pay-offs. Moreover, the most important consideration in management terms is knowing which technology push–market pull combination the new product development represents, so that the risks and pay-offs can be properly evaluated. Unless this is done, unncecessary risks are being assumed.

Another consideration is that many companies tend to focus too much attention on the efficiency of their technological efforts and too little on their business effectiveness. Many of the former efforts may be cost saving and are therefore worthwhile in their own right but, in terms of innovation and long-term wealth creation, they are essentially short-term and cosmetic improve-

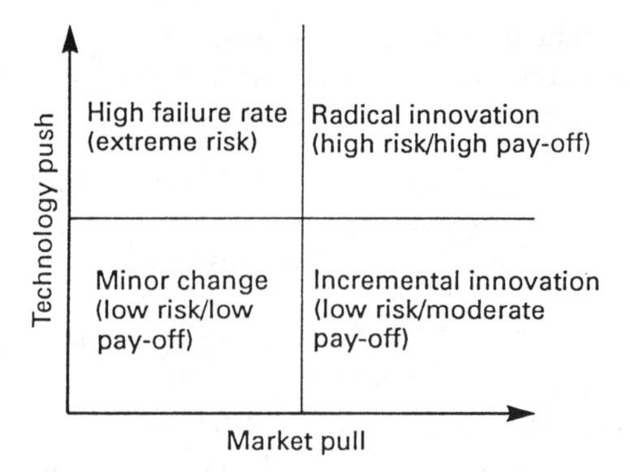

Figure 1.2 *Innovation matrix*
Adapted from (A. W. Pearson)

ments. They miss the critical factors in technological effectiveness which are concerned with:

- Which technology to pursue and when to pursue it?
- How to manage the transition from one technology to another?
- How to prepare the company for technological change?

These three elements determine a company's ability to manage technological transitions steadily in support of increasing business effectiveness and they are the basis of effective engineering management of innovation.

A further factor relates to the process of product design. One of the major consequences of technological change in recent times is the increase in product complexity. Engineers have always broken down complex product design tasks in some logical manner, but today's complexity requires the breakdown to be considered concisely and formally, particularly in terms of the needs of effective project management and coordination. This is essential when the design process involves the efforts of a large team, often involving many disciplines embracing other branches of engineering (for example, in microelectronics, the use of non-metallic materials and integrating the disparate skills of others such as industrial designers and non-technical specialists). High levels of analytical ability, imagination, realism and leadership are needed for this.

International competition

According to Dr Duncan Davies (1982) British industry overall is outstandingly innovative and still takes more quite new products through to profit than most of its international competitors. In primary innovation (winning large shares of new markets) Britain is unequalled. Furthermore, there is successful follow-up in the process industries (such as pharmaceuticals, agrochemicals, glass and textile products) with market share retained as the market itself grows: this is good secondary innovation. However, British firms do not follow up as successfully in the product, mechanical and electronic industries, where market share tends to be lost as the market grows (the transverse-engined hatchback car is an example). Substantial post-development gaps are left in these areas. Filling those gaps is becoming more imperative, because successful process industries are becoming increasingly dependent on mechanical and electronic engineering. Good secondary innovation (the Japanese skill) is needed. But this is expensive, and Dr Davies suggests the need to be more selective.

Superficially it appears to have been appreciated for many years that excellent research and development do not alone result in readily marketable products. Yet little constructively seems to have been done collectively by the engineering profession to encourage the pursuit among able engineers of those activities vital to the provision of products which can be successful against increasing competition on an international basis. To sell into a captive market might provide short-term financial success, but real engineering achievement must be measured in terms of the continuing ability to compete successfully in an intensively competitive international market. This lack of visible action by the profession might be based on a general ignorance of the problems involved or a hope that they will be resolved by others. At best this can only be a formula for missing the boat; at worst (and more likely) one for complete failure.

Product realization

Not only must engineering management pick the right products to work on and have the right people to engineer them, it must ensure that the entire system of the product realization process from earliest research and market analysis to final disposal is effectively managed and followed through. Engineers must look at performance reports not only to avoid current failures, but especially to foresee future improvements and opportunities for cost reductions. Engineers in manufacturing must look for process improvement. Quality control engineers must minimize the risk of failures. Service engineers must not only find ways to diagnose faults and execute repairs quickly and cheaply but also ensure that products are designed at the start with cheap and easy servicing in mind. Standards become the speciality of a few engineers and the money-saver of many. Preferred sizes and standard components offer immense cash benefits.

Engineering management of the entire product realization process includes all these factors. All of them require people to work closely and harmoniously with others and often outside the company in order best to serve the real needs of customers and, consequently, of the company.

So, managing the product realization process, linked as it is to complexity and rapid change, introduces problems of how to identify and achieve broad management objectives covering all the above linked activities. Management by detailed objectives alone is often inadequate because in a period of rapid change the interaction between objectives and methods of realization is often complex and not amenable to such a simple approach. Management by threat is hardly appropriate since the realization of first class, as against mediocre, innovations does not necessarily result from such an approach. Equally, gone are the days when the so-called management hierarchy can take over any task and perform it more adequately than those directly involved and thereby manage by direct example. The concept that strict financial control alone can lead to engineering success is equally absurd, even allowing that lax financial control can all too readily be shown to lead to ruin.

To sum up, this brief analysis of the management of the product realization process highlights very clearly the earlier statement that, more than ever, the necessity for engineering management to exercise a conscious leadership role aimed at setting and achieving the right broad objectives; this will include creating a corporate atmosphere or climate where those involved feel both committed to the objectives and motivated to success. For all this top class engineering managers are essential.

References

Babcock, D., 'Is the engineering manager different?', *Machine Design*, 9 March, 1978.

Davies, D., Innovation in UK manufacturing industry. Lecture to the Engineering Management Division of the Institution of Mechanical Engineers, 16 June, 1982.

Pearson, A. W., Project management in engineering R and D. Lecture to the Engineering Management Division of the Institution of Mechanical Engineers, 8 November, 1982.

2

Engineers in management

Norman A. White

When engineers and other technological specialists are assigned managerial responsibilities for the first time, they begin a career transition which is characterized by many new aspects (see Figure 2.1). First it requires new knowledge and skills, a shift in commitment from their professions to their organization, and the rekindling of their confidence – that they can be successful in a new activity. Because a person is a first class engineer it does not necessarily mean that he/she will become a first class engineering manager (even given both the will and the aptitude). At the least, the aspiring manager requires the opportunity to learn the new role and to acquire and practice new skills.

Part of the reason why engineers are often uncomfortable as managers – at least at first – is that most of their training emphasizes dealing with problems using objective measurement and established formulas. But the world of management is less predictable than the world of engineering.

Alternative career progression paths for engineers

Part of the dilemma faced by the profession in fully recognizing engineering management as an integral part of the engineering profession relates to the

- Requires new knowledge and skills
- Shift in commitment from their *profession* to their *organization*
- Rekindling of their confidence
- First class engineers do not *necessarily* make first class engineering managers
- The time and financial dimensions
- Navigating areas of ignorance
- Need for achievement

Figure 2.1 *Engineer's transition to management*

remuneration policies pursued by engineering organizations. These policies tend to reward management appointments at higher levels than technical engineering posts. Consequently there is a danger that, say, a competent engineering designer will seek to move to general management because he sees this as the only way that he can get the higher status and material rewards. In many cases this means that a competent engineering designer or other engineering specialist becomes a mediocre manager simply because his aptitude is more in engineering or design than management. Not all managers of technical activities have the wish or the aptitude to become managers of broader activities such as top management.

One solution to this dilemma is for engineering organizations to consider the concept of alternative career progression lines for engineers in industry. One career progression line could be for engineers carrying out technical activities and could be distinct from other channels of advancement on the more general management side of the organization. Such specialist progression ladders are designed to recognize technical engineering abilities and accomplishments and to reward those professional employees who concentrate on technological work and prefer not to have management responsibilities. This approach is often adopted in large-scale research establishments (see White, 1964).

An example of a dual progression ladder for a research establishment is shown in Figure 2.2. In this example, the two separate sets of rungs stem from a broad common base. The route taken depends on the individual's ability and inclinations as well as on opportunity. Lateral moves should also be possible, and corresponding rungs should have the same salary-promotion values and similar status symbols. A company which wishes to attract and retain first class scientists and professional engineers should not under-estimate the importance of maintaining parity in applying these status symbols.

Managerial	*Technical*
Director	Engineering or scientific adviser (or equivalent)
Assistant director or general manager	Senior research or engineering associate
Division manager	
Group Leader or Project Engineer or Project Chemist	
Engineer or chemist or physicist	

Figure 2.2 *Alternative career progression lines.*

Although there are potential advantages for an industrial company in offering a dual system of progression for engineers, there are also several possible sources of difficulty. Assignment to the specialist ladder may isolate the engineer from the main body of the organization and there may be a shortage of rungs on the specialist ladder. Also it may be difficult to arrange for the ceiling for advancement to be close enough to the top of the management ladder.

In short, although this approach has many apparent advantages, the adoption of such a scheme must be based on a careful evaluation of the organizational needs and be subsequently nurtured and sustained, if it is to realize the many potential advantages and produce the expected results.

Preparing engineers for management

Ideally, the engineer earmarked for senior management should have had the opportunity to acquire the new knowledge and necessary experience before the senior appointment is made. If this cannot be arranged, it is important that these are acquired as quickly as possible thereafter. The essential knowledge for all members of senior management and the board can be divided into four areas:

1 Knowledge of the business.
2 Understanding of the company's financial structure.
3 Knowledge of the company's senior managers.
4 Awareness of the economic, political and social background of those countries in which the company operates.

By a 'knowledge of the business' it is not meant that every member of the team should have a detailed knowledge of, for example, the product design or the manufacturing technology. The engineer will have this knowledge but the finance manager or personnel manager is unlikely to have it or need it. What is meant is that all members of the management team or board need to have a knowledge of what makes for success or failure in their particular business, which markets it serves and their relative importance to the success of the company, which markets are showing growth and which are standing still, or the state of the company's market share and the nature of the competition.

By 'an understanding of the company's financial structure' it is not a detailed knowledge of the company's accounts that is meant but an understanding of the strength and availability of the company's financial resources, the ability to interpret management accounts, an understanding of what levels of investment risk are acceptable, how much exposure the company can afford with regard to currency exchange and the cost of money and matters of a similar nature.

'Knowledge of the company's senior managers' takes into account the business capabilities of the people in the first two or three levels of the organization structure below the board of directors. Such knowledge is an essential prerequisite for an effective contribution to management succession planning, as well as to an understanding of the strengths and weaknesses of the company's principal asset – its people.

An awareness of the economic, political and social backgrounds of the countries or regions of countries in which the company invests or does business is a *sine qua non* for a responsible assessment of the risks involved.

Although the grafting on of knowledge and experience in these four essential areas will take time, it must be undertaken urgently in order that the unique contribution of the senior engineering manager can be identified and realized as soon as possible. What is this unique contribution? And how does it differ from that which has previously been outlined in respect of the management of technical activities? All aspects of the latter apply also at the more senior levels but the important difference is that senior engineering managers have the responsibility of anticipating the need for such contributions and ensuring that they are made and the related business decisions are taken and followed through. In essence, at the senior levels of engineering management the contribution required relates to engineering leadership of the business, but this is unlikely to be forthcoming until adequate orientation in the four key areas outlined has been achieved.

Management training

What is the most appropriate timing for management training? In other words, how much non-technical content should there be in the basic engineering degree or diploma course? This is a big topic, with many dimensions. For example:

1 If management subjects are included in an engineering degree or diploma course, it invariably means that some technical content has to be reduced because no one wants to see the course length extended in time
2 Is it reasonable to expect undergraduates to comprehend the significance of some aspects of management before they have had an exposure to engineering operations of one kind or another.

This subject is being actively debated in academic circles and in the various institutions. This author's view is that present courses for degrees and diplomas should satisfy the following requirements in combination:

- Sound engineering.
- Creativity.
- An appreciation of engineering economics.
- The ability to communicate.

Such combined skills are rarely, if ever, taught academically and they are equally rarely developed during the formal industrial training of engineers.

Having discussed briefly undergraduate work on engineering management, consider post-graduate studies. How should they interrelate? It seems that the preparation of engineers for a career leading to management can be divided into three parts:

1 Technical mastery of knowledge, practice and design in a specific engineering field.
2 Management of engineering resources (financial, economic and marketing communications; engineer in society).
3 Management of engineering operations (organization and personnel; corporate engineering leadership).

Part 1 and selected sections of Part 2 are currently part of an undergraduate course. The remainder is currently part of a post-graduate programme.

Yet another view is that education needs grafting on to work experience (i.e., the 'management' knowledge required by an engineer who becomes a manager, should be regarded as a continuing education requirement, something acquired by those who need it after they have become corporate members of professional institutions, and something which will vary in content between individuals depending on their field of activity and the industry in which they are engaged).

Part of the reason for these opposing views arises because insufficient systematic research has been carried out into engineers' managerial tasks. Figure 2.3 gives some results of a survey of engineers in the UK (Wearne and Faulkner, 1979). These engineers' answers show that most of them felt inadequately prepared for their present and prospective jobs, but that what is needed varies during different careers.

It is doubtful if all would agree that we are yet able to establish a clear basis for distinguishing between what preparation could be valuable in undergraduate courses and what at later stages. However, this work and similar reports are consistent with conclusions given in the Finniston Report (1979 that priority should be given to preparing engineering graduates for interdisciplinary project roles in manufacturing and other industries.

It might be questioned at this point whether the success of engineers as managers and the quality of management education and training they receive are related. If there is a shortage of engineers in the board room is it because too few engineers have the personal basic qualities, drive, ambition and extroversion required by such a career, or is it because they lack management training? Whatever the general broad basic quality level of our engineers is, a key factor certainly is the present lack of a recognized body of knowledge devoted specifically to engineering management; this lack inhibits even top

Skills and expertise	*Required in present job*		*Instruction received in basic training*	*Instruction received in subsequent training*	*Additional requirement in future career*
	Frequently	*Sometimes*			
Costing estimating	33	47	26	32	10
Evaluate projects	45	33	16	28	18
Analyse economic risks	19	25	5	20	18
Plan and control budgets	45	21	10	28	21
Company accounting	7	14	5	17	14
Corporate (business) planning	13	16	2	18	14
Plan and schedule project	39	34	17	28	12
Plan new product	11	14	5	11	4
Patenting	2	11	1	4	1
Plan research	7	16	4	6	3
Plan design and development	25	22	20	17	8
Plan construction	20	23	18	14	5
Plan production	13	10	11	11	4
Plan maintenance	14	20	10	15	6
Plan stocks and materials distribution	10	11	4	11	4
Plan services	14	17	6	8	4
Marketing of products	6	7	1	10	7
Marketing of consultancy, technical services	8	12	–	6	6
Use of company law	3	18	5	12	12
Use of health and safety law	23	44	6	35	18
Use of employment law	8	25	3	17	15
Use of consumer safety law	3	11	–	6	6
Draft contracts	17	27	12	18	12
Negotiate contracts (with client/customer)	18	23	4	13	14
Negotiate with supplier/contractor	28	31	5	15	10
Negotiate with employees	17	24	4	17	15

Skills and expertise	*Required in present job*		*Instruction received in basic training*	*Instruction received in subsequent training*	*Additional requirement in future career*
	Frequently	*Sometimes*			
Negotiate with trades union representatives	9	14	2	13	16
Negotiate with public authorities (not as customers)	13	20	2	7	6
Negotiate with senior management	34	24	1	13	12
Operational research	3	11	6	9	4
Work study (methods study)	5	16	12	14	4
Statistics	6	31	20	16	6
Data processing (including computers)	12	28	12	22	12
Systems analysis	4	18	5	9	6
Organizations and methods	4	17	7	13	6
Direct supervision of others	67	16	9	30	16
Motivation of others	66	15	8	30	15
Plan manpower requirements	39	26	6	16	16
Select personnel	29	36	6	23	18
Employee training: manual workers	6	13	2	8	6
Employee training: supervisors	10	18	3	10	7
Employee training: management	11	17	2	14	8
Others	7	2	1	4	1

Figure 2.3 *Engineers' managerial tasks*
Adapted from Wearne and Faulkner

class engineers from bringing their basic ability fully to bear on the important broad issues in their organizations.

The Diploma in Engineering Management

In 1982, in the UK, the Institution of Mechanical Engineers (IMechE) decided that many chartered engineers would benefit from a management qualification at post-graduate level which is specific to the needs of the professional engineer engaged in engineering management. It was designed to cater both for the individually motivated member and for those who are supported by their organizations and entry to the examination for the diploma is open to all chartered engineers and not restricted to corporate members of the IMechE.

The diploma, available on a modular basis, is taken in three parts:

1 Finance and human relations.
2 Use of Quantitative Methods and Information Systems plus one paper from a list of options.
3 Business policy.

The part 2 options have developed over time to cover the special requirements of engineers of other disciplines. They now include the following subjects.

- Managing Innovation
- Managing Engineering Operations
- Commercial Aspects of Engineering Operations
- Project Management
- Management and Public Administration.
- Civil Engineering Law andContract Procedures.

The diploma will be awarded on the successful completion of Part 1, Part 3 and the Part 2 core subject and any one of the options.

The remaining options may be taken as subsequent endorsements. Other options in Part 2 may be added as the need arises.

The assessment and examination procedures are designed to ensure that candidates can demonstrate their understanding of management processes and practice and also appreciate the skills needed for success in engineering management. Parts 1 and 2 will in the main be assessed by formal examination. Part 3 will be assessed either by set case study or by report on a management project, the choice being with the candidate. The management project will give the candidate an opportunity to investigate a management problem in depth and in so doing broaden his experience and also enable him to demonstrate his understanding of management processes and practices. It will be arranged jointly between the candidate's organization and the Institution.

Selection of engineers for management

The engineer's transition to manager will be most successful if the engineer has a basic need and ability for achievement. Recognition of this is a prime requirement in the selection of engineers for managerial responsibilities. Such engineering graduates with a broad-based education, which aids them in assimilating new techniques and responsibilities, will be more successful in effecting the transition than those who are trained in narrow specializations. Flexibility and adaptability, as well as social and management skills, are essential in a world of rapid change.

Engineering leadership

Management of engineering based activities can never be simple or easy, combining as it does ordinary business problems with the special problems and opportunities, presented by dynamic engineering technology. However, it will not be enough, if the country and industry in general are to be really successful in this ever more rapidly changing world, for engineering managers to be seen and valued only as managers of specialized technical functions.

One looks forward to the day when sufficient engineering managers have made their mark on the overall business plans and results of the enterprises for which they work, and when engineering leadership is a recognized and highly valued component of business leadership in all technologically based enterprises. And this, together with continuing improvement in performance by engineers, at all levels and in whatever capacity, towards meeting the needs of society (rather than just the desires of the engineering profession) will make irrelevant any further acrimonious discussion about the status of engineers.

References

Engineering our Future, report of the Committee of Inquiry into the Engineering Profession, Chairman, Sir Monty Finniston, Cmnd. 7794, 1979.

Wearne, S. H. and Faulkner, A. C., *Professional engineers' needs for managerial skills and expertise*. Report TMR 15A, School of Technological Management, University of Bradford, November, 1979.

White, N. A., *The human aspects of industrial research and development work*. Lecture to British Institute of Management Symposium, 30 September, 1964.

PART TWO

People and Organization

3

Organization

Edmund J. Young

There is an art of organizing that requires knowledge of aims, processes, men and conditions, as well as of the principles of organization.
Russell Robb, American electrical engineer, manager and pioneer organization theorist, 1909

All engineering managers need to be concerned with organization. The task of organizing is an art in practice. The study of organization can be approached as a science, which is multidisciplinary because technical, human, social, cultural, legal, economic and other aspects have to be considered. There are good organizers and bad organizers. All leaders should be good organizers. Engineers who aspire to become competent engineering managers need to be good organizers. A basic knowledge and understanding of organization, organization structures and the principles of good organization is essential for efficient engineering management.

The development of organization study

The Chinese philosopher Confucius (551–479 BC) once said 'The beginning of learning starts with the precise meaning of words'. Two words in the English language, *management* and *organization*, have been used so often, and with such diverse meaning, that much confusion, controversy and abstruse writings have been caused.

Management has become such a fancy, fad word that we are inundated with 'management experts' of all types. Civil engineers have conferences on coastal and river management. Medical practitioners are concerned with disease management. Farmers are involved with farm management. Garbage collectors are experts in waste management and health studios offer courses in health and diet management. Most of the meanings of 'management' refer to one or more of several words: planning, direction, regulation and control.

It was the French mining engineer-cum-top manager Henri Fayol (1841–

1925) who first defined management as a *process* consisting of planning, organizing, coordinating, directing and controlling, which subsequently laid the foundation of modern management theory. Fayol (1949) evolved his ideas on management mainly on the basis of his own personal observations and experiences as an engineer who became chief executive of a large French mining and metallurgical combine early in this century. He became so dissatisfied and concerned with the education and training of young French civil engineers that he was prompted to write a paper on his ideas on concepts and principles of administration to help the young engineers – a paper first delivered at the Mining and Metallurgical Congress in Paris in 1900. This paper became the basis of his classic but uncompleted book, *Administration Industrielle et Générale*, published in 1916. Posterity may well recognize Fayol as the 'founding father' of engineering management as well as of 'modern management theory'. Fayol viewed the management process as pervasive and operative through all types of organizations, and his view is often referred to as 'the universal approach' to management and organization.

Management and organization are not new. They have existed since the dawn of civilization. Primitive tribes, armies, ancient civilizations, nation states and empires of the past have had management and organization in order to survive, defeat their enemies, grow and prosper throughout human history.

Moses, the Biblical leader of the Israelites after their escape from slavery in Egypt, was advised by his father-in-law Jethro to set up an organization among his people by choosing 'able men out of all Israel, and made them heads over the people, rulers of thousands, rulers of hundreds, rulers of fifties, and rulers of tens'. (*Exodus*, c. 18, v. 25). So to quote the proverbial phrase, verily I say unto you, Moses was thus able effectively to 'manage' his tribe by using the process of delegation in his organization. He dealt only with the important, critical or *exceptional* issues, leaving his subordinate leaders to attend to 'every small matter'. This process is today often called the *principle of exception.*

Another Chinese philosopher Hsun Tzu (312–230 BC), now famous for his classic *The Art of War*, observed that 'men must organize themselves into a society. But if they form a society without hierarchical divisions, then there will be quarrelling . . . the ruler is one who is good at organizing men in society,' and advised 'plan before any undertaking, and carry it out with circumspection. (Hsun Tzu, 1963).

The earliest engineers had to be good organizers in their design and construction of temples, fortifications, walls, castles, bridges and water supply schemes. The Roman engineers were efficient, capable organizers and managers of projects for roads, bridges and viaducts. Engineering first evolved in two main streams: military and civil. The Roman military engineers were concerned with walls and forts, while their civil engineers constructed roads and town water supplies. Throughout history, the work of military and civil engineers has been interchangeable. An example of this is seen in Figure 3.1, which is an organization chart for a Chinese military bridge construction team under Colonel Yang, in the Ming Dynasty. Subordinate officers are shown with responsibilities for specialized operations, and there are dotted arrows showing the liaison lines of contact where particular advice and functional authority may be operative. Such organization charts are, therefore, not new. The ancient Sumerians, Egyptians, Persians, and Mayans as well as the Chinese all used them to indicate key personnel and their authorities and responsibilities in military and civil organizations.

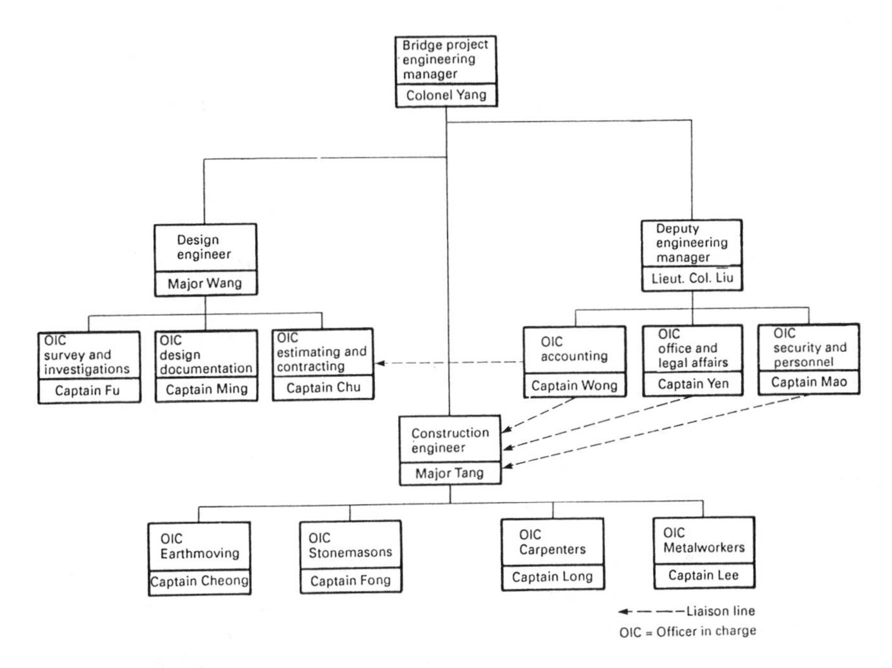

Figure 3.1 *Organization chart of officers in a Chinese military engineering battalion for bridge construction (Ming Dynasty)*
(Courtesy of the Chinese Association of South Australia)

Large-scale organizations, such as the British and Dutch East India Companies, the Great Wall and Grand Canal of China, the Suez and Panama Canals, and projects for railways, bridges and canals in Britain and North America in the nineteenth century used organization charts to show responsibilities and lines of authority. Among the earliest industrial organization charts was that by the English electrical engineer and manager J. Slater Lewis, who in 1896 produced a staff organization diagram in his classic book *The Commercial Organization of Factories* (Figure 3.2).

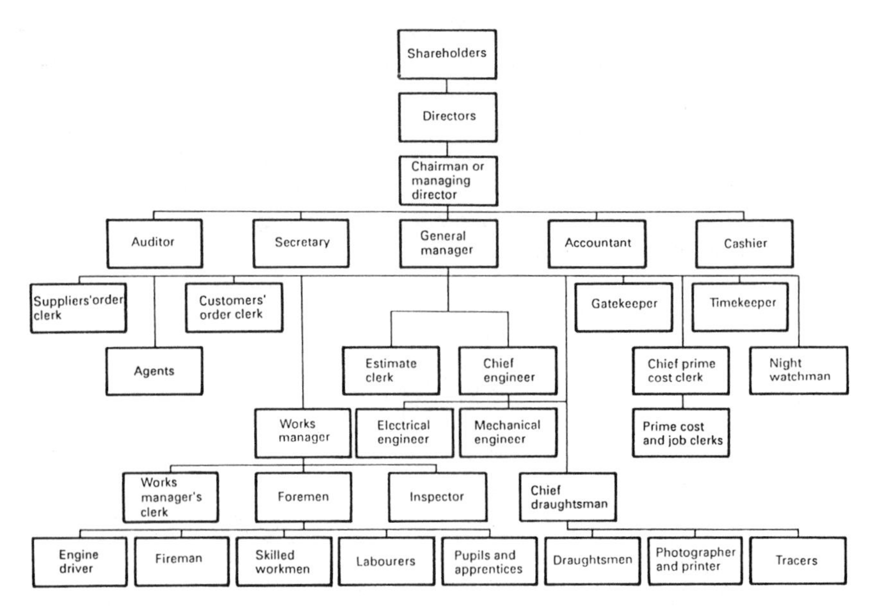

Figure 3.2 *One of the earliest printed organization charts, taken from J. Slater Lewis's book* The Commercial Organization of Factories *(1896)*
(Courtesy of E. and F. Spon, London and UMIST Library)

However, it was the American electrical engineer and manager Russell Robb who stands out as the pioneer organization theorist. He was a contemporary of the German academic sociologist Max Weber, who made some observations of the characteristics of bureaucratic organizations and to whom sociologists, behaviouralists and writers on organization mostly refer, generally ignoring (or being ignorant of) Robb's remarkable writings. Robb, who lived from 1864 to 1927, saw the problem of organization as similar to technical engineering problems. He noted that it required '. . . analysis, involves balancing and setting values, a constant estimation of the relation of parts to the whole' (*The Limits of Organization*, 1909). He also stressed the importance of human relations in organizations by observing that 'the organization that happened to have on its staff a good student in human nature, an expert on the stability of fads and fancies, was the one that had the really effective organization' and 'organization is but a means to an end; it provides a method' (Robb, 1909).

Many of today's approaches to organization study tend to be behavioural orientated which, although interesting, can be speculative or short-sighted, tending to overlook other aspects of organization.

But engineers have historically been involved with organizations and generally the most prominent and successful engineers have been good organizers of their enterprises. All engineering work requires efficient organization and competent management. This is true for design, development, installation, construction, manufacture, production, operation and maintenance of products, structures, systems, machines or equipment. It is also valid whether the engineered product is a bridge, a motor vehicle, a computer or a mousetrap.

Two American observers of the engineering profession noted that engineers had traditionally been associated with organization, and that organization and engineering were inseparable. Schott (1978) maintained that there was a symbiotic relationship between the engineering profession and the organizational environment in which engineers move, which explains the tendency for engineers to be continually moving into management roles. Layton (1969) called the engineer the 'original organization man'. All engineers need to be concerned with organization because engineering work entails organization of both human and material resources. It is preferable in engineering and technically orientated organizations that engineers are the managers and organizers, rather than having these roles undertaken by people without an engineering or technical background.

Defining the word organization

The term 'organization' generally has four main meanings, which are different, but related. (Altogether there are more than a dozen meanings, some with more subtle differences than others.) Staying with the four principal meanings, organization may refer to:

1. An amalgam or aggregation of human and material resources combined as a distinct, separate and integrated entity or system that may loosely be termed an enterprise. This might be a company, a government department, a factory, a construction project or even a ship with its crew. Such definition would include the buildings, land, machines, equipment, materials and products as well as the personnel working in the enterprise.
2. A distinct social group with formally defined relationships between its members. This could include a family, an engineering design team, a football club, a group of boy scouts or a women's association. Moses formed an 'organization' from members of his tribe.
3. A structure or system of authority and responsibility relationships among members of a cohesive social group or of an enterprise that is a distinct, separate and integrated entity set up to achieve the group's objectives.
4. The logical and systematic arrangement of work and activities, division of duties and tasks between individuals and groups, and the necessary allocation of authority and responsibilities in combination with the processes and material resources to achieve common objectives.

Organizations are usually set up to achieve certain objectives more effectively and more economically than individuals acting by themselves. For example, one person working alone might take three years to build a bridge across a river. A team of 100, or even ten skilled men working under a managerially competent engineer may build a better quality bridge, more economically and in a matter of weeks. Organization needs teamwork, a spirit of cooperation, coordination between individuals and integration of effort. This implies the need for leadership and management of the organization.

The four definitions of organization are interrelated. Definition 1 may be viewed as the total organizational enterprise, which must involve the relationships defined in 2, and which requires both 3 and 4.

In any study or analysis, it is important to know which meaning is used by various writers on the subject of organization, and by practitioners (such as managers, organization planners, and others involved in organizing).

The internationally famous British management authority, the late Lyndall F. Urwick (1966) noted this problem of semantics on 'organization' which still plagues the subject in spite of the hundreds (if not thousands) of papers and books written on the subject. Urwick (1937) himself once saw organization as a technical as well as a human problem. This is understandable in view of his early military background: in warfare organization becomes predominantly a technical problem of weapons, tactics, equipment, supplies, logistics and strategies besides the quality, training, morale and number of men. Woodward (1958) drew attention to the interrelationship between management, organization and technology from her studies of a number of manufacturing firms in South-East England, and significantly noted that the most successful firms were 'organization conscious'.

Some practical constraints

In practice, depending on the purpose and form of organizations, there may be a number of aspects such as technical, human and social, cultural, economic, legal, ethical and political to be considered. These are contingent upon what Robb once called 'purpose and conditions', and what we would now term the 'objectives and environment', some factors becoming more important than others depending on the circumstances.

An engineer planning to set up a business in construction, manufacturing or as a consultant would be constrained by the legal forms of business organization generally set down in statute law of the state in which he/she operates. The business enterprise may need to be a sole proprietor, a partnership, or a proprietary or public company. A Swedish or an Australian engineer, for example, accustomed to involving subordinate technical personnel in techniques of worker involvement and participation in managerial decision making may be frustrated in attemping to use similar methods in some Moslem, Asian or African construction projects, which traditionally have been used to autocratic, authoritarian and paternalistic management. Likewise, the use of modern construction equipment, requiring importation of highly skilled technical personnel into an underdeveloped country with low-level infrastructure, may be economically and politically unsuitable (because of the comparatively low labour costs of employing indigenous local personnel and the high unemployment conditions of the country in which the engineering project is to be carried out).

Thus, depending on circumstances, certain aspects of organization may outweigh others. Although the human or social aspects are always important and must always be considered, these may be overshadowed by other factors in some situations. It may be that technical factors such as automation, robotics, operation methods and technology, as well as legal, political and ethical factors will predominate over the human, social or behavioural aspects of the organization.

Analysis of organizations

Any basic analysis of organization requires precise definition of basic terms. Here, analysis is assisted by the use of the traditional organization chart.

Authority refers to the right or prerogative of requiring action over others or, in military usage, 'the right to command'. This is *formal* or *organizational authority*, so called because it is formally or officially conferred by the organization. There is also *informal authority*, which is not officially stated or recognized but is nevertheless real: this can be quite influential, sometimes even outweighing the formal authority. Informal authority could be technical or personal authority.

Technical authority derives from superior expertise, technical knowledge, skill and experience. *Personal authority* arises from certain attributes of personality, age, or from that intangible characteristic, charisma. It may be bearing, determination, willpower, appearance or height which gives one person more informal authority than another: it is the authority due mainly to personal influence.

Ideally, an engineer in charge should combine formal authority with the informal kind, technical and personal.

Responsibility refers to accountability, or to having an obligation to undertake a task, and to being held accountable for it and its successful completion.

Delegation refers to the release, allocation and assignment of both authority and responsibility to subordinate personnel within the organization. It is an essential process for effective organization. But although *authority* can be delegated to subordinates, the ultimate *responsibility* still resides with those at the top of the organization.

Specialization is important in organizations, for organization is necessary with specialization. Personnel need to specialize to carry out certain tasks more competently and efficiently. This was well recognized by the early Sumerians, Babylonians, Egyptians and Chinese. Our example of the Chinese Imperial Army bridge project organization in Figure 3.1 shows the various specializations and subdivisions of responsibilities. Thus design has the subdivisions of survey and investigation, documentation, and estimating and contracting. The construction and deputy engineering manager's roles are similarly divided among other specialists. Specialization is also reflected in Slater Lewis's staff organization diagram (Figure 3.2). A more recent example is shown in the chart of the Australian Grand Prix organization, Figure 3.3.

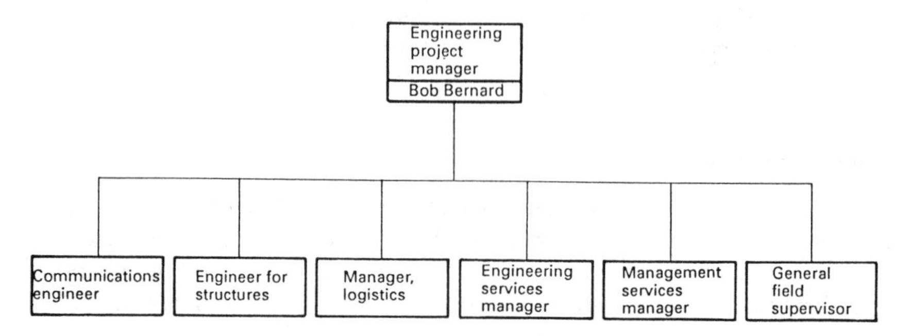

Figure 3.3 *Organization chart for the Australian Grand Prix, Adelaide, 1986* (Courtesy of Bob Bernard)

With increased specialization and levels of responsibility there is greater need for effective communication and efficient coordination. Without adequate coordination, both the purposes and working of the organization may not be achieved. Coordination, as Mooney (1949) stressed, is the essence of organization, and efficient coordination is a continuous internal objective of organization but is dependent on competent leadership, effective communications and the spirit of cooperation among the personnel involved. A dramatic illustration of the need for effective coordination was seen after the British ship *Endeavour* ran aground on part of the Australian Great Barrier Reef in 1770. Several long boats, with lines attached, were attempting to refloat the ship. But it was only after Captain Cook issued orders, at high tide peak, telling the boats' officers that they should all pull together at the same time that the ship came off the reef.

Urwick defined organization in terms of a formal communications network (Urwick, 1966) but this view tends to overstress the role of communications. Although this aspect is vitally important, it is only one facet of organization.

Organization charts

Organization charts (sometimes called organigrams) are most useful devices for organization analysis. These charts are simply diagrammatic representations of the people in organizations, showing their subdivision into various specializations, departments or sections, and depicting the formal lines of authority, responsibility and communications. A chart may also show liaison links, lines of staff advice and consultation, and lines of influence and status.

Please note that organizational status, which relates directly to the level of responsibility within the organizational hierarchy, is different from societal status. The status conferred on a person by society depends on such factors as occupation, personal wealth, education, class, heritage and various prejudices. Thus it is possible for a person with high organizational status to be regarded as of relatively low societal status (as a result of racial prejudice, for example). Conversely, if a medical practitioner were to be engaged as part of a construction project team, the doctor's role might not be regarded as one of high organizational status owing to the small number of people working directly for him (perhaps a small medical group): yet in Western society at least the doctor's status would be seen as being professional and highly regarded.

Most organization charts are drawn vertically, which means that the higher levels of authority and responsibility are shown at the top of the hierarchy. Hierarchical organizations are inevitable because in all cases there are leaders and followers (usually with far more followers than leaders). Matrix organizations (described later) endeavour to do away with some of the attributes of hierarchies but, in practice, these are often only temporary or auxiliary organizations attached to a larger hierarchy.

Sometimes it is convenient to draw organization charts laterally, to avoid the connotation of organizational status for some personnel who may be sensitive to this issue. In other instances, lateral or horizontal presentation can conveniently be combined with the traditional vertical format, simply to facilitate drawing and presentation. The organization chart shown in Figure 3.4 is an example that includes both horizontal and vertical presentation.

A less usual form of presentation is the circular or concentric organization chart. Although more difficult to draw than the more conventional chart, the lines of authority and management strata can be shown clearly and in a relatively small space. An example, which was used by BP Australia Ltd to depict their head office organization in 1967, is shown in Figure 3.5.

An organization chart presents a static picture, true only for a given point in time (in the same way as a company balance sheet, or like a cross-sectional view of a water-filled pipeline). No movement of personnel can be shown. The dynamic aspects of the organization are not apparent from a static chart,

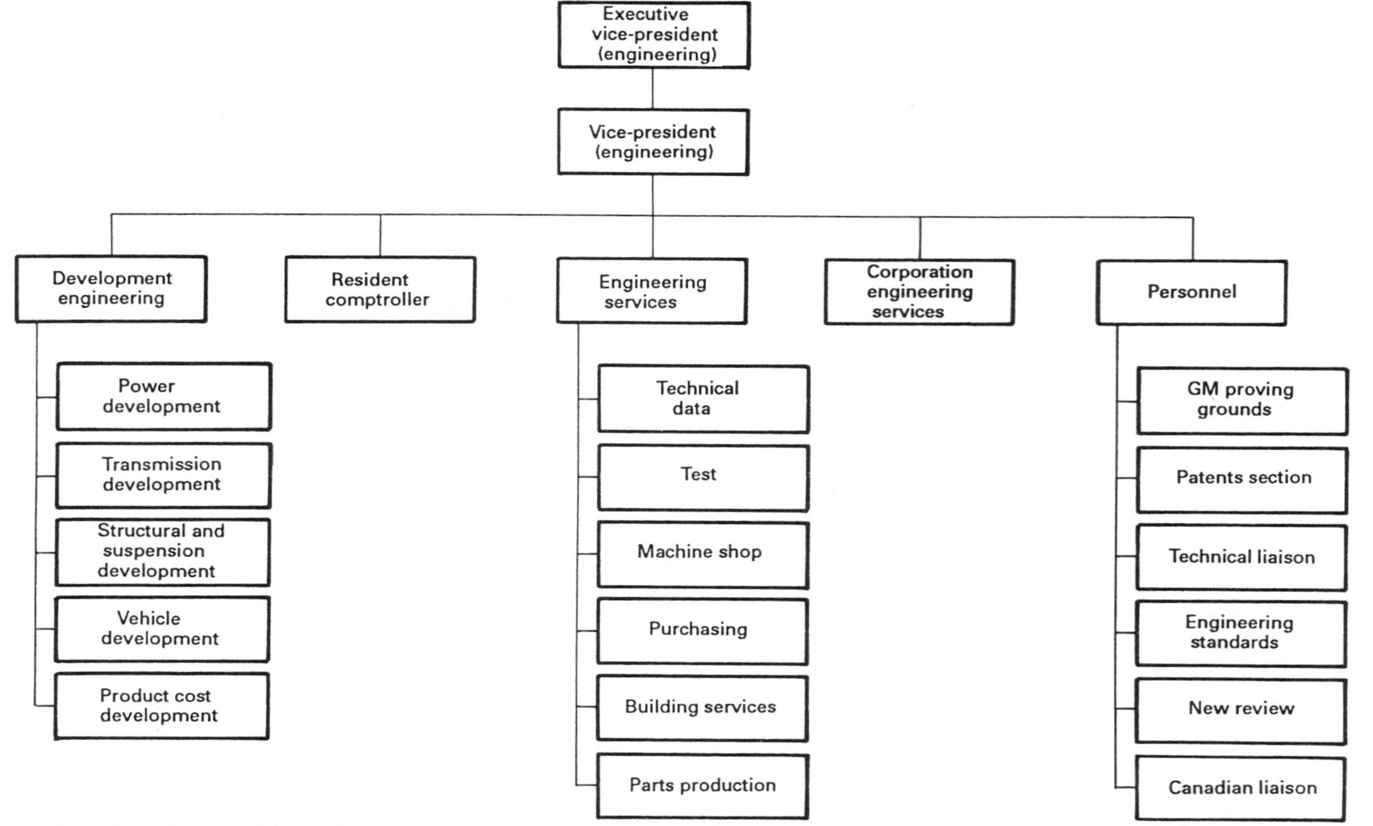

Figure 3.4 *Organization of General Motors engineering staff, General Motors Corporation, USA, 1963. This illustrates combined vertical and lateral presentation of an organization chart*

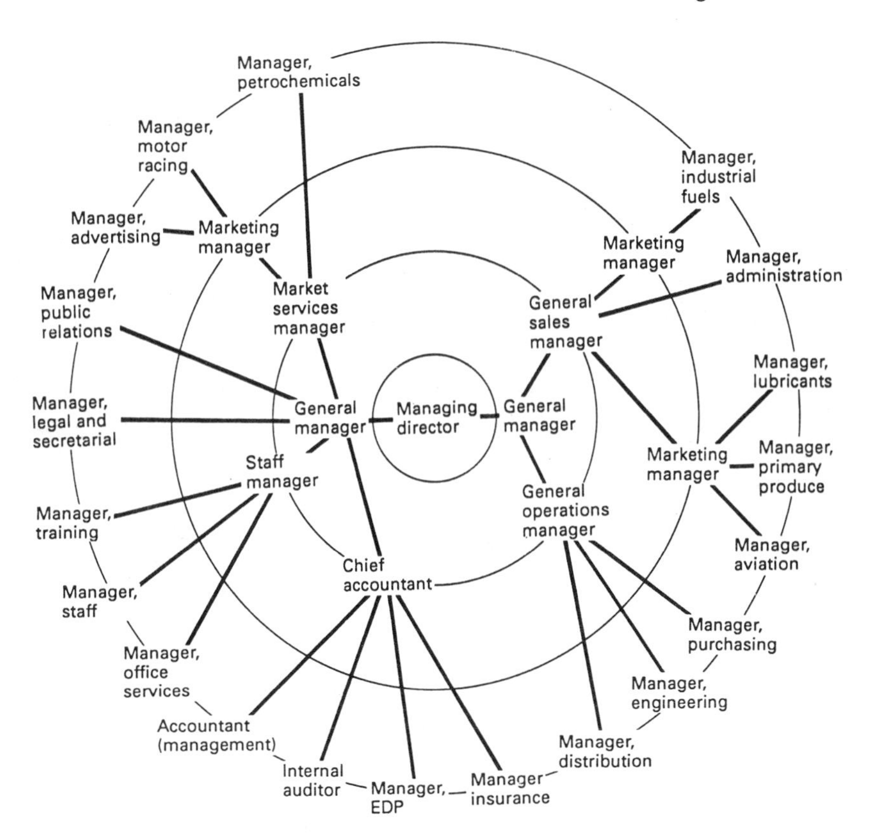

Figure 3.5 *A circular organization chart. This unusual chart was once used by BP Australia Ltd to show their head office organization. The lines of authority and management strata are presented clearly and in compact form*
(Courtesy of BP Australia Ltd)

although it is possible to produce a three-dimensional view or to use stepped images in a computer graphics system to show progressive changes. Just as a balance sheet needs to be updated, therefore, so must organization charts be kept up to date. And, just as a balance sheet must always bear the date at which the balance was made, so must every organization chart bear the date when the particular organization shown became valid (usually, but not always, the date when the chart was drawn). Dating is particularly important when a chart includes the names of individual job holders.

Generally, the advantages of organization charts are that they can show personnel, sections, departments, specializations, lines of authority and responsibility, and communications. But there are limitations. They do not show the informal lines of authority, and they do not always indicate the real bases of influence or power. Interactions between personnel or real status (apart from the organizational status) may not be apparent. Many behavioural writers have criticized the formal organization chart, and even discarded its use. But, when designing or redesigning an organization, the chart is a valuable tool. Like all tools, charts have their uses, advantages and limitations.

Types of organization structure

Line organization

The oldest and most traditional type of organization which is widely used today is the line organization. This depicts direct lines of responsibility and authority. An engineer setting up a consulting business employing a few young graduate engineers and draughtsmen to prepare designs, drawings, plans and specifications would use a line organization. Each member of the enterprise is engaged in carrying out prime tasks of the organization.

Line and staff organization

The line and staff form of organization can be traced back to the earliest civilizations and armies. Here, the term 'staff' has a specific organizational meaning quite apart from loose meanings synonymous with 'personnel' or 'employees' having certain privileges or status conferred on them by the organization. In the context of a line and staff organization, the word 'staff' refers to those who are specially appointed to advise, counsel, render service to, or otherwise assist the 'line'. For example, if the consulting business with a simple line organization expands, the engineer founder may need to employ an accountant and an administrative officer. Their duties will be to keep records of costs, prepare financial statements, manage clerical staff, and so on, none of which are prime 'line' tasks of the business. Both the accountant and the administrative officer are in staff roles. They should not give orders and directions to the engineering and draughting personnel, unless these are delegated by the engineer founder. Even then, the instructions should be written, and signed by the chief executive (the engineer founder).

Still considering the example of the small engineering consultancy, it may be that the next stage of growth for the engineer founder would be to form a partnership or a proprietary company, with the founder acting as chairman of the company as well as being chief operating executive. The laws of most countries generally require that such companies, engaged in professional engineering consultancy, shall have unlimited liability (just as the organization would have done when it was a sole proprietorship or a simple partnership). Here, the nature of the organization has important legal aspects.

A method of distinguishing between the line and staff roles in an organization is to ask the question 'who performs the prime tasks or basic functions of the organization?' Those are the line roles. Thus, in a manufacturing company, the purchasing, production and sales departments are generally considered to be line departments. Finance, accounting, personnel department and administrative services are regarded as staff. In the armed services the distinction is more clear cut: the line part of the organization includes those actually doing the front line fighting. But in non-military organizations there can be confusion, owing to the existence of line relationships between personnel in all sections with their direct superiors (for example, line relationships within an accounts department). Conversely, there can be staff roles in a predominately line function, such as the production controller or an inspector in a manufacturing department. While there are line relationships between the personnel manager and his direct subordinates (such as employment, training, safety, welfare, industrial relations and security officers) the personnel department as a whole is in a staff relationship with the rest of the organization. An engineer employed on design or construction by a private company moves from a line to a staff role if he/she changes jobs to work for a large bank, where the duties are to advise the bank's financial managers on the technical aspects of proposed new capital projects.

In the Email Organization's Range Division, the secretary, finance and personnel managers and members of their staff are all in staff roles (see Figure 3.6). The other personnel under the general manager are in line roles. But it is sometimes difficult to tell which is which simply by looking at the organization chart. It is necessary to analyse further the prime tasks, functions and relationships in the organization. These difficulties in distinguishing between line and staff in non-military organizations have caused some academic behaviouralists to ignore the concept. In practice, however, when organizations grow larger than about 100 personnel it is necessary to distinguish between the line and staff roles in duty statements, job descriptions, job specifications and organization manuals. Otherwise there is danger of conflict over limits of authority and responsibility, and the proverbial 'passing the buck' when things go wrong.

The line should always accept full responsibility for operations. The staff can only advise, and render assistance and services to the line. The line members can accept or reject staff advice. Unlike the military, where staff officers can also direct on behalf of their commander-in-chief, non-military staff officers tend to be restricted in their duties. The military constantly interchange their line and staff officers, so as to obtain better understanding of the roles and provide for the situation when staff officers may have to take command if the line officers fall in battle. In business and industry, such rotation of line and staff roles is rare, and is confined only to the more progressive companies.

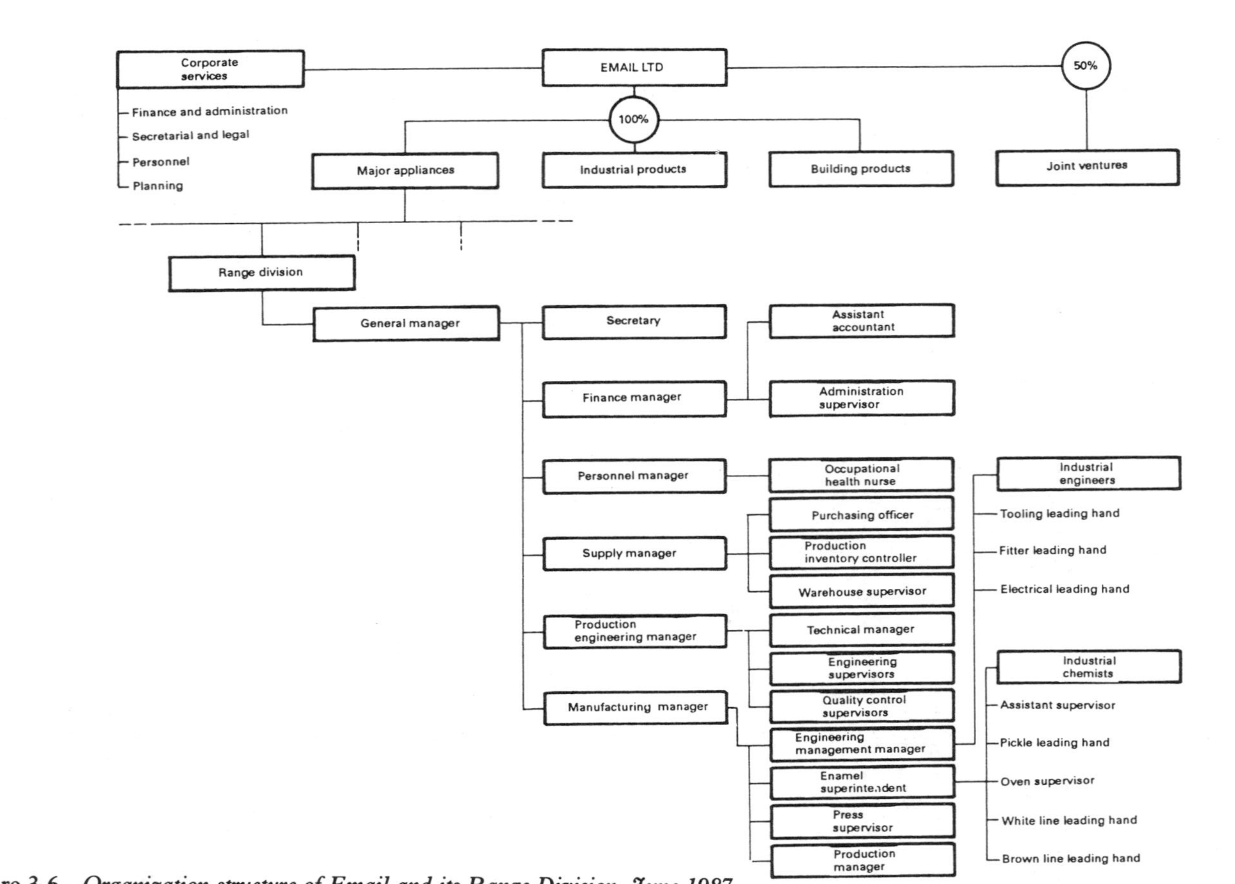

Figure 3.6 *Organization structure of Email and its Range Division, June 1987* (Courtesy of Email Ltd)

Line and functional organization

Many American texts on organization outline Frederick Winslow Taylor's functional organization (or 'functional foremanship' organization) as a distinct type of organization structure. Taylor (1856–1915), American engineer and pioneer of the 'scientific management' movement, advocated eight specialized functional foremen, each able to exercise authority over operatives under each other's control. However, Fayol (1949) was the first to recognize the difficulty of operating this type of organization. While he praised Taylor's use of specialized functional foremen, he could not see how it would work in practice because it violated the age old principle of 'unity of command'. There would be too many 'chiefs', and in any conflict of orders it would be more likely that the 'Indians' would do nothing or else follow the instructions of the most powerful or dominant superior. This principle of unity of command (one man, one boss) proved to be the Achilles heel of Taylor's original functional organization. After some unsuccessful applications in American factories it was eventually discarded as impracticable and too difficult to operate.

Today the term 'functional organization' is used for the subdivision of organizations into main functions or groups of similar types of activities (such as production, marketing, engineering, finance or others). It is also a form of departmentation or a pattern of organization, rather than simply a type of organization structure.

Committee organization

A committee is formed when two or more persons are appointed, elected or selected to work together as a single unit, to meet as necessary so as to come to joint decisions and act as an integrated single body on matters assigned or delegated to it. It is a conjoint body whereby the knowledge, skills and experiences of all may be utilized – the advantage of having 'two heads better than one'. A committee has the marked advantage in gaining cooperation, coordination of effort, and utilization of resources of all members involved.

Many organizations show separate committee organization charts, but some often combine both the committee and hierarchical formats. The board of directors, executive committees and sometimes planning and production committees are in line. Committees for safety, industrial relations and public relations are mostly staff.

Committees have certain advantages in that they require and stimulate personal involvement and commitment from their members. They are useful for consultation, coordination, deliberation, resolution of conflicts or differences, stimulation of creative ideas and the utilization of the diverse resources of their members. On the other hand, committees have certain limitations or disadvantages. They can be costly, time consuming, wasteful in generating trivial and irrelevant discussion, slow in decision making and a good vehicle for shelving responsibilities or passing the buck. A committee is often accused of 'having neither a body to be kicked nor a soul to be damned'. It is possible for some committees to be models of decisive indecisions. While on the one hand a committee can comprise members with diverse interests, on the other they can be split into factions with vested interests, pushing their own narrow, or parochial views and causes.

Much of the success of committees depends on the competency and leadership skills of the chairman and the efficiency of the secretary. Committees are useful devices in organization, but they should be used sparingly with due regard to costs, time and the speed of decision making. Urwick's monograph for the British Institute of Management on committee organization (Urwick, 1952) sets out the uses, abuses, advantages and limitations of committees: it should be read by all students of committee organization.

Matrix organization

The matrix is a more modern type of organization, developed by the American aerospace industry in the late 1950s. The matrix superimposes a lateral project organization over the traditional functional organization.

Figure 3.7 shows a matrix organization for a large engineering enterprise. Project managers A, B and C exercise equal authority and responsibility over personnel who are normally responsible to their functional heads or superiors. Thus development engineers X and Y are responsible to project managers A and B respectively, as well as to their chief development engineer. Similarly, the civil designing engineers P, Q and R are each respectively responsible to project managers A, B and C as well as to their chief designing civil engineer.

The matrix type can be subdivided into two forms of operation, which are the *fixed matrix* and the *shifting matrix*. In the fixed matrix, the same functional personnel remain responsible to the same project managers, with the various projects going through the organization sequentially under the different project managers. In the shifting matrix, personnel within the matrix may work under different project managers according to the priority of projects. It is possible in a shifting matrix that an engineer could be under not only his functional superior, but also under two or more project managers seeking the engineer's expertise or effort. The shifting matrix is more common than the fixed matrix owing to its flexibility and its adaptability to changing technologies.

The advantages of matrix organizations are their flexibility, clearer fixing of responsibility on one person to monitor and be in charge of a project from start to finish, more efficient progressing, monitoring and control, and the adaptability necessary to switch highly skilled technical personnel from one project to another. On the other hand, the main problem introduced is violation of the principle of 'unity of command'. Personnel in a fixed matrix

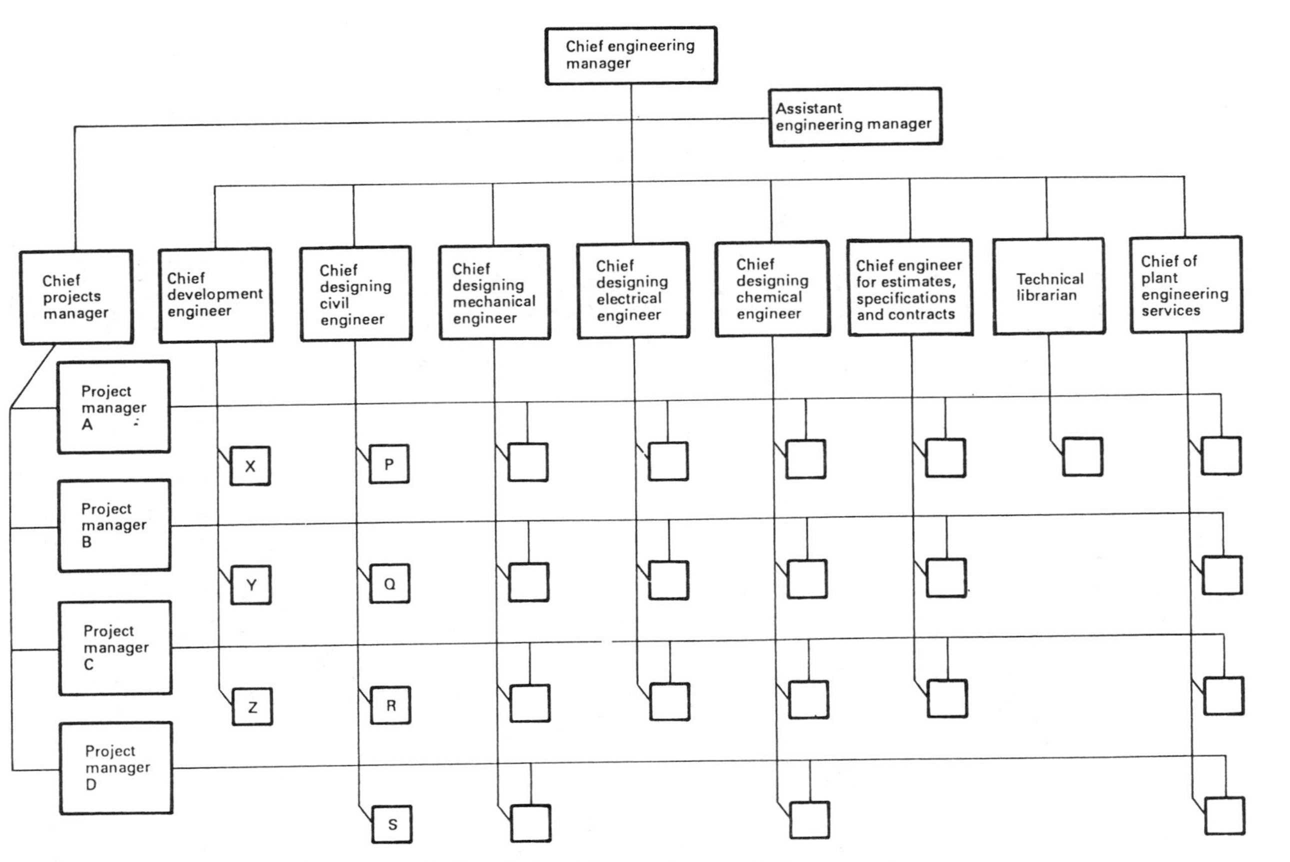

Figure 3.7 *The engineering department of a large industrial enterprise organized as a matrix*

have two superiors. In a shifting matrix there is the possibility of individuals having even more than two superiors. Such an organizational arrangement can lead to much confusion, role ambiguity and conflict on the part of the engineering (functional) and project managers involved. The personnel can suffer frustration, inertia and lack of motivation. Where such danger of conflict or disagreement exists between the functional and project managers, a useful device is to appoint a general coordinating or chief project manager over the various project managers. This person would consult with the functional heads, or appeal to more senior authority in the company in order to settle or arbitrate when differences arise.

An American consultant, Serge Birn, developed a *linear responsibility matrix* (LRM) or *linear responsibility chart* (LRC) to help resolve the questions of authority, responsibility and communications between all those involved with the matrix organization. Some people advocate a subdivision of duties and responsibilities, so that planning, scheduling and cost monitoring and control should be under the project manager, and the day to day allocation of tasks, the working methods and quality supervision should be under the functional heads.

In operating a matrix organization it is necessary to clarify all the tasks, duties and responsibilities needed for the projects being undertaken. All personnel should be adequately trained to operate within the matrix, and they should be clear on the person to whom they should turn for resolution of conflict, or the settlement of any confusion or disagreement. There has been much wishful thinking and speculation by organization theorists for the matrix and other 'free forms' of organization to replace the traditional hierarchical or bureaucratic line and staff organization. In practice, the use of matrix organization has increased in the high technology enterprises, but there have been many failures too. The matrix is usually designed and operated as an appendix or attachment to the normal line and staff organization, whether this be a business, industrial, engineering, military, political or academic organization.

General staff organization

A sixth type of organization, not usually discussed in books on organization, is the general staff organization. This was first developed by the Prussian military forces in the nineteenth century. It is an elaborate form of staff organization that combines the advantages of the committee form, functional specialization and line and staff organization. Mooney (1947) advocated application of the military general staff concept to his own enterprise, General Motors Corporation, in which he became Vice-President for Overseas Operations under the chief executive Alfred P. Sloan Jr.

In addition to the normal line and staff form, the general staff organization involves the use of a top level, continuously meeting, committee. This committee comprises top staff officers – the general staff group – headed by a chief of staff. The top staff officers are in charge of their own specialist staff officers. The general staff group will handle all details of staff work, assist in the formulation of plans, coordinate the various staff services, help in the issuing of instructions and assist in the overall supervision and control of the various arms of the organization. Figure 3.8 shows the adaptation of a military general staff type of organization to an Australian construction engineering enterprise with subsidiary companies and operations both in Australia and overseas.

Departmentation

So far in this chapter, six types of organization structure have been identified in terms of authority and responsibility relationships. Summarizing, these are:

1 Line organization.
2 Line and staff organization.
3 Line and functional organization, of which two types of functional organization can be classified:
   (a) Taylor's original 'functional foremanship'.
   (b) The modern view of functional organization.
4 Committee organization.
5 Matrix organization.
6 General staff organization.

Apart from analysing organization in terms of lines of authority and responsibility relationships, there is another approach. This looks at possible organization patterns in what some writers have called 'departmentation'. As seen in type 3 (line and function) organization, structures can be viewed and analysed as groups of similar and closely related activities (the modern view of functional organization). Accordingly, organizations can be subdivided into the following groups of activities or departments.

1 *Functional organization*. This is identical to the modern view of functional structure. It is the most common form of departmentation.
2 *Product organization*. Product organization is division along product lines. It is common in manufacturing, wholesaling and retailing.
3 *Regional or area organization*. Division into areas or geographical regions is commonly used for the decentralization of responsibility over certain areas.
4 *Project organization*. The organization of distinct projects is usually a combination of line and staff, functional and (sometimes) matrix organi-

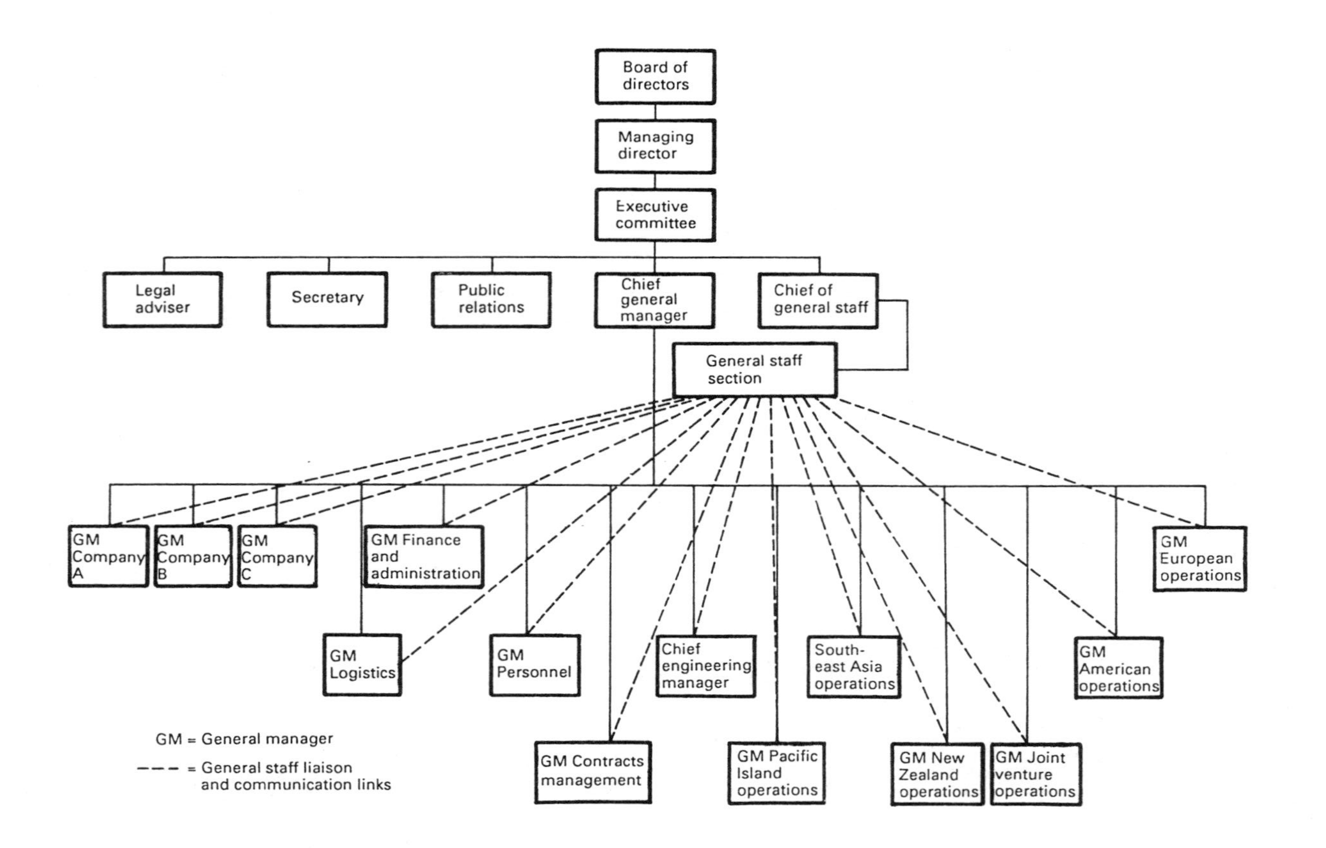

Figure 3.8 *Adaptation of military general staff organization for an Australian construction engineering organization operating on a large scale*

zation. Project organization is common for design and construction projects which have defined start and completion dates.

5 *Machine grouping*. Machine grouping is seen in factories and in construction work.
6 *Customer grouping*. Division into customer groups is used in wholesale and retail trading.
7 *Sequential grouping*. This identifies groups of sequential tasks or activities. It is often found in chemical processing, airfield, road and pipeline construction work.
8 *Service grouping*. This is division into groups of similar types of services, and is used in the manufacturing, retailing and service industries.
9 *Numerical grouping*. Numerical grouping is common where large numbers of personnel are employed, such as in Roman armies of the past, and in Chinese communes and Soviet work stations.
10 *Time grouping*. This is division according to time blocks. It occurs when shift work is undertaken in mining and manufacturing.
11 *Profit centre grouping*. Profit centres are sometimes identified as separate parts in a company for the purposes of setting financial performance targets and for monitoring the related achievement. They are used in the manufacturing, retailing and service industries.
12 *Personal attributes grouping*. This groups people according to attributes such as race, height, sex, religion, or particular skills (like language versatility). Such organizational groupings would be seen where tall men are employed as traffic policemen, or if short women were engaged as airline hostesses. Another example would be seen if workers from mixed ethnic backgrounds were grouped into teams of similar backgrounds to promote harmony and avoid conflicts.

In large organizations, a number of the above forms of grouping or other forms of departmentation may be combined with several types of organization structure.

Principles of organization

One of the most controversial debates in management and organization theory has been that of the existence or otherwise of the so-called 'universal principles of organization', which Fayol (1949) and Mooney and Reilly (1934) advocated for the efficient working of various organizations.

We have seen the operation of the principle of unity of command, which is as old as military history. Another principle is that of the 'principle of span of control'. The term 'span of control' refers to the number of *direct* subordinates of a supervisor or manager, and the principle is that this span should be limited to seven or eight subordinates. This principle was first advocated by the Lithuanian American engineer V. A. Graicunas (1937), who practised as a management consultant in Paris. Graicunas proved mathematically and graphically that the total number of relationships between an executive and his immediate subordinates became excessive when the span of control reached about seven. The debate centres on the precise number of subordinates. However, owing to the variability of human nature and circumstances, this principle must be interpreted broadly, since the maximum recommended 'span of control' may be affected by several factors, such as:

- The capacity and abilities of the particular executive.
- Whether the subordinates are highly skilled or relatively untrained.
- Whether the subordinates are concentrated within a confined area of a factory or are sparsely spread out along a few kilometres of a road or pipeline project.
- The nature of the work itself.

This means that the so-called principles are not rigid scientific principles or laws that can be repeated, tested and verified under all conditions. Rather, they are guidelines or rules which apply in most situations for the efficient operation of the organization, but with exceptions to the rule.

Simon (1945) was critical of the existence of 'principles of organization', which he called 'proverbs of organization', such that for every proverb there is a contradictory and opposing proverb. Simon maintained that the principles of span of control and unity of command are contradictory. But these two principles are really more complementary: the unity of command principle applies to vertical relationships, whereas the principle of span of control is more concerned with limiting the number of immediate subordinates.

The following principles of organization are pragmatic guidelines, generally culled from experience and found to improve the smooth working of organizations in practice.

1. *Definition*. There should be a clear line of authority and responsibility from the top to the bottom of the organization.
2. *Correspondence*. This means that there should be equal matching of authority with responsibility.
3. *Span of control*. The span of control should be kept within manageable limits, according to the prevailing circumstances.
4. *Unity of command*. Each subordinate should be responsible to only one superior.
5. *Balance*. There should be balance between all parts of the organization so that workloads are fairly apportioned.
6. *Clarity*. Each position and role in the organization should be clearly defined, and the duties, limits of authority and responsibility are made clear to all concerned.

7 *Unity of direction*. All parts of the organization should contribute equally to attaining the objectives of the organization and their efforts should be unified in one direction.
8 *Simplicity*. Aim for the simplest organization necessary to achieve the objectives.
9 *Flexibility*. Allow for flexibility in the organization to cater for changes in technology and changing circumstances.
10 *Ultimate responsibility*. The ultimate responsibility of those at the top of the organization cannot be totally delegated or absolved.

Some writers set out more principles for effective organization. Although organization principles may be criticized by some academic behaviouralists, they are nevertheless useful when designing, redesigning and operating organizations. The practical test of whether any principle is valid is to take the opposite course of action. Of course there are justifiable exceptions: if an engineer sees a workman fall down a hole orders have to be given quickly without reference to the principle of unity of command. Some managers have the exceptional ability of being able to cope with impressive spans of control. But, for most practical purposes, violate the principle of unity of command, exceed the sensible limit for span of control or refuse to define duties, lines of authority and responsibility and the results will probably be confusion, disorganization, and general avoidance of responsibility or passing the buck.

An organization manual is a useful tool which is used in large, progressive organizations to set out the organization structure clearly. The manual will show duties, lines of authority and responsibility, policies, procedures and relationships in the organization.

Change and stress in organization

In recent years there has been much focusing by organization theorists and behaviouralists on such topics as organization change, organizational culture and organizational stress. Changing technology, political and other factors such as union pressures, changing motivation and aspirations among employees and specific government legislation may force changes in organization and organization structures. Amos and Sarchet (1981) suggested that the following steps be adopted in making organizational changes:

1 Review or, if necessary, determine the organization's objectives.
2 Gather information on the existing status of activities, lines of authority and responsibilities, lines of communication, personnel and their relationships and duties (besides their abilities and skills). It is often necessary to draw up an organization chart and examine duty statements, job descriptions, job specifications, and flow charts of operations.

3 Prepare an action plan for the organizational renewal which will be necessary if the objectives are to be achieved with a specified time span. Here, duty statements, job descriptions, job specifications, an organization manual, flow charts, bar charts and even critical path network diagrams may need to be produced in addition to revised organization charts.
4 Gain acceptance of the plan by involving key personnel in the organization, by constant consultation, inviting questions and suggestions, and seeking cooperation through meetings and other effective methods of communication.
5 Prepare a schedule for implementing the changes. Here, written schedules, bar charts and critical path networks can be used for planning and controlling implementation.
6 Implement the planned changes with tact, determination and due consideration of the human factor, but continually stress the objectives and reasons for change to all concerned.

There is also Young's 'organizational modulus', which states that organizational stress is proportional to organizational strain. This is an engineering law that can be followed by organizational theorists.

There are hazards in organizational practice besides organizational stress. One of these is over-organization, which was recognized by Aldous Huxley in his book *Brave New World Revisited*. Then there are the practices of making individuals conform to the various mores and artificial customs of organizations, and the destruction of individual initiative, creativity and personality (as warned by William H. Whyte in his book *The Organization Man*). Organizations should be designed to achieve group or organizational objectives. They should serve men: not men serve the organization for the organization's sake. The general and overall purpose should override individual or factional self interests. As Russell Robb once observed, 'organization is but a means to ends: it provides a method'.

References and further reading

Amos, J. M. and Sarchet, B. R., *Management for Engineers*, Prentice Hall, 1981.

Blanchard, B. S., *Engineering Organization and Management*, Prentice Hall, 1976.

Brech, E. F. L., *Organization – The Framework of Management*, Longmans, Green and Co., 1957.

Child, J., *Organization: A Guide to Problems and Practice*, Harper and Row, 1977.

Dale, E., *The Great Organizers*, McGraw-Hill, 1960.
Dale, E., *Management: Theory and Practice*, 3rd ed., McGraw-Hill Kogakusha, 1973.
Fayol, H., *General and Industrial Management*, translated by Constance Storrs from the 1916 French original *Administration Industrielle et Générale*, Pitman, 1900.
Graicunas, V. A., 'Span of Control', *Papers on the Science of Administration*, Gulick, L. and Urwick, L. (eds.) Institute of Public Administration, 1937.
Handy, C., *Understanding Organizations*, Penguin, 1976.
Hsun Tzu, *Hsun Tzu – Basic Writings*, translated by Burton Watson, Columbia University Press, 1963.
Hunt, J. W., *The Restless Organization*, John Wiley, 1972.
Hutchinson, J. G., *Organizations: Theory and Classical Concepts*, Holt, Rinehart and Winston, 1967.
Koontz, H., O'Donnell, C. and Weihrich, H., *Management*, 7th ed., McGraw-Hill Kogakusha, 1980.
Layton, E., 'Science, Business and the American Engineer', in *The Engineers and the Social System*, Perucci, R. and Gerstl (eds.), John Wiley, 1969.
Lloyd, B., *The Organization of Engineering Work*, Macmillan, 1979.
Mackenzie, K. D., *Organizational Structures*, AHM Publishing, 1978.
Mooney, J. D., *Principles of Organization*, Harper and Row, 1947. Based on the original *Onward Industry* by Mooney, J. D. and Reilly, A. C., 1933.
Morrell, J. C., *Preparing an Organization Manual*, British Institute of Management, 1977.
PA Management Consultants Ltd, *Company Organization – Theory and Practice*, Allen and Unwin, 1970.
Pugh, D. S. (ed.), *Organization Theory – Selected Readings*, Penguin, 1971.
Pugh, D. S., Hickson, D. J. and Hinings, C. R., *Writers on Organizations*, 2nd ed., Penguin, 1973.
Rhenman, E., *Organization Theory for Long-Range Planning*, John Wiley, 1973.
Robb, R. in his papers published in part in *Processes of Organization and Management*, Catheryn Seckler-Hudson (ed.), Public Affairs Press, 1948.
Schott, R. L., 'The professions and government: engineering as a case in point', *Public Administration Review*, vol. 38, no. 2, month 1978, pp. 126–32.
Sloan, A. P., *My Years with General Motors*, Pan Books, 1966.
Shannon, R. E., *Engineering Management*, John Wiley, 1980.
Simon, H. A., *Administrative Behavior*, Macmillan, 1951.
Urwick, L., 'Organisation as a Technical Problem', in *Papers on the Science of Administration*, Gulick, L. and Urwick, L. (eds.), 1900.
Urwick, L., *Committees in Organization*, British Institute of Management, 1952.

Urwick, L., *Organization*, International Academy of Management, Nederlands Instituut Voor Efficiency, 1966.
Wearne, S. H., *Principles of Engineering Organization*, Edward Arnold, 1973.
Whyte, W. H., *The Organisation Man*, Penguin, 1960.
Woodward, J., *Management and Technology*, HMSO, 1958.

4

Engineering manpower planning

Peter M. Coughlin

People are a key asset in any organization. Most other assets, particularly the financial ones, are planned and controlled with meticulous care. Often this is not the case with manpower, even though this may be the organization's most valuable asset. The competitive environment of today's business demands a realization that company plans for at least five years ahead must include accurate and systematic manpower plans. Management should have a firm objective of ensuring that inadequate or badly developed manpower does not place a constraint on the achievement of corporate objectives.

The nature and purpose of manpower planning

Manpower planning is a management process that adds information about human resources to the corporate plan. It helps decision makers by indicating the consequences of their possible future actions, thus assisting in the search for the optimum strategy. Manpower planning can help line managers by enabling them to highlight existing or inherent problems in the way they are managing (or are proposing to manage) the human resources at their disposal.

The manpower planning process includes the development of policies, systems and procedures that will increase the chances of exercising effective control, thus helping managers to avoid making ad hoc decisions that could be detrimental in the long run.

Inevitably the development of managers, technologists, technicians and other specialists must be regarded as a long-term continuing process. Manpower planning helps future needs to be identified well in advance, also aiding the assessment of the long-term effects of today's decisions. An example of such a situation is the existence of surplus staff in a particular area of a business. The short-term view might lead to a decision to implement a redundancy plan, but investigation of the longer term workload forecast might reveal a substantial projected upturn. In such circumstances the short-term savings achieved by implementing redundancies would soon be eliminated by the cost of recruiting and training new staff to meet the future workload.

It would be a mistake to regard manpower planning as an isolated activity. Within the whole personnel field decisions are made on many matters including recruitment, training, manpower development, redundancies and working practices. The making of such decisions and their after effects form part of the manpower planning process. It is common to find that substantial employee-related investment decisions are made with little attempt at evaluation – before or after the event.

The personnel manager should press for the development of manpower planning procedures to ensure that staffing issues are considered in a comprehensive manner as part of the corporate plan. This will involve the establishment of flexible manpower policies, plans and control procedures to suit the changing environment of the organization.

Questions that manpower planning can help to answer

Manpower planning can identify areas of concern to management and staff. These may range from corporate level concern about organizational effectiveness down to individual worries about promotion, career prospects and the acquisition of skills through training. The planning process includes consideration of each of the problem areas. It facilitates the presentation of quantified options for management, who are then better able to decide on the course to be followed, the manpower planning process having shown the consequences of such decisions.

The commonest questions to which manpower planning can help provide the answers are:

1 *In an expanding or contracting workforce:*
How many should we recruit, and what sort of people?
Should they be young with potential or old without potential?
Should they be single-skilled or multi-skilled?
Which skills should they possess?
How many staff should we reduce by?
When should we implement the decision?
Will the method be early retirement, redundancy or natural wastage?
What criteria should we adopt for redundancy?
How can we end up with a leaner, fitter organization?

2 *In a situation of high labour turnover:*
How much turnover is controllable?
What is the total cost of turnover?
Would revised employment terms and conditions reduce turnover and, if so, by how much?
How would lower turnover affect plans for promotion and recruitment?

3 *Poor availability of staff for promotion:*
How has this happened?
What action can be taken for the future?
What short-term measures can be taken, bearing in mind the long-term needs?

4 *Experience:*
What combination of experience will be needed for future management?
Will present plans provide this experience?
How should we change the present plans?

5 *Policy on graduate and specialist recruitment:*
How many?
What sort?
Age range?
Should we recruit this year?

6 *Age structure:*
What is the most appropriate age structure for the organization?
How far from this is it at present?
How can we correct it?
What effects will such action have on the remaining staff?
How much will it cost?

7 *Manpower analysis and reporting:*
What information is needed by management and staff?
How can it be provided?

8 *Career structure management:*
Which promotion policies should we follow to maintain a viable career structure?
Do we currently promote staff at the wrong age?
Are staff contributing well in their jobs, or have they held them for too long?
Can we alter current career patterns and what implications would this have for numbers, ages, grades, skills and morale?

Definition of the manpower plan

Before the manpower plan is drawn up the organization must first have decided its goals. These include ensuring that the organization:

1 Obtains and retains the quantity and quality of manpower it needs.
2 Makes the best use of its manpower resources.
3 Is able to anticipate the problems arising from potential manpower surpluses or deficits.

The overall manpower plan will be a summation of plans drawn up for each group or department within the organization. For these groups information is collected, classified and analysed. Forecasts are made of demand, and also of internal and external supply. Each of these forecasts is edited and fed into the overall plan, which will consist of:

1 The succession plan.
2 The staff development plan.
3 Any recruitment plan.
4 Any redundancy plan.

The succession plan

The chief executive of the organization should ensure that succession plans are in place to cover the next two generations for each key post. For this purpose, key posts will usually be defined as the chief executive, executive directors, general managers, and the next level down from these.

Each of the individual succession plans should be prepared on a standard form. An example is shown in Figure 4.1, which is intended to cover a five-year period. Emphasis must be placed on jobs likely to become vacant as a result of retirement or transfers.

When the posts and post holders have been recorded on the form, the next stage is to complete the columns under the heading 'probable year for move' by entering *R* for retirement and *T* for transfer (where such moves are identifiable).

The term 'emergency successor' means exactly what it says, and against each post holder only one name should be recorded.

The 'short list' is compiled taking account of when a change of post holder is expected, as follows:

1 *Where a change is expected within two years:*
A firm list is recorded of not more than three names, from which the next appointment is likely to be made. A further list of names is also recorded to show those from whom longer term appointments to the post may be made (i.e., the next but one appointment).
2 *Where a change is expected from three to five years hence:*
A list of possible successors is prepared, but this will be less firm than if the change had been expected within two years.

When the short list has been compiled, one name is marked *P* to denote the preferred successor. He/she will be the candidate chosen unless there are other priorities. It may be that the emergency successor is also on the short list. In this case he/she will be designated *E*. Of course, there is every possibility that the emergency successor and the preferred successor will be the same person.

Each group succession plan is reviewed in the first instance by the director responsible for the relevant area of the business. Following this, all the group

PERSONAL AND STAFF CONFIDENTIAL

________________ – MANAGEMENT SUCCESSION

Note: Age – completed years as at 31/12
Years in post – years and months as at 31/12

Issued on:

Approved on:

ǿ Not available within 12 months of taking up present assignment
E Emergency successor T Transfer D Designate
P Preferred successor R Retirement

Grade	Post Title	Holder	Age	Years in Post	Probable year for move					Possible Successors		Remarks
					88	89	90	91	92	Emergency	Short List	

Figure 4.1 *Form for recording succession plan*

plans are reviewed collectively by the executive directors. During this review, possible additional names are considered, and any names deemed unsuitable are challenged. The outcome of this review is an overall agreed succession plan, which must be accompanied by an action plan for its achievement.

When the directors have authorized the overall plan they should delegate responsibility for implementing the associated action plan to the individual directors and departmental managers. The personnel manager should monitor the progress of all activities on the action plan and submit quarterly progress reports to the executive directors.

The staff development plan

It must be the responsibility of individual managers to consider and promote the training, development and progress needs of the work groups for which they are responsible.

The annual staff appraisal exercise (see Chapter 6) will be designed to encourage and assist managers to undertake a formal review of each member of the work group. During this review the aspirations, experience and training needs of each employee should be identified, agreed and recorded. The personnel manager should collate, review and summarize this data in order to prepare a company training and development plan. This plan will include an implementation programme and a detailed cost estimate. The cost estimate will form the basis of the company's training and development budget, to be submitted for approval to the executive directors.

When the plan and its budget have been approved, it should be the responsibility of the personnel manager, in close consultation with the line managers, to drive the plan to a successful conclusion. In practice it is found that a good deal of drive may be needed to persuade line managers, when the time arrives, that they did in fact propose and agree that members of their staff should be released from normal duties to undertake a period of training or a promotional move to another work group.

One method for ensuring the accuracy and effectiveness of the staff development plan is to establish a staff development committee which will formalize the relevant procedures. In large organizations it may be necessary to create, in addition to the main committee, a committee with specific responsibility for dealing with the development of technical staff.

The staff development committee

The staff development committee should comprise the executive directors and the personnel manager. This committee considers all senior vacancies and key personnel for placement, drawing up the appropriate action plans as required.

The work and terms of reference for such a committee would typically be described as follows:

Whilst the responsibility for staff development and succession planning rests initially with line managers, assisted by the personnel department, coordination of individual development and succession plans can best be achieved by the senior executives working together in a staff development committee.

The committee will have four main roles:

1 The collection of members' views on the performance potential and development needs of individuals identified for senior positions in the future to ensure that common standards of assessment have been applied.
2 To draw up and review succession plans for each department:
   (a) Identifying successors for management, specialist and other senior posts.
   (b) Initiating corrective action by internal interdepartmental transfers, or by external recruitment where there are insufficient internal candidates.
   (c) Assessing the readiness of the listed individuals to take over particular posts and to highlight the training and/or experience that they might need to fit them for these posts.
3 To identify staff with high potential for possible accelerated progression.
4 To review and advise on planned organizational changes in individual departments which affect other parts of the organization, especially where these involve the transfer of responsibilities or posts.

The staff development committee should meet quarterly to consider succession plans and the career development of selected staff (whose individual programmes will be reviewed annually).

The technical staff development committee

If necessary, on account of the large size of an organization or, more particularly, where it operates in several areas or countries, it may be desirable to establish a technical staff development committee. This will recommend candidates for appointment to all senior technical posts and ensure that a supply of trained technical staff is available to enable the organization to meet its objectives. Such a committee would be chaired by the technical director, and its members would be the technical managers and the personnel manager. Typical terms of reference for such a committee might be on the following lines:

1 To promote the interchange of technical staff between various company locations (worldwide) in order to maximize the availability and use of talent for the benefit of the company.

2 To consider nominations for appointments to senior technical posts and to advise the staff development committee, where appropriate.
3 To assess the implications on technical manpower of strategic developments within the company, and to cause appropriate action to be initiated where necessary.
4 To consider the supply of trained technical manpower to support current and forecast projects and operations.
5 To consider policies to correct any perceived imbalances of technical staff.
6 To ensure that technical staff are given adequate development training in order to maintain the desired level of trained staff.
7 To review and monitor the manpower resourcing of projects and operations.
8 To monitor the progress of any graduate recruitment programmes.

The numbers game

The starting point in developing a quantified manpower plan is the corporate revenue and overhead budget for the forthcoming period to which the manpower plan will apply.

The revenue budget will have been compiled from a combination of work in progress, known forthcoming projects and sales plans.

The overhead budget represents an assessment of the cost of the indirect activities and facilities required to run the business. For those readers who are unfamiliar with the terms direct, indirect and overheads, explanations will be found in Chapter 19.

These budgets are divided into the constituent components that will be allocated to each department.

The next stages of the operation are undertaken within the individual departments, the first such stage being the conversion of each departmental budget from financial terms into the equivalent workload. Workloads are typically expressed in terms of manhours per grade, but on the scale required in manpower planning it is more realistic to use the units of man years.

The next departmental planning stage is to convert the man years into people. The procedure for achieving this conversion must be adapted to suit the scale of the problem. Very often, the resource schedules resulting from project network analysis provide most of the information for the mainstream project or other engineering and design work (see Chapter 22). Allowances must then be made for nonproject activities, rework, supervision and departmental administration.

It is usually found convenient to prepare bar charts or histograms, depicting the estimated manpower needs on a timescale over (perhaps) two years. Examples of these forms of presentation will be found in Chapter 22. The charts should show the estimated staff requirements in terms of numbers needed in all the various disciplines and grades, from which the expected total staffing of each department can be quantified.

To move from mere departmental numbers to specific requirements needs a consideration of promotions, expected departures through retirements, resignations and other reasons, and training moves. The departmental succession and staff development plans are important factors in this study. The current staffing of each department, when superimposed on this information, will reveal where shortfalls or surpluses of staff are expected to occur.

On completion of all these studies, each department is able to submit its input to the corporate manpower plan, comprising the succession plan, staff development plan and the recruitment or surplus estimates (leading to recruitment or redundancy plans).

The manpower planning process has now reached the point where it moves from a departmental to a company wide operation. All the departmental elements are put together for review by senior management. This is likely to be the stage where the staff and technical staff development committees become closely involved in the overall succession and staff development plans. This is, of course, also the stage where the combined effect of departmental recruiting bids and forecasts of surpluses have to be reconciled, and internal solutions (such as staff transfers) considered to avoid unnecessary recruitment or redundancies.

When the proposed manpower plan has been drafted, it must be evaluated and checked back against the original corporate revenue and overhead budgets. If there is a serious discrepancy, the proposals must be re-examined and modified to reconcile the projected manpower cost with the budgets. When this has been done, it is possible for senior management to give their formal approval to the manpower plan, and to the action plans for its achievement.

Regular evaluation and control of the action plans is essential, and this should be combined with feedback to the departmental budgets to ensure that the whole process is leading to achievement of corporate targets.

The recruitment plan

Resulting from the formally approved manpower plan, the recruitment plan for the corresponding period will set out:

1 The numbers and types of staff required, and why and when they are needed.
2 Any perceived problems in meeting the recruitment needs, indicating how these are to be dealt with.
3 The recruitment programme.

The processes and procedures for recruitment are described in Chapter 5.

Permanent or temporary staff?

It is well worth giving very careful consideration to the basis of employing staff, since it may be possible to establish a balance between permanent staff, fixed-term contract and agency staff.

It is obviously unlikely that management and other senior staff could be employed on any basis other than permanent, and this will apply to the nucleus of experienced staff who are vital to the operation of the business. On the other hand it is common practice to take on draughtsmen, secretarial, clerical and other appropriate grades on an agency basis, so that they can be released as soon as work is no longer available for them.

Often a project requires specialist staff for only limited periods (particularly on construction activities) and it may be appropriate to engage them on fixed-term contracts.

Planning for a sensible balance between the numbers of permanent, fixed-term contract and agency staff can minimize the need for redundancies, so avoiding a sense of insecurity among the permanent staff and obviating the costs of retrenching permanent employees.

The redundancy plan

At one time the prospect of redundancy was regarded with horror, not only by employees who might be affected but also by the employers. Both parties were attuned to the concept that a job with a reputable employer was seen as a secure fixture, from the time of recruitment right through to retirement. Any suggestion that this tidy situation needed to be disturbed was difficult to contemplate. Redundancy carried a perceived stigma that was difficult to live with.

More recently, changed business circumstances have left employers with no alternative to trimming their workforces, sometimes drastically. Because redundancy has become commonplace it has lost much of its past stigma and employers have developed realistic termination packages to soften the blow.

It is essential that careful consideration is given to all relevant aspects, future as well as current, before any decision is made to implement a redundancy programme. Assuming that the manpower plan reveals a projected surplus of human resources in certain areas of the business, it is important to identify exactly where these surpluses will occur, the period of time during which the excesses are expected to exist and, finally, to quantify the surpluses precisely – on a name by name basis.

The next step is to examine possibilities other than redundancy. For example, can the surplus staff (with appropriate training if necessary) be redeployed to other work? It is important to review forthcoming retirements and potential resignations, since these might occur soon enough to allow some or all of the surplus staff to be 'carried' until they can fill the vacant places. Obviously the recruitment plan will require urgent and detailed review, since the suspension of recruitment might go at least part of the way towards solving the redeployment problem.

If, when all options have been examined, it is clear that there will be surplus staff who cannot be redeployed in the near future, then redundancies may be inevitable. If this is the case it is essential to lower a veil of secrecy while top management, as a matter of extreme urgency (to minimize rumours and resultant panic) address themselves as a close working group to the following:

1 Creation of a detailed redundancy programme.
2 Development of criteria to be used in redundancy selection.
3 Establishment of severance terms.
4 Arrangements for outplacing, including financial and pre-retirement counselling, to be offered to redundant staff.
5 Preparation of a provisional list of named staff to be discharged.
6 Detailed costing of the redundancy programme.

Careful consideration must be given to each item so that, once agreed by management, they can be 'frozen' prior to implementation.

The redundancy programme

The programme should start with the 'freezing' of the list of redundant staff, followed by an announcement to staff (by means of a general meeting addressed by the chief executive if at all possible). The announcement should include a statement along the lines of 'every employee will be advised of his/her position by, say, 5.00 p.m. tomorrow'.

Senior managers should interview each of their staff to explain the general situation and specifically to reassure the employee, or to hand over a formal notice letter. The notice letter should set out the severance terms, and there should be an additional letter explaining the counselling facilities that are to be offered. As soon as redundant staff have been advised of their position by their senior managers they should be seen by the personnel manager, who will be able to explain any matters that may be unclear.

There are several opinions as to the period that should elapse between an employee being advised of redundancy and his/her departure date from the company. Generally, it is found to be best to minimize this period – possibly allowing sufficient time only to arrange for the counselling facilities to be provided before departure. A long period between the giving of notice and departure could lead to the opportunity for creating dissention among the leavers and, more importantly, among those who are to remain with the company.

Criteria for selection for redundancy

In the past much was made of the need to apply the 'last in – first out' principle to selection for redundancy. Another method was to choose only staff within a few years of their normal retirement age. Attempts to adopt such broad-brush approaches are unlikely to achieve the desired result, because they assume that the newest or oldest employees are those of least value to the company. It would be surprising if this were the case.

Except, perhaps, in situations where redundancies have to fall among large groups of employees engaged on identical work, it is unlikely that calling for volunteers will achieve the results required. In many cases the volunteers would be the younger, most capable, and accordingly the most marketable employees. The company would wish to refuse offers from such people, and they might then have to nominate non-volunteers, so bringing the volunteer principle into discredit.

In any event the criteria to be used in selecting for redundancy must be written down and formally adopted by top management. Thereafter these criteria can be seen by all who need to know of them, which could include an industrial tribunal should any claim for unfair dismissal arise. The tribunal will wish to measure any such claim against the criteria that have been formally adopted. It is not recommended that the criteria are generally published within the organization, since this is likely to lead to arguments among redundant staff as to whether or not they were fairly measured against the criteria.

A realistic set of criteria might be developed along the following lines:

1 The overriding concern in selection, especially when business conditions are difficult, is to ensure that the company retains the workforce best suited to securing its future prosperity. The main criteria therefore, in selecting one or more staff for redundancy from a number employed in a similar capacity, are the company's view of:
   (a) The relevance of the individual's skills and knowledge to the future needs of the business.
   (b) The relative performance quality of staff with broadly equivalent skills and knowledge, and of their likely contribution to the company's future.
   (c) The extent to which the age of the individual assists in providing continuity of operation.
2 The following will, however, be taken into account:
   (a) Whether any staff in the relevant categories are over the normal retirement age, or in temporary or non-pensionable employment.
   (b) Eligibility for an immediate pension under pension scheme rules, and proximity to the normal retirement age.

   (c) Registration as a disabled person.
   (d) Known facts about any dependents or any special hardship.
   (e) Other factors being equal, length of service with the company compared with others in the particular category involved.

3 It is the normal presumption that a member of staff declared redundant will have been chosen from the company unit where a reduction of posts is to occur. There may, however, be cases where an employee so selected is a suitable replacement for an employee in another post in another unit, and the grounds for his being offered that other post are strong enough to demand that consideration should be given to declaring the existing incumbent redundant instead.

Severance terms

Statutory redundancy payments must be given to eligible employees but it is unlikely that an employer would regard such payments as being in any way adequate to compensate the employees. A package of severance terms usually comprises the following elements:

1 Pay in lieu of contractual notice.
2 Statutory redundancy payment.
3 An immediate enhanced pension for those within, say, ten years of normal retirement age.
4 An ex gratia payment.

It is not within the scope of this book to quantify pension or ex gratia payments. These items should be calculated at levels that will fall within the boundaries generally regarded as acceptable in the type of business concerned. Care must be taken to establish pension payments within the restrictions imposed by the Inland Revenue. It also has to be borne in mind that the early payment of enhanced pensions is likely to need substantial additional payments into the pension fund from the company, so that the fund shall not be depleted below a level approved by the actuary.

Ex gratia payments should not be advised in advance (in writing) to employees, to avoid the Inland Revenue regarding them as contractual and taxing them accordingly. It is usual to graduate ex gratia payments in favour of those likely to be hardest hit (i.e., those around the age of 45, who are not yet old enough to receive an immediate pension, yet find themselves too old to find a new job easily). It is sensible to limit value of ex gratia payments where necessary to ensure that no employee receives more money (severance payments plus pension) in the period between leaving employment and normal retirement date than he/she would have received by remaining in employment to that date.

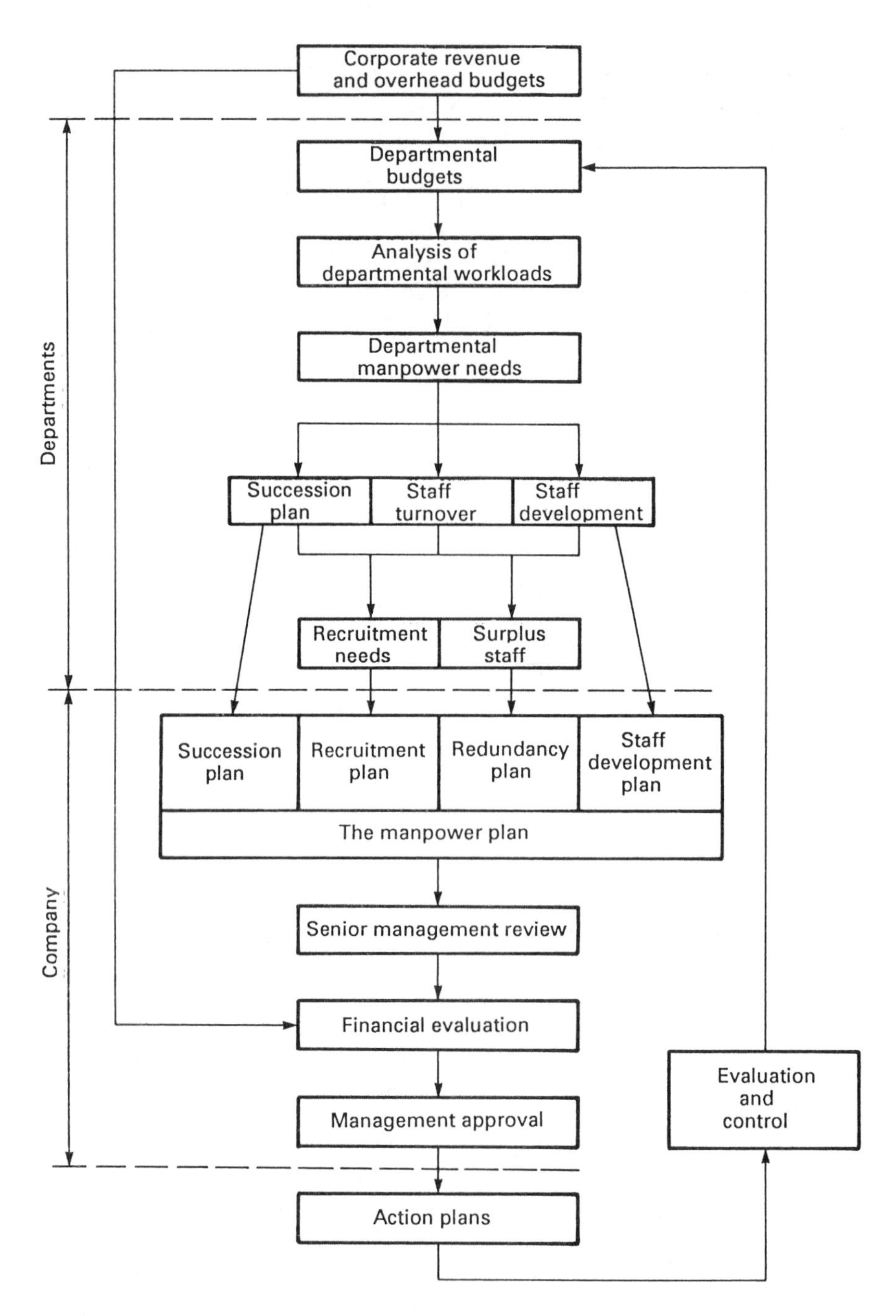

Figure 4.2 *The manpower planning process*

Counselling facilities

Redundant employees are at first bewildered and disorientated. They benefit greatly from expert counselling covering all aspects of job hunting (provided by a reputable outplacement consultant) and financial management (using experienced professional advisers who will provide group and private counselling to the ex-employee and to his/her spouse). For staff who are to take immediate retirement, pre-retirement counselling (involving their spouses) can be invaluable in helping them to prepare for the future.

When the redundant staff have left the company it is important that line managers devote as much time as may be needed to build up confidence for the future among the survivors. If it is honestly possible to do so, these staff should be given positive reassurance about the security of their jobs in the foreseeable future.

Summary and conclusion

The manpower planning processes are summarized in Figure 4.2. Details of the processes applicable to a particular organization will, of course, depend to some extent on the kind of business undertaken, but the principles illustrated in the figure are relevant to engineering businesses.

If management adopt a determined and consistent approach to manpower planning, particularly to succession planning, recruitment and staff development, it is likely that the manpower resource will be the major asset of the company. This asset will contribute greatly to the attainment of corporate objectives, while at the same time being cost effective and well motivated.

Further reading

Armstrong, Michael, *A Handbook of Personnel Management Practice*, Kogan Page, 1984.

Bramham, John, *Practical Manpower Planning*, The Institute of Personnel Management, 1982.

The Institute of Personnel Management, *Perspectives in Manpower Planning*, 1974.

5

Recruitment of engineering staff

Peter M. Coughlin

If an organization is to be staffed with competent, well-motivated staff it is imperative that senior management establish clear policies for the authorization of vacancies, for the selection of candidates and for the approval of engagements. Within these corporate policies the personnel manager must lay down and firmly control precise procedures to be followed for all recruitment throughout the organization, without exception. Finally it is essential that the personnel manager ensures that the progress of new employees is monitored by line managers so that performance and skills can be developed, or in the circumstances of inadequate performance the employment terminated at an early stage.

The recruitment process begins with the decision to fill a vacancy and the process cannot be regarded as complete unless and until the new employee has performed satisfactorily for a period of at least six months after starting employment.

Job descriptions

At the risk of stating the obvious it must be said that before embarking upon a recruitment process it is necessary to know exactly the duties to be performed by the new recruit. Far too many appointments turn out to be a failure because the duties actually required of the new employee are materially different from what was expected when the appointment was made. The means of avoiding this problem is the establishment of job descriptions.

The planning of the organization's structure will have included a statement of the job titles and principal functions for all posts in the organization. Armed with this basic data the personnel department should initiate the development of job descriptions for every post. A word of warning is appropriate here to the over-zealous administrator. It is counterproductive to aim for job descriptions that are meticulously detailed and purport to cover every facet of every activity that may be required of employees. Such documents are always out of date and may restrict the line manager from obtaining the flexibility he will require of his staff.

A workable job description ought to show:

1 The job title.
2 The title of the immediate line manager (for accountability).
3 A list of the principal duties and responsibilities required of the job-holder.
4 A clause indicating the flexibility of the job description (for example, 'to carry out such other tasks as are authorized from time to time').

The personnel department should arrange for line managers to draft job descriptions for all posts immediately subordinate to them. These descriptions should be drafted on standard forms – or at least in standard format. The draft descriptions must be checked and approved by the department managers. Finally, the personnel department should edit the descriptions, have them typed (preferably on a word processor for ease of storage and retrieval) and filed in an easily accessible manner.

The personnel department must ensure that if a new post is created anywhere in the organization, or if the functions of any post are changed substantially, the job description is prepared or amended. The objective should be to hold on file a complete set of up-to-date job descriptions. The recruitment process can then be put into operation smoothly and quickly.

Establishment of vacancies

If clear procedures do not exist there can be much confusion (accompanied by waste of time and money) if a line manager, believing that he has a vacancy, has the ability to trigger the recruitment process only to find at a later stage that a more senior manager vetos the engagement of another employee. It is vital, therefore, to define clearly the procedures for obtaining positive approval to recruit – before any effort is expended on recruitment.

Since labour costs form a very high proportion of an organization's expenditure it is essential that the staff complement is held at the lowest level consistent with the ability to achieve corporate objectives. It follows that all recruitment must be rigidly controlled and that every vacancy must be justified by the line manager and authorized at the highest practicable level – never lower than the member of corporate management responsible for the department claiming the existence of a vacancy. In a small organization it is sensible to arrange for the chief executive to authorize vacancies – whether for new posts or for replacements.

The preparation of the organization's annual budget will set the framework for the staff complement required for the coming year, and it is desirable that the labour costs incorporated in the budget are developed from a clear statement showing every post assumed to be filled during the budget year. Approval of the entire budget by the board of directors will therefore result in the existence of a listing of posts expected to be filled, allowing the personnel department to compile a listing of anticipated (but not yet approved) vacancies.

Caution ought to be exercised when reviewing the list of expected vacancies since budgets are prepared well ahead of the control period and circumstances may have changed by the time the anticipated vacancy was expected to arise.

In any event budget managers have been known to submit inflated bids for funds (and staff). Therefore the listing of expected vacancies should be regarded only as a guide. It must not be treated as authority to recruit.

The onus is on the line manager to initiate recruitment for each post – whether additional or replacement. He should prepare, on a standard form, a recruitment specification listing:

1 The job title.
2 The title of the immediate line manager.
3 Desirable qualifications.
4 Desirable experience.

At the same time the line manager should verify the content of the job description. The department manager should agree the recruitment specification before it is passed to the personnel department to initiate the recruitment process.

It should not be assumed that every post must be filled by a permanent employee. The organization's manpower plan should be reviewed in order to assess whether a new engagement is likely to be needed indefinitely. If this is not the case and if the eventual imbalance will not be redressed by a retirement or resignation, it may be preferable for the new engagement to be taken on for a fixed-term or on a temporary basis through an agency. It may be possible to fill the vacancy by internal transfer. These aspects should always be considered by the personnel manager for review with the department manager before the recruitment process begins.

Recruitment is initiated by the personnel department preparing a staff requisition. This document should show:

1 The job title.
2 The department.
3 Employment status (permanent, fixed term or agency).
4 Effect on staff complement (addition or replacement).
5 Availability of budgeted funds.
6 Salary grade.

An example of a standard staff requisition form is shown in Figure 5.1. This example is for the recruitment of permanent staff; it is usually convenient to use a different, slightly abbreviated, version for the engagement of agency staff.

The staff requisition is signed by the personnel manager prior to the department manager seeking authorization from the responsible corporate manager (or the chief executive in a small organization). The authorized staff requisition, the job description and the recruitment specification together provide the personnel department with all they need to proceed with recruitment.

Recruitment procedures

Obtaining a short list of candidates

It is important that all recruitment is conducted in a consistent and coordinated manner and nobody outside the personnel department should be permitted to respond to enquiries from candidates – particularly in respect of terms and conditions. The establishment of terms and conditions to be applied to a post is the clear responsibility of the personnel manager. He should do this at the outset of a recruitment programme and if it is necessary to obtain approval for a salary level or for the provision of benefits this should be done immediately so that clear answers can be given when dealing with candidates.

A recruitment campaign is likely to bring the organization into contact with many outsiders apart from candidates. These may include recruitment consultants, agencies and referees. It is important therefore that all parties are treated with efficiency, competence and courtesy. For example, every application and enquiry should be acknowledged on the day it is received.

It is scarcely possible to administer a recruitment campaign properly unless a progress schedule is prepared for each vacancy. The schedule should be updated with meticulous care. A progress schedule (see the example in Figure 5.2) should list the following:

1 Names of candidates.
2 Dates of enquiry acknowledgements.
3 Dates application forms sent and returned.
4 Initial decision – interview/hold/reject.
5 Interviews – dates/times/results.
6 Dates offers approved/sent/returned.
7 Processing – medical examinations and references.
8 Start date of successful candidate.

If it is decided that a post should be filled on a temporary basis it is appropriate to simplify the recruitment procedure – simply sending the recruitment specification and job description to suitable agencies, asking them to submit CVs and interview reports for potential candidates. Staff taken on in

Staff requisition (for permanent staff)
Form no 15. (Use form 16 for agency staff.)

Serial number: Date of issue:

Details of vacancy
Job title

Section

Department

Special requirements

Details of candidate
Educational requirements

Experience required

Signed ________________ Date __________
Head of department

Approval
This appointment is a replacement for __________
This post is/is not allowed for in the budget
I approve recruitment for this post

Signed ________________ Date __________
Personnel manager

Authorization
Recruitment for this post is authorized

Signed ________________ Date __________
Executive director

Signed ________________ Date __________
Managing director

Appointment
Vacancy filled by ________________ Start date __________

Figure 5.1 *Staff requisition. This is an example of a requisition for a permanent staff recruitment. The use of such standard forms ensures proper adherence to authorization procedures (the approval and authorization levels shown in this example would obviously be different from one company to another)*

Recruitment progress schedule										
Job title								**Requisition no.**		
Department										
Names of candidates	Enquiry rec'd	Application forms		Enquiry/ form to....	Reply reject hold int.	Interviews				
		Sent	Ret'd			Date	Time	Result	Advised	
NOTES										

Figure 5.2 *Recruitment progress schedule. A schedule such as this, used within the personnel department, is invaluable for keeping track of events in a recruitment programme*

this way are employees of the agency. The agency charges an hourly rate for the person, from which it deals with salary payment, National Insurance and other formalities.

Procedures for assembling candidates for a permanent or fixed term contract may include one or more of the following options:

1 The company 'grapevine' (personal contacts).
2 Direct advertisements.
3 Agency candidates.
4 Advertisement by a recruitment consultant.
5 Preliminary screening by a recruitment consultant.
6 Use of head hunters.

In most cases satisfactory results can be achieved by direct advertisement. However, time and cost can be minimized by careful selection of the advertising medium to be used. Care should be taken also to choose the appropriate publication day when using newspapers.

Agency introduction fees are expensive but it can be worthwhile using an agency if the post is specialized and the agency has a good reputation in the relevant field. Some agencies are very well known in particular areas and always have worthwhile candidates on their books. In such cases, if rapid recruitment is required, it may be worth using an agency despite the cost. It is common to advertise directly and simultaneously to use an agency in order to widen the field. It is always sensible to agree with an agency on their terms and conditions before asking them to offer candidates.

Sometimes it is worth asking a recruitment consultant to advertise the post, either because inclusion in a large panel advertisement by a well-known consultant may be more eye-catching, or because it is wished to conceal the employer's name (from candidates or existing staff).

Preliminary screening by a recruitment consultant is only worthwhile if the vacant post is very specialized and there is a shortage of resources within the employing organization for selection purposes.

Head hunters would be used only for very senior or particularly sensitive appointments.

In all cases involving advertising it is well worth having the text of the advertisement thoroughly vetted to ensure that the message will be clear. For direct advertising it will minimize internal effort if the advertisement specifies that candidates must submit CVs to a named member of the personnel department (no telephone number should be given in the advertisement). Perhaps, for junior posts where candidates may not have CVs, the advertisement could invite potential candidates to write to the named person requesting job details and an application form. Before the first responses are received it should be agreed as to who will have authority to reject candidates. Resources must be available for the rapid handling of applications. Normally, the personnel department should be capable of dealing with initial screening and rejection of candidates.

The progress schedule should be brought into use as soon as the first enquiry is received. Each application should be acknowledged on the day of receipt, and it is ideal if the preliminary screening can be carried out at once. If this is the case the first letter to the candidate will either enclose an application form or will incorporate a rejection.

When rejecting candidates, no reason should be given other than a statement that more suitable candidates have come forward. This method of responding should obviate possibly acrimonious communications from unsuccessful candidates wishing to dispute the reasons for their rejection.

It is useful for application forms to be sent to candidates whose CVs show them to be of possible interest. The use of forms ensures that all the required information is presented, and in a consistent manner. Application forms should contain, in addition to the usual personal and career details, the following:

1 A statement on the organization's equal opportunity policy.
2 Ethnic origin of candidates.
3 Details of disabilities.
4 Restrictions on geographical mobility.
5 Details of achievements and successes.
6 Names and addresses of two referees who have knowledge of the candidate's work.

The first three items will enable the equal opportunity policies to be monitored – an increasingly important facet of employment.

As the completed application forms are received a short list for interview is prepared and unsuccessful candidates should be advised as soon as possible. The short list must be prepared by the line manager in consultation with the personnel department. It is a very important part of a manager's role to be responsible for his staff and this responsibility should begin with the manager nominating the short list candidates. A manager should never be made to feel that unwanted candidates are being forced into his team.

Interviewing

It is wise to limit the number of interviews to a maximum of three per day and it is important that interviewing is properly organized and smoothly administered in a courteous manner. It is vital that interviewing is really effective. This implies that all interviewers should be capable of operating efficiently. In this regard it is highly desirable that managers should be trained in the acquisition of interviewing skills so that they will be fully effective in this important activity.

Clearly there are many methods of conducting employment interviews, but possibly the most effective interviews are conducted in two stages. The first should be handled by a senior member of the personnel department whose task is to assess the candidate in an overall sense, particularly in terms of personality and ability to fit into the organization. It is desirable that this interviewer has the power (to be exercised only in extreme cases) to reject a totally unsuitable candidate on the spot. Normally of course the candidate will go on to the second stage – an interview by the line manager (who may be assisted in the interview by not more than one member of his staff).

It is always necessary to make thorough preparations for interviews and it is recommended that arrangements include the following.

Physical arrangements

The interview room should be set out fairly informally. Probably the best arrangement is for the interviewers and the candidate to be seated at a round table – at least with the candidate being given a comfortable chair! If interviewers and candidates are seated on opposite sides of a desk, an unwanted barrier separates them. A circle of low easy chairs presents problems in locating papers, hands and coffee. Attention should be given also to heating, ventilation and lighting in the room.

Preparation of interviewers

Interviewers should be given (and should have studied in advance) copies of the job description, the recruitment specification and the candidate's application. As part of their preparation, the interviewers should agree between themselves on allocating areas of questioning. Often it is a good idea for the interviewers to ask every candidate for a post a set of standard questions, so that notes of their responses can facilitate comparison between them.

Interview structure

The senior interviewer should open the proceedings by introducing himself (or herself) and any other interviewers by name and title. He should then go on to set the scene as part of a wider introduction incorporating reference to the company, the department and the vacant post. The senior interviewer should then explain to the candidate the way in which the interview will be conducted. The bulk of the interview structure is very much a matter of personal choice and will depend to some extent also on the number of interviewers.

Irrespective of the interview structure, it is imperative that the interviewers ask open-ended questions (to encourage the candidate to talk) and that they *listen* to what the candidate has to say. As a rough rule of thumb guide, the candidate should contribute 80 per cent of the discussion.

At the end of the interview the candidate should be advised of the next step, being given some idea of the likely timing of a decision.

Control of the interview

The candidate should be put at ease at the beginning of the interview. Thereafter, in order to extract as much useful information as possible, the interviewers should not hesitate to ask probing questions, within their prearranged subject areas.

Points arising from the candidate's statements and answers should be followed up but interviewers should beware of digressing, and should take care that they do not find themselves talking excessively.

There is no reason why interviewers should not use notes, if these would help to organize the questions.

It is very important to prevent external distractions during the conduct of an interview. Strict instructions must be given that the interview may not be interrupted for any reason other than:

1 The building is on fire.
2 The arrival of coffee!

Coffee timing should be prearranged, so that a few moments' break can be taken at a convenient point, as an alternative to interrupting the interview dialogue.

Interview reports and follow-up
The personnel department interviewer should complete an interview report, preferably using a standard form, incorporating comments on:

1 Physical aspects.
2 Attainments and capabilities.
3 Intelligence and communication skills.
4 Disposition and personality.
5 Interests and activities.
6 Circumstances/availability/expectations.

The line manager and any other interviewer should each complete an interview report commenting on:

1 Suitability in terms of knowledge and experience.
2 Suitability in terms of disposition/personality/acceptability.

All reports should show the interviewer's overall assessment of the candidate using a scale constructed on the following lines:

- First class – would fill the post admirably.
- Capable of performing the job satisfactorily.
- Doubtful if he could perform the job satisfactorily.
- Unsuitable for the position.
- Worth considering for a different position, for example. . . .

The line manager should conclude his report with a summary of the interview and the action proposed. The personnel department should agree with the line manager as to the action to be taken in regard to each interviewee. Figure 5.3 shows an example of an interview summary report form.

When the successful candidate has been chosen, the terms of the offer of employment should be discussed between the line manager and the personnel department before agreement is sought from the personnel manager. Finally authorization is given by the appropriate member of corporate management and personnel department are able to make the offer of employment.

Having chosen the successful candidate it is necessary to consider carefully

<table>
<tr><th colspan="4">Interview summary and recommendation</th></tr>
<tr><td>Name of candidate</td><td colspan="3"></td></tr>
<tr><td>Post for which considered</td><td colspan="3"></td></tr>
<tr><td>Interviewed by</td><td></td><td>Date of interview</td><td></td></tr>
<tr><td>Physical
First impression, appearance, manner, speech, poise and confidence. Anything special about health or illness</td><td colspan="3"></td></tr>
<tr><td>Attainments and capabilities
Anything additional to the facts given on the application form about academic achievements, experience at work or other attainments or capabilities</td><td colspan="3"></td></tr>
<tr><td>Intelligence and communications skills
Bearing in mind the post for which the candidate is being considered, indicate level of intelligence (pointers are given by clarity of thought, mode of expression, speed of understanding, powers of reasoning, etc) and ability to communicate</td><td colspan="3"></td></tr>
<tr><td>Disposition and personality
Comment on the candidate's reliability, likely acceptability to others, ability to influence and lead, self-dependence, initiative and responsibility. Does the candidate have a personality which fits the job in question?</td><td colspan="3"></td></tr>
<tr><td>Interests and activities
Comments additional to those made on the application form</td><td colspan="3"></td></tr>
<tr><td>Circumstances/availability/expectations
Comments additional to those made on the application form. Indicate problems that could arise, e.g. movement of home, travelling difficulties, salary expectations and matters relevant to likely acceptance of an offer from us.</td><td colspan="3"></td></tr>
<tr><td>Summary of interview and general comments</td><td colspan="3">Mark candidate's position on scale
First class – would fill post admirably
Capable of performing job satisfactorily
Doubtful if could perform the job satisfactorily
Unsuitable for position
Worth considering for a different position – please specify</td></tr>
<tr><td>Recommendation</td><td colspan="3">Signature</td></tr>
</table>

Figure 5.3 *Example of an interview report and recommendation form*

how to deal with those who were not first choice. It is never certain that a chosen candidate will accept an offer of employment. Hence it is sensible to pick one or two other candidates, considered to be capable of performing the job satisfactorily, as second choice. Holding letters should be sent at once to these and rejection letters to the others. It is as well at this stage to check the recruitment progress schedule to ensure that rejection letters have been sent to all candidates other than the chosen two or three.

Offer and contract of employment

Conditions precedent

It is important to ensure that all offers of employment are subject to the candidate satisfying two conditions:

1 The issue of a medical report by a doctor nominated by the organization.
2 The receipt of satisfactory references from the nominated referees (and any previous employers).

When the offer of employment is made the candidate should be instructed to attend for medical examination. This is important because verification is necessary of the candidate's physical ability to undertake the required duties. In any event the employer should be aware of any serious disabilities or medical problems so as to be prepared for future difficulties that might arise.

It is sometimes argued that references are not worth the paper on which they are written but it is always worth obtaining them, particularly from the most recent past employer, because they may well reveal past problems. If a reference is vague it is wise for the personnel department to contact the referee by telephone – often people will say what they would not write. A letter should always be sent to thank a referee for his courtesy in writing the reference.

The contract of employment

Every employee has the basic right under the Employment Protection (Consolidation) Act 1978 as amended by the Employment Acts 1980 and 1982, provided he works sixteen hours or more a week, to be given a written statement or a written contract of employment setting out the following details concerning the employment:

1 The identity of the parties.
2 The date on which the employment began.
3 The date on which the employee's period of continuous employment began (taking into account any employment with a previous employer which counts towards that period).

4 The expiry date of the contract (if for a fixed term).
5 The rate of pay and the pay period.
6 Any rules as to hours of work, including normal working hours.
7 Entitlement to holidays, including public holidays, and rates of holiday pay.
8 Rules on sickness or injury absence and sick pay.
9 Details of the pension scheme and whether the employment is contracted out of the State Additional Pension Scheme.
10 The length of notice the employee is entitled to receive and must give.
11 The employee's job title.

The contract must include notes on the following:

1 Any disciplinary rules affecting the employee (there should be clearly stated rules known by all employees).
2 The person to whom the employee can apply if dissatisfied with any disciplinary decision.
3 The person to whom the employee can take any job grievance.

The organization's disciplinary and grievance procedures must be included. It should be noted that subsequent changes to any of the above items must be notified in writing to employees within one month of the change.

A contract of employment that contained only the items set out above would not be adequate from the employer's point of view since he will wish to have the specific right to impose additional requirements. Thus, within a single printed contract all the above matters should be included together with the following:

1 Name of department (indicating that the employer has the right to move the employee to another department).
2 Place of work (but adding 'or such other place as the company shall, on giving reasonable notice, determine').
3 A list of the employee's obligations including confidentiality of information concerning the business or other persons.
4 Restrictions on undertaking other work.
5 The ownership of rights over inventions and discoveries.
6 Rules on dealing with the media, writing books and the like.
7 Any restrictions on dealings in the employer's shares.

Only the personnel manager should be authorized to sign employment contracts on behalf of the employer. Duplicate signed copies should be sent with a covering letter to the successful candidate, inviting him to sign one copy and return it to the personnel department. When the signed contract has been returned by the new employee, the personnel department should write to any candidates 'on hold' to advise them that they have been unsuccessful.

Induction and probationary period

Induction

It is essential that every new employee should be introduced into the organization by means of a carefully planned induction programme which should begin immediately on arrival for the first day of employment. Proper induction will engender a feeling of welcome which will be valued by the new employee, helping him to feel part of the organization at a very early stage.

Depending on the seniority and nature of the new employee's work, up to a week should be set aside for the induction programme. Irrespective of the programme's length it should incorporate four elements, namely:

1 An initial session with the personnel department.
2 A meeting with the head of department (not the immediate superior).
3 An introduction to the department by the immediate superior.
4 An introduction to all the other departments in the organization.

The receptionist should be briefed in advance of the new employee's arrival, and should take him to the personnel department for the initial session, which should include provision of answers to questions raised by the new employee. This session should include:

1 A review of the company's organization and management structure.
2 The provision of information on benefits, clubs and the like.
3 Completion of initial administrative procedures for salary payment, income tax and similar items.
4 Issue of the organization's health and safety statement.
5 Issue of the employee information handbook.
6 Issue of the detailed induction programme.

The new employee should then be taken to meet the head of department who, thus, will be seen as a less remote figure in the future. The head of department should welcome the new employee and give an outline description of the department's work. The head of department should then hand over the new employee to the immediate line manager for detailed induction, aimed at the following:

1 To put the new employee at ease.
2 To interest him in the job and the company.
3 To provide basic information about working arrangements.
4 To indicate the expected standards of performance and behaviour.
5 To advise details of any training arrangements.

To avoid bewildering the new employee, no attempt should be made to cover all the above matters during one session. The programme should contain a balance of departmental induction, visits to other departments and free time (based at the new employee's working place). If the company's facilities so permit a lunch could be arranged during the induction programme to enable the new employee to meet his colleagues in an informal manner.

Probationary period

It is often seen as unfashionable formally to incorporate a probationary period into the terms of employment. Nevertheless in practice such a period is essential. During the probationary period the new employee's performance progress is monitored to ensure that he is settling in and to check on how well he is doing. If there are problems they must be identified at an early stage so that they can be corrected. If they are ignored, the longer term repercussions may be serious.

The monitoring of new employees is also important as a means of checking on the organization's selection procedures. If a mistake has been made it is useful to find out how it happened so that the selection procedure can be improved in the future. Possible causes of poor selection may include inadequacy of the recruitment specification or the job description. Other causes may be poor advertising arrangements, inadequate interviewing techniques or prejudice on the part of the interviewers.

A probationary period of six months is appropriate in most cases and, during this period, the personnel department should ask the new employee's line manager to provide written reports of the results of progress checks. Such checks should be made, formally, after two months and again after four months. If the new employee is not proving satisfactory, he should be interviewed by his line manager and advised of the reasons for any dissatisfaction. In such cases the new employee should be given either:

- The necessary guidance or training to improve the level of competence.
- A warning that, if his performance does not improve, continuing employment will be at risk.

Any written warning given during the probationary period should be copied to the new employee's personal file (kept by the personnel department).

After six months employment, the new employee's head of department should complete a report commenting on:

1. Ability
2. Intelligence.
3. Further training or experience required.
4. Working relationship with colleagues.
5. Attendance.
6. Punctuality.
7. Recommendation as to future employment.

If the head of department reports that the new employee is unsatisfactory, it is likely that the wisest course of action will be for the personnel manager to terminate employment. Occasionally, it may be worthwhile giving a marginally unsatisfactory employee a further period of time, say, two or three months, during which to bring his performance up to the required standard. However this is rarely a practical solution since an employee who has not met the required standards, despite progress reviews, is unlikely to improve over a period longer than six months.

In most cases, if the selection and recruitment processes have been properly carried out, the first months of employment will be valuable to both the employer and the new recruit. This is a period when, with the help and encouragement of the employing organization, the new member of staff attunes his performance to the needs of the employer – thus developing a mutually beneficial relationship that should form the basis of a long and enjoyable career.

Further reading

Higham, M., *The ABC of Interviewing*, IPM, 1979.
Plumbley, P., *Recruitment and Selection*, 2nd ed. IPM. 1976.
Sidney, E (ed.), *Managing Recruitment*, Gower, 1988.

6

Remuneration

Peter M. Coughlin

The remuneration policy of a company must be aimed at attracting, motivating and retaining staff of the calibre needed to enable the organization to meet its objectives. A further important aim must be the control of payroll costs. It will be apparent, therefore, that employee remuneration is a much wider topic than it might appear at first sight. Certainly it should not be regarded simply as answering the question 'how much must we pay them?'

Scope of remuneration

Remuneration covers the whole spectrum of an organization's commitments to its employees in terms of salaries, other payroll related items, pension schemes, insurances, and other benefits. Within this range of items, remuneration should be geared to a properly designed and controlled pay and benefits structure, tailored to suit the complete employee hierarchy in the organization. Once the need for this across-the-board matrix approach is appreciated, it becomes apparent that a more or less scientific procedure is essential in order to minimize anomalies and discontent. Indeed, avoidance of discontent is an important measure of a remuneration policy's success.

Developing a remuneration system

Outline

Assuming that the remuneration system under consideration is to be implemented in an existing organization, it can obviously be assumed that a remuneration system of some sort must be in operation, although this may be inadequate or badly defined. The planning of a new system usually starts, therefore, with a careful analysis of the existing situation. The steps needed to complete the planning are:

1 Analysis of the existing situation.

2 Remuneration policy development.
3 Planning the implementation.
4 Job evaluation.
5 Market surveys.
6 Salary structure development.

Analysing the existing situation

Detailed analysis of the existing arrangements should begin by collecting the following basic data:

1 The organization structure, including the scope and numbers of staff.
2 The current method for evaluating jobs.
3 Payments, including salaries, bonuses, benefits, and payroll associated costs (such as employer's National Insurance contributions).
4 The current policies, and past results.
5 Influence exercised by trade unions or other organizations.

Once the facts have been gathered, the next step is to consider all the reasons prompting the need to change the system. These may include:

1 Ineffective controls.
2 Existence of anomalies.
3 Payroll costs are out of line with market conditions.

Another area requiring preliminary examination is the effectiveness of non-salary items of remuneration, such as bonuses, benefits and the value of the total remuneration package.

Developing a future remuneration policy

When considering a new remuneration policy, with all its proposed systems and procedures, account must be taken of the type, size and style of the organization, including the nature of its management style (how formal?) and the influence of any unions or other negotiating bodies. As a general guide, the aim should be to develop a policy which will not be excessively rigid or bureaucratic, but which will cover all staff in a properly coordinated, controllable and demonstrably fair manner. At the same time, the arrangements must lead to costs at a level appropriate to the economic situation of the organization.

The remuneration policy should take account of:

1 Comparable external rates.
2 Fairness within the organization.
3 The degree of formality required in the salary structure.
4 Provision for cost of living and performance related increases at salary reviews.

5 The development of a staff appraisal system incorporating a means of determining increments at salary reviews.
6 The place of incentives and benefits in the remuneration structure.
7 The development of procedures for controlling and reviewing remuneration.

Planning implementation of a new policy

When the future policy has been decided, it is necessary to prepare an action list for its implementation. This document will show all the steps to be taken, set down in chronological order with their target dates and responsibility nominations.

Job evaluation

Aims

A fundamental prerequisite to the establishment of a remuneration policy is the determination of the comparative values of jobs throughout the hierarchy. These values are used as the foundations on which to build the pay and benefits package. Armed with the range of job values, comparisons must then be made with other companies operating in the same or relevant industries in order to establish the prevailing market rates. The procedures used must be easy to understand and administer, and the system must be perceived as fair by the employees.

The aims of job evaluation can be summarized as:

1 Providing a basis for ranking jobs, so that they can be graded and grouped for the purposes of creating a salaries scale.
2 Ensuring that job values are assessed on objective grounds, from an analytical study of the job content, as far as possible.
3 To contribute to the establishment of a database from which job values can be reviewed and updated.

Techniques

Job ranking

Job ranking is the process of identifying and listing all jobs in the organization, and then placing them in their assessed order of importance.

In a job ranking scheme, jobs values are generally assessed by considering them as a whole, rather than by analysing them into their constituent parts. Each complete job is compared with others in the organization. Even though the complete job is under review, it is important to provide the evaluator with a list of job aspects to which attention must be given when making comparisons. If such guidelines are not laid down when the system is designed, there is a risk that an evaluator will pick aspects that are not key parts of the job, resulting in incorrect ranking.

Factors on which comparisons should be based include:

1 The nature of decision making expected in the job, including the importance of decisions, the degree of discretion allowed (bearing in mind the level of authority), the innovation or creativity required and the relevant time factors. This mention of time factors cannot be allowed to pass without reference to Elliot Jaques who, in his pioneering work at the Glacier Institute of Management, placed particular emphasis on the timespan of discretion as a measure of a job's worth. His definition of the timespan is, approximately, that this is the longest period which can elapse in a job before the jobholder's manager can be sure that his subordinate has not been exercising marginally substandard discretion continuously in balancing the pace and quality of his work (see Jacques, 1964).
2 The resources controlled, such as the number of staff, value of the budget and volume of turnover.
3 Complexity of the work and the variety of tasks to be performed.
4 The qualifications and skills needed to enable the required tasks to be performed. Management, administration and communicating skills must be included in addition to technical requirements.

Job descriptions (see Chapter 5) should exist, but these will need amplification in order to highlight those factors which are particularly relevant to the ranking comparisons.

The next stage is to designate certain benchmark jobs, arriving at the most important, the least important, and another which is approximately in the middle. All the remaining jobs are then fitted in around these benchmarks in their order of importance. The technique used in this part of the programme is that of paired comparisons, in which each job is compared with others on a one-by-one basis. This technique is illustrated in Figure 6.1.

The final stage is to group the jobs into grades by bracketing jobs of broadly equal responsibility, so that salary ranges can subsequently be decided and allocated to the grades.

Job classification

With job classification, the grades are defined first. Jobs are then assigned to their appropriate grades.

Job classification is best used in conjunction with job ranking. It involves an

Department and section		MECHANICAL DESIGN GROUP											
Name of evaluator		J SMITH ENGINEERING MANAGER											
Reference		A	B	C	D	E	F	G	H	I	J		
Reference	Job titles	GROUP LEADER	PROJECT ENGINEER	DRAWING OFFICE CLERK	MECHANICAL ENGINEER	SENIOR ENGINEER	DESIGN DRAUGHTSMAN	MESSENGER				Job score	Ranking
A	GROUP LEADER	X	0	2	0	0	2	2				6	4
B	PROJECT ENGINEER	2	X	2	2	2	2	2				12	1
C	DRAWING OFFICE CLERK	0	0	X	0	0	0	1				1	6
D	MECHANICAL ENGINEER	2	0	2	X	0	2	2				8	3
E	SENIOR ENGINEER	2	0	2	2	X	2	2				10	2
F	DESIGN DRAUGHTSMAN	0	0	2	0	0	X	2				4	5
G	MESSENGER	0	0	1	0	0	0	X				1	6
H									X				
I										X			
J											X		

Figure 6.1 *Job ranking by means of paired comparisons. This is a simple example to illustrate the method. Each job listed on the left is compared in turn with the jobs listed along the top. Marks are awarded according to whether the left-hand job is more important (two marks), of equal importance (one mark) or less important (no marks). These scores are added up along each line, and the jobs are ranked according to their total scores*

initial specification of the grading system to be adopted, including a detailed definition of each grade by reference to the factors which were referred to in the job ranking approach. Each benchmark job description is compared with these criteria in order to slot the job into its appropriate grade. At this point it is useful to refine the grade definition by reference to the job description. Remaining jobs can then be placed into grades by reference to the ranking hierarchy. Once this has been done, there may be little point in retaining the job classification system, since the grading system will include all the ranked jobs.

Points systems

In points systems, the idea is to award points to different aspects of each job, add them all up, and value jobs according to the total points scored.

Points systems are complicated, and involve a great deal of time-consuming and expensive analysis. As the analysis proceeds the points scored have to be reviewed and weighted according to the assessed importance of each factor. It is then necessary to check on how the various jobs stack up against each other.

The system appears to be scientific and impartial, but judgement must still be applied in deciding the weighting factors and in reviewing the scores. Several cases have been observed of detailed analyses being presented to the 'review committee', only to have them adjusted drastically by a senior manager because the results did not accord with his perception of the practical realities of the situation. On balance, unless particular circumstances dictate otherwise, it may be sensible to avoid the complexities of a points system and concentrate instead on job ranking, job classification, or a combination of both.

Making market surveys

A competitive salary structure will take account of salaries paid by comparable employers. It is obviously essential to investigate the market before attempting to set salary ranges for the various grades. It is equally important in the future to check regularly that the organization's salaries remain in line with those paid by competitors.

It is rarely possible to obtain exact parallels when seeking data on market rates. It is likely, therefore, that data will be assembled from several sources and reviewed critically with the objective of compiling a composite picture of salaries in areas which are close to those of the organization.

Useful salary data is obtainable from surveys published by various organizations. Those produced by Inbucon, the British Institute of Management, and other professional institutions are particularly helpful because they are presented with some statistical analysis. The upper and lower quartile levels and the median are given for each case. Other useful sources are trade associations and employers' organizations. Regular reports are also published by Incomes Data Services and *The Industrial Relations Review and Report.*

Often, other companies in a similar line of business are prepared to exchange salary data on a fully reciprocal basis. For such exchanges to be meaningful, it is desirable to conduct them by reference to job descriptions for benchmark jobs. This ensures proper comparability.

Recruitment advertisements can yield useful market data on salaries for certain jobs, particularly those which appear in professional and trade journals. In the same way, recruitment agencies are able to advise on market levels of salaries for posts lying in their areas of specialization.

When all the accumulated data have been analysed and assembled it is possible to make a first estimate of the likely mid-point salary level for each proposed grade.

Developing a salary structure

The salary structure should be designed so that there is scope for flexibility, and for adequately rewarding good performance. It is convenient to install a unified grading and salary structure for all levels in the organization, since this leads to, and is seen to involve, a consistent approach.

The structure should provide minimum and maximum salary levels for each grade. It is important to consider the span of salaries within each grade as well as the differentials between grades (measured from the mid-points). Clearly the span of salaries possible within a grade ought not to be too narrow, or there will be insufficient scope for rewarding performance. A wide spread not only allows this, but also gives the opportunity to slot new employees into a grade without difficulty. Usually a span of 20 per cent on each side of the mid-point will give an adequate spread (in which case the top salary payable in the grade will be 50 per cent above the lowest salary in the same grade).

The differentials between grades should be wide enough to make a clear distinction between one grade and the next, but not so wide as to cause difficulties in grading borderline jobs. A differential of about 15 per cent between the mid-points of adjacent grades is likely to produce acceptable results.

The number of different grades obtained from a job ranking exercise embracing all employees will obviously depend on the size of the organization and its structure and purpose. For illustration, suppose that a scheme has thirteen different grades (which is a likely number). Assume that the market survey has indicated a mid-point of £6000 for the lowest grade (grade 1) and about £32 000 for grade 13. If each grade has a salary range allowing a variation of ± 20 per cent about the mid-point, the resulting salary structure might be that shown in Figure 6.2.

In practice the best course of action is to prepare a structure using criteria such as those outlined above, and then to review the results critically, modifying them if necessary to suit the needs of the organization.

Once a salary structure has been implemented, it is necessary to review it periodically (usually annually) to take account of cost of living increases. The appropriate increases are made to the mid-point levels, and the upper and lower limits for each grade are then recalculated by applying the 20 per cent (or other operative) criterion.

It is, of course, possible to take a salary structure one stage further by

Job grade	Annual salary level		
	Minimum	Mid-point	Maximum
	£	£	£
1	4 800	6 000	7 200
2	5 520	6 900	8 280
3	6 348	7 935	9 522
4	7 300	9 125	10 950
5	8 395	10 494	12 593
6	9 654	12 068	14 482
7	11 102	13 878	16 654
8	12 768	15 960	19 152
9	14 683	18 354	22 025
10	16 886	21 107	25 328
11	19 418	24 273	29 128
12	22 331	27 914	33 497
13	25 681	32 101	38 521

Figure 6.2 *Example of a salary scale structure. In this example, the mid-points have been set to give intergrade differentials of 15 per cent, and the upper and lower range limits in each case lie at ± 20 per cent about the mid-point*

dividing each grade into a number of increments (which is the case in public service organizations, for example). However, this can lead to mechanistic, rigid and inflationary salary reviews. Consequently, most organizations do not consider this further refinement to be necessary.

Salary reviews

Principles

It is standard practice to review employees' salaries on a regular basis, taking account of changes in the cost of living and their performance. Many organizations review salaries twice a year, the purpose of one of these reviews being to award a standardized percentage increment judged to be relevant to all or part of the increase in the Retail Prices Index over the previous twelve months. At the second review, each employee is considered for a possible further award, based on performance over the past year.

The concept of two reviews each year is losing favour, particularly since the annual rate of inflation in the UK has become much lower than in the recent past. In these circumstances two reviews in a year will have an inflationary effect on the salaries bill, since even poor performers will receive a significant increase attributable to the cost of living, and many employers are reluctant to offer no increase at all on the occasion of the merit review.

On balance, it is probable that a single annual salary review will prove to be satisfactory. Such a review can be performance based, the value of individual increases being established as part of an annual staff appraisal procedure.

Performance appraisal

The annual staff performance appraisal must necessarily extend over a protracted period. In a well-managed scheme, these appraisals should include at least one interview between each member of staff and his/her manager, during which such matters as past performance, points needing correction or improvement, future performance objectives, possible training, and the employee's future career aspirations can all be discussed and documented for follow-up. For the purposes of salary reviews, the appraisal scheme should be designed to produce numerical performance ratings on a standard scale. These ratings must obviously be available in good time for the annual review, which will probably mean ensuring that they are extracted from the appraisal programme as a priority item, well ahead of the entire appraisal process completion.

An example of a performance rating scale follows, in which each level of rating carries a definition intended to help the managers responsible for carrying out the assessment, and to let the employees know the factors against which their ratings have been adjudged. The factors in the example range from one to five, as integers, but it should be made permissible for the process to be refined somewhat by the award of intermediate ratings, such as 1/2, 2/3, 3/4 or 4/5.

A performance rating example

1 *Outstanding*. The performance is distinctive. A major contribution is being made by the employee, setting him in the front rank of peers.
2 *Superior*. The employee consistently surpasses standard targets for the job.
3 *Fully effective*. A sound, all-round effective performance. The standard of work is satisfactory, targets and deadlines are met regularly, effective use is made of time, and working relationships are good. Difficulties and emergencies are minimized and coped with adequately, without unnecessary reference to superiors. It is expected that most staff will be operating at this level of performance.
4 *Below standard*. Performance is below the expected standard in one or more respects. With additional learning, experience and effort it is likely that performance can be improved to become fully effective.
5 *Poor*. The overall level of performance falls well below that desired. There

is consistent weakness in several major parts of the job. If urgent action by the employee to improve is not evident, serious consideration must be given to demotion or termination of employment.

Salary review procedure

The overall salary review process is likely to be a fairly complex affair (the degree of complexity depending upon the size of the organization and its policies). A formal and detailed programme should be prepared. Early steps in the programme should be:

1 First assessors list provisional performance ratings for staff under their control (the first assessors are the line managers directly responsible for the relevant staff).
2 The first assessors discuss the provisional ratings with their superiors or other second assessors, and they agree jointly on the ratings to be put forward to management. This is part of a process designed to ensure uniformity across the company (individual managers may hold very different views on what constitutes outstanding performance!) and to reduce the possibility of prejudice or other personal unfairness to any individual employee.
3 The company management (or its representative committee) consider all the proposed ratings, checking for consistency of approach across the organization, and amending the proposed ratings where necessary.
4 Management notifies assessors of the finally approved ratings. These are entered on the individual staff appraisal forms for discussion at subsequent individual appraisal interviews between the first assessors and their staff.
5 A listing is made of the numbers of staff who fall into each rating category, together with their current basic salaries.

Armed with the data derived from the last of these steps, it is easy to evaluate the total additional salaries cost that would arise for various rating/percentage increase combinations. The starting point for preparing such combinations is to select a percentage judged to be appropriate for all staff with fully effective performance (rating 3). This should account for most employees. Significantly greater percentage increases should be awarded to those with 1 and 2 ratings, while a substantially lower increase should be given to those with a 4 rating. Poor performers (rating 5), whose continued employment is in question, should normally qualify for no increase.

Some salary anomalies may be found during the review, with some staff being paid less than their peers. In such cases the fairest approach is to correct each anomaly before applying the appropriate performance related increase. In this way, the proposed post-review salary will be in line with others of equal merit (since the employee will have the benefit of the percentage increase being applied to a higher base salary).

A random example of correlation between performance ratings and salary increases might be as follows:

Rating	*Percentage salary increase*
1	10
1/2	9
2	8
2/3	6
3	5
3/4	4
4	2
4/5	1
5	nil

It should be borne in mind that there are unlikely to be many (if any) employees rated 1 or 1/2, so that the costs of their higher percentage increases are unlikely to be significant in terms of the total salaries bill.

The cost of pensions and other payroll items are related to (and will increase with) base salaries. It may be considered desirable, therefore, to minimize the increase in the total base salaries bill, while not depriving the employees of a reward for better than average performance. One way of achieving this is to set aside a small part of the total sum available to the company for pay increases, and pay this element in the form of a once only, non-pensionable bonus to those with performance ratings of 2/3 or better.

The option of a once only bonus may also be found useful in the case of an employee who would otherwise be taken over the top of his/her salary scale by a high performance increase, where the job responsibilities have not changed sufficiently to warrant immediate promotion to the next higher grade. In this case the base salary would be increased to the maximum scale limit, with the bonus used to provide the excess.

As far as possible, promotions to higher grades and their associated salary adjustments should take place at times other than the general salary reviews, in order that these two distinct functions are not confused.

Total payroll cost

The total payroll cost will be very much greater than the sum of all basic salaries. The details will depend to some extent on the accounting methods used and company policies on pension schemes and other benefits, but readers may be interested in the following example, which could apply to an engineering company operating a non-contributory pension scheme.

Item	*Percentage of basic salaries*
Basic salaries	100
Pension contributions	28
Employer's National Insurance	10
Permanent health insurance	1
Benevolent fund, etc.	9
Total payroll cost	148
Allowance for sickness absence	4
Total allocated cost	152

In this example, no less than 48 per cent has to be added to the basic salaries costs as a payroll burden in order to arrive at the total payroll cost. In practice, a further allowance has to be made for absence through sickness (in this example bringing the effective total payroll costs to over 150 per cent of the basic salaries).

After each salary review, the basic salaries payable to each grade of employee must be totalled, and enhanced by the payroll burden. The total, divided by the potential working hours available in a year (after allowing for leave and public holidays) will indicate the average or standard hourly costs of employees in each grade. These data are required for estimating, cost accounting, and budgetary control purposes.

Bonuses and other payments

Individual bonuses

The possible use of individual bonuses to reward staff with performance ratings above average has already been mentioned. Typically, a sum equivalent to about 1 per cent of the total basic salaries bill could be made available for such bonuses. This sum can be apportioned over staff rated 2/3 or better, using a suitably graduated scale that is designed to award the highest percentage bonuses to the best performing employees.

It is unlikely that profit related bonuses, which would be payable across-the-board to all employees, would have much (if any) incentive effect. Such bonuses are not recommended. Most of the potential recipients would feel that their ability to have a direct influence on corporate profits is remote.

Bonuses for senior executives

Whereas across-the-board bonuses for all staff are not seen as effective motivators, some organizations may feel that the situation is quite different when it comes to their senior executives. The main problem is to decide whether or not the performance of the executives in question can be shown to have a real influence on profitability. If a direct performance to profits link is apparent, there may be a case for introducing a bonus scheme.

A formula can be developed such that bonuses become payable if the organization's profits exceed a specified threshold level. The formula would allow calculation of individual bonuses according to the amount by which profits exceed the threshold. Each senior executive should, perhaps, receive a bonus of not less than 10 per cent, and not more than 40 per cent of basic salary. It is not practicable, however, to lay down guidelines for such a scheme, since the conditions, profits, salaries and numbers of qualifying executives vary so widely from organization to organization.

It is questionable whether senior executive bonus schemes have much motivating effect in the long term. Possibly the participation of the executives in any individual bonus scheme (if their performance so merits) at the time of salary review is all that is required. There is no general answer to this question. Each organization must consider the matter in relation to its own situation.

Bonuses for sales staff

Bonuses for sales staff engaged in obtaining orders for off the shelf goods, or for fairly simple manufacturing orders (such as double-glazed replacement windows) are likely to be payable in the form of commission on the value of orders taken, usually measured and paid against monthly figures. Other incentive schemes involve competitions, with valuable prizes for those who achieve the highest sales in a period.

In engineering organizations it is less usual to find sales staff who can claim to have generated any particular sale single handedly, and it will probably be more difficult to relate the sales volume achieved to any particular individual without being unfair to the other staff, both in and out of the sales department, who worked to obtain the orders. This being the case, there is unlikely to be much point in introducing an incentive scheme for these sales staff.

London weighting and relocation allowances

The cost of establishing a home in the London area, and the costs of living in the area, are markedly greater than elsewhere in the country. It is almost impossible to attract employees from other parts of the UK without making special provisions in the remuneration package.

A London weighting allowance is (usually) a non-pensionable addition to the monthly salary. Reference to the *Department of Employment Gazette* or to the *Industrial Relations Review and Report* will indicate the current London weighting values.

For new recruits, or for existing staff asked to relocate to the London area, allowances should be provided to cover the expenses of buying, selling and moving home (legal and agents' fees, survey charges and furniture removals). Normally, a daily living allowance is provided for an initial period, to bridge the gap between selling one home and moving into another.

The organization's regulations for the payment of relocation allowances should specify which allowances are payable to house owners, as opposed to tenants. They should incorporate time limits for house hunting, and they should be approved by the Inland Revenue to avoid the possibility of employees being asked to pay income tax on the allowances.

Share schemes for employees

Private share ownership is growing rapidly in the UK and this is partly due to the development of employee share schemes. Such schemes certainly increase the allegiance of employees and enhance their interest and long-term commitment to the organization.

All share schemes must conform with government regulations, and they must be approved by the Inland Revenue. The schemes are dependent upon a trust being set up by the employer. The employer provides money to the trustees for the purchase of shares, and the shares are held by the trustees on behalf of the individual employees. Two common types of scheme will now be described.

Profit sharing schemes

In these schemes, the Government specifies an upper limit to the value of shares which a company can set aside for any employee each year. Within this limit, the company decides on the total value of shares to be allocated for the year. After the trustees have purchased the shares, the employees receive any dividends in the normal way.

Under current regulations, the shares are held in trust for five years, after which they are given to the employees without deduction of income tax. An employee may ask that the shares be given to him after two years, but he will then become liable for income tax on their value according to the period held:

2 to 4 years 100 per cent taxable
4 to 5 years 75 per cent taxable

Although it is not mandatory for employees to contribute to such schemes, it is usual for employers to require them to do so. Typically this is arranged on a one for one basis, with the company providing a sum equal to the employee's contributions.

Savings-related share option schemes

In this arrangement, employees can opt to enter into a five-year Save as You Earn (SAYE) contract with the Department of National Savings or an approved building society. The fixed monthly payments can be chosen up to a specified maximum.

At the beginning of the five-year period, the share option price is established. At the end of the period, the accumulated savings, together with bonuses, are used to buy shares at the original option price.

Company cars

The provision of cars for employees who need them for their work is not really part of the organization's remuneration system. Such cars are tools of the job and, although they may be valued by the employees, they should not be considered in the same light as 'perk' cars.

In the 1970s, high income tax levels led employers to substitute tax-effective benefits for highly taxed pay. The most valued of these benefits was the company car. Although cars and company-provided fuel for private motoring are now taxed fairly severely, they remain tax-effective and are therefore attractive to employees. Apart from the tax aspect employees benefit significantly by not having to find the capital for car purchasing. On financial grounds the company car is a worthwhile benefit.

The provision of company cars is so widespread that employees over a certain seniority level consider them to be an essential part of their remuneration package. Today it is less common for cars to be provided as part of salary. It is usual to allocate a company car to senior staff without asking for any salary sacrifice in return. Indeed, it is now unlikely that an organization would be able to recruit suitable senior staff without the inclusion of a car in the offer package.

It is essential to have clear rules, firmly and impartially applied, to cover:

1 Definition of the seniority level at which a car will be allocated, together with definitions of the higher levels of seniority which qualify staff for more expensive cars.
2 Definition of the types of car to be offered (by make and model or by reference to cost limits and other criteria).
3 A statement of the associated motoring expenses for which the employer will pay.

Company cars are a very emotive issue. To minimize the risk of dissatisfaction and jealousies it is sensible to use staff grades to define the levels at which cars will be allocated. In the grading example given earlier in this chapter, it might be decided to provide cars for those in grade 10 and above. It might be further decided that better cars would be allocated to staff when they reach grade 13 status. In any event, the basis for allocation should be decided by the board of directors, and the resulting rules must be made known to all staff in the affected grades.

Unless a car user will drive very heavy mileage on business, the sensible replacement frequency for cars is either three or four years (although a shorter period may operate if the cars are subject to a contract leasing or other special fleet arrangement).

Where employees are given some freedom of choice in the type of car, it is obviously necessary to specify the permissible limits. One method is to nominate a 'marker' car in each class (normally two categories should be sufficient) from which the permitted upper on-the-road costs limits can be set. Provided these limits are not exceeded, and provided also that the company is not associated with a particular part of the motor manufacturing industry, there is probably little reason to restrict an employee's choice of car.

Occasions may arise when a member of staff would wish to have the use of a better car, offering to contribute part of the cost. Such arrangements are generally unwise and not recommended, one of the many possible difficulties being the risk of arguments about the disposal of eventual sale proceeds.

The employer should insure the cars comprehensively, and meet all maintenance and repair costs. Providing membership of a motoring organization or recovery service can also be considered.

It is expensive to charge all fuel used for private and business purposes to the employer's account, but it would be difficult to monitor and control a split allocation of fuel costs between the employer's and the employees' accounts. Unless the company is willing to bear the full cost, probably the easiest method for dealing with fuel expenses is for the employees to purchase all fuel themselves. They can then claim reimbursement for fuel used on the company's business (on a fuel cost per mile basis).

Unless a company car is a true pool vehicle, used by several employees and garaged at the workplace, its provision will lead to an income tax charge on the employee. The car is regarded by the Inland Revenue as a taxable benefit, and the charge is effected by the application of 'scale benefits' to the employee's income tax code.

The car benefit scale is published annually by the Inland Revenue. There is a basic range of benefits, applied according to several categories of engine capacity, for those who drive at least 2500 business miles during the tax year. Those driving less than 2500 miles suffer an increase of 50 per cent in the scale benefits, while 18 000 business miles in a tax year will reduce the benefit by 50 per cent. Benefits are reduced as from the tax year in which a car becomes four years old.

If any fuel is provided for private motoring a fuel benefit will be added to the car benefit, again on a graduated scale according to engine capacity. The scale benefits are similar to the car benefits, and they are reduced by 50 per cent if more than 18 000 business miles are travelled in a tax year.

It is usual to allow employees to acquire their allocated cars on favourable terms if they retire on pension or are made redundant. In other cases, departing employees would return their cars to the employer.

Private medical insurance

Private medical insurance, to cover the costs of specialist treatment, falls into a category which is similar in some respects to the provision of company cars. The benefit has become so common that it is almost essential to offer it as an inducement when recruiting, and as a means of giving peace of mind to existing employees. Both employer and employee benefit from the ability of a sick person to receive specialist consultation without delay, followed where necessary by treatment at a convenient time. These benefits are particularly helpful in parts of the country with long National Health Service waiting lists.

It is sensible to offer private medical insurance to all permanent staff in the organization (with their dependent spouses and children). It is seen to be divisive to offer this facility only to senior staff. The scheme can be set up in two ways:

1 A free scheme, in which the employer pays all the premiums.
2 A company scheme where the employees meet part or all the costs of the premiums (the cost advantage to the employees being the significant discount secured by the company).

For employees who are defined by the Inland Revenue as 'higher paid' the amount paid by the employer towards the premium is treated as a taxable benefit.

If an organization is contemplating the introduction of a scheme, discussions should be held with several of the companies offering the service. It is possible to negotiate substantial discounts from the standard premiums for group membership schemes set up by employers.

Retirement pensions

The provision of pension benefits is an area in which legislative changes are likely to continue for many years. It is essential that any employer considering the implementation of a pension scheme should refer to current legislation, in consultation with a first class professional adviser.

All pension schemes are subject to approval by the Inland Revenue, which lays down strict limits that may not be exceeded. These limits apply to the following:

1 *The amount of annual retirement pension payable* which, at the time of writing, must not exceed two-thirds of the final salary after ten years' service.
2 *A lump sum, payable on retirement.* Part of the normal pension may be commuted for a cash lump sum. This (again correct at the time of writing) is allowable to a maximum of three-eightieths of final salary for each year of service up to forty years. Thus the maximum is one and a half times final salary, but only if service is twenty years or more.
3 *Spouse's benefits in the event of an employee's death in service.* A cash benefit of up to four times the employee's annual salary plus a pension equal to two-thirds of the employee's notional pension at normal retirement date (based on current salary).
4 *Spouse's pension on an employee's death after retirement*, the limit being two-thirds of the retired employee's pension.
5 *Provision for inflation.* Pensions may be inflated in payment up to limits established from time to time by the Inland Revenue.

Increasingly pensions are becoming regarded as deferred pay rather than as benevolent hand-outs from generous employers. Another development has been the enactment of legislation that gives employees the right to make personal pension arrangements to facilitate the preservation of their pension benefits as they move from employer to employer during their careers. It is anticipated that employers will continue to provide schemes for those who would prefer to join them rather than setting up their own personal pension plans.

Matters to be considered when preparing to establish a pension scheme are:

1 Whether the scheme is to be administered directly or insured. Trustees will be required in both cases.
2 Whether there should be a single scheme with a common scale of benefits for all employees (generally this is the preferable arrangement).
3 The cost of the scheme as a proportion of basic salaries, and a decision as to the proportion (if any) of salary to be contributed by employees.
4 Entry and normal retirement ages. These should be the same for men and women (for example, an entry age of 18 years with a normal retirement age of 60).

The following details must be considered in relation to details of the proposed benefits (always bearing in mind the limits set by the Inland Revenue):

1 If the scheme is not to be contracted out of the State Earnings Related Pension Scheme, the formula to be applied for integrating these benefits into the company scheme benefits.
2 Definition of final salary to which the pension formula will be applied.

3 The proportion of final salary to be accrued for each year of service (for example, one-sixtieth).
4 Provisions, if any, for voluntary early retirement, or for early retirement at the company's request.
5 Pensions for permanent disability.
6 Preserved pension benefits and transfer values for early leavers.
7 The formula, if any, for incorporating the effects of the basic State Retirement Pension into the company pension.
8 Dependant's benefits on the death in service of an employee (lump sum payment, spouse's pension and any children's allowances).
9 Dependant's benefits on the death of a pensioner (spouse's pension, position in the event of the surviving spouse's remarriage and any children's allowances).
10 The formula and limits for commutation of part of the pension for a lump sum payment.
11 Arrangements for employees to make voluntary contributions in order to enhance their benefits.
12 The method to be adopted in deciding on annual pension increases.

Pension provisions have a high national profile, with much legislation existing and more to come. Inland Revenue regulations and limitations are strict and liable to variation. It cannot be emphasized too strongly, therefore, that the services of a first class professional adviser are essential. A competent adviser will be able to:

1 Prepare outline schemes for evaluation and consideration.
2 Prepare documentation for the chosen scheme, including the trust deed, application for Inland Revenue approval and handbooks for employees.
3 Cost the scheme and advise the required funding rate.
4 Undertake full administration of the scheme, including the maintenance of a database, provision of benefit statements, annual reports and the like.
5 Carry out regular actuarial investigations.
6 Provide regular help and guidance to the employer.

The investment of pension funds, outside the scope of this chapter, is clearly another matter for the maximum possible expertise. It is likely that small schemes will use the services of insurance companies or properly qualified investment organizations. In either case it is wise to invite proposals from a number of competing organizations for consideration by the employer, the fund trustees and the professional pensions adviser.

When establishing a pension scheme it is important to include a statement in the announcement to employees to the effect that:

> Whilst it is the intention of the company to maintain the scheme for the long-term benefit of employees, the company reserves the right to amend

or terminate the scheme at any time in accordance with the formal documents constituting the scheme.

Death, accident and disability insurance

Death in service benefits are incorporated into company pension schemes (up to a maximum of four times salary) but it is worth considering the additional provision of personal accident insurance. This type of insurance is reassuring to employees and the cost to the employer is low, rarely leading to claims. Cover should be on a world-wide, 24 hours per day basis, to include:

1 A multiple of salary (say, five times) payable on accidental death.
2 A scale of benefits payable in the event of major disability due to accident.

Apart from the peace of mind given to employees by personal accident insurance, there are administration savings since there is no need to arrange special cover when employees travel on company business.

At one time companies provided permanent disability insurance for their employees in order to safeguard (at least a proportion of) salaries, and to maintain pension contributions in the case of people absent from work for continuous periods of more than six months. Such insurance tends to be complicated in administration and its provisions are difficult to explain to employees. It is better to incorporate a provision into the main pension scheme so that, for example, after six months continuous absence on account of ill health or injury the employee will be granted an immediate pension of two-thirds of salary (or such lesser amount as the Inland Revenue may allow).

Subsidized meals

Provision of a staff restaurant or canteen is a benefit that is greatly appreciated in many companies. If it is decided to provide such a service, it will usually minimize the burden of administration if a professional catering organization is used, preferably paid on the basis of cost reimbursement plus a management fee.

The first decision is whether or not to provide meals for all employees. Usually it is best to make provision for all, if internal catering facilities are to be established. A related decision is whether or not to provide separate facilities for senior managers, or for visitors, or for both. It is usually desirable for some private facilities to be available. Sometimes senior managers consider it necessary to eat together in an atmosphere conducive to private discussion. It is often appropriate to entertain visitors privately.

To avoid unnecessary allocation of space, it may be satisfactory to have a compromise, whereby the norm is for all staff to use a common restaurant.

One or two small rooms can be set aside, to be used when required for entertaining visitors, by management, and for group working lunches.

It is important to agree a detailed annual budget with the caterers, and to insist on prompt monthly reports of expenditure against the budget. This is essential in order to control expenditure, whilst delegating the specialist management of the operation.

To make use of the restaurant attractive to staff, the costs will be heavily subsidized. It may even be considered that staff should not be asked to pay at all. Whatever the case, it is wise to provide staff with some form of identity as proof of their entitlement to use the restaurant. It is not unknown, where such precautions do not exist, for the number of diners to exceed the organization's staff complement, explained by the presence of messengers, contractors' workmen, casual visitors, other occupants of the building, relatives, girl-friends, boyfriends and even passers-by who have come to hear of the free or cheap meals available.

As an alternative to providing a staff restaurant, consideration may be given to the issue of luncheon vouchers. These, however, are a taxable benefit. In view of the high cost of meals in public restaurants, the annual cost per employee of issuing luncheon vouchers to a realistic value is likely to be of a similar order to the cost of modest internal catering. This may indicate the internal arrangement to be preferable, provided that the necessary space can be found. Certainly the use of a staff restaurant is likely to lead to shorter lunch periods than would be the case if staff had to seek their meals outside.

Effect of the total remuneration package

In the days of very high taxation levels, employers were continually seeking devices that could attract, motivate and retain staff by substituting tax-free goodies for part of salaries. A thriving industry grew up, peopled by advisers eager to advise employers of the latest schemes for circumventing the Inland Revenue. In that environment, a good range of perks acted as a strong magnet for the acquisition and retention of employees.

With falling tax levels, and with rising activity by the Inland Revenue to combat tax avoidance, the number of benefits being offered to employees has been much reduced. It is now more appropriate to set up direct pay arrangements that will be seen as competitive in the market place, while at the same time providing good financial rewards and structured promotion prospects based on proper appraisal of performance. If the direct salary is accompanied by a small number of ancillary benefits (sound pension scheme, share scheme, private medical insurance, cars for senior executives and a staff restaurant) plus the ability to earn bonuses for superior performance, it is likely that the complete remuneration package will prove satisfactory to staff, and it will therefore be effective.

References and further reading

Beacham, R., *Pay Systems: Principles and Techniques*, Heinemann, 1979.
Cowling, A. G. and Mailer, C. G. B., *Managing Human Resources*, Edward Arnold, 1981.
Greenhill, R. T., *Employee Remuneration and Profit Sharing*, Woodhead-Faulkner, 1980.
Jacques, Elliot, *Time-Span Handbook*, Heinemann, 1964.
Kenaghan, F. and Redfearn, A., *Determining Company Pay Policy*, Institute of Personnel Management, 1979.
Vernon-Harcourt, A., *Rewarding Management*, Gower, published annually.

7

Employment legislation

Peter M. Coughlin

Employment legislation exists to regulate matters concerned with the working environment, with who may or may not be employed, with various employment related benefits and with the protection of the employee. Most of the legislation in these areas has been subject to change in recent years and there is no reason to suppose that future changes will not occur. Textbooks are inevitably overtaken by such events, and any reader needing to know definitively the current position on any aspect of legislation should always consult a reputable publication, such as *Croner's Reference Book for Employers* (updated monthly).

This chapter concentrates on legislation enacted specifically for employment matters, but there are laws covering wider fields which also have significance in the employment field. Examples of such laws are the Finance Acts, Social Security Acts, Data Protection Act and some European Economic Community (EEC) legislation.

The importance of good communications

While it is essential that all employers comply with the legislation that affects them, compliance does not guarantee immunity from industrial relations problems. Such problems will be minimized only by adopting a positive attitude towards good communications and sensible consultation with employees at all levels.

Often a staff consultative committee can be a useful forum for disseminating information – in both directions. Care should be taken to set up any such committee under the chairmanship of senior management. This ensures that quick and authoritative answers can be given to employees' questions, and it has the additional benefit of preventing the committee from becoming a breeding ground for 'barrack room lawyers' who might be inclined to use it for subversive discussion of matters such as employment terms and conditions.

The existence of recognized trade unions can have a major influence on employee-related issues – often enshrined in legal provisions. This is not the place for detailed discussion of relations between employers and trades unions, but it may be worth bearing in mind that the engineering office which fosters good communications and effective consultation between management and staff is likely to engender a culture without undue pressure for union recognition.

The Health and Safety at Work Act 1974

The Health and Safety at Work Act lays obligations on employers and employees. It covers most work situations and its provisions are supported by codes of practice and guidance notes.

Employers' duties

Every employer must endeavour to ensure the health, safety and welfare (at work) of his employees. To that end attention must be given to the safety of plant, to the proper use of materials, to the provision of safety instructions and training. In particular proper attention must be given to the maintenance of the working environment, including the provision of safe access and means of escape. The employer must take account of the public and visitors when dealing with these matters.

Employers of five or more people must prepare, and keep up-to-date, a written policy statement about health and safety at work. The statement must show a firm commitment to the policies, describing the organization and arrangements for carrying them out. The statement must be brought to the attention of all employees, preferably by the issue of individual copies.

The policy statement should remind employees of their duty to care for their own health and safety as well as that of others who might be affected by their work. Employees have a duty to cooperate with the employer in helping him to comply with the relevant legislation.

Duty of designers and manufacturers

The Health and Safety at Work Act lays on designers and manufacturers the duty of ensuring that their products are designed and built to be safe and without risk to health when correctly used. This duty extends to testing and examination of products to check their adequacy for the specified duty. It is a requirement that information shall be provided that will enable products to be used at work safely and without risk to health. Similar considerations apply to the construction and installation of manufactured equipment.

Health and Safety regulations

Regulations promulgated as a result of the Health and Safety at Work Act 1974 apply to all employers. The more important regulations are described in the following paragraphs.

Safety representatives and committees

The Safety Representatives and Safety Committee Regulations 1977 provide for recognized trades unions to appoint safety committees in specified circumstances. The purpose of a safety committee is to work with the employer to ensure the health and safety at work of employees.

Reporting of injuries, diseases and dangerous occurrences

Regulations, issued in 1985, impose a duty on employers to report to the appropriate authorities, *inter alia*, fatal and major injury accidents, and dangerous occurrences. Records must be retained in the prescribed form for at least three years after each event.

First aid at work

The Health and Safety (First Aid) Regulations 1981 require employers to make first aid provision for their employees. Factors influencing the required provisions include the number of employees, the degree of danger and the nature, size and location of the business. The Health and Safety Executive has published a code of practice and guidance notes to assist employers.

First aid equipment

Every employer must provide clearly marked first aid boxes containing prescribed items. The locations of first aid boxes must be publicized, as must the names and locations of those in charge of them. In large companies or in hazardous conditions additional first aid equipment may be needed. If there are more than about 400 employees, a first aid room should be established.

First aid personnel

If the organization is small (less than 150 office staff or fifty industrial staff) a person must be appointed to take charge of the first aid box and to organize help in cases of illness or injury. This person need not be qualified in first aid. In larger organizations a trained first aider must be available at all working times. It is good practice to have one trained person for every fifty employees, whether in an office or in an industrial environment.

Enforcement

The body responsible for enforcing the provisions of the Health and Safety at Work Act is the Health and Safety Executive, which is responsible to the

Health and Safety Commission (which in turn is responsible to the Secretary of State for Employment).

The Executive can appoint inspectors to carry out its enforcement and advisory duties. Inspectors have the right of access to an employer's premises at any reasonable time, or at any time where the inspector considers that a dangerous situation exists.

Inspectors can issue notices which require an employer to take action necessary to comply with the Act or its regulations within a specified time limit (known as improvement notices).

Where an inspector discovers an activity that could lead to serious injury he may issue a prohibition notice, which forbids the employer from carrying on the dangerous practice.

Compulsory display of notices

Every employer is required to display certain items for all employees to see. The health and safety policy statement and first aid notices (both referred to already in this chapter) fall into this category. Further items relevant to all employers are the employer's liability insurance certificate (see later) and instructions on the procedure to be followed in case of fire. Other items, described below, are appropriate to specific locations.

Factories

The Factories Act 1961 requires the display of an extract from the Act together with information about the inspectorate, the medical adviser and the working hours of women.

Offices

The Offices, Shops and Railway Premises Act 1963 provides for the display of a thermometer on each floor. It also requires the publication (or distribution to all employees) of a copy of the Act.

The Factories Act 1961

The Act defines a factory as premises where people are employed on manual work for specified processes. The Act applies also to various building works and to works of engineering construction. The Act is long and detailed: hence the following is a brief summary of its provisions.

Notices and records

The factory inspector must be advised a month before a factory is opened and a 'general register' specified by the Act must be maintained.

Health and welfare provisions

In addition to matters regulated by the Health and Safety at Work Act, the Factories Act makes provisions concerning cleanliness, overcrowding, temperature, ventilation, drainage, lighting, sanitary accommodation, drinking water, washing facilities, accommodation for clothing and facilities for sitting.

Safety provisions

Additional to the requirements of the Health and Safety at Work Act, this Act prescribes regulations about:

Fencing and guarding dangerous machinery and restrictions on cleaning machinery by persons under the age of 18 years.
Training and supervision of people using dangerous machines.
Hoists and lifts.
Chains, ropes and lifting tackle.
Cranes and other lifting machines.
Floors, passages and stairs.
Safe means of access and safe places of employment.
Dangerous or explosive substances and fumes.
Steam boilers, receivers and containers.
Air receivers and gasholders.

Special provisions

The Factories Act makes special regulations about the health, welfare and safety of workers, the employment of young people and normal factory working hours and overtime.

Offices, Shops and Railway Premises Act 1963

This Act applies, *inter alia*, to all offices and requires that the local authority should be advised in advance of the first use of premises as offices. The local authority will automatically inform the relevant fire authority. The Act covers much the same ground as the Factories Act in terms of health, safety and welfare.

Employers Liability (Compulsory Insurance) Act 1969

Insurance cover for at least £2 million must be taken out under an approved policy against liability for personal injury and disease sustained by employees on account of their employment.

Fire Precautions Act 1971

A fire certificate, issued by the local fire authority, must be obtained for every place of work where more than twenty people are employed. Such certificates prescribe fire precautions for the premises. Even where less than twenty people are employed, some basic fire precautions must be observed, such as outward opening doors for heavily occupied rooms, clear and marked escape routes and the provision of fire extinguishers.

Employment of non-UK nationals

EEC nationals

Other than nationals of Portugal and Spain, EEC nationals may enter the UK to work without a work permit. Admission is normally allowed for six months and, if the person takes up employment, the Home Office will issue a residence permit. Such permits can be issued for up to five years, and after four years' continuous employment the person may live and work permanently in the UK. Any employment must be on equal terms to those applied to UK nationals.

Commonwealth citizens and non-EEC foreign nationals

Except for Commonwealth citizens who can prove that one grandparent was born in the UK, the prospective employer must obtain a work permit from the Department of Employment. Permits are issued under stringent conditions for a maximum initial period of twelve months for specified categories of workers. Changes of job must be approved by the Department of Employment. After four years in continuous employment the time limit may be removed.

The Disabled Persons (Employment) Acts 1944 and 1958

The Manpower Services Commission maintains a (voluntary) register of disabled people who meet certain criteria and the Acts provide that employers of not less than twenty people must engage a quota of disabled people. The level at the time of writing is 3 per cent.

Where a company employs more than 250 people, the annual directors' report must describe the policy in relation to the employment, training and promotion of disabled people.

Employees with convictions

In accordance with the Rehabilitation of Offenders Act 1974 convictions are treated as 'spent' after periods of between five and ten years, according to the severity of the sentence. Lesser periods may apply to young offenders. Employers may not ask job applicants whether they have any spent convictions, and they may not discriminate against employees or dismiss them because of spent convictions.

When giving a reference, an employer must take great care if he wishes to reveal a spent conviction. Such disclosures must be relevant, correct and made without malice.

Racial discrimination

The Race Relations Act 1976 affects employment by forbidding discrimination against job applicants, employees and members of the UK Government's Youth Training Scheme. An industrial tribunal may make awards to such people if they are discriminated against. The Act says that discrimination occurs when, on racial grounds:

1 A person is treated less favourably than others would be treated, or is segregated from others.
2 An unjustifiable regulation is made which is such that the number of people in a particular racial group able to comply is proportionately smaller than the number outside that group that can comply.
3 A person is victimized for becoming involved in discrimination proceedings.

In the recruitment field it is unlawful for employers to discriminate in recruitment arrangements, in the employment terms or by not offering the job at all.

It is unlawful for employers to discriminate between employees in employment terms, in matters of promotion and transfer, or by penalizing employees on racial grounds. The employer also carries a liability for the discriminatory behaviour of his employees. Dismissal of an employee solely on racial grounds would be ruled unfair in terms of the Employment Protection (Consolidation) Act 1978.

An industrial tribunal is empowered to make a monetary award up to a specified maximum to a job applicant or employee who succeeds in a discrimination claim. However, such compensation may be awarded only if the complainant has suffered loss (and this includes injury to feelings). The tribunal also may make an order against the employer, declaring the complainant's rights or recommending the employer to stop the discrimination.

Sex discrimination

Matters concerning sex discrimination are covered by the Sex Discrimination Acts 1975 (amended) and 1986, the Equal Pay Act 1970, the Equal Pay (Amendment) Regulations 1983 and various EEC Council directives. All of these combine to outlaw sex discrimination in employment. They also establish the rights of men and women to equal treatment in their terms and conditions of employment if they are employed on work of equal value.

Except where a person's sex is a genuine occupational qualification it is unlawful for an employer to discriminate against women in the areas of recruitment, employment conditions, promotion opportunities or benefits (other than pension benefits). Similarly a woman may not be subjected to any detriment or be dismissed because she is a woman.

All contracts of employment are deemed to include an equality clause for women. This applies to like work, to work rated as equivalent to men's work and to work of equal value to that undertaken by a man. Claims related to alleged contravention of equality clauses may be taken to an industrial tribunal, which is empowered to award arrears of pay and damages for a period not exceeding two years before the proceedings began. No claim may be referred to a tribunal more than six months after employment has ended.

Most provisions relating to sex discrimination against women apply equally to the treatment of men. The legislation also forbids discrimination against married people in employment.

Statutory sick pay

Employers must pay statutory sick pay (SSP) to employees who satisfy rules about incapacity for work, periods of entitlement, qualifying days and notification of absence. When SSP has been properly paid, the employer may reclaim it by deduction from the company's National Insurance contributions.

Employees are not entitled to SSP in specified circumstances such as being over the state pension age, being on an employment contract for a fixed term of three months or less, earning less than a prescribed amount or working outside an EEC country. In addition, an employer may withhold SSP if the employee was not ill or if he/she failed to meet the employer's notification regulations.

Inspectors from the Department of Health and Social Security (DHSS) are empowered to check that employers are applying the SSP rules correctly and maintaining the prescribed records. The Social Security and Housing Benefits

Act 1982 prescribes a range of penalties that can be imposed on employers for non-compliance with the SSP scheme provisions.

Statutory maternity pay

Employers must pay statutory maternity pay (SMP) for up to eighteen weeks to employees who meet certain qualifying conditions, including having been employed for not less than twenty-six weeks. It is outside the scope of this chapter to describe the detailed operation of the SMP scheme but it should be noted that a woman's right to return to work after her baby is born is a separate issue from her entitlement to SMP.

If SMP is properly paid employers may deduct it from their National Insurance contributions.

Employers must maintain SMP records for three years. The Social Security Act 1986 lays down a number of penalties for breaches of regulations prescribed by the Act in relation to SMP.

Holidays and holiday pay

There is no statutory requirement that employees must be granted holidays. Entitlement to holidays and holiday pay is governed either by agreement with a recognized trade union or by the employment terms laid down by the employer. In any event the Employment Protection (Consolidation) Act 1978 (amended) lays down that holidays and holiday pay are among the terms and conditions of employment to be included in the written statement given to employees not later than thirteen weeks after starting employment.

Pension benefits

When considering pension matters employers are strongly advised to obtain assistance and up-to-date guidance from a reputable and competent consultant.

At the time of writing several changes to occupational pensions were in the pipeline. These derive mainly from the Finance Act 1986 and the Social Security Act 1986. Further changes are likely in the years ahead, and it is not practicable here to do more than mention some of the more important benefit changes. These include the following (which became effective on 6 April 1988 unless otherwise stated):

Membership of occupational pension schemes

Employers may not compel employees to join or to remain members of occupational pensions schemes. With effect from 1 July 1988 all employees acquired the right to join instead a personal pension scheme (PPS).

If an employee does decide to join a PPS he will have to leave (or may not join) an occupational scheme. If an employee joins a PPS his employer is not compelled to make pension contributions beyond the level of National Insurance contributions.

Additional voluntary contributions

An additional voluntary contribution (AVC) contributor need no longer contribute for a minimum of five years, and the level of contribution need not be constant. Contributors who began to make AVCs after 8 April 1987 are not allowed to commute for cash any pension arising from the AVCs.

Employers must provide in-house AVC schemes for members of their occupational pension schemes. On the other hand, employers must allow such employees to join 'free standing' AVC schemes if they wish to do so (schemes in which the contributor chooses the investment medium outside the company arrangements).

Retirement age

Since November 1987 it has been unlawful to force women employees to retire at a younger age than their male colleagues. The law does not, however, require that women's pensions shall be paid at the same age as men's pensions.

Leavers

Pension scheme members who leave an employer after less than two years' service may take a refund of their own pension contributions. With more than two years' service a preserved pension or a transfer value is provided.

Employment protection

Employment Protection Act 1975, as amended by the Employment Act 1980

This Act was set up to improve industrial relations and the Advisory, Conciliation and Arbitration Service (ACAS) was established under its provisions. One function of ACAS has been to lay down a consultation procedure to be followed before redundancies may be implemented.

Trade unions and employee involvement

Several Acts specify matters concerning the relationships between trade unions and employers. These include trade disputes, picketing, ballots and union dues. ACAS is involved in the production of relevant codes of practice.

Where a company employs more than 250 people the annual directors' report must include a statement showing how the company has consulted its employees about decisions that would affect them, how information has been provided to employees and how employees have been encouraged to become involved in the company's performance.

Employment Protection (Consolidation) Act 1978 as amended by the Employment Acts 1980 and 1982

Not later than thirteen weeks after their employment has begun, all employees who work for sixteen or more hours per week must be given *either* a written statement setting out their terms or employment *or* a contract of employment in which these terms have been incorporated. The information required was described in Chapter 5.

It is sensible to incorporate the company's disciplinary and grievance procedures into the statement or contract of employment. There is a code of practice to assist in preparing these procedures. Examples of typical procedures are given in Figures 7.1 and 7.2.

A contract of employment becomes effective as soon as a job had been offered and accepted. Employers may change the terms provided that employees are notified in writing within one month of the change.

Miscellaneous employment items

Itemized pay statements

Itemized pay statements must be given on or before pay day.

Guarantee payments

Guarantee payments must be made for a day on which the employer sends the employee home for business reasons.

Medical suspension

Full-time employees are entitled to be paid for up to six months if they are suspended on medical grounds.

Termination of employment

Minimum period of notice

Once an employee who works for sixteen hours or more per week has been employed for at least one month he/she is entitled to be given notice as follows:

- Not less than one week, if employed for less than two years.
- Not less than one week for every year of employment, if employed for between two and twelve years.
- Not less than twelve weeks if employed for twelve years or more.

Disciplinary Procedure

General

This document is to be taken to refer to both male and female members of staff notwithstanding that for reasons of convenience expressions used refer to one sex only.

The policy of the Company is to ensure that, where possible, staff who show signs of failing in their duties and standards of conduct are, without invoking formal disciplinary procedures, given the guidance necessary to correct themselves. The intention of this procedure is to ensure that any who fail to observe the standards of performance and behaviour necessary for the efficient operation of the Company, the safe performance of work, and satisfactory relations between all persons at work, are treated in a fair and orderly manner.

Rules and standards

Staff are expected to comply with:

1 Rules and procedures as publicized by the Company or their Department from time to time.
2 Such other standards of behaviour as clearly must be observed in the interests of safe efficient working and of satisfactory relationships between staff, or between staff and outside parties.

Investigation

Where a member of staff may have breached Company or Department rules or procedures or standards of performance or behaviour, he will be so informed and the complaint fully investigated, before any formal disciplinary action is considered. The investigation will be undertaken by the immediate supervisor, or more senior management, according to the apparent gravity of the offence.

An investigation may range from a discussion between the member of staff and his supervisor to a formal 'board' comprising senior management, but the same general principles will apply whatever the form of the investigation.

In every instance the member of staff concerned will be given the opportunity to state his case and to refer to witnesses, if this is relevant. During the investigation he may be accompanied by a fellow member of staff of his choice, if he so desires.

Every endeavour will be made to conduct the investigations as speedily as possible with due consideration for the need for thoroughness and fairness.

In all cases an investigation will be kept confidential within the Company, in so far as this is practicable.

Nature of disciplinary action

According to the nature and seriousness of the matter and depending on the frequency with which it has occurred, disciplinary action may take one of the following forms:

(i) an informal unrecorded warning;
(ii) a warning delivered orally but recorded;
(iii) a written warning, which may include being placed on special report, and may be a final warning;
(iv) suspension with pay;
(v) disciplinary transfer/demotion;
(vi) dismissal with notice;
(vii) summary dismissal without notice.

Figure 7.1 *Example of a disciplinary procedure*

The disciplinary action to be taken will depend on the circumstances, but the following will normally result in dismissal:

(i) serious infringements of safety regulations and requirements;
(ii) serious damage to Company property, or to other property on Company premises, resulting from deliberate act or from negligence;
(iii) theft, or unauthorized removal, of property from Company premises (including property of other staff);
(iv) physical violence;
(v) unauthorized disclosure of Company information, declared or known to be confidential;
(vi) wilful disregard of Company or Department rules;
(vii) falsification of evidence or references;
(viii) continued failure after reasonable warning to maintain satisfactory standards of performance or behaviour.

This list is not exclusive or exhaustive.

Suspension with pay may be applied while investigations are taking place, in circumstances where it is clearly undesirable that the member of staff should return to work before investigations have been completed.

In the case of Court proceedings which establish an offence, Company action will depend on the relevance of the offence to the position of the individual as a member of staff.

Authority for disciplinary action

Oral warnings will normally be administered by the immediate supervisor, whether they are to be recorded or not.

Written warnings may also be administered by the immediate supervisor, except that final letters of warning as well as letters concerned with suspension, disciplinary transfer, demotion or dismissal, must be issued by the appropriate Head of Department (but see final paragraph of this procedure). Dismissal requires the authority of the Head of Department and the personnel manager.

When determining the appropriate disciplinary action to be taken, every consideration will be given to the need to satisfy the test of reasonableness in all the relevant circumstances and to take mitigating factors into account.

The format of the disciplinary letter

Every written warning must contain clear statements about the following:

(a) a brief description of the matter involved;
(b) the action required of the member of staff to correct his shortcomings;
(c) the nature of the penalty, if any, imposed;
(d) where appropriate, a statement of the possible penalty in the event of a recurrence of the offence or a further breach of disciplinary rules;
(e) his right of appeal and details of the person, to whom the appeal is to be addressed, and the time limit for appeal;
(f) the period of validity of the warning (this will not be greater than two years).

Every notification of dismissal on disciplinary grounds will cover items (a), (c) and (e) above.

Right of appeal

A member of staff who wishes to appeal against any disciplinary decision concerning himself should, in the first instance, discuss the matter with his immediate manager or supervisor. He may then lodge an appeal in writing within three working days to the manager or supervisor who signed the

Figure 7.1 *Continued*

written confirmation of disciplinary action, or to the personnel manager. Normally an appeal will be heard by a line manager senior to whoever made the disciplinary decision. The member of staff has the right to be accompanied at an appeal hearing by a fellow member of staff of his choice, if he so desires, and to have opportunity to contact other staff who might be able to give evidence in his favour.

If, on appeal, it is decided that there is insufficient or no basis for the disciplinary action affecting the applicant, all papers relating to the case will be removed from his personal file.

Records

Every disciplinary action taken by the Company apart from an unrecorded oral warning, will be recorded in the member of staff's personal file under 'confidential' classification. The record must cover details of the nature of the matter, the action taken and the reason for it, whether an appeal has been lodged, its outcome and subsequent developments if any.

The record must also show evidence that the warning has in fact been delivered to the member of staff concerned, either in the form of a signed acknowledgement by him or, if he is unwilling to sign, by testimony of a person present when the warning was delivered.

Records will be retained in personal files until:

(a) in the case of a written warning – the period of validity stated on the letter of warning has expired without need for any further disciplinary action;
(b) in the case of a warning delivered orally but recorded – after one year has elapsed without need for any further disciplinary action.

Personnel function

The Personnel Department is available to provide advice and guidance to individual staff or to supervisors and managers, and must be given adequate advance notification and opportunity for consultation in case of intended written warning, suspension, disciplinary transfer or demotion, or dismissal.

Figure 7.1 *Continued*

An employee need give an employer only one week's notice. Despite the foregoing, if contracts of employment specify notice periods greater than the legal minima, the contracts take precedence.

An employer may choose to give pay in lieu of notice. If he does so, it may be paid without deduction of tax or National Insurance contributions.

It is not necessary to give notice to an employee at the end of a fixed term contract since, in effect, notice was given when the contract was made. Similarly formal notice need not be given to an employee retiring at normal retirement age.

Summary dismissal

An employee may be summarily dismissed in a case of gross misconduct, i.e., a matter of such gravity that no reasonable employer could tolerate the continued employment of the person.

Grievance Procedure

1 If a member of staff wishes to raise a question on any matter affecting his employment or relationship with the Company, he should, in the first instance, raise it with his immediate superior.
2 If the member of staff is not satisfied with the outcome of stage 1, he should raise the question with the manager at the next senior level.
3 If the grievance still remains unresolved and the Head of Department has not hitherto been involved, the member of staff may raise the question with the Head of Department.
4 At stages 2 and 3 a written report on the discussion will be prepared by the person hearing the grievance. Reports will be lodged with the personnel manager.

Nothing laid down in this grievance procedure is intended to preclude members of staff from raising problems of a personal nature with the personnel manager or other senior member of the Personnel Department.

Figure 7.2 *Example of a grievance procedure*

Written statement of reasons for dismissal

An employee with six months' continuous service is entitled to request a written statement giving details of the reasons if:

1 He or she is given notice of termination of his contract or
2 His or her contract is terminated without notice or
3 A fixed term contract is not renewed upon its expiry.

Such a statement must be given to the employee within fourteen days.

Unfair dismissal

An employee who works for sixteen or more hours a week and who has been continuously employed for two years or more has the right not to be unfairly dismissed and if dismissed can complain to an industrial tribunal. This does not apply if:

1 The employee works outside the UK
2 He/she reaches state retirement age.
3 A fixed term contract expires and the employee has previously agreed in writing not to make a claim at the end of the term.

A complaint to a tribunal claiming unfair dismissal must be made not later than three months after the dismissal. The employee must prove that he/she was dismissed, in which case the employer must prove the reason for the dismissal. There are only five permissible reasons for dismissal:

1 Inadequate capability or qualifications of the employee to do the work for which he/she was employed.
2 The employee's conduct.
3 Redundancy.
4 A legal restriction.
5 Some other substantial reason.

Once the reason for dismissal has been established the tribunal has to determine whether the employer was reasonable in treating it as grounds for dismissal. The tribunal must also satisfy itself that the manner of the dismissal was fair.

There are circumstances where a dismissal is automatically unfair. Most such circumstances relate to union membership or to the employer acting against an agreed redundancy procedure. Usually the dismissal of a pregnant employee will be deemed unfair.

Unfairly dismissed employees can choose to have their jobs back or to take compensation.

If reinstatement is chosen and the employer does not comply with the corresponding order from the tribunal, compensation may be awarded plus an additional award of between thirteen and twenty-six weeks' pay. A higher award may be made if the dismissal was an unlawful act of race or sex discrimination.

If the employee does not want his/her job back the tribunal will order compensation.

Redundancy definitions and payments

Chapter 4 dealt with redundancy procedures and severance terms. This chapter completes the picture by covering definitions and other statutory matters, all of which apply to men aged between 18 and 65 years, and to women aged between 18 and 60 years. Further requirements are that the employee must normally work for sixteen or more hours per week, and must have done so for at least two years.

Redundancy is defined by the Act as dismissal either because the employer has ceased to carry on the business for which the employee was engaged or because of diminution in the relevant workload. It is important to note that the workload of other employees engaged on different work has no bearing on the grounds for declaring redundancies in a particular work group.

Employers can avoid liability for redundancy payments if they are able to offer suitable alternative employment to surplus employees. In such cases employees are entitled to a trial period in the alternative job. If an employee unreasonably refused an offer of alternative employment a tribunal would be unlikely to find in his/her favour.

If a business changes hands and an employee continues to work (but for the new owners) he is not considered to be redundant. If he is made redundant at a later date, the new employer must count the whole employment when calculating redundancy payments.

When an employer intends to make redundant 100 or more employees within a ninety-day period he must notify the Secretary of State at least ninety days in advance. If ten to ninety-nine employees are to be made redundant within a thirty day period the Secretary of State must be given thirty days' notice.

Statutory redundancy payments are:

- Half a week's pay for each year of employment between the ages of 18 and 21.
- One week's pay for each year of employment between the ages of 22 and 40.
- One and one half weeks' pay for each year of employment between the ages of 41 and 64 (men) or 41 and 59 (women).

The UK government announces each year the maximum 'weeks pay' that can be taken into account.

Employment is deemed to end as follows:

1 Where notice is given, the date on which the notice expires
2 Where no notice is given, the date when the termination takes effect.
3 The date on which a fixed term contract ends, if it is not renewed.

Data Protection Act 1984

The Data Protection Act regulates the use of information about living people recorded on computers or similar processing equipment. The Act imposes specific duties on data users and computer bureaux while giving certain rights to individuals (including employees) about whom the data is held (known as the data subjects). Data users and computer bureaux must apply for registration to the Data Protection Registrar, on a special form. The Registrar has published a series of guidelines that are helpful in interpreting the Act. The address is: Office of the Data Protection Registrar, Springfield House, Water Lane, Wilmslow, Cheshire SK9 5AX.

There is a set of data protection principles, which state that personal data shall be:

1 Collected and processed fairly and lawfully.
2 Held only for the lawful purposes described in the register entry.

3 Used only for those purposes and only disclosed to those people described in the register entry.
4 Adequate, relevant and not excessive in relation to the purpose for which they are held.
5 Accurate and, where necessary, kept up-to-date.
6 Held for no longer than is necessary for the registered purpose.
7 Surrounded by proper security.

The principles also provide for data subjects to have access to data held about themselves and, where appropriate, to have the data corrected or deleted.

If the data user fails to provide the requested information, the data subject may seek a court order to obtain it. The court may award compensation from data users or computer bureaux if the data subject has suffered distress or damage on account of the loss, destruction or unauthorized disclosure of data.

The Act does not apply to all personal data and there are certain exemptions from the whole of the Act. Data users who hold only exempt personal data do not need to register, and the data subjects have no rights in respect of the exempt data.

So far as employees are concerned, the exemptions from the Act are:

1 Personal data used only for calculating and paying wages and pensions, for keeping accounts or for maintaining records of purchases and sales in order to ensure that payments are made. This exemption does not apply if the data are used for wider purposes (such as a personnel record or for marketing).
2 Personal data used for distributing articles or information to the data subjects. In practice this exemption refers only to names and addresses. If a data subject objects to the holding of such data about him or herself, then the exemption does not apply.
3 Information processed only for preparing the text of documents – the 'word processor exemption'. The effect of this is that the Act does not apply to information entered into a computer for the sole purpose of editing the text and printing out a document.

The Act precludes data users from disclosing data to any person other than those described in the register entry, but there are the following exemptions to the non-disclosure rules:

1 Disclosure to an employee to enable him to do his job.
2 Disclosure to the data subject or his representative or to others with the consent of the data subject.
3 When disclosure is urgently needed to prevent injury or other damage to any person.
4 Disclosure required by law or in connection with legal or court proceedings.

Employees have no right of access to data held about them in written form (their personal files for example).

Policy on business conduct

Companies are required to establish a policy on business conduct. This policy must be brought to the attention of all employees. Of necessity, the content of the policy statement will be specific to each organization, but the example given in Figure 7.3 is comprehensive and includes all matters likely to be required.

Further reading

Croner's Reference Book for Employers, Croner Publications, updated monthly.
Croner's Employment Law, Croner Publications, updated monthly.

Business Conduct

The Company's policy on contributions, payments to persons and ethical conduct is set out below and must be complied with at all times.

Laws and regulations

Employees must at all times observe the laws and regulations of every country and jurisdiction in which the Company operates or has any contractual association. The use of the Company's funds or assets for unlawful or improper purposes is prohibited.

Political contributions

The Company makes no political contributions in the UK. Group companies operating outside the UK may make contributions in support of political parties in the countries in which they have activities, provided that such contributions:

1 are not unlawful
2 conform to local practice
3 are of a modest amount
4 are properly recorded in the accounts
5 are approved by the Board of the Group Company

Relationships with government officials, customers and suppliers

The relationship of the Company and all employees with government personnel, customers, suppliers and other persons with whom there is a contractual relationship should at all times be such that neither the Company's integrity nor its reputation would be damaged if the details became a matter of public disclosure.

In this respect extraneous payments may not be made to or received from such persons regardless of amount and whether directly or indirectly.

Furthermore:

1 no gifts should be made or received if they are of material value; and
2 no entertainment should be extended or received if it could be considered extravagant;

in each case having regard to all relevant circumstances. The amount expended on such gifts or entertainment shall be properly entered in the accounting records.

Figure 7.3 *Example of a company's statement of business policy*

Commissions, fees and similar payments
If commissions, consultants' fees, retainers and similar payments are required to be made and can be justified in the normal course of business, the sums paid shall be clearly related to and commensurate with the services to be performed. No payment that is not so related or that could be seen to be an improper inducement should be made unless the approval of the Board of the Company concerned has been obtained before it is agreed to be paid.

Proper control and accounting
Compliance with prescribed controls, accounting systems and rules is required at all times. All Group companies must devise and maintain a system of internal control sufficient to provide reasonable assurance that management are controlling adequately the operations for which they are responsible. All accounts must accurately reflect and properly describe the transactions they record, and all assets, liabilities, revenues and expenses must be properly recorded in the books of the appropriate company or entity. No secret or unrecorded fund of money or other assets is to be established or maintained.

Company resources and personal interest
The Group has valuable resources, both in terms of tangible assets such as materials, equipment and cash, and intangible assets such as computer programs, trade secrets and confidential information. These resources may not be used for personal gain, nor otherwise than for the proper advancement of the business of the Group.

Ethical behaviour
Ethical behaviour is always a matter of spirit and intent, characterized by the qualities of truthfulness and freedom from deception and fraud. Conduct exhibiting these qualities is expected of the Group worldwide and of its employees. Ultimately it will be a case of our own good judgement in any given instance. If there is a situation which poses difficulties employees should always seek counsel, and managers should always be available to give consideration and advice on ethical behaviour generally. Employees should be encouraged to raise for consideration any matter of conduct that causes them concern, no matter how small or insignificant it may seem to be.

Figure 7.3 *Continued*

8

Training and development

Darek Celinski

This chapter aims to inform companies how their training and development should be organized and conducted, and what methods and procedures should be used.

Introduction

Training and development are two distinct learning processes. Although they have some areas of mutual overlap, they are treated in this chapter as different because they are used to satisfy different kinds of needs in companies, use different methods and procedures, and produce different results.

This chapter gives a great deal of attention to policy statements. This is because no training or development can be effective unless it is correctly organized and conducted. Policy statements are nowadays considered to be essential for achieving this. By specifying the correct steps to be taken in each case, and how each step is to be carried out, policy statements enable companies to ensure that their training and development are correctly organized and conducted.

Training and development that are conducted strictly in accordance with well-formulated policy statements practically cannot fail to produce the required improvements in job performance. This represents a tremendous improvement on the results that training and development usually produce in companies where such statements are not in use. This is because without the discipline that these policy statements impose, all kinds of shortcuts and simplifications of the essential methods and procedures are invariably taken.

For example, ever since the earliest days of companies providing any kind of formal training, it has been well known that it should not be undertaken until a need for it has been identified beforehand and precisely defined. In practice, it is almost unknown for this requirement to be properly carried out. Instead, it is either completely ignored, or the need is so vaguely stated as to be of no practical use. The effect of this is that a great deal of training that is nowadays provided by companies is needless. Needless training does not produce any improvement in the subsequent job performance: thus, by definition, it is ineffective.

Needless training is so widespread, that it is only a myth that training increases the efficiency and productivity of companies, thus making them more competitive and successful. This cannot be anything other than a myth as to this day there seems to be no example of a company which actually produced any of these increases. Instead, there is much more evidence that it does not do anything of the kind. Some of this comes from companies which noticed that however much training they were providing, it was making no difference to their efficiency and productivity. Quite simple follow-up studies invariably confirmed that their observations were completely correct. Similar evidence comes from companies which, after years of providing a great deal of training and development, were surprised to find that it made no difference to their efficiency and productivity when all training and development had to stop completely.

Even more convincing, because it was scientifically obtained, is the evidence provided by the Bath University report on management training. This particular report was published in June 1986, under the title *Management Training – Context and Practice*. Its purpose was to report the results of a pilot survey on management training that was conducted by the School of Management of Bath University, sponsored by the Economic and Social Research Council and the Department of Trade and Industry. Among its many important findings (based on the information obtained from nearly 2300 companies that responded to the survey) one amounts to saying that management training is a completely futile activity. Summarized, this reads: '*Incidence of training was not related to performance. Companies doing no training were as likely to be successful as those doing a great deal*'.

Initially it may appear to be quite unbelievable that trained managers could be no more effective than those who were not trained. However, this has to be recognized as unavoidable for as long as the present-day methods and procedures used to conduct management training remain unchanged.

The most widely used form of management training is attendance at management courses. Because it is particularly difficult to make this form of training fully effective, it is important that every step that should be taken in these situations is properly carried out. The most fundamental of these steps is the identification and definition of the needs that the course is to satisfy.

Anyone who has had the experience of attending a management course is likely to be aware that this is not usually preceded by any attempt to identify the manager's actual needs. Without knowing what needs the training is supposed to satisfy, it is not possible to establish what is relevant, and therefore the content of most of the management courses is based on assumptions. The training itself is frequently not like training at all – this is the case every time that the stated objectives are expressed in terms of providing a knowledge or an understanding of some principles, practices, or techniques which it is assumed that the managers need in order to improve their performance. Instead of evaluating this kind of training immediately on its completion, the participants are usually asked to assess it under a number of headings, and when this appears to be generally satisfactory it is assumed that the course was effective. This is the reason why management training that was not improving performance has been running for so many years without any steps being taken to increase its effectiveness. Courses can be made effective only when there is a reliable feedback about the results that they actually produce. Unfortunately, the widely used practice of having courses assessed by their participants produces a completely misleading feedback. For example, there can be no doubt that many of the courses that were the subject of the Bath University survey had been praised by their participants for their relevance and high quality and assessed as excellent. Yet the scientifically conducted investigations found them to be completely ineffective.

This introduction is distinctly critical about the results that the training currently provided by companies produces. This is in order to discourage the companies that intend to start providing their own training from adopting the practices that are currently widely used. Also, to encourage the companies that are already providing their own training to re-examine their practices in order to ascertain that their results are fully satisfactory.

Every criticism that is included in this introduction can be corrected by companies taking steps to ensure that:

1 Every kind of training is provided only after a need for it has been identified and precisely defined.
2 The training aims and objectives show that it is capable of satisfying the above defined needs.
3 The training itself is by instruction and a great deal of relevant practice, with immediate feedback so that its progress can be reliably monitored.
4 The training is always followed-up to ascertain that the initially defined need has been satisfied.

Training policy statements like the examples included in this chapter can be a great help to companies to ensure that each of the above steps is properly carried out. Then the companies can be certain that their efficiency and productivity is being increased. An additional benefit of this is that, on balance, such training costs companies nothing to provide. In practice, its total costs should be only a fraction of the value of the improvements produced in efficiency and productivity. This can often be proved in specific financial terms.

Development activities in companies tend to suffer from similar problems and difficulties than those of training. These problems and difficulties can be overcome by a somewhat similar approach to that recommended for training. The section on development included in this chapter contains the relevant details.

Training young people

Although this chapter is about the training and development of adults, a brief outline of the methods used by engineering companies to train young men and women for their first jobs in life may be appropriate. In the UK, the schemes described below may be part of the Youth Training Scheme.

Companies recruit and train young people in order to ensure a continued supply of well-qualified technical, specialist and other key personnel to satisfy their future needs. Because different categories of jobs require various qualifications and training, a number of different training schemes for young people are operated by engineering companies. The following are the principal schemes:

1 Craft.
2 Technician.
3 Technician engineer.
4 Professional engineer (technologist).
5 Business technician.
6 Business professional.

Most of the above schemes take about four years to complete. Companies, according to their needs, decide which and how many of these schemes to operate. In each case, periods of in-company training complement and are integrated with periods spent in establishments of further education. In-company periods, in addition to learning the job, include attachments to various departments so that the newly qualified young person has a considerable understanding of the whole company. The time spent in the places of further education is in order to follow appropriate study courses which, when completed, will lead to formal qualifications.

All the training schemes for young people are fairly flexible, so that they can be adapted by companies to suit their particular needs. This includes variations in requirements for the entry into a scheme, variations in courses of further education to be followed and variations in the final qualifications to be obtained.

In order to make these explanations more specific, listed below are the kinds of qualifications and typical first appointment jobs that young people obtain on completion of their various training schemes:

Craft

Qualifications: Certificate of Craftsmanship awarded by the Engineering

Industry Training Board and the appropriate City & Guilds of London Institutes' Certificate.

Jobs: jig borer, tool maker, model maker, skilled turner, miller, grinder, assembly fitter, maintenance fitter.

Technician

Qualifications: Business & Technical Education Council (BTEC) Certificate and/or BTEC Higher Certificate.

Jobs: Estimator, planner, progress planner, mechanical maintenance technician, electrical maintenance technician.

Technician engineer

Qualifications: BTEC Diploma and/or BTEC Higher Diploma.

Jobs: Jig and tool draughtsman, test engineer, electronic maintenance engineer, quality controller, illustrator.

Professional engineer

Qualifications: Engineering degree.

Jobs: Design engineer, research engineer, development engineer, sales engineer, industrial engineer.

Business technician

Qualifications: Appropriate BTEC Certificates or Diplomas.

Jobs: Accountant, buyer, stock controller, administrator, personnel officer, industrial relations officer.

Business professional

Qualifications: Appropriate degree.

Jobs: Accountant, buyer, programmer, marketing executive, sales executive, personnel executive, training executive.

Further details about all these training schemes are available in the UK from the Engineering Industry Training Board (EITB). The EITB is a statutory body that was set up by the provisions of the Industrial Training Act 1964. As almost all engineering companies are within the scope of the EITB, their help is readily available. This can be through personal contacts with EITB advisers, as well as from the many kinds of training booklets that they publish.

Training adults

The training which companies provide for their employees can be defined as the process of bringing a job holder to a desired standard of job performance by instruction and practice. The job holder in this definition can be absolutely anyone who works in the company, regardless of his/her status or function. This includes all levels of managers and supervisors, as well as all kinds of manual, clerical, technical, commercial and professional employees. In order for training to fulfil its purpose, as defined above, the following conditions must be met:

1 Training must be a planned and methodical process, conducted by people who have been trained to train, and who are competent in using the appropriate methods and procedures.
2 The process of bringing a job holder up to a desired standard of job performance requires that the standard has to be known and defined before any training is undertaken.

Why train?

It seems that training is very rarely presented as a means of reducing operational costs of companies. Yet, when training is correctly organized and conducted, it makes a significant difference to these costs. Companies as the places of work are also the places where a great deal of learning is going on most of the time. Where no training is provided, this learning is expensive because it is slow. It wastes time and resources and is costly in terms of the adverse effects that it has on efficiency and productivity. Training produces many improvements that reduce the operational costs significantly, which often can be proved in precise financial terms.

It is unavoidable that a great deal of learning is going on in companies most of the time. Absolutely everybody on joining a company has to learn his new job. The initial rather intensive learning, when usually no useful work is performed, is always followed by months or even years of further learning from experience until full competence is reached. Every major or minor change in product or services, practices, methods, procedures, technology or markets necessitates more learning. Still more learning is unavoidable whenever somebody's job has been reorganized or in any way altered, or the job holder has been promoted or transferred to a different job or to a different kind of work. Yet more learning is unavoidable whenever an improvement in someone's current job performance is required. It is only by learning how to do a better job, or some parts of it, that the required improvement can be in practice achieved.

In companies where no formal training is provided, most of this unavoidable learning is by trial and error. Although this is often thought to be less expensive than the formal training, because of the problems and inefficiencies that it produces it is in practice much more expensive.

Learning by trial and error is much slower than when it is by a properly prepared and well-presented instruction. This kind of instruction is practically never available in companies which do not provide formal training. In such companies the managers and supervisors are responsible for providing any training that those who report to them may require. However, because managers and supervisors are practically never trained in training, the effectiveness of their instruction is usually well below that of qualified instructors. Furthermore, training is time consuming and it is invariably given low priority by the busy managers and supervisors. For this reason the learners in companies are left to themselves for most of the time, or the more experienced colleagues at work are asked to help.

There are several potential problems when the more experienced colleagues at work are asked to help. First of all, they are certainly not trained in training: therefore their instruction is usually piecemeal, unsystematic, and sometimes difficult to follow. Furthermore, not all experienced colleagues at work are willing to take the necessary trouble, or to share their know-how. Some may enjoy the advantage that their superior knowledge gives them and may on purpose withhold some important information. Also, some more experienced colleagues may themselves not be well informed, or very good at their jobs, passing on incorrect methods that should be eliminated instead of being perpetuated.

Learning is usually difficult in companies where the immediate managers, supervisors, and the experienced colleagues provide the training. This is because such training is unplanned, often haphazard, and affected by the pressures of the moment. Also, managers, supervisors and experienced colleagues, unlike trained instructors, tend to just show or explain how something should be done and then leave the learners to learn on their own how to do it. This means that the learners are responsible for their own progress, and are expected to ask for help when they find themselves in difficulties.

There are several reasons why the learners are often reluctant to ask for help. They may simply not want to impose on the good will of their busy managers or supervisors. They may fear being criticized or even being reprimanded for any mistakes made. Or, they may not want to admit that they are having difficulties in learning something that has already been explained to them. Because of these kinds of pressures, learners often think that they have no practical alternative but to try to learn on their own, by trial and error. Learning by trial and error almost always leads to the acquisition of less than the best work habits. This tends to be confirmed by the findings that whenever an unsatisfactory quality or quantity of work is investigated in a company, inadequate training is usually identified as its principal cause. Also, many of the mistakes that occur at work are due to insufficient training.

In view of all these problems and difficulties, it cannot be surprising that when this kind of learning is replaced by planned and methodical training conducted by qualified instructors, many valuable improvements are obtained. The most important of these are:

1 At least 50 per cent reduction in the time taken and the costs of learning to do a job, a task, or a group of tasks.
2 At least 10 per cent higher quality and quantity of work produced on completion of training.

Less specific, but nevertheless quite valuable additional improvements that such training produces are:

3 Better utilization of machines, materials and equipment.
4 Elimination of the mistakes, scrap and rework that are usually produced during learning by trial and error.
5 Improved compliance with the safety requirements and corresponding reduction in accidents at work.
6 Managers and supervisors are not personally involved in training and therefore have more time to spend on their more important work.
7 Generally higher morale, better cooperation and increased loyalty are usually found in companies that are efficient, and where those who work there are well informed and properly trained.

The savings produced by the 50 per cent reduction in the time needed for unavoidable learning are usually sufficient to pay the costs of providing formal training within a company. When these reductions are greater than the 50 per cent (as it is often the case) and the value of the increased quality and quantity of work are added together, the overall cost savings produced by formal training become quite significant. This figure becomes still higher when the value of the less specific improvements is also taken into the account.

It is important that no improvement or saving produced from training is taken for granted or assumed. In every case the results produced should be measured and, where possible, expressed in financial terms. This is important for two reasons.

1 Training is demanding in terms of the effort and resources needed to make it effective. This commitment is unlikely to be allowed to continue unless company managements are given reliable proof that the promised improvements are being achieved.
2 Those responsible for organizing formal training in companies need accurate feedback about the results that they are actually obtaining. Only through such feedback can they find out when the right level of effectiveness has been reached. Even when training has proved to be effective, continued feedback is necessary in order to maintain success. If at any time any aspect of training becomes less effective than it should be, reliable monitoring and feedback alerts those responsible, and appropriate action can be taken to correct the shortcomings.

Where to train?

There are just two kinds of environments where all the training that is provided by companies takes place: one is called 'on-the-job' and the other one, 'off-the-job'.

Training is on-the-job when it is conducted in precisely the same environment as that in which the particular job is normally performed. This includes not only the same physical location, but also the same conditions and surroundings, the same tools, materials, standards, and facilities used for the actual job.

Off-the-job training is that which takes place outside the day-to-day pressures of the job, and away from the location where it is normally performed. There are three main kinds of locations where off-the-job training is conducted. These are:

1 Companies' own training bays or training areas and workshops, which may or may not be adjoining production departments.
2 Locations outside the company that belong to somebody else (e.g., premises of the suppliers of equipment that the company already owns or plans to install).
3 Lecture rooms, which can be either in or outside the company.

On-the-job training is by far the most effective form of training that is available to companies. It is unsuitable for training managers and supervisors and other employees while they are attending to the company's clients and customers. However, whenever there is any choice, on-the-job training should be used instead of off-the-job.

On-the-job training is so effective because unlike off-the-job training (especially when conducted in a lecture room) it cannot fail to be:

1 Relevant – learning by doing the actual job automatically makes it strictly relevant.
2 Effective – the instructional technique is by far the most effective training method.
3 Correct – progress in doing the job provides immediate and accurate feedback, which is so essential for learning to do something correctly.
4 Applied on the job – on completion of training, the learner continues doing the job without any interruption, and thus does not experience the difficulties of transferring the learning from one environment to another.

When to train?

The need for training is indicated when a job holder needs to:

1 Learn to do something that he is unable to do now – this is when the job-holder has been newly recruited, or newly transferred to a different job, or newly promoted into a higher level job.
2 Learn to do his job, or some parts of it, differently than at present – this is when any kind of changes that affect the particular job-holder are to be introduced.
3 Learn to do his job better than he does it at present – this is when some

specific improvements in the job-holder's current job performance are required.

Listed above are all the learning needs that arise in companies. Any training that is not designed to satisfy one of the above needs is certain to be needless. By providing formal training that satisfies everyone of the above needs, companies ensure that any need for learning by trial and error is eliminated.

When deciding to train a job-holder, the timing should be carefully considered. People have an almost unlimited capacity for forgetting what they have learnt, when it is not immediately applied in practice. This is unlikely to present a problem as far as the timing of the on-the-job training is concerned. However, it is a distinct problem when the off-the-job training is being considered. The best results are obtained when it is so timed that the new learning is immediately applied in practice. However, if it is not, the delay should be no longer than about two weeks. If this is longer, a great deal of the learning is usually forgotten, and thus the benefits of the training are largely wasted.

Who is involved in training in companies?

Apart from the job-holder who is to be trained, his immediate manager or supervisor, and a member of the company's training staff are involved.

Managers and supervisors are involved because it is they who define their subordinates' desired job-performance standards that the training is to produce, and afterwards ascertain whether or not the desired standard has actually been achieved. This is one duty that managers and supervisors cannot delegate to anyone else. Only the manager or supervisor knows precisely what improvements in his department are expected: therefore it is only he who can properly define the training needs of the job holders who work there.

The other group of people who are involved in training are the company's training staff. No company can possibly derive proper benefits from any training that it provides unless there is someone who is able to make it effective. This applies even when the company uses a great deal of outside expertise. There are many different duties that such a person carries out. The most important of these is to help managers and supervisors to identify training needs within their sections and departments, and afterwards, to take the necessary steps to ensure that these needs are satisfied.

In companies which provide formal on-the-job training there are also departmental part-time instructors who are involved in training. Such an instructor is an experienced job-holder who has been trained in using an instructional technique, spends most of his or her working time on doing his normal job, and instructs when the need arises within the department.

How to train?

General procedure
There is a procedure which should be used whenever any kind of training is conducted. This is a four step process, based on that which is widely used in systematic problem solving. In its basic problem solving guise, the procedure comprises the following four stages:

1 Identify the problem and establish its causes.
2 Decide what should be done to remove the problem.
3 Do it.
4 Check the results.

When this procedure is applied to training, the four stages of the process can be retitled:

1 Identify the training needs.
2 Choose, design and prepare the training.
3 Carry out the training.
4 Evaluate the results.

The first and fourth steps in this procedure are carried out by the managers and supervisors. The second and third steps are conducted by training specialists.

Training on the job
It is easy to comply with the four step procedure when training is carried out on the job. The situation at work makes it quite apparent who must be trained to do what. Also, this kind of training is easy to evaluate, because when the training has been completed the job performance itself should give clear indications of the results. Even the training needs do not have to be precisely defined, since the job itself ensures that the training is strictly relevant.

Formal on-the-job training is conducted by the departmental trainers who have been trained in using an instructional technique. This requires them to prepare an instructional breakdown sheet before giving the instruction. The instruction itself is highly structured and requires the instructor to explain and demonstrate one stage of the job at a time. This is immediately followed by the learner practising it until eventually the whole job has been mastered. Throughout all this time, the instructor remains with the learner and gives any help and advice as and when this is required.

Training off the job
When training is conducted off the job, in a lecture room, it is necessary that the four step procedure is complied with strictly, without any shortcuts or

simplification. The identification and definition of training needs has to be precise, so that the trainers can design training that is strictly relevant. It is also essential that the managers and supervisors of the trainees are able to recognize the results of training, and know whether or not the needs which they originally identified have been met in practice.

Training typically has three elements: explanation, demonstration and practice. Demonstration and practice may be difficult or unsuitable for the instructional techniques used in lecture rooms: therefore a number of different methods may have to be chosen to ensure that adequate explanation, demonstration and practice (the so-called hands on experience) are provided in order that the training shall be effective. The practice element is particularly important for a number of reasons:

1 It enables the learners to learn by doing the work.
2 It provides the feedback which is always necessary whenever anyone is learning to do anything.
3 What people do they understand.

Training policy statements

The term 'policy' can be defined as 'an agreed framework within which specific actions can take place'. Thus, a training policy statement is a document which details who should take what actions when operating one of the company's training schemes. This means that a company needs one such statement for each of the training schemes that it operates.

It is usual that, after deciding to operate a training scheme, the company's training specialist drafts the appropriate policy statement. This should incorporate every essential method and procedure that is needed in order to make the particular training scheme fully effective. Also, when drafting this document, the company's organization, facilities, and constraints have to be taken into the account.

The draft, when completed, is submitted for approval to the appropriate policy maker, usually the company's chief executive. After the approval it becomes the official policy document, compliance with which is mandatory.

The use of detailed training policy statements by companies is a comparatively recent development. The practice of formulating such statements and then insisting that they are strictly complied with started in companies that were providing a great deal of formal training and wanted to find out how much they were benefiting from this. Their findings were deeply disappointing. They found a great deal of waste and many inefficiencies, mainly due to the haphazard and unsatisfactory methods and procedures used to organize and conduct their formal training. For example, attendances at courses often made no difference to the trainees' subsequent job performance because the methods used for their selection were not correct. These findings showed the benefits of identifying the best practices to be used in each case and then, by means of specific policy statements, of ensuring that these practices were consistently applied throughout the company.

The following pages contain several examples of policy statements. Each of these was formulated for use in a medium-sized engineering company that employed a training manager and two training officers (see Figures 8.1 to 8.5).

Understanding development

There are several different explanations of the term *development* in use. One such explanation is that development is the process of increasing the effectiveness of an employee through planned and deliberate learning that widens and extends his existing knowledge and experience. The employee is in this case absolutely anybody (including managers and supervisors) who is employed by the company, regardless of level or function.

This section describes just one form of development. This is management development that uses an in-company project as a vehicle for widening and extending the managers' existing knowledge and experience. The reason for selecting this form here is that management development is by far the most widely used kind of development that companies provide. Also, the in-company project methods can be used for developing all other levels of employees, besides the managers. Finally, the in-company project method is probably the most effective means for developing managers that at present exists.

The companies that use the in-company project method invariably started to do so after finding that the much more popular form of management development, which is attendance at courses, did not result in increasing the effectiveness of managers. The problem with courses is that it is almost impossible to identify the skills and knowledge that the managers need in order to increase their effectiveness. Thus, it is no more than an assumption that managers need to learn about the processes of management, financial control, styles of management, etc. Also, the skills that managers need in order to become more effective are further examples of such assumptions. These include the widely popular problem solving and decision making, leadership, negotiating, communicating, and many other topics.

The most reliable evidence that management development courses based on such assumptions do not increase the managers' effectiveness, comes from the companies that conducted special follow-up studies. Other sources that support these findings are the companies which arrived at the same conclusions as a result of their own observations. The most noteworthy of these were the published views of the management development manager of one of the best-known companies in the country. His view, which attracted a good deal

XYZ Company Ltd	TRAINING POLICY
	PAGE NO:
	ISSUE NO:
	DATE:

TRAINING PLANNING AND BUDGETING (ANNUAL)

AIM: To identify training needs in every department that the Company's next year's business plans generate; to formulate a training plan for satisfying these needs; to allocate budgets for fulfilling the plan.

THE POLICY IS:

1 THAT the company provides formal training to satisfy the following four categories of needs:
   - 1.1 NEW APPOINTEES – training of those who are newly recruited, transferred or promoted to fill vacancies.
   - 1.2 CHANGE – training of those whose jobs will be affected by the impending changes of any kind.
   - 1.3 IMPROVEMENT – training of those whose current work performance should be improved.
   - 1.4 LONG TERM – training of young people for their first adult jobs in life through a range of appropriate training schemes.

2 THAT all training is provided to satisfy previously identified and precisely defined work performance needs of individuals, groups of individuals, departments and the company.

3 THAT training plans are formulated each year and cover the same period as the company's financial year.

4 THAT the following procedure is used to formulate training plans:
   - 4.1 IDENTIFYING TRAINING NEEDS. The general principle is that directors, managers and supervisors are responsible for work performance of those who report to them directly and therefore for identifying their training needs. Training manager and training officers only assist in this process and advise how these needs can be satisfied in the most cost-effective manner. Thus, in November of each year:
     * Managing director meets with the training manager in order to identify the training implications of the Company's Business Plan for the ensuing year.
     * Training manager and training officers make appointment with directors, managers and supervisors in order to assist them in identifying training needs within their respective areas of responsibility.

Figure 8.1 *A company policy statement for annual training planning and budgeting*

XYZ Company Ltd	TRAINING POLICY
	PAGE NO:
	ISSUE NO:
	DATE:

TRAINING PLANNING AND BUDGETING (ANNUAL)

4.2 FORMULATING TRAINING PLANS

* On completion of the interviews with the directors, managers and supervisors, training manager and training officers draft training plan for each section and department where the training needs have been identified.
* Draft training plans are submitted to respective directors, managers and supervisors for their approval and are amended as necessary.
* In early December, each of the six directors review all the draft training plans for their own areas of responsibility, and when in agreement give their approval.
* Training plans for the ensuing year are finalized, and redrafted where necessary.

4.3 PREPARING TRAINING BUDGETS

* Training department calculates the cost of fulfilling training plans in each function and the whole company;
* Training department negotiates the final training budget for the ensuing year.

5 THAT the training plans and the budgets are formally presented by the training manager to the executive board of directors during their December meeting.

6 THAT each director is responsible for the fulfilment of all the training plans within his own area of responsibility.

7 THAT the training manager is responsible for organizing, coordinating, and controlling all the activities involved in training planning and budgeting and for monitoring the fulfilment of the whole training plan during its life.

Figure 8.1 *Continued*

of attention at the time, was that it makes no difference to companies whether they provide any management development or not. Thus, the companies that provide this do not appear to be any more successful than those which do not. Conversely, the companies which do not provide any management development do not appear to be less successful than those that do provide it.

In view of the above, it seems that the companies that wish to increase the effectiveness of their managers are unlikely to achieve this by using manage-

XYZ Company Ltd	TRAINING POLICY
	PAGE NO:
	ISSUE NO:
	DATE:

INDUCTION TRAINING

AIM: To enable newly recruited adult employees to become considered by their immediate managers and supervisors as fully competent in one-third of the time that it takes without formal training.

THE POLICY IS:

1. THAT every newly recruited adult employee, regardless of function or department, receives formal induction training.
2. THAT the Company induction training provides Company knowledge and skills, departmental knowledge and skills and the job knowledge and skills.
3. THAT the nominated training officer organizes, coordinates and controls all induction training activities.
4. THAT all training is conducted in accordance with requirements of an instructional technique, and its instructional breakdown sheets are formulated after training needs for each knowledge and skill area have been precisely defined.
5. THAT the newly recruited employees report to the lecture room for the first hour of each working day during the first week of employment to be trained by the training officer on the subject of the company knowledge and skills.
6. THAT the departmental knowledge and skills and the job knowledge and skills are provided by the departmental part-time trainers.
7. THAT even if only one new employee joins the company on a particular Monday, the whole induction procedure is followed as usual.
8. THAT the induction of newly recruited managers and directors is purpose designed to suit the individual requirement.
9. THAT the nominated training officer is responsible for the effectiveness of all the trainers who provide induction training in their respective departments and from time to time conducts follow-up studies to ascertain whether or not the aim given at the top of this page is being satisfied.
10. THAT the nominated training officer provides further coaching of departmental trainers, whenever their results are less than expected.

Figure 8.2 *A company training policy statement for induction training*

XYZ Company Ltd	TRAINING POLICY
	PAGE NO:
	ISSUE NO:
	DATE:

ON-THE-JOB TRAINING (BY TRAINED INSTRUCTORS)

AIM: To obtain the following improvements on the results produced when this form of on-the-job training is not provided:

1 At least 50 per cent reduction in the time and the costs of learning to do a job or a task,

2 At least 10 per cent increase in the quality and quantity of the work produced.

THE POLICY IS:

1 THAT all on-the-job training is conducted by departmental instructors who have been trained to instruct using an instructional technique.

2 THAT annually, during training planning, the number of instructors required by each department is reviewed and adjusted as necessary.

3 THAT departmental instructors instruct only when there is a need for it within the department and spend their remaining working time doing their usual jobs, i.e. they are part-time instructors.

4 THAT departmental instructors are selected and trained in accordance with the Company's training policy statement headed 'Training of Departmental Instructors'.

5 THAT departmental instructors remain responsible to their departmental managers and supervisors and are subject to the same conditions of employment as other employees within the department.

6 THAT departmental instructors are paid the wages and salaries that apply to their main jobs, without any special additions for their instructional duties.

7 THAT departmental managers provide adequate facilities for their instructors to be able to do their work effectively, especially accepting instructing as a priority over all other work of instructors.

8 THAT departmental managers monitor the effectiveness of their instructors, and from time to time, investigate their results in order to ascertain that the aim stated at the head of this is being satisfied.

9 THAT departmental managers and supervisors cooperate with the training manager and his staff, and seek their help and advice on all matters related to the work of the departmental instructors.

10 THAT departmental managers inform the training manager whenever any additions or replacements of their departmental instructors are required.

NOTE: Whenever possible the training manager should be given four months' notice as this is the time required to select, assemble sufficient number of trainees to run a viable programme and train departmental instructors.

Figure 8.3 *A company training policy statement for on the job training using trained instructors*

XYZ Company Ltd	TRAINING POLICY
	PAGE NO:
	ISSUE NO:
	DATE:

TRAINING OF DEPARTMENTAL INSTRUCTORS

AIM: To enable departmental instructors to use an instructional technique when conducting on-the-job training within their departments.

THE POLICY IS:

1 THAT vacancies for departmental instructors are advertised within the departments and the applications to respective departmental managers are invited.
2 THAT the applicants are, first of all, given a copy of this policy statement to read as well as that headed 'On-the-Job Training (by Trained Instructors)'.
3 THAT every applicant who accepts the requirements laid down by these two policy statements is interviewed by his departmental manager with the training manager, and no more than a few days later is informed whether successful or not.
4 THAT the preferred applicants are those who are well experienced in the work of the department, do not have any supervisory duties, and are interested in training and helping their colleagues at work.
5 THAT the selected applicants are allowed by their managers to attend an in-company run course in instructional and related techniques.
6 THAT the course is run by the Company's training manager and its total duration is ten days: four consecutive days during the first week, and two days during each of the three subsequent weeks.
7 THAT a sufficient number of courses is run each year to meet the needs of all the departments, considering that the optimum number on each course is eight participants.
8 THAT departmental instructors are encouraged by their managers to seek help and advice from the training manager and his staff.
9 THAT the training manager and his staff monitor the standards of instruction by the departmental instructors and provide further training and coaching whenever they consider this to be needed.
10 THAT departmental instructors may ask their managers, at any time, to be released from their instructional duties.

Figure 8.4 *A company policy statement for the training of its departmental instructors*

XYZ Company Ltd	TRAINING POLICY
	PAGE NO:
	ISSUE NO:
	DATE:

ATTENDANCE ON EXTERNAL COURSES

AIM: To define the procedure to be followed to ensure that the attendance at external courses results in learning that is immediately afterwards applied on the job.

THE POLICY IS:

1 THAT the three-part 'External Course Attendance Control Form' is completed for each individual course attendance.
2 THAT the nominating manager holds a pre-course briefing with his nominee in order to discuss and agree the aim of the attendance and to complete Part 1 of the Control Form.
3 THAT the training department select a course that is likely to satisfy the need stated within the Part 1 of the Control Form, inform the manager and/or the nominee, book a place and make all the necessary arrangements
4 THAT the attendee, immediately after completion of the course completes Part 2 of the Control Form.
5 THAT the nominating manager holds a post course debriefing with his nominee, and completes Part 3 of the Control Form.
6 THAT the training manager and the training officers are available to assist the nominating manager at any stage in the above procedure.
7 THAT the nominating manager assists his nominee in the transfer of learning from the course to on the job application.
8 THAT the plans for the course attendance are cancelled, if the nominating manager is not able to state on the Part I of the Control Form specifically what his nominee should be able to do after the course that he is not able to do before the course.

Figure 8.5 *A company training policy statement on the use of external training courses*

ment development courses. Management development that uses an in-company project as a vehicle for widening and extending the managers' existing knowledge and experience provides a far more effective alternative.

In-company project-based management development

The essential information about this form of management development is presented below as answers to the questions which begin with the words what, why, where, when, who and how.

What is it?
It is an arrangement where a small group of managers, in addition to their usual duties, work together on a project that is needed by the company.

Why use the project work?
1 Because this is distinctly the most effective method of developing managers.
2 Because management development that really increases the effectiveness of managers provides companies with excellent means for becoming more successful.
3 Because it is exceedingly cost effective as usually there are no additional costs. Therefore the value of the completed project alone presents a net benefit.

Where is it conducted?
Mainly in the company's own working environment, often in the office of one of the participating managers.

When to conduct it?
There is no specific answer to this question, apart from the need to avoid holiday periods.

Who organizes and conducts it?
The company training manager is the person best suited to organize every in-company project-based management development event. There is no one from outside the group responsible for conducting it, as the groups are self-governing.

How is the project work organized?
Every project has its 'owner' – this is usually a senior manager or a director who has identified a need for some kind of investigations within his area of responsibility, and is interested in resolving it. After agreeing to be the owner of the project, his duties are:

1 To formulate the project brief.
2 To brief personally the 'project team'.
3 To attend the project team's verbal presentation of their findings/solutions/recommendation, on completion of the project work.
4 To receive and study their written report and meet the team again in order to announce his decision regarding the acceptance of their findings/solutions/recommendations.

Figure 8.6 provides an example of a policy statement which details the actions that an engineering company decided to take when conducting its in-company project based management development.

XYZ Company Ltd	TRAINING POLICY
	PAGE NO:
	ISSUE NO:
	DATE:

IN-COMPANY PROJECT-BASED MANAGEMENT DEVELOPMENT

AIM: To enable the managers to increase further their effectiveness both as individual managers and team members.

THE POLICY IS:

1. THAT the Training Manager is responsible for planning, organizing, coordinating and controlling, and the overall effectiveness of management development.
2. THAT participation is open to every manager, but is strictly voluntary.
3. THAT the company senior managers and directors select live problems and projects and are their 'owners' to whom solutions and recommendations are made.
4. THAT the problems and projects selected are real, challenging and require creative solutions.
5. THAT whenever possible the problems and project selected should extend beyond the boundaries of a single department or function.
6. THAT the work is conducted in 'project teams' usually comprising four managers with different backgrounds, expertise and functional responsibilities.
7. THAT the owner prepares written project brief containing its title, background information, statement of objectives, how will the success of the project be assessed, and the timescales required.
8. THAT the owner briefs personally the project team, and provides them with copies of the project brief.
9. THAT the project team meets for one day a week during the life of the project, which should be no longer than three months.
10. THAT during the first meeting of the team it is decided how is the work to be organized, who is to be responsible for what and an action plan is drawn with provisions for recording the progress of the project work.
11. THAT the training manager keeps in touch with each team, and only when asked, advises and/or arranges help from other specialists.
12. THAT each project team writes a brief report on their work and findings and recommendations.
13. THAT each project team makes a verbal presentation of its findings/ solutions/recommendations to the Owner of the project and other teams.
14. THAT within two weeks from the date of presentation, the owner of the project meets his team in order to discuss and announce whether or not he or she accept their findings/solutions.

Figure 8.6 *A company policy statement for in-company project-based management development*

It is usual that the training manager assists the project owner in carrying out each of the above duties, as well as organizing the composition of the project team, and generally assisting them in the successful completion of their project.

Counselling at work

Counselling at work is included in this chapter because, although it is neither training nor development, it is a management technique used to improve job performance. This technique is used in situations where a job-holder's performance is adversely affected by any kind of problems and difficulties. The purpose of counselling is to help the job-holder to overcome these problems and difficulties, and thus restore his work performance to an acceptable standard.

More specific information about counselling is given below in the form of answers to questions which begin with the words: what, why, who, when, where and how.

What is counselling?

A glossary of training terms defines 'counselling' as 'a direct personal relationship in which the counsellor's friendliness, experience, and knowledge are made available to another person in order to assist the latter in solving his/her problem'.

Why counsel at work?

Counselling, as a management technique, is used to assist job-holders to eliminate any apparent obstacles to their good standard of job performance which, if not corrected, may develop into even more serious problems.

Who counsels whom at work?

In companies, it is always the immediate manager or supervisor of the job-holder. Unlike other forms of counselling, counselling at work does not require understanding of psychology – managers and supervisors willing to comply with a few simple rules of counselling at work can obtain excellent results.

When to counsel?

Problems at work tend to escalate if no actions are taken to correct or eliminate them. For this reason, counselling should take place as soon as it becomes apparent to a manager or supervisor that someone who works in his department has a problem of any kind that hampers work performance.

Where to counsel?

Counselling has to be in private and confidential. This requires that it should be conducted in a quiet office, free from interruptions and preferably not open to view.

How to counsel?

Counselling takes the form of an interview – it is therefore a conversation with a purpose. It is one-to-one conversation, held in privacy and confidential. As is the case with any kind of interview, it is the counsellor who is completely responsible for its outcome. The job-holder being counselled is in a subordinate role and can only respond to the counsellor's handling of the interview.

Work performance related counselling requires the manager or supervisor to be sympathetic towards the job holder. Although the manager has the power to dictate what should be done, and how, in order to correct the situation, when counselling it is essential that this power is not used. Instead, the manager should help the job-holder to find his own solutions and to solve his own problems.

A counselling session can be said to have been successful when the job-holder is left with the feeling that he works for a good boss; that the boss has helped him out of a situation which, if not solved during the counselling interview, would have produced unpleasant and undesirable consequences. Thus, a counselling interview is distinctly different from the disciplinary one. If it did ever happen to generate feelings like those that are usually generated by a disciplinary interview, this would indicate that it was completely mishandled.

The first step in the procedure used for counselling at work, is when the manager or supervisor notices that one of the job holders in his department has a problem or difficulties, and decides to hold a counselling interview. In order to carry this decision to its successful conclusion, the counsellor will need to do the following:

1 Prepare for the interview:
   (a) Review the problem, difficulties, situation and collect any facts, data, evidence that may be needed during the interview.
   (b) Write down the specific outcome that the interview will aim to achieve.
   (c) Decide when and where to hold the interview, and inform the job-holder concerned accordingly.
2 Start the interview:
   (a) Explain the purpose of the interview.
   (b) Explain the method of approach that will be used.
   (c) Make the interviewee relaxed and at ease.

3 During the interview proper:
   (a) Explain his/her own views/understanding of the problem.
   (b) Outline likely consequences of the problem.
   (c) Invite and listen to the explanations and views of the interviewee.
   (d) Invite suggestions as to how the problem may be solved.
   (e) Assist the interviewee in selecting the best solution.
   (f) Summarize the progress made during the interview.
   (g) Restate the action that was mutually agreed to take.
4 Ending the interview:
   (a) Write down the actions that were agreed.
   (b) Make arrangements for monitoring progress and checking results.
5 After the interview:
   (a) Assess the interviewee's feelings generated by the interview: motivated? cooperative? neutral? sullen? crushed?
   (b) Compare the results achieved against the aim stated in 2(a) above – if the aim is not achieved, identify the reasons why and decide what improvements to make when holding the next counselling interview.

Further information

The Engineering Industry Training Board has its offices at 54 Clarendon Road, Watford, Herts., WD1 1LB.

There is a useful publication, *The Personnel Yearbook*, published annually by Kogan Page in association with the Institute of Personnel Management. This is available from IPM Distribution, George Philip Services Ltd, PO Box 1, Littlehampton, West Sussex.

9

Personal communication skills

Gordon Bell

> One cannot expect a committed and effective workforce unless they understand the reasons behind the decisions which affect them. Accepting this principle, it is hard to see how one can have too much properly planned communication within a company.
> *Sir David Plastow*, Chairman and Chief Executive, Vickers plc

A qualified engineer has to put in years of hard, serious slog before achieving a degree. With some other so-called disciplines it is enough to attend a few lectures, to be reasonably intelligent and agreeable for three years or so and a degree drops into the lap as easily as badly managed spaghetti. Balls, picnics and parties are the order of the day (and night) for these lightweights: but not for the engineer, whose exacting course forces his nose to the grindstone throughout. The intensity of his workload tends to isolate the engineer from the frivolity going on elsewhere in the university and he is often, unjustly, written off as an oily swot and a dull dog. Socializing and the development of communication skills form little part of his education. The engineer suffers for this when he goes into management. Management has been aptly described as the art of getting work done through other people which, in turn, demands expertise in:

- Writing.
- Speaking.
- Handling meetings.

No one capable of becoming an engineer need lack such expertise. The image makers, the shiny ones, should not have a monopoly of effective communications. The engineering manager also needs them, and will probably use them to a better end.

Why should managers bother about communications? We can all write. Communications? Is that not just talking and writing? Why all the fuss? 'I'm too busy,' says the rut man, 'with important matters such as cost accountancy, production problems and getting some sort of profit for myself and the company. Talking and writing? Phooey! They come naturally. Techniques my foot! I haven't got time. Meetings? They happen every day. Letters and reports? I get by. As for speaking and presentations, I always dodge them if I can – I don't want to expose my limitations. Other people may be clever at that sort of thing. They have a gift for it. Let me switch off my mind and rationalize. Success stems from who we know, not what. Let me pull the walls of my rut closer around me and snuggle down into it.'

Successful managers at all levels are good communicators, but this skill does not come giftwrapped to anybody. You have the talent, but your talent has to be thought about, worked on and developed. Techniques are simply tricks of the trade that make a job easier. As a professional communicator (and that is what a manager must be) the time you invest in mastering techniques pays valuable dividends.

Effective communications stimulate the desired reaction. There is no other criterion. How to stimulate the required reaction: that is what we shall be looking at. How to get the right reaction when you speak, the right reaction when you write and agreement to your reasonable wishes.

Reports

Some companies behave like slow-witted dinosaurs. Information never reaches the action-stimulating targets in time – if at all. Management blames the workers. The workers blame the management. Reports and letters become unreadable bumph. A great divide develops between sales, research, production and administration. Customers go elsewhere. Who is to blame them?

One of the first duties of a newly appointed manager is to make a study of the organization's nerve system. Does the right information get to the right people at the right time for action? Do customers declare enthusiastically that your company deals with them better than any other firm in your type of business? Make this a general survey to start with and then investigate your own department in detail. Note with stark honesty the strengths and weaknesses revealed within your special area of authority. The starting point for improving matters is yourself. You might have to make changes in your thinking and in your attitude towards those above you in the hierarchy, to those below you and to the people sideways to you.

It is a hard fact of life that your full-scale technical reports will receive a limited welcome, except from your fellow engineers (and not all of them will be overjoyed with routine stuff). The company chairman and his fellow directors rarely read more than three pages if they can help it. Your sixty-eight page report might well drop on their desks with similar wodges from the geologist, the labour relations manager, the sales and personnel departments, the accountant, the public relations man, the production manager and many, many others from perhaps ten different countries. Your tome will not be read.

The furniture in any busy director's office always includes a yawning wastepaper basket ready to receive unfiltered, unedited gobbledegook. So, be warned. Find out what he needs and wants: give him no less but, above all, no more.

Engineers often inflict upon themselves a misguided feeling of inferiority in the presence of the glossy ones in the boardroom – as Cleopatra puts it in *Antony and Cleopatra*, like 'mechanic slaves with greasy aprons, rules and hammers'. This sort of nonsense is grossly out of date and must be thrown out at once.

The UK had long been written off as the 'sick man of Europe'. Strikes became known as the 'British disease' and Britain was regarded as a thing of the past among trading nations – until the dramatic resurgence which has recently developed. The UK is again among the league leaders and going to the top. It cannot be pure coincidence that so many British boardrooms are nowadays manned in increasing numbers by qualified engineers – people trained in logic, trained to plan and to design; practical people who create the conditions for manufacturing what the world needs and will buy.

We must listen to what these proven leaders say, for example, about reports. What do such men require from senior people reporting directly to them? This question was among those discussed with some distinguished engineers when writing began on this chapter (their cooperation is acknowledged at the end of the chapter).

Sir William Barlow (Chairman and Chief Executive, BICC) has been kind enough to let me quote his Executive Instruction (see Figure 9.1). Sir William adds, in a letter:

> I am also very keen when people have made visits, particularly overseas or to other companies, that immediately they return they write a report so that those who have not been on the visit can benefit from the experience of the man who has made the visit.
>
> Unfortunately there are far too many people in the engineering industry who have not developed, or been trained, to have the ability to write clear concise and short reports, so that if your book is encouraging that, it will indeed be well directed.

Please adjust your thinking to suit your intended readership. Write especially for them and their needs. If your reader wants a needle, do not bury it in a haystack and expect him to search for it.

Definition: *A report is a working document that helps the other fellow to do his job*. This applies in all directions (Figure 9.2).

Those you lead will write better reports for you if you brief them properly. Sir William Barlow gave us an excellent example: we know precisely what he wants and expects from his managers. You must likewise clarify your own requirements and discuss them with your subordinates – before instructing

Executive Instruction
Monthly Reports

Executives in charge of business units are required to write a monthly report to accompany the monthly accounts.

The following rules should be observed:

1. The reports should be kept to a minimum number of words and the narrative should not exceed five A4 pages.
2. The report should be written by the person responsible and not delegated.
3. It must give a clear picture of what is going on compared with budgets.
4. Problems and difficulties should be mentioned together with the steps being taken to improve matters and by when. Once a problem has been mentioned in a report reference to it should be repeated until it is no longer an issue.
5. Outstanding decisions required from headquarters or other divisions should be mentioned.
6. Simple, clear English should be used and jargon avoided sc that a stranger could read and understand the report. Initial type abbreviations should be limited to those in general use in the Group.
7. Each report should be understandable without reference back to earlier reports for the simple reason that the reader will rarely have previous reports to hand when reading the report.
8. The report is from the manager to his superior, thus in the case of an operating unit it would be to the chief executive of his subdivision. In the latter case it would be to the chief executive of the division, and finally the chief executive of the division makes a report to me.
9. In addition to comment on performance and prospects, information which helps us to keep abreast of the particular business activity should be included. Headings would normally cover:

   Sales
   Profit
   Cash flow
   Orders and prospects
   Pricing
   Competitors' strategy
   Exports
   Capital expenditure progress, and
   Industrial relations

   If there is nothing to report under any of the foregoing, the heading should be excluded. Similarly if there is information worth imparting under some other heading, please include it.
10. The intention of the monthly report is to enable all necessary knowledge about the business to be available to those responsible for controlling and monitoring the business. If the monthly report is kept simple but comprehensive, most of the questions that would otherwise have to be asked should be answered and it should have the effect of reducing meeting time and the amount of other ad hoc forms of reporting.

Figure 9.1 *Example of a reporting instruction* (Sir William Barlow, Chairman and Chief Executive, BICC)

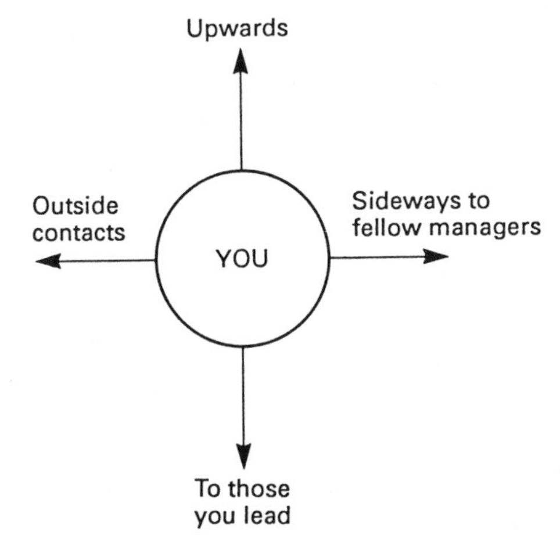

Figure 9.2 *Communication directions*

them to write a report. Sloppy reporting indicates a sloppy attitude at the top, and it is contagious. Your own reports must set a good example.

Some specialists regard report writing as a task thrust upon them by an uncaring world, or as a form of drudgery that interrupts other work which excites them more. Such an attitude wrecks many a career. Particularly in fact-finding research, a report is often the true purpose and the only end product of the work. Also, a report sometimes acts as your sole representative. People know you and judge your value from no other evidence than black marks on paper.

Communication is not a one-way process. It develops from interplay between you and those you wish to influence. Keep reaction in your mind – their reaction. Their reaction is the yardstick for your success.

Astonishingly few report writers bother to discuss their efforts with their readers. A fool has as his motto 'I shot an arrow in the air. It fell to earth I know not where'. He loses an awful lot of arrows that way. Equally, many a report writer loses friends because he neither establishes a target nor aims at one. A wise man seeks customer satisfaction and takes the trouble to listen at least as much as he talks or writes. Before you write, insist on a precise briefing, and thoroughly research your readership. After they have read your work, take the trouble to ask questions and to learn from their reaction. Galileo said 'Nature abhors a vacuum'. Readers abhor reports with empty spaces (those which exclude themselves). You are writing for them, after all.

The main factors to be considered are:

- Circulation and distribution.
- Layout.

- Visual aids.
- Numerical information.
- Language.

Circulation and distribution

Circulation means that a report moves from one person on a list to the next. Distribution means that each reader gets his own copy for action or for reference.

Who will receive the report and how will they use it in their jobs? This must be in your mind from the outset. In some companies the deplorable custom exists of tacking on the circulation list after the report has been written. How on earth can the writer consider the readers if he does not know who they are – their technical level, their needs in relation to the subject, their preferences about layout, what sort of language they can absorb, and so on. Once the need for a report, its subject and the objectives have been established, the users must be your prime consideration.

Who needs to know? (This will determine the distribution of your report).

What and why do they need to know? (This will decide the scope of your report).

How can you best convey the information to them?

These are key questions requiring positive thought before you write a paragraph. Brief yourself clearly and fully and have your readers firmly in mind. A clinical but human attitude to report writing is half the battle and becomes a personal asset.

Report contents and organization

You have advantages with your training and experience. Adapt engineering techniques to the job of writing. Know what the finished product has to do, plan and design it, gather the materials, and then build it to achieve the objective.

A first class report is conceived and starts its development right at the beginning of the project. It is not an afterthought. Suppose that a national leader were to drop dead this morning. Do you imagine that journalists reporting the event would need to scurry in all directions and make panic telephone calls to find out details of his career and other basic facts about him? Of course not. The material is on file and readily accessible. So, too, must the raw material for your reports be organized – long before you think of writing. But, remember, it is raw material and you probably have less pressure to meet a deadline than a journalist. This does not mean that you can sit back and dream. According to Parkinson's first law, 'Work expands to fill the time available for its completion'. So, without hurrying, get down to your job of writing, systematically. Old adage: *Plan your work. Then work your plan.*

Stage one: gathering material
You will find it helpful to clear the decks for action. Start with an uncluttered desk. Then go through every document about the project that you can muster. Read everything. Create three piles:

1 Completely relevant material
2 Material that you feel has no bearing on this report. Skim through this, but do not discard anything until you are certain that it contains no pertinent facts. Then get rid of it out your way. You are now confident that nothing has been forgotten
3 Facts and figures which might require validation or other treatment. Include in this pile notes for immediate action to check and, perhaps, to process this information. In short, do nothing more until you have all the checked facts and figures at your disposal.

Stage two: organizing the material
Before drafting your report ask yourself the following questions and discipline yourself to write down precise answers to them. Writing, even in note form, often forces one to make the thought clear. Answer these questions on paper:

- What is the subject and scope of this report?
- What is the purpose of this report?
- Who will read this report, and what are their needs in relation to this matter?
- Have I collected and checked all the information I may need?
- Have I selected from this material the information that my readers need?

If you spend time and care on these questions you will make the writing much easier for yourself. Boil down the answers to these questions and have them on a card in your sight all the time until you have completed your report.

Now that you have reminded yourself of all the facts, now that you have considered your readers, try to obtain a distinct picture of the report structure by making a table of contents. Set out the main headings and subheadings clearly (eventually on one sheet of paper).

If it is to be a full scale technical report, familiar traditional headings may serve. These can also be adapted to suit other reports. The headings are:

Summary – the busy man's page
Table of contents
Introduction – object and reason for the work done
Methodology – tests used, tolerances allowed, etc.
Results – facts and figures arising from the tests
Full summary
Conclusion – conclusions which you draw from the results
Recommendations for action

Appendices, acknowledgements, etc.

You will realize that not all your readers need, say, details of methodology. Your fellow specialists certainly will, but not the chairman. He (and the rest of us) trust you on such matters.

Layout

The first impression that a report reader receives is from the cover. Does the report look as if it wants to be read? Does the title tell the story in a headline? When the reader turns to the first page does he see a busy man's page, or an executive summary, as some companies call it?

The busy man's page (not necessary for a report of five pages or fewer) is the last thing you write. It gives all your readers the guts of your report in two or three minutes, so that even laymen can discuss it at their own level. The language that you use here must be understood by every reader. The busy man's page should provide the following information:

1 The report title. If your company uses a keyword retrieval system, try to use such keywords in the title.
2 The author's name. The writer might be a junior and need further authorization. A courtesy much appreciated by some readers will be the author's telephone and extension number. This saves an enquirer a good deal of trouble in locating him in a large company.
3 The date. This tells us whether the work is current or old stuff.
4 The library file number for reference. In many companies, the top people receive only the busy man's page. They can ask for a full copy from the library if they need it.
5 The distribution list (if confined to six names or fewer) so that we know who else has a copy, for discussion.
6 Keywords (again, if not too many).
7 A digest of the whole report, giving (in from fifty to not more than 150 words) a well-pruned outline of your story, including the object and reason for the report, the main results, the main conclusions and the main recommendations for action.

Your reader now has the main headlines. If you have captured his interest he will turn to the next page. This is a table of contents, setting out the various sections of the report and telling him where to find what particularly concerns him (the financial considerations, for example). A good table of contents clearly indicates that you have taken the trouble to help your reader to find his way. Do not flatter yourself that all your readers will want to read every word of your work. Most of them have other claims on their time, and will not do unnecessary digging. Make your layout as easy as possible for them.

If your company demands a standard layout, use it. People know from experience of the normal practice where to look, for instance, for graphs, charts and detailed figures. If these are usually placed at the end of reports, put them there. Many executives look first at the pictures and diagrams before reading the words.

The standard layout, however, is not Holy Writ. If you feel that it can be improved, make a suggestion to the appropriate people. As an example, a sales director was having trouble with his travelling representatives. He was losing expensively trained men. Their main grouse was that, at the end of a long day, they had to write reports covering every call, and each call report was usually three of four pages long. This chore, worked on at home or in hotel rooms, occupied several hours each evening. The sales director found the daily task of ploughing through these reports equally tedious.

On examination, the data were found to be repetitious, usually consisting of numbers, prices, quantities, etc. A system of cards was devised, so that the sales team merely had to fill in the spaces for routine data, and make observations on the back of each card about anything unusual. Thus, the representatives were happier, filing was simplified, and the paper saved paid for the change in about two weeks. So, use the standard practices but be on the lookout for improvements.

A useful exercise is to make a study of six or seven journals, and discover how the layout of each suits its regular customers. Readers of *The Times* know exactly where to look for the city news, the sports pages, the letters and the leader columns. Observe the clever headings in *The Guardian* and note how they lead one on to read the story beneath. Examine the sentences in bold type which follow the headings: these lead sentences (as they are known) outline the main facts of each story and encourage the reader to seek the details which follow. However serious you might be about your own speciality, no harm will come from finding out why millions of people are persuaded daily to pay good money for general news. Your own professional material would normally be of more direct personal interest to your readers. The company pays you to provide it, and it need never be dreary.

Headings, subheadings, paragraphing, the placing of illustrations – all such matters of layout – demand thought which should lead to your reports being more readable than most. Nobody wishes you to be facetious, but a report should be a pleasure to read: not just a dollop of solemn obscurity. Let your readers welcome your reports. Remember, they need to know and sharing your knowledge should be agreeable for them as well as a duty.

Visual aids

If someone presented you with 10 000 words explaining in detail a concept which you could easily understand from an engineering drawing on one sheet of paper, would you thank him? It would be equally thoughtless for an engineer to offer a technical drawing to a layman who could not even recognize the symbols.

The first essential in a visual aid is that the reader can grasp the point at a glance. The point must be seen. Second it must be understood at once. The objective is to convey your message as quickly as possible without leaving any doubt in the reader's mind as to precisely what you mean. Informative captions help your reader: a naked 'Figure 1' rarely does so. Cross reference your illustrations to the appropriate passages in the manuscript and *vice versa*. If all your detailed tables are placed in an appendix, consider whether a brief extract would facilitate the reading and clarify a point without interrupting the flow. Then refer your reader to the full table in the appendix.

The type of aid used, again, depends on the readership. A pie chart (see Figure 9.3) provides a simple means of offering crude information. To satisfy an expert seeking more detailed comparisons, you might have to use something more esoteric such as a statistical graph, logarithmic graph, or even a log-log graph (so long as you are sure he can understand it). The same consideration applies to all illustrations. There are many options: tables, graphs, charts, diagrams, drawings, photographs and others. Select the most appropriate for your reader.

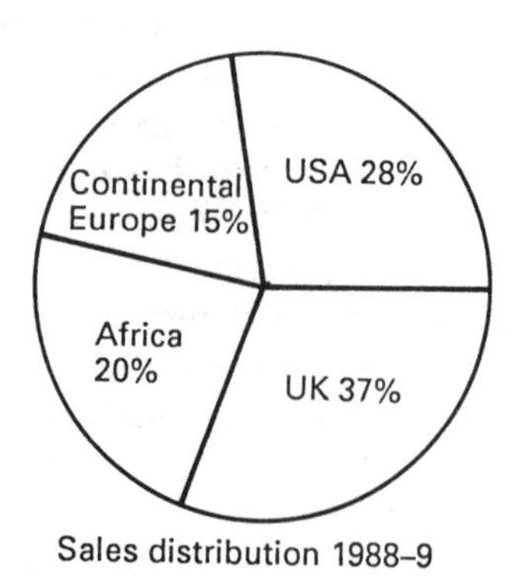

Figure 9.3 *A pie chart*

Your visuals should be integrated into your manuscript at the first planning of the report. You may well find that this saves you a considerable amount of writing and gives the report a well rounded look.

Numerical information

Most reports are fundamentally about numbers. With numbers, significance is everything. Unless figures are significant, or highlight other information that is significant, leave them out. Does your reader really need precision to three decimal points, or would rounded figures be more acceptable? If you are rounding up or down tell your reader, who will probably be grateful for simplicity.

The presentation of numbers demands the same forethought as visual aids. Do not offer a whole page of numbers without a clear indication of what the reader has to take in. Use headings, captions or an explanatory paragraph. Formulae can confuse people not strong in mathematics. Readers will accept your findings if shown a simplified formula. The experts can challenge you if they wish by referring to the appendix, where the formulae are worked out and explained in detail. Significance is all – or nearly all. Shine the light on the significant numbers. The others are for the nit-pickers who, of course, have their value.

On the same principle that a long report has less chance of being read than a succinct one, numbers lose power from being too numerous and smothered in a crowd.

Language

Please imagine that you are a personnel director. You are sifting through applications for a job you have for a new manager when you come across a *curriculum vitae* containing the passages reproduced in Figure 9.4. This extract represents a mere one-fifth of an application actually received by a company. It continues for several more pages in the same vein. Count the full stops. There are only three in the whole passage.

You have probably found several other factors in the application which would deter you from engaging this man. Reading between the lines, one can feel that this chap is a worthy sort, but he is writing to impress us rather than to express his thoughts.

His writing needs surgery. Even the most elementary surgery demands that the surgeon knows the various parts of the body, how they work and what to do when the trouble has been diagnosed. It is equally so with writing. One must at least master the main elements, their uses and their values. This chapter includes a few of the techniques of clear writing. These might help you to begin with. The list of recommended books at the end of this chapter will give you more information for later study.

A survey of 5000 graduates revealed that many of them took a holiday during English lessons. The trouble, and sometimes a lifelong fear of grammar, starts with a schoolmaster who teaches the English language as if it were Latin and also the exclusive property of the grammarian. Such impertinence takes the breath away and must be vigorously opposed. The language belongs to us – all of us. The purpose of grammar is to ensure that the thought becomes clear. For example, 'I ain't got none'. Does it mean I have not got none (I have some) or I have not got some? So we have a sensible

In my capacity as Personnel Manager I have always been totally involved and committed to all the important, vital and numerous areas of the Personnel Function and Services and, with my last employers, I was part of the Senior Management Team involved in the whole area of the business which included being involved in the Planning, Development, Administration and Management of various projects of different sizes and complexity in the United Kingdom and other parts of the world which were in the Division's Sphere of Operations which included the setting up of Offices and Sites in these different areas away from base for the successful operations to be carried out with all the associated problems.

A part of my responsibilities were (sic) the overall Industrial Relations and I have always found that I enjoy good communication skills at all levels with all nationalities and, I believe that I have set an enviable record of achievement in this vital area and, I have found that most any problem can be solved to everyones (sic) satisfaction if enough effort and thought is put into it enhancing a good working relationship and thereby improving the performance of everyone, including myself.

Although holding no formal qualifications, which I have found to be no deterrent to my career and progression, I feel extremely confident in applying for this position based on the strength of my 15 years wide and varied experience in a Personnel, Safety and Training environment and I am proud to have reached my present position which was gained through hard work and dedication to my employers and their employees as I started out at the bottom but my energy, enthusiasm, versatility, flexability (sic) and success has enabled me to hold down some extremely responsible and demanding posts which I have found both enjoyable and rewarding, dealing with a vast range of problems, people of various nationalities and needs, and I have always enjoyed the full confidence of my employers and their employees to handle all situations in my sphere of expertise to the best of my ability which has involved working for long and extended periods of time outside the normal working day at the office, home and away, which I believe has made me an all-round Professional Personnel Specialist.

Figure 9.4 *Three sentences from a job application*

rule about double negatives. One needs rules to prevent such ambiguity: and that is the sole function of grammar – to make things clear. You must respect grammar, but treat it as your servant. You are not its slave.

If you were lucky enough to have had a good teacher please skip the next page or so (we must not teach grannie to suck eggs). Unfortunately, many people need to be at least reminded of basic aspects of control in their writing.

Some principles of clear writing

- Write to express – not to impress.
- Keep sentences short.
- Provide variety for your reader. For example, vary the length and structure of sentences.
- Prefer the concrete – especially for nouns describing the subject of each sentence.
- Prefer the active.
- Do not use a phrase if a preposition will serve.
- Tie in with your readers' experience and needs.
- Avoid pompous cliché words and phrases.
- Write as you would talk (tidied up).
- Clarify the thought before you write it.
- Remember that there is a human being at the other end of your communication.

Let us take the sentence as our module. A sentence can be defined as a single thought, completely expressed. A sentence has two main parts:

1 The subject of the sentence – the topic, the who or what we are talking about. For example, the dog chewed a bone. The subject: the dog.
2 The action – what happened(s): chewed a bone.

If you make these two main parts of your sentence clear, your reader is happy. If you do not, your reader suffers.

Go back, please, and study the job application in Figure 9.4. Take the first sentence and note ten ways by which you could make it more acceptable. For example:

- The sentence does not convey one thought, but several.
- The subject, apparently, is *I*, the dullest word in the English language. Several other subjects loiter.
- The sentence is spattered with adjectives: important, vital, numerous, complete, comprehensive, etc. – as vulgar as sequins on a mourner's trousers. (Just as a reminder, adjectives modify and give more information about nouns. A noun is the name of something, someone or a place.)
- Unnecessary adverbs cling to the verbs: always, totally. (Another reminder, adverbs modify verbs. Verbs are the action words).
- Only one full stop gives the reader pause – after 120 words.

Keep sentences short. Each sentence one thought. Use your (and your reader's) best friend, the full stop. The word 'and' can often be replaced by a full stop. Otherwise sentences babble on like Tennyson's brook. Be sparing with commas. Too many of them should warn you that you are becoming verbose.

Nouns

Concrete nouns are the names we give to real things and people. Words like cat, book, cigarette, petrol-pump, policeman, girder, wall, are names for real, physical things. They conjure up in your reader's mind what is in yours. As Gertrude Stein remarked, 'A rose is a rose is a rose'.

Abstract nouns are the names we give to ideas – non-physical concepts such as achievement, reliability, consideration, departmentalization and economics. They build into abstract phrases, such as 'the fundamental realities in relation to the engineering aspects of this matter'. With such abstract nouns and phrases readers can, and do, choose their own interpretation. It might not be the one that you intended.

So give your reader a chance to agree with you, at once, as to what the subject of each sentence is. Prefer concrete nouns, especially as the who or what, the topic of each sentence.

Verbs

Verbs are the action words in the 'what happens' part of a sentence. There are three main types:

1 The verb 'to be' – I am, you are, he is, I was, you were and so on. The verb 'to be' simply indicates existence, or acts as an auxiliary. If overused it becomes dull, because it lacks the very stuff of a verb – action.
2 Passive verbs, where the subject of the sentence does nothing, but suffers the action. That is to say, the subject is acted upon. Also, the passive construction always includes a part of the lifeless verb 'to be'.
3 Active verbs, where the subject does the deed of the verb.

Let us take a horrid example and work backwards, noting the pitfalls and how to avoid them.

> In so far as the lubrication with respect to certain mechanical appliances, *viz* engines, is concerned, this process is carried out by means of the careful application of selected oleagenous products to the appropriate various moving parts areas of the appliances.

The galloping verbal diarrhoea indicated here is, unfortunately, not rare. Remember, your reader wants to know in each sentence: What is the subject? Who or what are we talking about? Read the horrid example again. See whether you can disentangle the subject from the maze. Try it before you read on.

You will have decided that 'this process' is the topic of the sentence: but you have to backtrack to confirm that 'this process' really means 'the lubrication with respect to certain mechanical appliances, *viz* engines'. So that is what the writer is talking about. Or is it?

If we look at the 'what happens' part of the sentence we find that the subject does nothing. But it gets 'carried out' by another confusing set of notions – a passive sentence.

Let us trim it a little. We do not need 'in so far as the . . . is concerned' at all. 'Selected oleagenous products' can go hang. So can 'the appropriate various moving parts area of the appliance'. So, we can cut to 'The lubrication with respect to engines is carried out by means of using oil'. This is still passive.

There are two phrases which can be replaced by simple prepositions. Prepositions are little linking words such as of, in, at, by, towards, with, above, which thoughtless writers pad out into phrases. Here are some examples:

Preposition	*Pads out to*
About	In so far as
Above	In a superior position to
By	By means of
In	Within the boundaries of
Near	In close proximity to
Of	With respect to
Towards	In the direction of
With	Through the medium of

Never use a phrase if a preposition will serve. Instead of the horrid example we now have 'The lubrication of engines is carried out by using oil'.

The sentence is still passive. The verb 'carried out' is nonsense. The real action is 'lubricates'. Should we say 'Engines are lubricated by oil'?

The sentence remains passive. If you study the phrase 'The lubrication of engines' you will see that the verb 'lubricates' has been smothered in an abstract phrase. Not only that: the action word 'lubricates' has crossed over and no longer works on the action side of the sentence. Why waste a powerful verb and bury it in a phrasal subject?

The verb acts as your most potent weapon. In general, if you get the verb right and make it work properly, it is difficult to write a bad sentence around it. So, 'lubricates', a power verb conveying the action in a word. Who or what does the lubricating – oil. Who or what does the oil lubricate – engines. And so we get:

Oil lubricates engines.

This, quite incidentally, uses only three words in place of the original thirty-nine.

Never strive for brevity. You might become curt or leave out information. Make every word work and brevity will come naturally.

Make active power verbs work for you and your reader. Fuddy-duddy professors, especially chemists and engineers, still force their students to write entirely in the passive. They assert that the only alternative to the passive is the personal, using the word *I*. This is simply not true. Here are a few sentences in the active (the subject performs the deed of the verb):

This generator develops 100 megawatts.
Diamond cuts glass.
Acids neutralize alkalis.
Windspeeds over 100 mph will overload the structure.
The crusher handles up to 100 tonnes of ore per hour.

Not a personal note anywhere. If you had this 'write only in the passive' nonsense inflicted upon you, shake it off. Use the passive when appropriate by all means: it often serves best. Use the verb 'to be' too: it has its uses. Variety is the spice of life. Vary your sentence structures. Vary the lengths of your sentences. Your readers will be grateful.

Tie in with your reader's experience. Avoid technical concepts and language unless you are certain that your readers will be able to follow you. Either use your readers' language or explain your own. Write as you would talk (tidied up). Remember that a report is a working document that helps the other fellow to do his job.

Letters and other written documents

When writing letters, memoranda, orders, instructions, manuals and public relations material, keep the same principles in mind as those for reports.

A letter should serve as an ambassador to both you and your organization. Even when you write to big companies such as ICI or BP you are not addressing a building. At the other end is a man or a woman with a job to do. Their reaction measures your success. An agreeable relationship does no harm. So let in a little humanity. The big *C* acts as a guideline. Your letters should be:

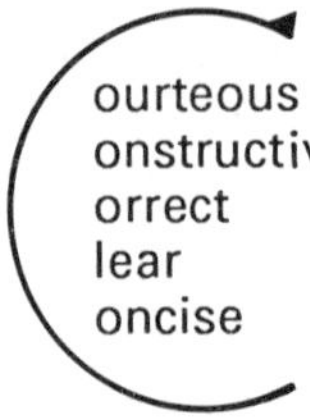

C ourteous
onstructive
orrect
lear
oncise

Many people mistakenly believe that courtesy is a soft quality. Courtesy demands discipline and stems from consideration of the other fellow – sometimes an abnegation of oneself. Courtesy is not always easy, but it is essential for effective communication. Courtesy does not mean that one uses the cant phrases of so-called business English:

Assuring you of our best attention at all times.

Do not hesitate to contact me.
Regretting the inconvenience caused.

Would you say things like these to your husband or to your wife? Be yourself, not merely a funnel for tired old clichés which show that you are too lazy to find your own words.

Be constructive. Help your reader. Even if you cannot supply something, perhaps you could tell him where else he might get it.

Be correct. Get your facts right, your language right.
Be concise. See the principles of clear writing.
Be clear. See the principles of clear writing.

Clear writing is simply clear thinking on paper. Clear writing is habit forming and keeps your mind alive. Thoughtless writing also becomes a habit and can drag you down to its own level.

Speaking

> Communications skills are essential for all engineers aspiring to be managers. To understand the existing knowledge of the recipient, to formulate the message, and then express views in a clear, succinct manner. This applies equally to writing reports and letters or giving oral presentations at business or public meetings.
> *Dr Kenneth Miller*, Director-General of the Engineering Council

Many engineers say that they can write, but that speaking in public terrifies them. Why? The requirements for a first class speaker are completely straightforward:

- He must have something worth saying.
- He must have something worth listening to.
- He must know his subject thoroughly.
- His subject must be relevant to the listener.
- He must prepare to give his listeners full value in return for the time they give him.
- He must deliver with ease and clarity.
- He must make the occasion memorable and profitable for all concerned.

When asked what terrifies him, the worried, would-be speaker always includes among his reasons: nerves, lack of experience and no time to prepare – all to hide the real cause of his troubles – his own vanity.

The main cause of ineffective speaking is that the speaker focuses on himself and on his own problems instead of working on the needs of his audience. The worried speaker will, of course, disagree. He will declare that he is not vain; that he is shy, timid, and introvert; as though looking inwardly at himself represented a virtue. Sensible people will face the fact that being more interested in 'How am I going to do?' rather than 'How are they going to do?' will inhibit any progress they might make as speakers. *A successful speaker is one who gives his audience a success.*

As with all communications, the reaction of the recipients is the criterion of success. If they react in the desired manner, the speech is effective. If they do not, however glorious the speaker's voice, however clever his words, however much he litters his presentation with expensive visual aids, the talk will be a flop.

Reaction measures the value of any communication.

Let us look in more detail at the main requirements for a first class speaker.

He has something worth saying. If you have nothing worth saying, please do not say it. Save your breath and other peoples' time.

He must have something worth listening to – from the audience's point of view. Your subject must be relevant to the needs and wishes of the audience on this occasion. A brilliant analysis of Beethoven's piano sonatas will not do if your audience thirsts for knowledge about the Industrial Training Act. To a group of music students, the reverse would be true.

So, assuming that you feel you have something worth saying, assuming that you truly know your subject, you must make a thorough study of your audience and mesh in their needs in relation to your topic. Do this right at the beginning. Who and what are they? On what basis will they be able to relate: as experts, as learners, as financial people, as laymen, as men, as women, or as a conglomeration of all sorts? You cannot begin to prepare a talk intelligently and obtain the desired reaction unless you study the source of the reaction. Do not simply transfer the problems to your hearers. Solve your own problems by thorough preparation. There are few real excuses for ineffective speaking and most are self-inflicted. An audience is on your side. They want you to succeed – to have a success themselves – through you.

History demonstrates that the human animal changes but little. As they always did, the same causes and effects establish and condition man's relationships with other people. 'For whatsoever a man soweth, that shall he also reap' (Galatians). One does not plant an apple tree expecting a yield of gooseberries.

Preparing a talk

Step one

Eliminate excuses and self-denigrating attitudes to speaking.

Step two

Establish the purpose of the speech and the exact reaction you seek.

The first question that a speaker must ask himself is 'What result do I want from this talk?' And he must work (yes work) on this question until the answer becomes clear, positive and dynamic. He must not settle for a vague generalization. He must know precisely what effect is required, so that he can create a suitable cause with his speech to gain the precise effect.

Step three

People: take active steps to know well the source of the reaction – your hearers. Who and what are they? What are their needs in relation to your subject? What would they like to take away from the occasion? What examples close-home to them can you weave into the presentation? How do they relate to your subject now and what changes do you wish to induce? What is in it for them? If they see a gain for themselves in your proposition they are likely to agree with it.

Step four

The proposition, the theme. Consider, and then write in a single sentence the main message you wish them to receive.

Step five

Power points. Pick out the three most telling facts that support your theme and build your talk round them. Make sure that they stand out prominently from the lesser facts.

So, now we have the five Ps: purpose, people, proposition, power points and, finally, profit (profit for them). If all these elements in your preliminary thinking have solid bases the rest is a matter of organizing the material and of delivery.

You will have noticed that these bases echo the initial preparation for a written presentation. You are right. The system works for all types of human communication, from a telephone call to winning a parliamentary election or selling a million packets of Bloggo. The professional at work knows with absolute clarity his:

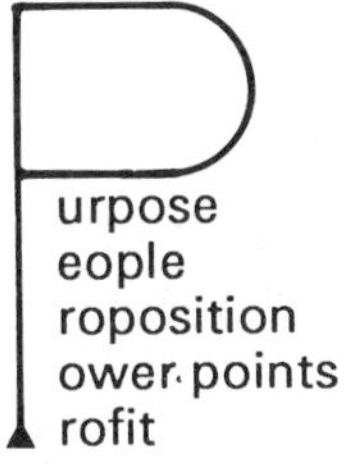

Step six

At this juncture, a table of contents, a view of the structure of your talk, will give you added command. We need a good beginning, a good middle and a good end.

You will be familiar with network analysis or progress charts. We can begin at the end, knowing with complete clarity what the end product should be, then work backwards to the steps needed to achieve the end product.

Step seven

Gather every scrap of information about the subject you can muster. Make rapid notes about your own knowledge and experience of the subject. Look elsewhere and add to these notes. Review the whole subject from every source available. Work at speed and aim for a quantity of ideas and facts until you have at least ten times more material than your talk demands.

Include in your notes thoughts about the audience and their present knowledge of the subject. Sketch in examples that they will understand and appreciate from their own experience. If possible use an example close to them for every link in the chain of your proposition. Never leave them out. It is their talk. It is yours alone at your peril.

Ensure that you have several options for both the content and structure of your talk. Then you need only choose what is best to serve your purpose. The rest can be discarded.

Now you are confident that you have not overlooked anything important. Also you have probably refreshed your memory about matters you might have forgotten or thought too obvious and elementary. Experts, like old men, do forget.

You now have plenty of material and some to spare. When you have confirmed to yourself that your own thinking about the occasion is clear, work positively to create a first class presentation. Weave in:

1 Your impact at the beginning. Make your opening remarks lively. Use an interesting thought, not a bromidal cliché.
2 With a technical talk or a boardroom presentation your audience will be helped by some signposts indicating the purpose and scope of your talk. Know where you are going and let them share the outline of your route so that they also know. Do not dwell too long on the signposting – just make the indicators clear.
3 Develop your theme, the proposition, by working towards the first of your power points.
4 Use visual aids if they are appropriate to the occasion. They often make a point more quickly and with greater effect than words alone.
5 Use check points. From time to time, summarize to ensure that your listeners have grasped the main facts and that you have said what you intended, so far.
6 Do not worry about your voice. Just make sure that everybody can hear you, and that you vary the pitch, pace, volume and sentence length. Each new heading should be introduced with a variation from the tone that

ended the previous point. Also change your physical position from time to time.

7 Use a few notes by all means, but spare us a dreary reading. if you are not in complete command of your subject this is no time to tell us.

8 Go on developing your theme and the main facts that support it.

9 Work towards your final, persuasive ending. Summarize. Make your main facts memorable. If you want action from the audience, stimulate it, especially at the ending.

10 When you have finished *stop talking* and let us be delighted to have heard you.

Checklist for speakers

Structure

Opening: Is it alive? Does it have impact? Does it involve the audience at once? Does it start an agreeable relationship?

Signposts: Do they:

- Show where we are all going together?
- Indicate the scope?
- Give the audience footholds, a map of the route?

Main facts: Are they justified, authoritative, significant? Do they strongly support the theme?

Summings-up: As check points. Are your listeners with you so far? Are you following your declared signposts? Do you need internal signposts?

Final summary: Make main facts unforgettable. Not too long. Build up the climax.

Ending: Curtain line creating the afterglow, or getting action on their part. Is your ending truly powerful?

Delivery

Narrative: Will they understand your language? The story line? Examples concerning themselves – include them in everything. Pictures in their minds?

Contact: Can they see your own mind through your eyes? Are you watching, interested in them and their reactions?

Visual aids: Are they appropriate? Clear? Simple? Memorable?

Voice: Have you planned for variety of pitch, pace, volume? For an occasional controlled silence?

Time: Are you sufficiently within the time set to allow for the extra time needed when working the audience?

Will your listeners enjoy a worthwhile experience? Will they have a success through you – and you through them?

Tis not in mortals to command success.
But we'll do more Sempronius;
We'll deserve it.

Business meetings

'Journeys end in lovers meeting'. Lovers' meetings can result in bliss or disaster, depending on the lovers, the controls applied and whether the consequence matches what was aimed for. Just like business meetings.

Incredibly few organizations regard meetings as an aspect of their business worth serious study. Companies can tell what their raw materials cost and how much the labour force costs. They watch and record the profits and the losses year by year, month by month, even week by week. But ask a director 'What do meetings cost in your organization?' and the response is a blank stare, as if that were a stupid question. 'Sorry', he says, 'I'm too busy to fret about that sort of thing. Right now I'm rushing off to a meeting'.

If a workforce sits on its bottom with nothing to do, that is serious. If materials lose their value because they have been exposed to bad weather, bad workmanship or sheer waste, that is serious. Slack stock control, strikes, customers going elsewhere: all such deficiencies cost money, and the loss can be measured in terms of cash. But few companies consider the gaping hole in the money bag caused by dud meetings. Few companies even know the price of their meetings or recognize it as a major cost.

Let us define a business meeting as two or more people getting together for a specific business purpose.

If you were to ask a group of managers to list all the meetings that occur in any large organization, their own perhaps, the result would run into hundreds of events – board meetings, management meetings of many sorts, production meetings, personnel interviews, sales interviews, union meetings, conferences, training, foreign trips, technical meetings, scientific research seminars, progress meetings – just to start with. Many more will occur to you.

Now consider the cost. Managers, directors and staff all cost money, even when they sit bored stiff and inactive at meetings. Rail fares, air fares, hotel bills, expenses, cars, conference rooms, training premises (including rent, rates, heating, lighting and cleaning), secretarial preparation and the work of, say, accountants and others, literature, reports, handouts, etc. Companies who have been wise enough to study the cost effectiveness of their meetings are usually aghast when confronted with the facts. After an in-depth survey one Norwegian company estimated the cost of its meetings as one and a quarter times its gross profit. A British giant quoted 'about £1 million a week' as its figure.

How can any sensible management afford to be casual about meetings?

Apart from all the money thrown away, fruitless meetings diminish all who take part in them. They are more than a waste of time. They are a waste of life.

Any meeting that you attend should be the better for your presence. No meeting that you organize should be less than businesslike. We measure the success of a meeting by the profitable action that ensues (Figure 9.5). Please keep these proportions in mind. Be positive in your attitude towards meetings. In general, never call a meeting merely to discuss something. Never call a meeting just to obtain a decision. Call your meetings to discuss, to decide and to get profitable action actually started as a result. The meeting will use up less time if the preparation has been thorough. More meetings fail through skimped preparation than for any other reason.

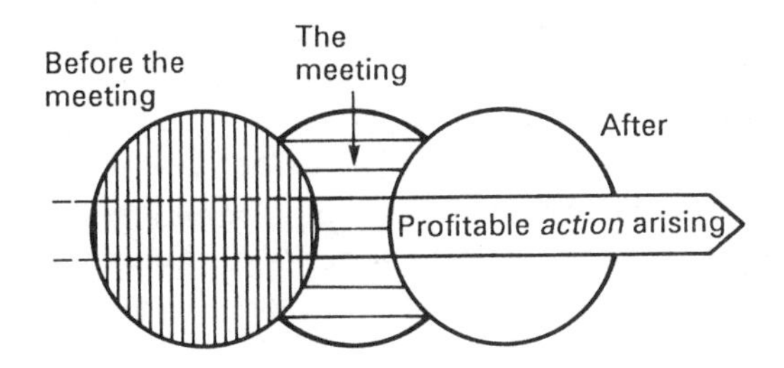

Figure 9.5 *The purpose of a meeting is to generate profitable action*

Before the meeting

Assume that, as head of department, you are not only organizing the meeting, but will act as chairman. You would be wise to have trained a first class secretary or personal assistant. The good ones are worth their weight in platinum, and a shrewd manager makes every effort to deserve such a treasure. He will know what follows and see to it that you think about these questions:

1 Is the meeting necessary? Or can the business be done without dragging people away from other work? Would an exchange of letters, a few telephone calls, or some other personal exchange serve as well? If the answer is 'We must have the expertise of others and a face-to-face examination of the facts' you have a meeting on your hands.
2 What is the precise subject of the meeting and its scope?
3 Are my terms of reference clear?
4 Have I clearly defined the objective of the meeting? What business is it meant to achieve? What action would be desirable as a result of the meeting? What other options are there? Weigh up the pros and cons of each possible outcome of the meeting and anticipate their effects.

5 Am I sufficiently briefed on the subject (I must not pose as an authority but I need not be a fool).
6 Does the agenda cover the subject? Is the agenda too weighty? Can the proposed business be accomplished in the time available?
7 Have all (and only) the right people been invited to the meeting? You will of course be careful to avoid the professional meeting-goers, whatever their status, unless they can be expected to make a real contribution. Would some people lower down the scale bring practical experience to, say, a production problem?
8 Do I know enough about the members and their needs? Is the venue the most convenient for all concerned? Must we start at 9.00 a.m. if this means that five people have to travel late and stay overnight in a hotel? Would a start at 11.30 a.m. enable them to spend the evening with their families – and save five hotel bills? Have I got their names and titles correct? Have I arranged for them to be met? Has the receptionist been given a list of those arriving, with instructions to help them as much as possible?
9 Do not forget that the chairman or secretary must arrange that:
   (a) Invitations or notices are sent out neither so far in advance that they might be forgotten nor so late that members might have accepted conflicting appointments.
   (b) Any handouts are prepared and taken to the place of the meeting.
   (c) All members are fully briefed, and have been sent the agenda in advance together with any reports or other literature which they should study beforehand.
   (d) Each member knows, if possible, who else will be attending.
   (e) Arrangements have been made to take the minutes, if necessary.
   (f) The room and seating accommodation are adequate and reasonably comfortable.
   (g) All equipment (such as the projector) is in place and in working order.
   (h) There is a clock in the room, visible to all.
   (i) Telephone calls will be intercepted and not put through to interrupt the meeting (arrange for messages to be noted so that they can be given during suitable breaks in the meeting).
   (j) Ashtrays are provided (if smoking is allowed).
   (k) Refreshments are provided at the right time.

Starting the meeting

A good chairman creates a team with a common aim, to get the business done with the greatest profit for all concerned. The profit need not be money, although that is usually the ultimate aim.

You will recall that an effective speaker establishes a good relationship with his audience at the outset. A chairman has the additional responsibility of establishing good relationships between the members of the meeting. The manner in which you begin can colour the entire proceedings.

Be the last to be seated. If you sit with egg on your face waiting for them, you have missed a trick. Have them seated ready to get to work. The moment your backside reaches the chair is an important moment. The effect should be that the meeting is now officially under way.

The group will respond to your manner at the start. If you casually loll in the chair and dither, that becomes the key attitude of the members towards the business in hand. Be firm, be pleasant, be businesslike. If introductions need to be made, make them. Briefly sketch in the background of the problem under discussion. Then, with a sure touch, build in the controls which will sustain your authority and help to keep the business moving at the right pace and in the right direction.

Remind the members of the exact purpose of the meeting. Declare the scope and the limits within which this subject is meant to be explored at this meeting. If, later on, someone with an axe to grind strays outside these limits you have a legitimate choke-chain with which to pull him back into line. Discourage time wasters, now. Remind everyone that the meeting and its members have a time limit. Also, discourage the prima donnas. Make it clear that every member of the meeting is important, and that we want their views: that is why they were invited.

Get the nod from the members about these control points at the start. They will sound arbitrary if you try to impose them later.

So now your meeting has a clear, common objective. The scope and limits are specific. Every member knows that his contribution is needed. They are aware that time is finite and valuable. They realize that the chairman knows what he is about and is worthy of their respect. One cannot demand respect: it has to be earned (mainly by showing respect for others).

The larger the meeting, the greater the need for formalities. Even a small gathering works better within suitable rules. Get these agreed at the start. To make them up as you go along just does not work.

If someone is making a formal presentation for discussion, decide and announce beforehand whether questions will be permitted during the presentation or left until the end. Otherwise matters could get out of hand.

Conducting the meeting

The last opinion we want is that of the chairman. A senior member can often stultify discussion by declaring his view too soon. If he is sensible and reserves his own opinion the group might, in any case, come to the decision that he desires. Then all he needs to do is to agree with them. They will certainly feel better for having made the decision without pressure from the boss.

To get people talking, pose an open ended question of a general nature. Do not let the specialists take over too soon. One way of deadening a meeting is to use the cartwheel method. That is, to begin with the fellow on your immediate left hand, and then continue with the fellow on his left and so on round the table. The first man needs little voice to reach the chairman; the second will also mumble and keep the focus narrow. People on the right hand side feel remote from the discussion, and will probably interject out of turn. The chairman then has to decide whether to reprimand the interrupters or let them rip, spoiling his one-at-a-time-clockwise instruction. On the other hand, if nobody butts in we are faced with a dreary succession of low key remarks.

It is much better to keep the control in your own hands. You should place a lively mind at the far end of the table, and invite him to start the ball rolling. He will need to speak clearly to reach the chairman's ear, thus including the whole meeting in what he says. Next, bring in someone to your left, followed by someone from your right. Every member of the meeting will be aware that the chairman might drop on him at any moment, so they pay attention to what is being said.

Keep the members on the path leading to a successful conclusion of the business in hand. If anyone strays, remind him of the purpose. Ensure that all the members have an opportunity to express their views. Note the points which are agreed, and prevent further discussion of them if this would hold up progress. You might need to sum up from time to time. This is one way of bringing the ball back into your court if the meeting has become ragged.

When the meeting has reached a decision based on intelligent discussion, seek to get action agreed – and who is to do what and by when.

Minutes

And while the great ones depart for dinner,
The secretary stays, getting thinner and thinner;
Doing his best to record and report,
What the great ones think they ought to have thought.

Minutes are best created at the meeting. The chairman dictates and asks the secretary to record:

1 The main facts which emerged from the discussion.
2 The decisions arrived at.
3 Who takes what action and when.
4 A date for reporting progress.

Do not form subcommittees. Make named individuals responsible for specific actions. If they care to form committees, that is their affair.

Bumph breeds more bumph. Minutes should be agreed on by the meeting. After smaller meetings the secretary can type the minutes, preferably on one sheet of paper, photocopy them and hand a copy to each member before he leaves. If the members hurry off, the minutes should be posted and waiting on their desks when they arrive at their offices – with the action marked in red.

Recommended follow-up

If you are not already actively associated with your professional institution you should immediately seek information as to the advantages you can gain by being so. Such organizations provide a wealth of sensible literature pertinent to engineering. The specialized courses they sponsor create opportunities for self-development. Get to know more of your fellow professionals at regular local and national meetings. Find out what these organizations can provide for your benefit. Become actively engaged.

Action 1

Telephone or write today to your professional organization. Develop your skill as a communicator. Read more about speaking, writing and meetings. A few book titles will be found at the end of this chapter.

Action 2

Visit your local library or bookshop as soon as you have taken action 1.

Action 3

Prepare a checklist, and monitor the next ten meetings you attend. Do they follow the guidelines which you have just been reading? Note and put into effect every improvement that you can make, personally, and among your colleagues and staff. *Vis inertiae*, according to the Rev. E. Cobham Brewer (in *A Dictionary of Phrase and Fable*), is:

> That property of matter which makes it resist any change. Thus it is hard to set in motion what is still, or to stop what is in motion. Figuratively, it applies to that unwillingness of change which makes men 'rather bear the ills they have than to fly to others they know not of'.

Do not succumb to *vis inertiae*. Take action this day. It is in a splendid cause – yourself.

Useful organizations

The Engineering Council, 10 Maltravers Street, London, WC2 3ER. Telephone 01 240 7891.

The Institution of Mechanical Engineers, 1 Birdcage Walk, Westminster, London, SW1H 9JJ. Telephone 01 222 7899.

The Fellowship of Engineering, 2 Little Smith Street, Westminster, London. Telephone 01 222 2688.

Or the governing body of your own special discipline.

Further reading

Bell, Gordon, *The Secrets of Successful Speaking and Business Presentations*, Heinemann, 1987.

Cooper, Bruce M., *Writing Technical Reports*, Pelican 1987.

Goodworth, Clive, *The Secrets of Successful Business Letters*, Heinemann, 1987.

Gowers, E., *The Complete Plain Words*, revised edition, HMSO/Penguin, 1987.

Mitchell, John, *How to Write Reports*, Fontana/Collins, 1987.

Scott, Bill, *The Skills of Communicating*, Gower, 1986.

Acknowledgements

I am indebted to the following for their encouragement and cooperation during the formative stages in writing this chapter:

Sir William Barlow, Chairman and Chief Executive, BICC.
Sir Alistair Frame, Chairman, Rio Tinto-Zinc Corporation.
Michael Hoffman, Chief Executive and Managing Director, Babcock International plc.
Dr Kenneth Miller, Director General of The Engineering Council.
Donald Newbold, CBE, Chairman, Foster Wheeler plc.
V. J. Osala, CBE, Executive Secretary, Fellowship of Engineering.
Sir David Plastow, Managing Director and Chief Executive, Vickers plc.

10

Supervision, leadership and motivation

Edmund J. Young

The topics of supervision, leadership and motivation are interdependent. Effective supervision requires that the engineering manager exercises competent leadership, and has adequate understanding of how to motivate subordinate personnel (whether they be highly qualified engineers or unskilled workers). In this chapter, these three topics will be analysed separately at first. Then the principles and concepts will be examined together, to see how they apply to engineering management practice.

Supervision: an essential task in engineering management

The terms 'supervision' and 'supervisors', like the terms 'management' and 'organization', are troubled by a problem of semantics in the English language. A supervisor is any person who is in charge of and is responsible for the actions, efforts and behaviour of one or more other people. A supervisor exercises formal authority as conferred by the organization and is therefore in a formal leadership role, exercising authority and accepting responsibility for personnel in his group or organizational unit.

Owing mainly to historical usage and custom, there is a status connotation with the terms 'supervisor' and 'manager'. In many organizations a supervisor refers to a foreman, group leader, or some low-level manager who heads a group of workers, usually at the bottom level of the organization. Some organizations make a clear distinction between supervisors and managers, whereby the supervisors are the subordinates of the managers. As such, supervisors are not usually accorded the status, remuneration and privileges of managers. Managers have greater authority and higher responsibilities. A common distinction made in many organizations is shown in Figure 10.1, which depicts three levels:

1 Managers.

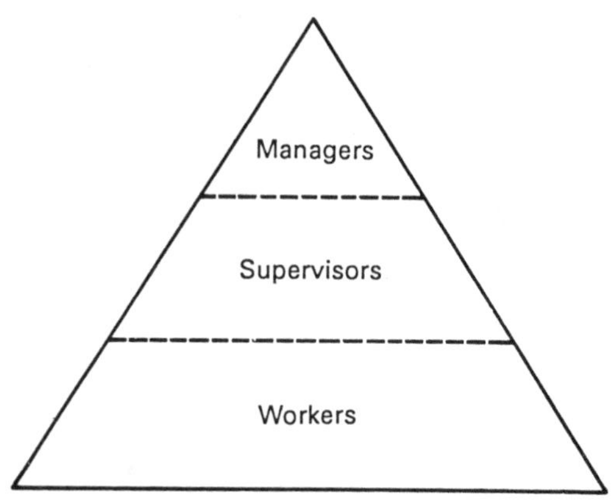

Figure 10.1 *Common arbitrary distinction between managers, supervisors and operative workers*

2 Supervisors.
3 Operative workers.

The term 'operative workers' here refers to those personnel directly involved in performing the operative tasks, whether designing, creating, researching and developing, installing, constructing, producing, manufacturing, operating and maintaining the product, machine, system or service. Many make a further arbitrary distinction between top or general management, middle management, and supervisory management (the last being the 'supervisors') in the classification of managerial levels as distinct from the operative workers at the bottom of the organization structure.

Another view which is gaining increasing acceptance is that *supervision* is a basic universal and pervasive function that is exercised *at all levels of the organization* from the top or chief executive to the lowest level manager, supervisor or chargehand, and is therefore applicable to everyone who is in charge of, exercises authority over and accepts responsibility for, one or more persons in his group or organizational unit.

These two views of supervision may be called the lateral view and the vertical view. The lateral view is so-called because it considers the word 'supervisors' as referring to those spread across the lowest level of management responsibility. The vertical view, on the other hand, considers that everyone who exercises supervision over subordinates is a 'supervisor': the task is seen as a pervasive one, operative from the top to the lowest level of managerial responsibility. Most progressive organizations regard the lowest level supervisors as part of the 'management team'. Generally it is only the traditional, more conservative managements who prefer the preservation of status differences between managers and supervisors.

We will initially consider mostly the vertical view. The principles and concepts are equally applicable to the lateral view, with the proviso that the distinction in practice is based more on differences in levels of authority and responsibility, status, nature of duties, remuneration, privileges and other fringe benefits. We shall view the supervisor and his subordinates as comprising the personnel in an organizational unit, whether this is a team, a section or a department.

The term supervision generally has three main meanings, pertaining to the task of:

1 Overseeing, planning, organizing, coordinating, directing, leading, evaluating and controlling *subordinate personnel* in the organizational unit.
2 Overseeing, planning, organizing, coordinating, directing and controlling *operations, personnel and material resources* of the organizational unit.
3 Monitoring, checking, evaluating, recording and reporting *progress* so as to ensure that operations and things are proceeding according to predetermined plans. When a government engineer 'supervises' a contract it means that he needs to ensure that the contractor is carrying out the work according to the standards and requirements of the plans and specifications agreed to in the contract, and that work is progressing satisfactorily within limits of agreed schedules and costs. If not, the engineer can reject defective work and withhold progress payments until the contractor remedies all defects.

In general, supervision refers to 1 and 2. Significantly, the supervisory task is a managerial function that can be defined in terms of Fayol's (1949) original analysis and most books on supervision (as well as on management) find it difficult to avoid the Fayolian functions.

Supervision is usually the *first managerial task* of most engineers but it is a task for which most engineers are little prepared. The author's own experience and observation as an executive engineer once responsible for supervisory and management training in a large Australian company found that the supervisory hurdle is generally the most difficult for young engineers to surmount.

There are several reasons for this. First, engineers' initial engineering education gives little guidance for assuming future supervisory and managerial responsibilities, as Badawy (1982) indicated. Second, the general attitude and temperament of many engineers, being concerned more with things than with people, are inhibiting factors that make the transition from the technical specialist engineer to an engineering supervisor more difficult. The late Professor Merritt A. Williamson, an American pioneer of engineering management education and founding editor of *Engineering Management International*, once noted that engineering tends to attract candidates whose initial orientation is more concerned with things than people. He also remarked upon the 'myopic technical specialization' engendered in engineering schools (Williamson, 1961). If the young engineer does not succeed as a competent supervisor in handling people, then he is less likely to be considered for promotion to higher levels of managerial responsibilities. Generally, supervision is automatically the first step after the engineer has demonstrated his technical competence. It is common for the most technically competent to be considered first for promotion. But, as Rubey, Logan and Milner (1970) observed:

> A poor engineer is not likely to develop into a good manager (nor is he likely to be given the opportunity to do so): on the other hand, a sound technical education does not automatically make a good manager.

Thus supervision of personnel is a task which every engineer who aspires to engineering management or general management should master. Supervision is the first step upwards on the management ladder of any organization.

The essence of success in supervision is in the ability to deal with and handle people in an effective, harmonious and fair manner, tempered by consistency, courtesy, humility, firmness, integrity and a sympathetic understanding of human nature and behaviour.

The engineer who aspires to supervisory and managerial roles should always be an avid student of human nature, be interested in people, trying always to understand why and how they behave in certain ways, be tolerant of human weaknesses and idiosyncrasies and always bear in mind that all individuals and groups differ. Their individual experiences, education and training, groups to which they belong, attitudes, personality and many other factors such as age, health, sex, ethnic background, philosophy of life, hopes and ambitions will all influence their pattern of behaviour. All these attributes and factors make the study of human beings much more difficult than the study of machines, structures or engineering systems.

The engineer aspiring to supervision and management should develop interpersonal skills in communicating, persuading, influencing, negotiating and empathizing with people. The engineer, being trained in analysing and solving technical engineering problems in a logical, rational, objective and systematic manner, should not expect people to behave in a similar fashion to machines or inert objects. Rather, people are more often swayed by their beliefs, attitudes, feelings and emotions rather than adhering to logic, rational analysis and objectivity. Their actions and behavioural patterns may often be governed by self interest but again their philosophy of life, belief and value systems, state of their health, emotions and feelings may influence their present behaviour. The old Chinese saying 'a boil on the back of a man's neck is of far more importance to him than the mortal illness of the Emperor' holds true in many cases of self interest. But if a person is influenced much by prevailing attitudes of comradeship ('mateship' in Australia), certain beliefs, moral and ethical codes or altruism then these factors may prevail over self interest. Material gains such as more pay, greater accessibility to goods and services and promotional opportunities may not be the main motivators or stimulus to certain behaviour because, in the words of the Bible, 'man does not live by bread alone'.

Often the manner of communication can evoke different responses. A simple order given in a certain manner may be interpreted or misinterpreted in a number of ways and could evoke a range of possible actions or reactions with different people. This human element makes the task of supervision and management more difficult that solving scientific engineering problems.

This emphasis on the human element in supervision and management, and the need for change in orientation from analysis of technical engineering problems to the more complex task of analysis and solving human problems, make the first hurdle in supervision particularly difficult for many an engineer.

In his Chairman's address to the Engineering Management Division of the Institution of Mechanical Engineers, Dr N. A. White (1983) noted that the most difficult thing for an engineering manager to learn is to live continually with areas of uncertainty, if not of outright ignorance. He quoted a former Master of Clare College at Cambridge University, Lord Ashby, as saying that 'the secret of good management lies not in the manager's vast and exact knowledge, but in his skill at "navigating areas of ignorance" '. However, most practitioners maintain that certain principles of supervision culled from experience are far more useful than operating in total ignorance, or simply learning from the wasteful process of trial and error. As Lyndall Urwick once said, 'a manager's or a supervisor's errors are other people's trials'.

In spite of what behavioural theorists and academics like Simon (see Chapter 3) and Mintzberg (1973) say about 'principles of administration' and their theoretical analyses of managerial functions, Henri Fayol's (1949) statement on principles is almost a truism: 'The fact is that the light of principles, like that of lighthouses, guides only those who already know the way into port, and a principle bereft of the means of putting it into practice is of no avail'.

It is the human element in supervision and management that makes for uncertainty in human actions, and thus renders the practice of supervision and management more of an art than a science.

Personal attributes of engineers aspiring to become supervisors

It is essential at the start of the engineer's career to decide whether he/she wishes to become a supervisor of others or become solely a technical specialist. Otherwise a wrong decision based mainly on the temptation of promotional opportunities, higher status, greater remuneration and more fringe benefits associated with supervisory and managerial roles could lead to future difficulties, job dissatisfaction, frustration and failure. An engineer who by personality, temperament and motivation is essentially an individualistic technical specialist and who dislikes dealing with people (especially having to supervise them) would find greater job satisfaction, and a happier career path, by remaining an engineering specialist concerned primarily with technical achievements and excellence.

The engineer aiming for supervision and engineering management must have what Bower (1966) called 'the will to manage'. Without this will, motivation and determination to manage, the engineer should not consider taking on any leadership role in a group. He may be a most knowledgeable leader in technical matters but this is different from being a leader of a group or organizational unit.

In their study of engineers and scientists at NASA (National Space and Aeronautics Administration in the US), Professors Bayton and Chapman (1972) classified three types of engineers in respect of their orientation towards management:

- *Type 1 engineers*, who had strong managerial motivation from the outset. These generally performed well in managerial roles.
- *Type 2 engineers*, who initially had specialist motivation but who, once in managerial roles, found them challenging, stimulating and rewarding, with accompanying job satisfaction.
- *Type 3 engineers*, who had only specialist motivation, with little or no inclination for management. These, when promoted to managerial roles, would find them disappointing, dissatisfying and frustrating.

It is the type 3 engineer who usually makes a good technical specialist, whether in design, research and development, consulting or teaching. But type 3 people should avoid aiming for management. It is unlikely that they would become other than a second or third class technical supervisor or manager. The sad experiences of many good technical engineers who failed dismally by becoming poor or bad managers have caused many non-engineering executives (as well as engineering managers such as Bone [1972]) to raise the perennial question: 'should engineers be managers?'

Much of the problem goes back to the formative education of engineers. Dr Don Williams advocated, when he was Chairman of the South Australian Division of the Institution of Engineers, Australia, that engineering schools should cater for the education and training of two distinct streams of graduates – the technical engineering specialists and the management oriented engineers, with the latter aiming for supervisory and managerial roles. Various surveys of the engineering professions in Australia, Britain, New Zealand and the USA have found that management, supervision of personnel, human relations and communications were the weak links in the education of professional engineers.

Supervisory role and duties

It is useful to analyse the supervisory role from the lateral view, in that the relationships between the supervisor, superiors, peers and subordinates can be seen. Figure 10.2 shows a senior civil design engineer in a supervisory role in respect to his subordinate design engineers. In turn, he is responsible to the principal civil design engineer who in turn is responsible to the manager of engineering. It may be that our senior civil design engineer is in charge of hydraulic design, and his peers at the same organizational level are also senior civil design engineers responsible for structural, bridge, roads and airfield design, all reporting to the principal civil design engineer. Similarly, the principal civil design engineer may have peers such as principal mechanical, electrical and chemical design engineers, who in turn are all responsible to the manager of engineering.

The senior civil design engineer (hydraulics) has been delegated the authority over, and responsibility for, personnel in his organizational unit. This comprises civil design engineers, chief draughtsman, draughtsmen, tracers, and junior, cadet and trainee engineers and draughtsmen. The chief draughtsman is himself a supervisor in charge of his draughtsmen and tracers. The cadet and trainee engineers may be assigned to various design engineers, who in turn exercise supervisory functions over them. In this case the span of control of the senior civil design engineer is limited to the chief draughtsman and the number of design engineers directly responsible to him.

As a supervisor, the senior civil design engineer has three formal sets of relationships within the organization – upwards to his superiors, laterally to his peers, and downwards to his subordinates. There may be extensive external relationships, such as those with outside consulting engineers, surveyors, geologists, quantity surveyors, architects, clients or customers and even members of the general public (although the last may be handled by a public relations or corporate affairs manager, if not by some member of top management).

In this supervisory role the senior civil design engineer exercises formal authority over his organizational unit and is thus responsible for the work, effort and behaviour of *all personnel in the unit.* The organization, or more precisely his superiors, delegated authority and responsibility to him for the operations of that organizational unit.

His duties would include planning and allocation of work to his subordinates, interpretation of the company's policies, objectives, plans and procedures from higher management, transmission and implementation of

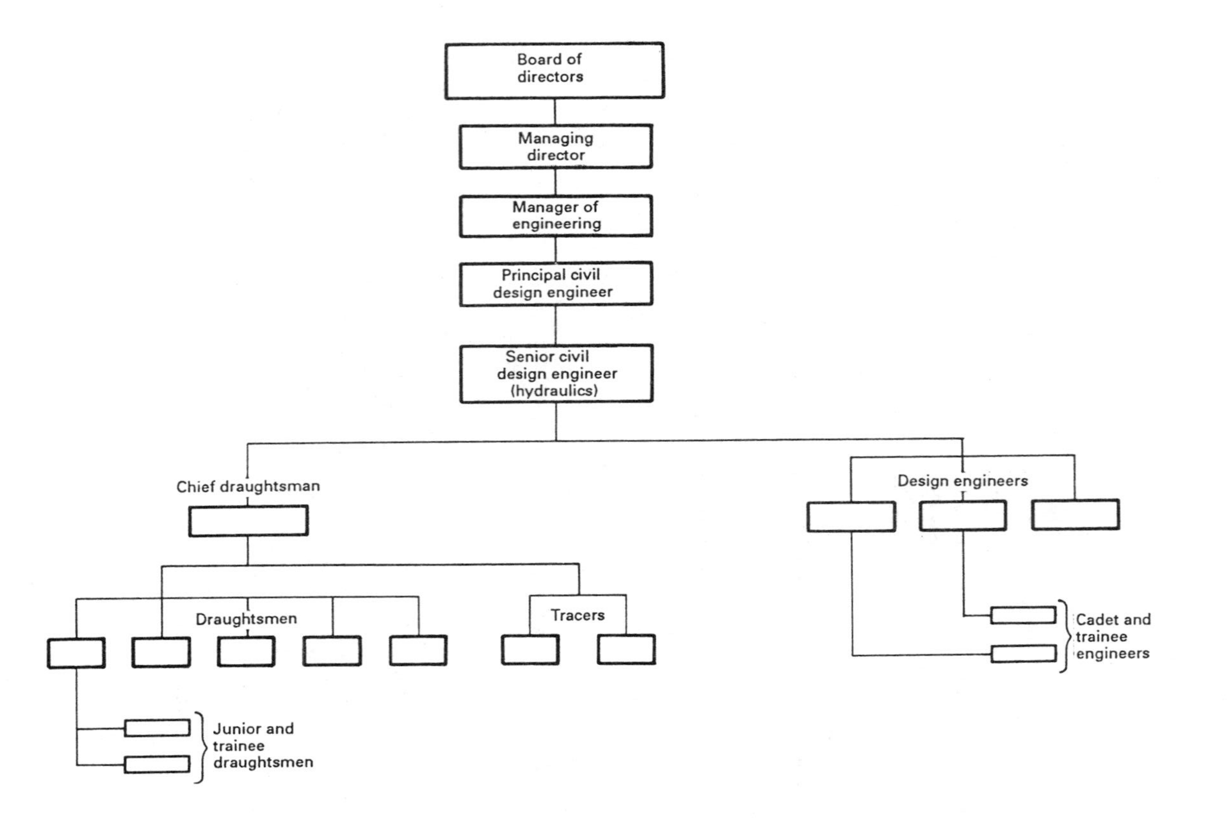

Figure 10.2 ***Managerial chain of command with the senior civil design engineer (hydraulics) in a supervisory role***

instructions, and general supervision of the daily operations of the organizational unit. To higher management, he should convey the thoughts, ideas, suggestions and other important matters that affect the whole organization. He needs to deal with grievances and complaints expressed by his subordinates. Where necessary he needs to exercise disciplinary measures, so that rules and procedures are followed. He needs to explain the reasons and necessity for such rules and procedures to his subordinates.

He needs to be concerned with both daily and short-term planning and scheduling of work, making day-to-day decisions so that predetermined plans are followed. Where necessary he must take corrective actions to ensure that operations go according to plans and not out of control. With his peers he works in cooperation to achieve joint actions, and obtains advice, suggestions and complaints so as to resolve any grievances and conflicts. The supervisor is vitally concerned with communications, written and verbal, upwards, laterally and downwards.

The giving of orders and instructions, the conduct of regular meetings with subordinates, attending to various grievances and complaints, use of disciplinary measures, initiating ideas, encouraging and inspiring subordinates, obtaining advice and suggestions, motivating personnel, maintaining a high state of morale, organizing and coordinating effort, directing, monitoring, evaluating and controlling operations besides recording and reporting to senior management, are all part of the duties of the supervisor. Sometimes he may be engaged in selection of personnel, but his continuous responsibility includes their training and development, safety, welfare, security and relationships with unions and other representatives concerned with industrial relations. He thus needs to be familiar with the awards and conditions under which his subordinates work.

Ideally, the supervisor should know each of his subordinates in his organizational unit, their backgrounds, their educational and training, their qualifications and experiences, their prime motivations and aspirations, their ambitions and even philosophies of life. To some extent the supervisor should show interest in the families of his personnel, for this demonstrates a personal interest in people's backgrounds. A supervisor who shows a genuine concern, understanding and interest in individuals and their personal background is more likely to inspire loyalty and willing effort than one who shows no concern and treats people like machines or automatons. Even the manner of greeting subordinates first thing in the morning, and courteous and friendly behaviour during working hours can affect the attitude, motivation to work and loyalty of subordinate personnel.

The supervisor should explain to his subordinates the whys and wherefores of various actions, to evoke their greater cooperation, commitment and effort. Where necessary he must make quick and firm decisions, but he should investigate and gather the facts beforehand if time allows. He should be able to inspire and influence people, setting an example where required and providing the necessary zeal, determination and confidence for getting the job done. He may need to reprimand and impose discipline. It is better to be a firm but fair supervisor than one who vacillates and is weak. The latter will not gain the respect and loyalty of his subordinates.

Many useful hints can be gathered by studying the more practically orientated books on supervision, and by reading various personnel and supervision journals: more so than by studying books on organization behaviour written by academic theorists who have never supervised or managed personnel outside their ivory towers.

The supervisory task is not an easy one. It requires much more than technical knowledge. It needs knowledge and understanding of people.

Principles of effective supervision

Various texts on supervision have come up with a number of principles deemed necessary for effective supervision. Most of these have been taken from the experiences of many practising managers and supervisors. The most useful principles found by this author are the following:

1 Treat each person as a distinct and valued individual. Show respect and courtesy. Get to know and understand him. Show interest in personal and family background, qualifications, experiences, education and training, attitudes, ideas and thoughts expressed, aspirations and ambitions, particular interests on and off the job, habits, customs, dress, etc.
2 Give praise and credit where due. It is far better to praise, give due recognition and encouragement for work well done than to criticize, belittle and discourage subordinates. Where possible, indicate what contribution that effort has made to the organization, and let other personnel 'especially superiors' know about the good work. Give adequate recognition and reward for superior effort. Everyone likes being praised and recognized for good work. A supervisor can instil loyalty and greater effort by such actions.
3 Continually stress and reiterate the organization's objectives. Point out to your subordinates the targets that your organizational unit is aiming for and when you expect such to be achieved. If possible, set out your targets in pictorial, chart or diagrammatic form. Bar charts and arrow or precedence diagrams are useful aids in visualizing targets, schedules and progress.
4 Remember the importance of self interest to all individuals. Attempt to integrate individual self interest and goals with the objectives of the organization. The question 'What is in it for me?' should be answered by

'If you play your part, then you will get . . . and we shall all benefit from your contribution'. Appeal to higher or group objectives. Make each individual feel that his contribution is important to the whole enterprise and that he has a vital part to play no matter how menial may be the task. Field Marshal Viscount Slim (1956) of the 14th Army in Burma went round to each military unit and used a watch to show his men (whether they were ordinary soldiers, cooks or clerks) how important their jobs were to the whole military operation. By showing that the watch would stop if the small wheels stopped and drawing the analogy of the big wheels being the officers and the small wheels being the men themselves, Slim stressed how important their jobs were to the whole enterprise and so raised morale and the will to fight back.

5 Try to aim for high ideals and standards of ethical conduct, in practical and realistic terms. Remember that there is always the tendency for subordinates to follow your own example.

6 Never reprimand any individual subordinate in front of others, especially his peers and subordinates. Reprimand privately, make sure of your facts and always listen to the subordinate's side of the story. It could be that you have made a mistake, or have overlooked certain actions that could have been taken to prevent the wrong occurrence. When reprimanding, always indicate where the mistakes were made, what the consequences were and what should have been done. Show trust and confidence in the subordinate's ability to avoid repeating the mistake. The only occasion to reprimand openly is when the whole group has been at fault, in which case the group must be treated as if it were the individual. Subordinates are more likely to respond positively by such actions and appreciate your trust and confidence that they will not make the same mistake again but will do better next time.

7 Always show confidence. Never show disappointment. Do not show that you are worried even if you are. Do not have regrets. Show hope, optimism and belief in the cause. As a supervisor you are in a formal leadership role and your group's confidence, trust and morale can be affected by your attitude and behaviour. Field Marshal Viscount Montgomery, as noted by the Australian biographer Alan Moorehead (1946), displayed one characteristic of leadership that spurred his men to greater effort under the most difficult conditions – he exuded a sense of confidence, optimism and the will to be victorious.

8 Learn to delegate, and train your subordinates accordingly. Give clear, simple and direct instructions, and then trust your subordinates. Encourage people to accept greater responsibilities and to undertake a variety of tasks. Not all individuals have equal abilities: select those who are more capable than others and give them greater responsibilities. But develop the others to attain higher standards of performance.

9 Always listen and heed the advice and suggestions made by your subordinates. Sometimes they can see things which you have overlooked. Have a flexible mind, and adapt your actions to changed situations.
10 Be loyal to your subordinates and, in turn, you will find that they become loyal to you. If you make a mistake, admit it. Do not try to cover up mistakes and shelve responsibility for them on to others, whoever they are. If you admit a mistake, your subordinates are more likely to understand and forgive. 'To err is human. . . .'.
11 The saying 'do as you would be done by' is a good principle to follow. But always beware of the few who, for sake of personal ambition, may try to deceive you, put the 'knife in your back', or otherwise try to weaken or destroy you. Then, the teachings of Machiavelli (1952), the Chinese philosopher Hsun Tzu and Japanese Prince Iyesu Tokugawa may be more appropriate as a means of self preservation.
12 Always reserve time to think about future plans and courses of action. Reflect on what should have been done. Learn from mistakes. Where you encounter obstacles, devise various ways of going around them rather than confront them. Devise various alternative or contingency plans. Both the Chinese philosophers Confucius and Hsun Tzu stressed the importance of adequate planning as the secret of success of most courses of action. Plan meticulously but always be flexible and adaptable to changed circumstances.
13 Always try to simplify the problem. Sort out the 'wheat from the chaff'. Do not become immersed in details. Sometimes, certain details may be important. Sort out the essential details, and concern yourself with the important issues. This is the old 'exception principle' mentioned in Chapter 3.
14 Always retain a sense of humour, especially when things look bleak. It is better to belittle oneself rather than belittle or mock others. Integrity and honesty are important to both your superiors and subordinates. But always have the determination, will, zeal and courage to undertake action which you consider just and right.

The author's own experiences as a civil engineer, first in charge of survey and investigation teams and later in charge of construction projects, found that adherence to the above principles generally works well. This is true of people from practically any nationality, provided that trouble is taken to understand their cultural differences. Some groups respond more to autocratic styles of leadership than others, while some (especially Australians) tend to be more individualistic, responding more to the democratic or participatory leadership styles.

The essence of effective supervision is to develop harmonious, cooperative teamwork, so that each individual in the organization unit or team strives willingly and enthusiastically to make his contribution. This entails competent leadership.

Leadership: an essential part of engineering management

All supervisory and managerial roles are formal leadership roles in the organization. As with 'management', definitions of 'leadership' abound. Simply, leadership is the capacity of an individual to lead others. It is the ability and power to influence others to follow willingly and enthusiastically to achieve group objectives. One of the early scholars of leadership, American industrial psychologist Ordway Tead (1935), defined leadership as 'the activity of influencing people to cooperate toward some goal which they come to find desirable'. Many years later, Field Marshal Viscount Slim (1962), defined leadership as 'that combination of persuasion, compulsion and example that makes men do what you want them to do'.

Tead (1935) in his studies of leadership in social, industrial, business, political and military groups found that leaders in general possess the following ten characteristics or traits which differentiate them from their followers:

1. Physical and nervous energy.
2. A sense of purpose and direction.
3. Enthusiasm.
4. Friendliness and affection.
5. Integrity.
6. Technical mastery.
7. Decisiveness.
8. Intelligence.
9. Teaching skill.
10. Faith.

Slim (1962) considered five important qualities necessary for leadership: courage, will-power, judgement, flexibility and knowledge. Note that with his military background, Slim included the aspect of 'compulsion' in his definition, and in today's definitions, Tead's qualities of 'friendliness and affection' and 'technical mastery' would be replaced by 'human empathy' and 'technical knowledge and skill' respectively.

Two leading writers on management, Professors H. Koontz and C. O'Donnell (1978) of UCLA, defined leadership as 'the art or process of influencing people so that they strive willingly toward the achievement of group goals'. Most definitions of leadership revolve around the concept of 'influencing people'. Two American sociologists, J. R. P. French and B. Raven (1959), analysed five sources of power and influence as follows:

1 Reward power – whereby the leader has power to reward followers.
2 Coercive power – where the leader has power to punish followers.
3 Legitimate power – power conferred by the organization or, as defined in Chapter 3, the power of formal authority.
4 Referent power – where followers like to be identified with and willingly follow the leader.
5 Expert power – whereby the leader possesses certain expertise, knowledge, skills and other attributes which the followers do not have but are prepared to follow the leader due to these qualities.

Reward and coercive powers are those of the proverbial 'carrot and the stick' and both referent and expert power are closely associated with informal authority as seen in Chapter 3. This classification of sources of influence for leadership is useful for engineers, especially in cases of the need to switch from one source of power to another so as to lead subordinate personnel in changing situations. Leadership entails not only the ability to persuade and influence people but (in some situations) to direct, command, cajole, goad and control followers towards the achievement of group goals. This would be in line with Slim's other attribute of leadership – compulsion.

In early studies of leadership, there arose two schools of thought:

1 The trait school.
2 The situational school.

The trait school, of which Tead is often considered as one of the early exponents, maintained that leaders possessed certain characteristics or traits which made them stand out as leaders. The situational school maintained that leadership was more of a function of the particular situation rather than traits possessed by the leader, and some people become competent leaders in one situation but are followers in other situations. The debate raged on for decades but today most authorities on leadership tend to take a compromise stand and maintain that leadership is a product of the particular situation and that leaders possessed certain traits that propel them to leadership roles in such situations.

The English scholar on leadership, John Adair (1968), maintained that the task involved must be related to the team for leadership to function. Today we note that not only do the leaders need to possess certain qualities for leadership, but the task, the situation and the nature of the group or followers as well can influence leadership.

An American sociologist, Kurt Lewin, classified three types of leadership: autocratic, democratic and laissez-faire (or free-rein). The *autocratic leader* is one who tends to dominate the followers. He makes most of the decisions and gives all the orders. He seldom considers the views of his followers and tends to override any opposing views or objections of his group. The *democratic leader* consults his followers, encourages them to participate in decision making, and may be swayed by a majority vote or consensus view of his followers. He tends to be more considerate of the views of his followers. The *laissez-faire or free-rein leader* exercises little direction but tends to suggest actions to his followers, give occasional guidance or counselling, and generally allows a high degree of freedom of action on part of his followers. Figure 10.3 is a diagrammatic representation of these three types of leadership.

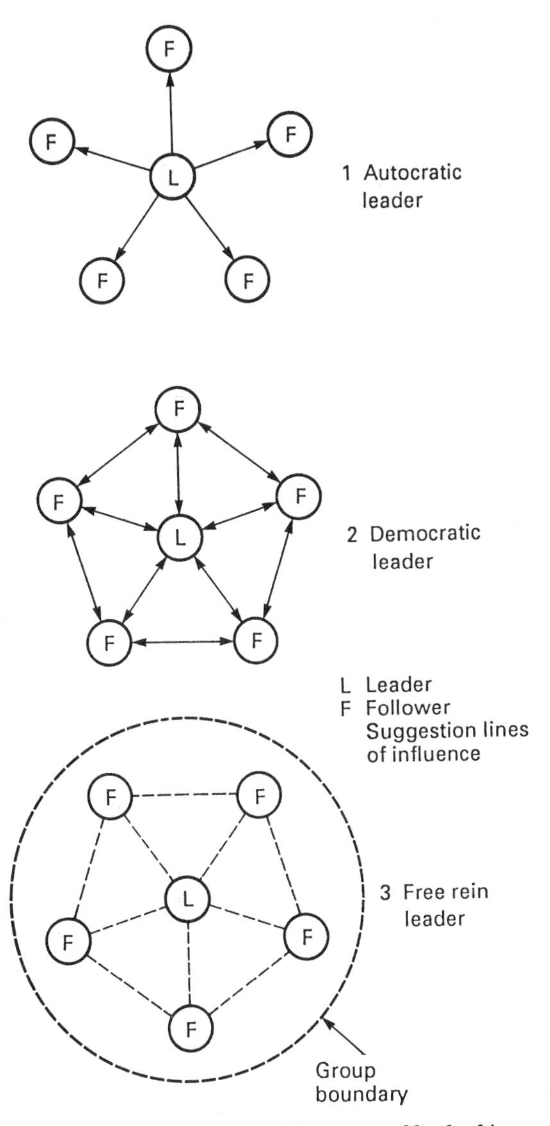

Figure 10.3 *Diagrammatic representation of three types of leadership*

Recent studies by behaviouralists maintain that, rather than simply three types or styles of leadership, there is a range or spectrum of leadership styles ranging from the most authoritarian and autocratic leader at one extreme to the most free and easy 'laissez-faire' leader at the other. Based much on Tannenbaum and Schmidt's (1973) analysis of leadership patterns, Amos and Sarchet (1981) present a range of leadership styles which an engineering manager may exercise in practice, as shown in Figure 10.4.

However, in recent years the situational school of leadership has gained more support. One of their most prominent scholars was Fred Fiedler in the USA. Fiedler (1967) is known for his contingency approach to leadership. After studying over 800 different groups ranging from military combat crews, factory and office groups to construction teams, Fiedler maintained that there were two main styles of leadership – autocratic or production orientated, and the democratic or human relations orientated.

To distinguish between the two types of leadership, Fiedler devised a questionnaire test which he called the LPC (least preferred co-worker) score test. The respondents, who were leaders in various situations, were asked to rate themselves on an eight point scale ranging from one extreme to another (such as cooperative to uncooperative) on about twenty characteristics. They were asked to think about their past experiences and assess how they thought their least preferred co-worker rated on these scales. The positive attributes were given high scores on the scale. Averages were obtained for all the characteristics. Fiedler found that high averages correlated with human relations orientated leaders, whereas low averages were associated with task-orientated leaders.

Leadership was seen as a function of both leadership style and the particular situation. Fiedler maintained that the group situation was affected by three sets of factors:

1 Position power, or the degree of formal authority.
2 Task structure, or definition of the task.
3 Leader-member relations.

These sets of factors could be favourable or unfavourable to the leader. Fiedler's empirical findings were plotted in relation to the three sets of factors. The two types of leadership style were correlated with his observations of leadership effectiveness, which he defined in terms of achieving certain goals or output that were measurable.

Fiedler found that the autocratic leader was most effective in situations where the factors were most favourable or most unfavourable, but in between the extreme situations, the democratic leader was more effective. He concluded that rather than try to change the personality and style of the leader it would be better to change the situation, i.e. to fit the job for the leader. He suggested that by changing some of the factors, job re-design and reforming

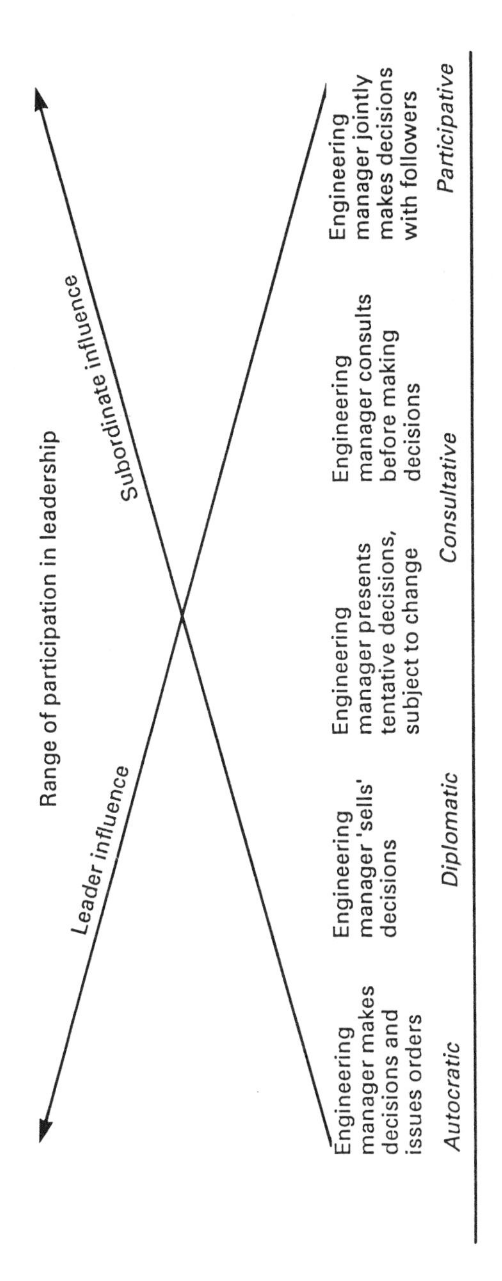

Figure 10.4 ***Leadership styles which the engineering manager may use***
(Courtesy of Amos and Sarchet)

the members of the group, more effective leadership (and thus improved productivity) would be achieved. For example, in his studies for NATO, Fiedler suggested that French speaking officers be placed in charge of French speaking ratings in the Belgian Navy, with Flemish speaking officers in charge of Flemish speaking ratings.

As applied to engineering management practice, Fiedler's studies have some significance. It may be that an engineer with a quiet, unassuming personality and with a democratic human relations-orientated style of leadership is better suited to lead a team of design engineers than a more boisterous, extrovert type of engineer (who tends to be autocratic and task-orientated and who may be more suited to supervise a construction project). The author's own observation in engineering practice is that human nature varies widely and that some engineers are more flexible and adaptable than others. Some can change their leadership styles to suit the particular situation, whereas others find it difficult to change.

Fiedler's generalizations may hold good in many cases, but there are also many exceptions due to the variability and complexity of human behaviour. Fiedler's assumptions have been criticized by other behaviouralists, who question his simplified classification of two types of leadership, the validity of his LPC score tests, the inadequacy of his three sets of factors and his neglect of the influence of culture. This last criticism would apply, for instance, in countries where autocratic leadership is the norm, and any attempt to exercise a democratic style of leadership (such as encouraging construction workers to participate in decision making) could result in many difficulties and eventual chaos.

The following concepts are important in leadership:

1 There can be no leader without followers.
2 Qualities that make for good leadership for one group may not necessarily make for good leadership of another group. The particular situation, nature of the task and the group itself, can influence the type or style of leadership and personal characteristics of the leader required.
3 Certain personal qualities such as decisiveness, technical knowledge and skill, job experience, communications skill, intelligence (above average of the group), ability to influence and persuade others, determination, will-power, ability to resolve conflicts, ability to crystallize and express the general thoughts of the group, ability to inspire others, ability to initiate ideas and actions, ability to engender loyalty, trust and confidence on the part of followers, honesty, frankness and integrity are important for effective leadership. In some situations the leader needs to be firm, persistent, domineering (even with democratic styles of leadership) and sometimes ruthless to ensure that the general good takes precedence over individual self interest. As Tead and Slim stressed, integrity, faith in the

cause and a sense of destiny (often coupled with ego by some charismatic leaders) are important for successful leadership.

4 Leaders are not born leaders, but leadership is developed through experience, purposive guidance, self development, education, training and the will to lead others.

Behaviouralists like to classify personality into different types, such as extravert and introvert. The American psychologist Maccoby (1976) listed four types of managers and supervisors after his study of 250 managers from twelve industrial companies:

Type	*Personal attributes*
The craftsman	Conservative work ethic and pride in the job.
The jungle fighter	Entrepreneur, ruthless manipulator, seeking power – not cooperation.
The company man	Fearing insignificance, people of this type become functionaries who accept bureaucratic roles.
The gamesman	Detached, but with drive to compete. Makes play of work, but must win by using imaginative gambles.

This classification may be applicable to American industry, and it is no wonder that many managers suffer from personal and organizational stress. In reality, the author's own experience is that human nature varies so widely that some people exhibit some or most of the personal attributes identified by Maccoby, whereas others may not satisfy any of the above criteria. Culture, of the group or organization, and of the society in which the enterprise operates, can influence the effectiveness of leadership in engineering considerably. All four of the above types may be applicable to American business society, but they may not be relevant to some European, Asian or African societies.

Giegold (1981) stressed the importance of leadership as essential in engineering management and suggested a useful technique for developing leadership skills in engineering managers. Much can be learned from studying case histories and biographies of both successful and unsuccessful leaders of the past, especially of their strengths, weaknesses and mistakes.

Motivation

Much has been written on the subject of motivation, including a book by Gellerman (1968) entitled *Management by Motivation*. Motivation is a psychological force inherent in the individual that propels him to act and behave in a certain manner. Psychologists maintain that people are motivated to satisfy certain needs which they come to see as desirable. It is often stated that managers and supervisors must motivate their subordinates to greater effort, productivity and better quality standards and performance on the job.

However, little has been written on the motivation of the managers and supervisors themselves: if they are not sufficiently motivated, it is difficult to see how their subordinates will be motivated. Slim's emphasis on example in leadership is relevant to motivation.

The American behaviouralists have developed a number of theories of human motivation since American psychologist Abraham H. Maslow (1970) proposed a theory of hierarchy of human needs. Maslow conceived a five step model of motivation which acted as a stimulus to satisfy human needs and viewed it as a hierarchy, as shown in Figure 10.5. Maslow proposed that:

1 There is a hierarchy of human needs. Once each need is satisfied, the individual seeks to progress to the next higher level.
2 A satisfied need is not a motivator of action.

He saw the higher level needs as more intangible and more concerned with the mind and spirit than with material things. At the base level were the physiological needs such as food, clothing, shelter, sex, etc. and he maintained that once these were satisfied, then the individual progressed to seek satisfaction of the next level of needs – safety and security.

There has been much criticism of Maslow's simple model. This centres especially on the automatic progression up the hierarchy, failure to consider the individual's philosophy of life, values and beliefs, and the influence of culture, customs and traditions. There is also the impact of other individuals, groups and the environment to be taken into account.

It is true that self interest largely governs human behaviour. But there are other potential influences on behaviour than are contained in the simple picture painted by Maslow. At the top of his hierarchy is self-fulfilment, self-realization or self-actualization (an American term), but some years ago, when the present author was project engineer on a construction project in Singapore, the Australian project manager and he both thought that the epitome of life, especially after a hard day on site, was to eat various Asian foods and listen to the Chinese sing-song girls with high slit skirts in a cabaret or restaurant. We were told by others that true self-fulfilment was 'to have an American salary' (American oil engineers were then the highest paid in South-East Asia), an Englishman's home, Chinese food, a Japanese wife and a French mistress. Alas, Maslow did not consider the multiple and multidirectional motivations of real life! Our security, social and status needs as construction personnel were minimal.

Another theory of motivation which has gained much prominence among behaviouralists is that proposed by Frederick Herzberg (1959) and his colleagues. This is the motivation-hygiene theory, motivation-maintenance theory, or two-factor theory of motivation. In his study of over 200 engineers and accountants, Herzberg found (by asking them what made them happy or unhappy, satisfied or dissatisfied in their jobs) two sets of factors. The first set

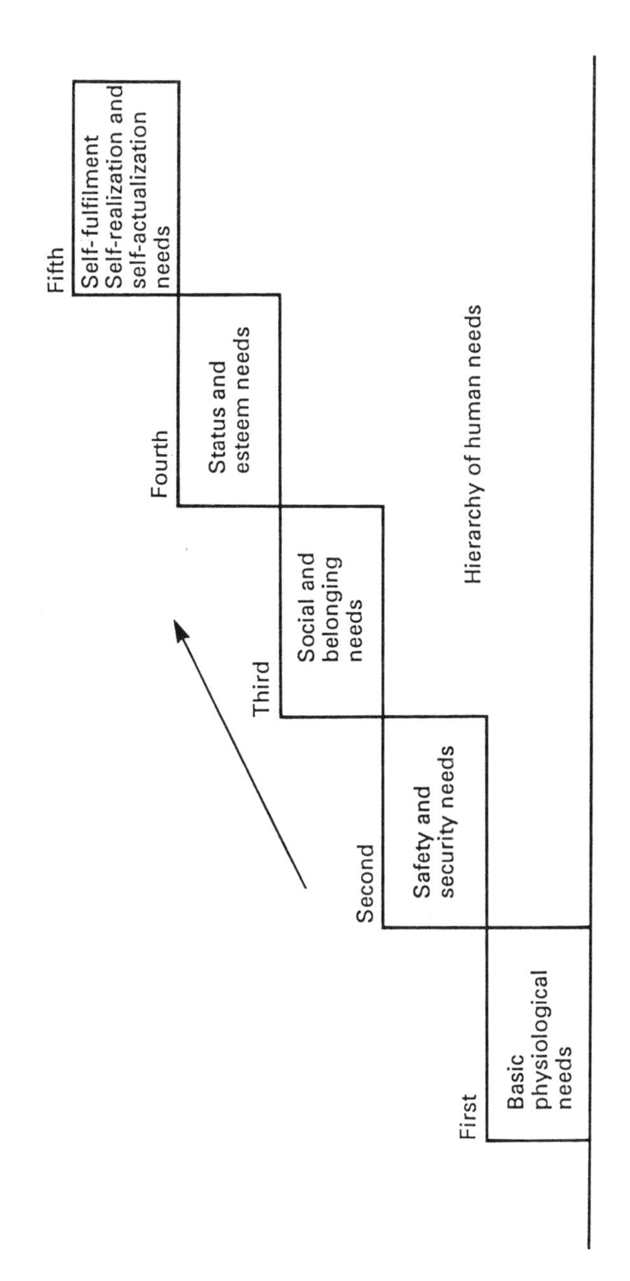

Figure 10.5 *Maslow's theory of human needs in motivation*
(Adapted from Davis, x, *Human Relations at Work*, 3rd ed., 1967)

he termed 'motivators' and the second set he called 'maintenance, hygiene or environmental' factors. Out of a total of sixteen factors, the first set – the motivators – included sense of achievement, recognition, the work itself, responsibility, status, and opportunities for advancement and personal growth. The maintenance factors included such things as salary, technical supervision, working conditions, personal life, job security, company policy and administration, and interpersonal relationships with subordinates, superiors and peers.

Herzberg found that the first set of factors were the true motivators of action and were more significant than the second set of factors. He saw both sets of factors as having a range or spectrum of satisfaction. With the motivators it ranged from satisfaction to simply no satisfaction. With the maintenance factors the range ran from dissatisfaction to no dissatisfaction. If the first set of factors were present, then they operated to build strong motivation. But if they were absent, then they did not prove highly dissatisfying, but simply resulted in no satisfaction. On the other hand, the maintenance factors were such that if they were absent then there would be dissatisfaction, but their presence did not build motivation. The motivators were more job centred, whereas the maintenance factors were more concerned with the environment of work.

Koontz and O'Donnell (1978) drew a comparison between Maslow's and Herzberg's theories of motivation with the motivators at the top levels of Maslow's hierarchy. But Herzberg's theory has been criticized on the basis that his initial study was made on professional workers, although Herzberg claimed that later studies made on factory and office workers verified his initial findings. Critics of Herzberg's theory also point out that in some situations maintenance factors could be more important than motivators such as salary, job security and personal life (especially in times of high unemployment) and also that he neglected the influence of culture. From some studies made outside the USA it appears that Japanese workers regard maintenance factors as of equal importance to the motivators, whereas Argentinian workers tend to follow the American pattern of motivation as perceived by Herzberg.

Another American theory of motivation especially favoured by many engineers is the expectancy theory proposed by Victor Vroom (1964). Vroom developed an equation of:

$$\text{motivating force} = \text{valence} \times \text{expectancy}$$

which mathematical expression seemingly appeals to many engineers.

Vroom saw motivating force as a measure of the strength of a person's motivation, and valence as the strength of an individual's preference for an outcome. Expectancy was the probability that a particular action would lead to a desired outcome.

Thus Vroom argued that a person's motivation towards an action at any time is determined by his anticipated values of all the outcomes of the action, multiplied by the strength of that person's expectancy that the outcome would yield the desired goals. A number of engineers have developed mathematical expressions for motivation based on Vroom's theory. However, it must be remembered that in practice people do not behave in a rational and logical manner as Vroom's equation suggests, and many actions are taken simply based on their emotions, moods, feelings, attitudes and idiosyncracies. Their health, age, sex, ethnic background, experiences and beliefs can govern their motivation at different times in their lives. A person may change his motivations, place different priorities on various motivational factors or even have a set of different motivating factors instead of one. The subject of motivation is difficult and complex in practice in spite of all the academic theories and studies made to date.

Significantly, in spite of all the criticism levelled against him by behaviouralists, it was the American engineer Frederick Winslow Taylor who first remarked:

> There is another type of scientific investigation . . . which should receive special attention, namely the accurate study of the motives which influence men (Urwick, 1962).

Taylor had stated this many years before the so-called 'pioneer of the human relations movement' George Elton Mayo had arrived on the industrial scene. And yet many behaviouralists consider Mayo as the founding father of the behavioural sciences (Trahair, 1984).

However, Hofstede's (1980) analysis of whether American theories on motivation, leadership and organization apply outside the USA, and Maloney's (1982, 1986) papers on motivation and supervisory problems in international construction are worthwhile reading for engineering managers.

The technique of management by objectives

A technique claimed by some to be most useful in supervision and management is that of management by objectives (MBO). The concept was first proposed by Peter Drucker. It was spelled out in his classic work *The Practice of Management* (1955) in the following statements:

- Business performance requires that each job be directed towards the objectives of the whole business.
- These objectives should lay out what performance the manager's own managerial unit is supposed to produce.
- The objectives of every manager should spell out his contribution to the attainment of company goals in *all* areas of the business.
- The goals of each manager's job must be defined by the contribution he has to make to the success of the larger unit of which he is a part.

- This requires each manager to develop and set the objectives of his unit himself.
- Higher management must reserve the power to approve or disapprove these objectives. But their development is part of a manager's responsibility – it is his first responsibility.
- Objectives are the basis of control. To be able to control his own performance a manager needs to know more than what his goals are. He must be able to measure his performance and results against the goal.
- A manager's job should be based on a task to be performed in order to attain the company's objectives.

What Drucker proposed was not new. American engineers like Russell Robb and James D. Mooney (see Chapter 3) in their writings on organization and Field Marshal Viscount Slim (1956) in his writings on the Burma Campaign, leadership and morale, had expressed similar ideas, but Drucker crystallized these ideas in his chapter on MBO.

John Humble (1972) defined MBO as a 'a system that integrates the company's goals of profit and growth with the manager's needs to contribute and develop himself personally'. In an earlier paper, Humble (1969) set out MBO as a continuous process of:

1 Reviewing critically and restating the company's strategic and tactical plans.
2 Clarifying with each manager the key results and performance standards.
3 Agreeing with each manager a job improvement plan which makes a measurable and realistic contribution to the unit and company's plan for better performance.
4 Providing conditions in which it is possible to achieve the key results and improvement plans notably by having:
   (a) An organization structure which gives a manager maximum freedom and flexibility in operation.
   (b) Management control information in a form and at a frequency which makes for more effective self-control and better and quicker decisions.
5 Using systematic performance review to measure and discuss progress towards results and potential review to identify men with potential for advancement.
6 Developing management training plans to help managers overcome their weaknesses, to build on their strengths and accept responsibility for self-development.
7 Strengthening a manager's motivation by effective selection, salary and succession plans.

Figure 10.6 is a systems diagram showing the cycle of MBO as viewed by

American authority of MBO, George S. Odiorne (1965). Canadian Professor Bill Reddin (1970) is another leading authority on the MBO technique.

The advantages claimed for MBO include provision of a logical and systematic technique for setting objectives by which performance can be measured, involving managerial participation at all levels in setting objectives and so engendering a greater sense of commitment, mutual setting of performance standards by managers and their superiors and subordinates, encouraging greater initiative at all levels of management, and exercising better self control and self discipline by all managers and supervisors. Humble claimed that it is an important technique for management development.

In the present author's experience, MBO schemes can fail for a number of reasons. These include ineffective leadership or the failure to make firm decisions by top management, inadequate supervision, insufficient allocation of resources to make the scheme work and low motivation or morale of the group or organizational unit.

Another practical problem with MBO is that much is left either to senior management (to prescribe the procedures) or to the subordinates (in attaining the agreed objectives). Where both parties have insufficient experience of MBO, or are otherwise uncertain or ignorant of the principles, then it may be a case of the blind trying to lead the blind. Again, one questions the Machiavellian philosophy of 'the ends justifying the means', especially when dealing with people. It all comes back to the question of competent or incompetent leadership at the top.

Some companies have spent large sums of money to pay management consultants to implement MBO, with little in the way of results (for the reasons already given). Many practising managers consider MBO as another gimmicky technique, propagated by consultants and others to promote their own business (along with such things as sensitivity analysis, managerial grid, and transactional analysis). Only time and posterity may prove whether they were right or wrong.

Further comments on MBO

This section is contributed by the handbook editor, who has had some experience of MBO in companies in the UK. Two cases will be quoted. The first was total disaster, leading to demoralization and considerable wasted expenditure. The second was reasonably successful, allowing us to end this part of the chapter on a more optimistic note.

A large company decided to implement MBO and undertook a detailed programme of training, led by academics from a well-known higher educational establishment. Groups of senior staff spent many hours in training sessions, and all professional staff had to compile their own job descriptions.

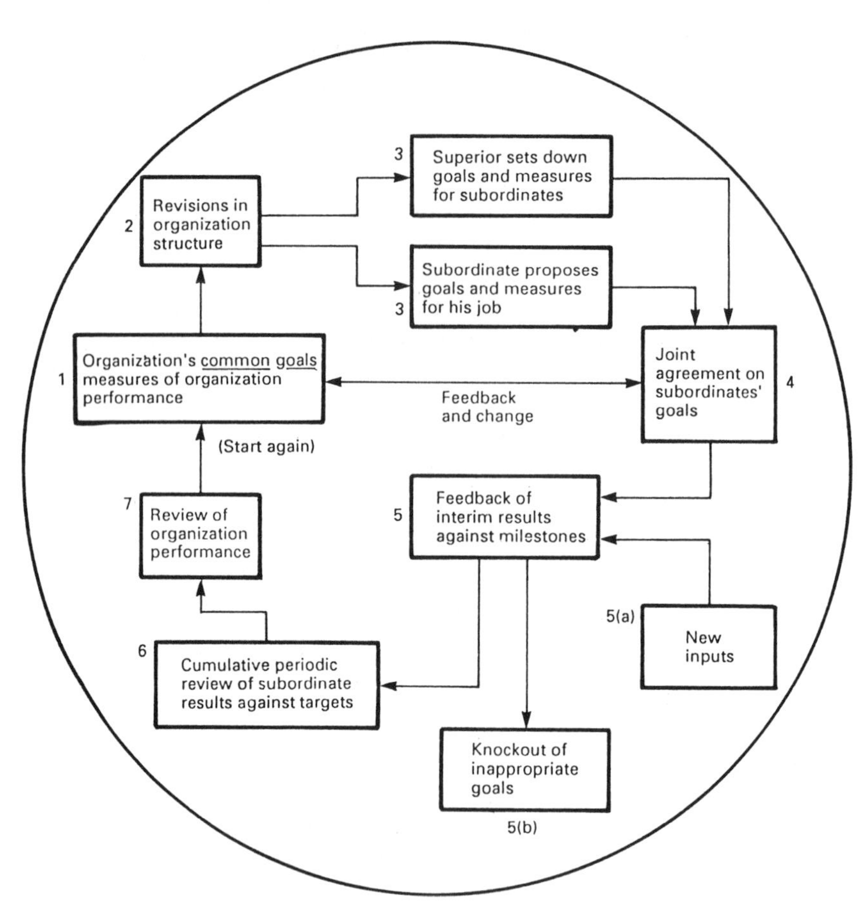

Figure 10.6 *A systems diagram for management by objectives. This shows the MBO cycle*
(Courtesy of G. S. Odiorne)

Later, the company's objectives were analysed and broken down into individual objectives, so that a pyramid of targets was established from top management right down through the organization. This was carried right through to the lowest level of professional engineers, involving perhaps some 200 people. After about six months of this effort, another group of consultants recommended a top to bottom company reorganization. The previous functional arrangement of the engineering groups was changed to a mixture of teams and functional groups and every single job title was changed. Management implemented this change without reference to the parallel MBO programme. Now consider the chaos and scepticism generated throughout the workforce. After many months of training and preparation, all the job descriptions and objectives were rendered obsolete overnight, because they related to the obsolete organization. Rigid application of MBO rules, which must be tied to the organization structure and its objectives, is bound to fail, simply because organizations are dynamic and subject to change. Formal MBO is too cumbersome.

Here is a brighter story. Honeywell decided to introduce a simple version of MBO some years ago. This was top-led, with a directive reaching Honeywell Controls Limited in the UK from the USA headquarters. The programme consisted of a simple briefing, given in clearly written notes, followed by individual target setting of about four or six measurable objectives agreed between people with their immediate superiors. The objectives had to be quantifiable, in order that any improvement could be measured. Every three months there was an informal review, and every six months new targets were set. This was a simple, effective aid to performance appraisal and development. It succeeded because it was kept simple, did not demand expensive involvement from consultants, and was sufficiently flexible to adapt to small organizational changes.

Integration of supervision, leadership and motivation

The topics of supervision, leadership and motivation are closely related and all depend on understanding the human factor and human relations in engineering management. Effective supervision requires competent leadership on the part of the engineering manager, who should understand human motivation and what makes people behave in certain ways. But unlike the analysis of structures, machines and products, human beings are not always rational, objective or logical. Rather they can be most irrational, most illogical and subjective depending on factors such as personality, attitudes, feelings, experiences, education and training, sex, age, state of health, beliefs and philosophy of life. These make the task of management and supervision a much more difficult and challenging job than solving technical engineering problems. The engineering manager should continually cultivate, develop and improve his leadership skills by study, reading, absorbing the best theories and the most practical principles, and by gaining experience in various leadership roles.

A factor often overlooked by managers, and seldom discussed by theorists, is morale, or *esprit de corps* (Fayol, 1949). This factor greatly influences the motivation of individuals and groups. Military commanders place great emphasis on morale. Field Marshal Viscount Montgomery set out four important rules which he found important for success (see Moorehead, 1946):

1 Morale. Study the individual soldier. Create an atmosphere of success. Morale means everything.

2 Simplify the problem. Sort out the essentials which must form the basis for all future action; and once you have decided upon them ensure that those essentials stand firm and are not swept away in a mass of detail. As a commander, lay down the general framework of what you want done – then within that framework allow great latitude to your subordinates. Explain the plan to them carefully and fully, then stand back and avoid being encumbered with inessentials.
3 You must learn how to pick a good team of subordinates and, once you have got them, stick to them and trust them. All men are different and all generals are different; so are brigades and divisions. But if you study human nature, you will be able to fit them into the right places.
4 Make yourself know what you want, and have the courage and determination to get it. You must have the will to win; it is much more important to fight well when things are going badly than when things are going well.

Translate the military terms into their civilian counterparts (worker for soldier, manager for general, will to manage instead of will to win, etc.) and these rules hold good in engineering management. The last statement, particularly, is a truism: the test of good leadership, supervision and motivation occurs when things go wrong. It is then that the engineering manager needs most to inspire his workers, getting them to produce their best effort in the most difficult circumstances.

References and further reading

Adair, J., *Training for Leadership*, Macdonald, 1968.

Amos, J. M. and Sarchet, B. R., *Management for Engineers*, Prentice-Hall, 1981.

Badawy, M. K., *Developing Management Skills in Engineers and Scientists*, Van Nostrand Reinhold, 1982.

Bass, B. M., *Leadership, Psychology and Organisational Behavior*, Harper and Bros., 1960.

Bayton, J. A. and Chapman, R. L., *Transformation of Scientists and Engineers into Managers*, NASA Document SP – 291. Washington, D.C.; NASA, 1972.

Bone, G. W., 'Should Engineers be Managers?' *Management for Engineers*, (ed. Semler, E. G.), Inst. of Mech. Engrs.

Bower, M., *The Will to Manage*, McGraw-Hill, 1966.

Brown, J. A. C., *The Social Psychology of Industry*, Penguin, 1954.

Christensen, C., Johnson, T. W., and Stinson, J. E., *Supervising*, Addison-Wesley, 1982.

Cleland, D. I. and Kocaoglu, D. F., *Engineering Management*, McGraw-Hill, 1981.

Cribbin, J. J., *Effective Managerial Leadership*, American Management Association, 1972.
Drucker, P. F., *The Practice of Management*, Heinemann, 1955.
Du Brin, A. J., *The Practice of Supervision*, Business Publications, 1980.
Eckles, R. W., Carmichael, R. L. and Sarchet, B. R., *Supervisory Management*, 2nd ed., John Wiley, 1981.
Fayol, H. *General and Industrial Management*, Translated from the French by Constance Storrs, Pitman, 1949.
Fiedler, F. E., *A Theory of Leadership Effectiveness*, McGraw-Hill, 1967.
Fraser, J. M., *Understanding Other People*, Pitman, 1953.
French, J. R. P. and Raven, B., 'The Bases of Social Power,' in *Studies in Social Power*, (ed. Cartwright, D.), University of Michigan, 1959.
Gellerman, S. W., *Management by Motivation*, American Management Association, 1968.
Gibb, C. A. (ed.), *Leadership*, Penguin, 1963.
Giegold, W. C., 'Leadership – The Essential of Engineering Management,' *Engineering Management International*, vol. 1, no. 1, July, 1981, pp. 49–56.
Giegold, W. C., *Practical Management Skills for Engineers and Scientists*, Wadsworth, 1982.
Herzberg, F., Mausner, B. and Synderman, B., *The Motivation to Work*, John Wiley, 1959.
Hofstede, G., 'Motivation, Leadership and Organisation: Do American Theories Apply Abroad?' *Organisational Dynamics*, Summer, 1980, pp. 42–63.
Humble, J. W., 'Management by Objectives', *The Director*, November, 1969, pp. 275–80.
Humble, J. W., *Management by Objectives*, British Institute of Management, 1972.
Hunt, J., *Managing People at Work*, McGraw-Hill, 1979.
Koontz, H. and O'Donnell, C., *Essentials of Management*, 2nd ed., McGraw-Hill, 1978.
Maccoby, M., *The Gamesman*, Simon and Schuster, 1976.
Machiavelli, N., *The Prince*, Translation from the Italian by Luigi Ricc, Mentor Books, 1952.
Maloney, W. F., 'Supervisory Problems in International Construction', *Journal of the Construction Division, Proc. American* Soc. of Civil Engrs., vol. 108, no. 603, September, 1962, pp. 406–18.
Maloney, W. F., 'Understanding Motivation', *Journal of Management in Engineering*, ASCE, vol. 2, no. 4, October, 1986, pp. 231–45.
Maslow, A. H., *Motivation and Personality*, 2nd ed., Harper and Row, 1970.
Maude, B., *Leadership in Management*, Business Books, 1978.
Mintzberg, H., *The Nature of Managerial Work*, Harper and Row, 1973.
Moorehead, A., *Montgomery*, Hamish Hamilton, 1946.

Odiorne, G. S., *Management by Objectives*, Pitman, 1965.
Pfiffner, J. M. and Fels, M., *The Supervision of Personnel*, 3rd ed., N.J.; Prentice-Hall, 1964.
Plunkett, W. P., *Supervision: The Direction of People at Work*, Wm. C. Brown, 1976.
Reddin, W. J., *Managerial Effectiveness*, McGraw-Hill, 1970.
Rubey, H., Logan, J. A. and Milner, W. W., *The Engineer and Professional Management*, Iowa State University Press, 1970.
Sartain, A. Q. and Baker, A. W., *The Supervisor and the Job*, McGraw-Hill, 1978.
Shannon, R. E., *Engineering Management*, John Wiley, 1980.
Slim, Field Marshal the Viscount, *Defeat into Victory*, Cassell, 1956.
Slim, Field Marshal the Viscount, 'Leadership', *The Manager*, January, 1962.
Tannenbaum, R. and Schmidt, W. H., 'How to Choose a Leadership Pattern', *Harvard Business Review*, May–June, 1973, pp. 162–164ff.
Tead, O. *The Art of Leadership*, McGraw-Hill, 1935.
Thurley, K. C. and Hamblin, A. C., *The Supervisor and his Job*, HMSO, 1963.
Trahair, R. C. S., *The Humanist Temper: The Life and Work of Elton Mayo*, Transaction Books.
Urwick, L., *Management, Current Affairs Bulletin*, University of Sydney, vol. 30, no. 8, August 27, 1962.
Vroom, V., *Work and Motivation*, John Wiley, 1964.
Westbrook, J. D., 'An Integrated Theory of Motivation', *Engineering Management International*, vol. 1, no. 3, December, 1982, pp. 193–200.
White, N. A., 'Engineering management – the managerial tasks of engineers', *Proc. Inst. of Mech. Engrs. – Management and Engineering Manufacture*, Part B, vol. 197, 1983, pp. 143–55.
Williamson, M. A., 'Stepping into Management', *Works Management*, vol. 14, no. 3, 1961, pp. 18–21 and p. 38.
Wortman, L. A., *Effective Management for Engineers and Scientists*, New York; John Wiley.

PART THREE

Facilities and Equipment

11

Creating the office

Peter Lebus

A professional engineer may be involved in new offices either as a user, as a professional consultant, or as the project manager for a client company. Essentially the problems and processes in each case are identical.

There are advantages in treating an in-house project for creating a new engineering office as if it were being managed for a third party. This will dictate a more rigorous and objective approach. The work involved is different from many other engineering projects. Office requirements relate mainly to people, rather than to things or processes. In addition, the intending occupier (the client) is usually a fellow engineering professional, with quite clear ideas as to what he wants to achieve. He is also able to judge how well his objectives are being met.

In an office this is often the first time that the client has handled this type of work. Owing to politics and to imprecise forecasting, the organization and staffing frequently change during the course of office projects. Unfortunately all too many people believe, incorrectly, that they are experts themselves. Guidance from an office client tends to be indefinite and imprecise. Therefore the project professional will need to take a leading role. He should be given responsibility for decisions and advice, which puts a high premium on past experience and knowledge in the field. The purpose of this chapter is to define a strategy for this type of project and to provide some of the knowledge that is essential.

The challenge for management

The main challenge for management arises out of the opportunities that can be taken – or missed. An office move is often a traumatic experience, but it does provide management with a rare opportunity to make dramatic changes to the way that the company is run or organized.

Unfortunately change is often uncomfortable. This can put pressure on maintaining the status quo. It is therefore of crucial importance for management to seize this chance and to examine the way that the company is run. They must also set long-term corporate objectives which will form the basis for the new accommodation.

There are a number of potential changes that should be examined at the beginning of the project.

Technology

A relocation can be a good time for the introduction of new technology and equipment into the company. If possible this should be carried out prior to, or during, the move. Many firms take this opportunity to introduce new data processing, CAD or telephone systems.

Management philosophy and methods

The shape and character of a building has a major impact on the way in which companies are managed and organized. The traditional long and thin building, which may also be rambling in geography, dictates a highly cellular layout. This results in long communication lines and a more formal management approach. This in turn tends to create extra layers of management.

Conversely the more modern office, which is deeper in plan, can allow a more open and egalitarian approach to management. Many firms take this opportunity to improve lines of communication and introduce a more informal management style.

It is important to recognize that the building should reflect the future management objectives, not vice versa.

Corporate identity

The relocation of offices can be an opportunity to promote a new or more visible corporate identity. This can reflect the new management style and be used positively to assist with the company's marketing efforts.

Improved efficiency

Another benefit of a relocation is the almost inevitable improvements that can be achieved in operational efficiency. The accommodation for individuals and groups can be designed to meet their specific needs. New services and facilities can be installed.

Although some overheads such as rent may increase, energy costs can be reduced through the use of modern building management technology. Other overhead costs can also be reduced by the provision of automated or simplified back-up services.

However care must be taken when introducing change. Where there is an opportunity for improvement, there is usually an equal and opposite opportunity for problems to arise. If the building is not properly planned and fitted out, it is quite possible to increase the overall costs, to reduce efficiency and to lower environmental standards. Just because a technology or concept is new, it does not follow that it is necessarily better or more appropriate for your particular company needs.

For instance, there has been an element of fashion in the open office environment. This has run through the German Burolandschaft approach of the 1960s, followed by the American systems furniture approach of the 1970s, and the new large floor area designs of the 1980s. These new large floors are aimed primarily at the international finance houses, which require immense dealing floors or clerical paper factories. However they are not ideal for most other types of organization.

The challenge, that management must recognize and meet, is to sort the worthwhile and efficient from the inappropriate, however new and shiny that may be.

Project organization

A major problem with project work is that it is temporary in nature. It is unusual to have the right type of organization in place. Furthermore, because the office project is so different from the conventional engineering project, the organization and skills may not be readily available. Costs and disruption are imposed on top of all the normal corporate activities, which must continue to function efficiently. The project is equivalent to a new division within the company, with an annual turnover of several hundred thousand pounds, or even several million pounds.

The project organization is likely to be necessary for a period of six to eighteen months for most typical jobs. In view of this temporary nature, it is tempting to give responsibility for this to 'old Joe' (who is not quite up to running a department any more) or to 'Fred' (who is about to retire). This can be disastrous, as the rest of the company must then live for many years with the results of his decisions.

The control of the project must be set up at the inception of the job with members who have sufficient authority and skills. A typical organization tree is shown in Figure 11.1.

Project committee

It is crucial that the project committee has sufficient authority to take decisions and to ensure that they stick. The chairman should be a senior director or partner who is involved with the overall corporate strategy. Despite their seniority, committee members must be willing to listen, often to junior staff, as well as to talk.

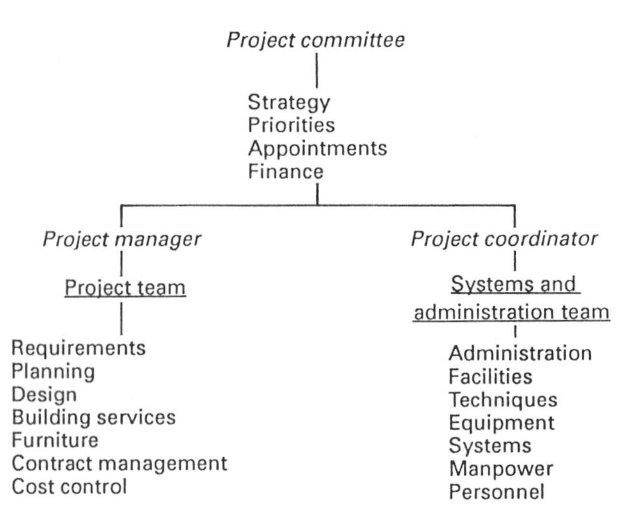

Figure 11.1 *Organization tree. This shows a typical project organization with the main responsibilities*

Ideally the membership is three plus the project manager and coordinator. The committee should be responsible for the project strategy. They should have authority to agree building layouts, space standards, provision of facilities, amenities, environmental standards, programmes, budgets and the appointment of specialists and contractors.

Meetings should be held at intervals of two or three weeks. Minutes, with action lists, must be issued promptly and the actions progressed. Sensible delegation can, with the right team, keep the involvement needed from committee members down to about one half day in every two weeks. Obviously there will be occasions when more intensive work is necessary in order to decide strategy or resolve policy questions. Visits to other buildings and meetings with advisers, suppliers or contractors will probably be required.

Project manager

The role of a project manager is familiar to most engineers. He is in charge of the project or technical team. In practice, the project manager is responsible for ensuring that the building and its contents are operational and efficient, on time and within budget. He will have within his team all the technical expertise that is necessary, such as design, building services, structures, costing, programming and contract supervision. He will also be responsible for employing contractors and for coordinating specialists, such as telecommunications, security, power supply, catering, furniture, etc.

The project manager will also need a requirements and space planning team. The members should have a commercial rather than a design background. This is because designers are usually trained to develop *solutions*. This team needs to identify and then investigate management *problems*. It is crucial that this role is carried out rigorously, as it will form the basis for all the design and layout that follows.

The planners must be familiar with office routines and responsibilities. They will investigate staffing and organization, growth, the need for facilities and services, and communication links. They must be competent to investigate the total company operation from the corporate identity to the filing systems.

The project manager and his team are specialists who need particular types of experience. Consequently, external consultants are often used in this role. Although outsiders have a learning period when they start, this should be compensated by the ability to provide impartial and up to date advice and techniques.

Project coordinator

The position of project coordinator should be filled by the in-house administration or office manager. He will work very closely with the requirements team, and provide data and contacts for them. He should be the secretary to the project committee, with direct access to its chairman. His role is to ensure that all the detailed operational needs in the new premises are properly thought out and planned. Therefore he will coordinate all the operational studies and decisions that may be necessary. This could include any or all of the following:

- Telecommunications systems.
- Data processing equipment.
- Word processing systems.
- Filing, storage and archive equipment.
- Catering systems and suppliers.
- Security systems.
- Personnel relocation policies.
- Future facilities management policy and systems.

The coordinator should liaise with departmental management and with equipment suppliers, and recommend the installation of new methods for the offices. If he can be appointed at the outset to be the facilities and administration manager for the new premises, this will ensure that he looks at all suggestions with an eye to their efficiency in the future. He should develop the systems and equipment that will be needed to back up the day-to-day running of the company.

Decision factors

There are four main decision factors in a project:

1 Objectives.
2 Quality.
3 Cost benefit.
4 Timing.

Each is important, but none can be considered in isolation from the others. A change in objectives or quality, for example, would alter the cost or timing – and vice versa.

Objectives

Objectives have to be defined at two levels. First there is the corporate strategy, which will include general as well as specific points. Some may appear obvious, but they should be spelt out nevertheless. All too often crucial factors can be forgotten in the pressure of events and special pleading. Typical objectives would include:

1 Improvement in space utilization.
2 Better interdepartmental communication.
3 Introduction of computer terminals within operational departments.
4 Reduction in layers of management.
5 Better management/staff relations.
6 Stronger corporate identity.
7 Better working environment for staff.
8 Better flexibility for change.
9 Better services and facilities.
10 Reduction in office running costs.

Second, a detailed user brief must be developed by the requirements team, working closely with the project coordinator.

The brief must be a comprehensive document (or series of documents). In a short project, a single document may well suffice. However, if the project stretches over many months it may be necessary to produce additional data as the company develops and changes.

On this basis it is possible to help the design team to achieve a flexible building. In the early stages, it is sufficient (and preferable) to define the range of options that should be catered for. For example:

1 The number of private offices on each floor will be between eight and twelve.
2 There will be twenty to fifty computer terminals.

3 The accounts department is twenty staff but may have thirty-five within five years.
4 The computer suite should allow for 100 per cent expansion.

Excessive detail at this stage can lead to a very inflexible building, which will not permit easy change in the future. This data will be used for block planning, outline design, and the scope of works.

During the course of the works more specific detail will be required, based on the most up-to-date forecasts. The definition of objectives and requirements is a continuing process with ever increasing detail. This ceases shortly before occupation, when each individual workplace is 'fine tuned'.

Quality

The quality objective is often described as 'good quality without extravagance'. This is too vague, however, when detailed design is being developed.

The future facilities manager for the new premises should have a major impact in this area. There will be special fixtures and finishes in some public areas such as reception, dining rooms, meeting rooms, etc. The facilities manager is more interested in performance than appearance. He will want to assess the advantages of every part of the new premises. For instance, he should check:

1 Types of wall coverings.
2 Relative merits of floor services outlet boxes.
3 Wearability of floor coverings.
4 Cleanliness of window blinds.
5 Replacement of ceiling tiles.
6 Maintenance of equipment.

He should lobby for the best possible quality-in-use that the overall budget will allow. He will also need to balance his priorities within the limited resources that are available.

Cost benefit

Cost is clearly a major factor, but it must not be considered in isolation. It is not merely a passive cost accounting process. A new office is certainly a cost, but it is also an investment. Therefore the project committee must always bear in mind what it is that the costs are providing. The objective should be to maximize the cost effectiveness of the whole building as well as its individual parts.

Cost benefit studies should be made during the requirement and design development phases. Sometimes it may be hard to quantify the potential benefits in terms of cash. Nevertheless these questions should be asked, and estimates made, to form a basis for decisions. Typical cost benefit studies could include the following:

1. The location of columns and window mullions. This, plus the depth of building, will have a major impact on the efficiency with which space can be used in the building. The additional costs of larger modules may be overridden by the better use of space leading to a smaller building.
2. The interrelationship of lighting, air conditioning and ceiling modules. This will influence the location of partitions and therefore the size of offices and the flexibility for change.
3. The location of cable outlets. This will dictate what has to be provided above the floor and within the office areas.
4. The design of windows. This will affect the method of cleaning and therefore the need to install external window cleaning equipment.
5. The design of entrances. This will influence the security systems and therefore the number of people who must be employed for this purpose.
6. Storage systems. New techniques and equipment can reduce the space requirements. A good working environment can improve staff efficiency and reduce absenteeism.

In some instances, the project committee, for overriding strategic reasons, may decide to proceed with decisions which are not the most cost effective. In these cases, the cost benefit study will help them to assess the costs of their decisions.

It is the responsibility of the project manager and the project coordinator to make sure that the committee are made fully aware of all the relevant facts and estimates before such a decision is taken.

The preparation of budgets is a process of ever increasing detail. When the building has been selected, the requirements identified, and the outline scope of work developed, an initial budget estimate must be prepared. This may be as broad as the cost per square foot but ideally it should include a breakdown into the main cost elements. This will help the project committee decide whether they must reconsider their strategy, the requirements or the quality.

It is tempting, but dangerous, to proceed in the hope of finding savings as you proceed. Nearly always this results in money running out towards the end of the project, so that late but high priority items may be lost.

It is reasonable at this early stage to allow a contingency of up to 10 per cent for budgeting purposes. This should be used to cover unforeseeable design or construction problems. It is not an allowance for further upgrading of the requirement.

Once this budget has been agreed, it should be used as an upper limit for design purposes. The committee and the project team must not allow any overrun against this. Any additional requirement that is identified must be costed separately and agreed as being essential. A further budget item can be listed for these user variations so that the overall cost position can be controlled.

Timing

The overall duration of an office relocation project is frequently underestimated. It is unrealistic, but all too common, to assume that because one wants something to happen quickly, it can be made to do so despite evidence to the contrary.

It is important to ensure that adequate time is allowed for each stage of the project. The search for new premises and the negotiation of the lease details are matters which are largely controlled by outside factors. They are unlikely to be shortened dramatically. Once a building has been chosen, it is rarely less than three months before the legal documents have been agreed and signed. The time taken to find the right building will depend on availability in the right location. In central London this may take many months or even years. Elsewhere it may be a matter of a few weeks.

The development of the company's requirement is all too often rushed. The result of this can be a succession of changes during the design and fitting out period, which lead to abortive costs and considerable aggravation. In other cases, it leads to the inclusion of features or facilities which are inefficient or inappropriate, but which cannot be changed economically. It is usually possible to develop an adequate design brief in one to three months (depending on the size and complexity of the company).

Many firms wait until they have signed a lease before committing themselves to establishing what the requirements are. This leads to unnecessary time pressure on the requirement and design phases. The requirement study should be started as soon as a decision has been taken to relocate the company. This gives adequate time during the search and negotiation periods to ensure that the requirement is properly thought through and detailed. Individual managers can be given adequate time to sort out what their long-term needs really should be. Figure 11.2 shows a typical project timetable for a relocation to a building of about 50 000 square feet.

Selecting a building

The selection of new offices must be based on an agreed set of requirements and priorities. Therefore a brief must be developed which is based on the corporate strategy. This brief will have two main purposes. It will be used by the agent to search for possible buildings and it will be used by the project team to assess the suitability of each building.

The brief should include data on a number of factors. In some instances this

Weeks	0 4 8 12 16 20 24 28 32 36 40 44 48 52 56 60 64 68 72 76 80
Initial study and building brief	0–4
Building selection	4–16
Negotiate and lease	16–30
Phase 1	
Requirements	16–24
Sketch scheme and block layout	24–30
Initial budget	27–31
Phase 2	
Detailed design and specification	30–38
Tender	38–42
Appointment of contractor	42–44
Phase 3	
Contract works	44–70
Snagging	69–71
Furniture and equipment	70–73
Move	74

Figure 11.2 *Typical project timetable for the selection and fitting out of a 50 000 square foot building*

will be a single requirement, in others it may be a range of acceptable options. The data should include the following:

1 Total space.
2 Floor sizes.
3 Types of layout (cellular, shared, open).
4 Shapes of space (deep, narrow, divisibility).
5 Services (lighting, power, cableways, heating and ventilating, plumbing).
6 Environmental standards (temperature, humidity, light levels, glare, acoustics).
7 Facilities (storage, catering, lifts, entrances, etc.).
8 Location.
9 Availability.
10 Appearance (image, quality, style).
11 Flexibility (subletting, expansion, change).
12 Outlook (windows, view, noise, dirt).
13 Costs (fitting out, running, energy, rent, rates, services).

All of these factors will not be of equal weight. Therefore one must differentiate between those factors that are *essential* to the success of the project, those which are very *desirable* but not essential, and those which would be *useful* to have but are not of prime importance. It is also necessary to define any factors which would be *unacceptable*, such as specific locations and limits to rent.

Once the agent has produced a short list of possible buildings the assessment process can proceed.

Several factors are easy to check in sufficient detail on the basis of the documentation. This will include such items as the overall size, number of floors, location, some costs, the environmental design standards, floor shapes, etc. Other factors will need a visit, and even the examination of architectural and engineering detailed design drawings. The suitability of the shortlisted buildings should be assessed on the following basis.

Layout

A few trial layouts will soon tell you a great deal about a building and its suitability. Areas of waste space can be identified as well as rooms that would be too large or too small. Access routes may be defined by column positions. Storage may be located in the wrong parts of the building. The shape of open areas will dictate how efficiently the space can be used and what the acoustic environment will be.

Space

The size of floors should be compared to the needs of departments and facilities. The trial layouts can be used to assess how well space in each building meets company needs. In particular the room sizes and amount of corridor space will vary from building to building.

Relationships

The size of individual floors, and the main access routes will dictate whether correct interrelationships between departments and facilities can be achieved. Essential communication links must be preserved in a natural and easily manageable fashion.

Services

The provision of cabling facilities, and the ability to expand them, have become major factors in the efficiency with which new offices can be used. Flexibility will certainly be required to reflect the constant change and updating of office technology. Many new buildings are now designed with false floor systems. Nevertheless, the degree of accessibility may still be quite low. It is important to ensure that additions can be made to cabling in the future without excessive costs. The type of floor outlet box should be carefully examined, because they have hard wear, and can create safety hazards due to sharp raised lids.

Flexibility

The modules for partitions, services, lights, and heating, ventilating and air conditioning should be checked for compatibility. Partitions will have to be moved in the future, so that the shapes and sizes of space can be changed to meet operational needs. The extent of future growth should not be underestimated as this can lead to prohibitive costs and premature relocation. The ability to sub-let some parts of the building will create flexibility for the future within economic constraints.

Acoustics

Noise levels on occupation are difficult to forecast. Particular problems occur at the ceiling and at the perimeter of buildings. These areas should be examined with care to see whether adequate levels of privacy can be achieved where this is essential. The modern tendency towards the use of linear diffusers for air conditioning and for lighting creates problems of sound leakage between adjacent rooms.

Lighting

This should be checked for both natural and artificial illumination, including the lighting level, glare, individual control and reflection.

Heating and ventilation

In many buildings heating and ventilation systems are designed with building operation and economy as the main, and sometimes the only, consideration. Frequently not enough attention, or original investment, is given to creating a good and flexible environment for staff. Proposed standards must be checked against the company requirements. The plant and duct capacity for further upgrading to meet equipment needs must be assessed.

Maintenance

The ease with which a building can be cleaned, repaired and maintained must be checked. The access to equipment and facilities is important and may have been scrimped in order to increase the net lettable area. No one likes to pay rent for the space which really needs to be re-allocated to plant and services. The type of finishes will have an impact on the maintenance bill and will need to be checked for wear and cleaning.

Cost

The relocation costs and the future running costs should be calculated at this stage. This should include rent, rates, service charges, fitting out costs, running costs, maintenance and the cost of future changes. The project team should be able to develop these figures in collaboration with the estate agent and quantity surveyors.

Other factors

When a building is being shortlisted, there are several other factors which need checking to ensure overall acceptability. This will depend on the specific needs of the company and could include the following:

- Security.
- Floor loading.
- Catering facilities.
- Access for equipment.
- Goods and staff entrances.
- Parking.
- Toilets.
- Appearance.
- Outlook.

It can be seen that this analysis requires a great deal of detailed checking. This is justified because the company is being committed to a major investment, and a significant continuing cost. The decision will be a major influence on whether the company can achieve a high level of efficiency over the life of the lease. With many buildings it becomes apparent at an early stage that they are unsuitable, or at best only second rate. Clearly there is no need to proceed with detailed analysis in these cases.

At the end of this selection phase the project committee should have a clear choice before it, based on a report, setting out the relative advantages, limitations and costs of each building.

Fitting out process

Once a suitable building has been chosen, the time pressures start to build up. A specific programme can be prepared, which may include the disposal of existing premises. This in turn creates deadlines which become very expensive to miss.

As can be seen from Figure 11.2, the fitting out process subdivides into three main phases:

1. Sketch scheme, block layout and budget.
2. Detailed design, specification of works and tender process.
3. Contract works on site.

Some overlapping of activities may occur, but it is important to ensure that each phase is completed to time.

Phase 1 Sketch scheme, block layout and budget

During the selection process, much preliminary work would have been done in identifying policies, priorities and requirements. This must now be updated and more detail established. If the early work has been done properly, this can be a quick process. Contacts will have been made, and management should already be planning for the future. However, now that there is a real building, it does concentrate the mind wonderfully, and a number of new requirements may well emerge.

Space standards can now be specific, bearing in mind the actual building module and design. The amount of waste and access space should be calculated, so that the net usable space is known.

The first step should be to prepare a block layout of the building, based on the need for space, types of layout and the interrelationship of departments. This should define the location of sections, main facilities and equipment. Partition layouts should be developed, bearing in mind the building limitations.

At this stage the project committee must take a number of important decisions. It is almost inevitable that the requirements of several departments and the overall corporate strategy will not totally coincide. There will be disagreements over priorities, standards and locations. In this case it is essential that the committee takes clear decisions which can be acted on by the project team. It is equally important that the committee use their authority to inform departmental managers and special interest groups about policy decisions that have been taken. It is not advisable to pass difficult decisions through the project team in the hope of avoiding unpopularity. The team members do not have the authority to make them stick.

Liaison within the project organization is very important at this stage. All parts of the team must work closely together so that strategy and detail can evolve in a logical sequence. Surprises to the project committee can frequently end in abortive work.

Once the block layouts have been agreed at corporate and departmental levels, a sketch scheme and an outline scope of work can be developed. This should include partitioning systems; alterations to heating, ventilation, lighting, power and other building services; building works; special works for computers, telecommunications, catering and reception; security systems; joinery; standards of finish; and the provision of signs.

Budget costs must be developed for each element of work. This can be carried out by a professional quantity surveyor or by a building estimator, depending on the size and complexity of the works. This costing work leads to a major policy decision by the project committee. It is important that they take the necessary and difficult actions to finalize the budget. If the cost is too high, they must decide what must be changed to allow a reduction. They must ensure that additional items are not added at a later stage, except when this is genuinely unavoidable. In that case, the budget should be altered correspondingly.

At this stage it is reasonable to reduce the contingency factor in the budget to about 5 per cent to cover unforeseeable expenses in the works. It is the project manager's responsibility to make sure that this is not spent without proper justification.

During this initial design development phase, contact should be made with the statutory authorities so that the preliminary plans can be discussed. The main points for consideration would include:

- Planning regulations.
- Means of escape in case of fire.
- Building regulations.
- Health and safety at work.
- Public health.
- Offices, Shops and Railway Premises Act.

These discussions and permissions may take a considerable period of time, so it is important to start them early.

Phase 2 Detailed design, specification and tender process

Once the outline scheme and budget have been agreed, the planners and designers can prepare the detailed designs and documentation for the contract.

The working drawings must be developed for all works, services, finishes and fittings. The written specification of works can be prepared in parallel. This will describe the works and refer specifically to the individual drawings. Standards of workmanship should be defined.

The services specification is likely to be prepared separately by the building services engineers. Very close liaison is needed with the other members of the design team so that the services can be incorporated into the overall design and achieve a coordinated result.

When these documents have been finalized they should be agreed by the project committee before they are sent out to tender.

It is normal, in most projects, to use a competitive tender process to select the contractors and suppliers. If possible, at least five contractors should be sent copies of the documentation for quotations. In some circumstances, however, a negotiated tender may be used if time is very short or if the work is so specific that only one contractor can carry out the work.

The returned tender documents need careful analysis and checking. It is common to find that items are excluded or that extra costs are referred to in accompanying letters. It is important to insist on receiving a detailed cost breakdown against each item in the specification of work. This should highlight any arithmetic errors or unexpectedly expensive design elements.

The contractors should be interviewed to assess how efficient their administration, supervision, programming and cost control are. References must be checked and examples of previous work visited. It is only on this basis that value for money can be judged. The cheapest contractor may not turn out to be the best value in the end.

It is desirable for the selected contractor to meet the project committee before the final appointment is made. This can help to achieve a good working relationship.

Phase 3 Contract works on site

A senior member of the project team must be appointed as supervising officer or contract manager. He must be technically experienced and competent to supervise the works on site.

When the contractors have been appointed, the supervising officer will work very closely with them. A detailed programme of work should be prepared by the contractor and agreed at the outset. This will form the basis for the continual monitoring of progress on site.

Before the works start it must be announced that the supervising officer is the only person with the authority to issue instructions to the contractor. Without this it becomes impossible to control costs and timings.

Regular site meetings should be held and minutes issued, so that appropriate action can be allocated to individuals and then followed up.

At the end of the contract, time must be allowed in the overall programme for the snagging inspections and for remedial works. This will be followed by a final builder's clean, and the installation of furniture and equipment.

The increasing use of systems furniture with complex wire management has lengthened the installation process. Adequate time must be allowed for this, and for full testing of systems and equipment.

The building is now ready for occupation.

It is a fond, but unlikely, hope that the offices will be 100 per cent right on the first day (if for no other reason, than that the company may have changed during the course of the project, too late to modify the designs).

In order to control change and to sort the trivial from the important, it is advisable to refuse any alterations during the first three months of occupation – unless they are completely unavoidable. Frequently requests for change are withdrawn once staff become accustomed to the new premises, which should be an improvement if the project has been well run.

Conclusion

This chapter has described the process of fitting out an office. However, even in the best organized world, mistakes are made. It is worth noting some of the most common errors in the hope of avoiding them in the future.

1. Allowing non essential changes to requirements during the course of the works.
2. Preparing timid forecasts of change which are overtaken by events within months.
3. Giving insufficient authority to the project committee, the project manager and coordinator.
4. Setting unrealistic targets for time or cost that clearly cannot be achieved.
5. Working with inadequate expertise or experience in a field that is strewn with minefields.
6. Allowing false expectations to remain when inevitably they will be disappointed.

Despite these potential pitfalls, the setting up of a new office presents an opportunity which must be grasped by management. Despite the frequent strain, the change should be used as a catalyst for progress.

It must be remembered that these new offices will be a primary factor in the ability of the company to achieve its objectives for efficiency and cost control over the next ten or more years. One cannot hope for a better challenge than that.

Further reading

The Boisot Waters Cohen Partnership, *Business Property Handbook*, Gower, 1981.

CIBSE Guide (3 vols.) CIBSE, 1986. This covers air conditioning, heating and ventilation standards.

CIBSE Lighting Guide, CIBSE, 1984.

The CIBSE publications are available from CIBSE, Delta House, 222 Balham High Road, London, SW12 9BS. Telephone 01–675 5211.

Robichaud, B., *Selecting, Planning and Managing Office Space*, McGraw-Hill, 1958.

12

Drawing office equipment

Dennis Whitmore

The design department, as a very important link in the chain leading to product or project fulfilment, must be made as efficient as possible in order that the overall work is not held up through lack of drawings and specifications. Consequently, much effort has been put into making design work less costly and less time consuming. There are several ways in which this can be achieved. For example:

1 Aids to drafting (such as drawing boards of improved design and drafting machines) can be given to draughtsmen to facilitate the drawing of plans.
2 Computer aided design (CAD) radically reduces time spent on drawings.
3 The ergonomics of drawing office design and layout will have a marked effect on the efficiency of the draughtsmen, both in terms of the time taken to carry out work and of their psychological well-being.

The principles of ergonomics and office layout have been dealt with elsewhere in this handbook (Chapters 11 and 36). In the drawing office the important things include:

1 The furniture and equipment provided.
2 The levels and form of lighting.
3 The layout of furniture and equipment.
4 The ambient temperature and ventilation.
5 The decor and colours used on walls, floor and ceiling.
6 Freedom from unacceptably high levels of distracting noise.

Many of the above factors affect the individual psychologically and, as a result, are overlooked by management when they provide accommodation and facilities. A comfortable room with decent, if pricey, furnishings can be viewed as a luxury that cannot be afforded. But the increases in productivity which these conditions could bring make them economical in the long run. Unfortunately, psychological improvements are often difficult to attribute because they are not obvious.

Drawing units

Professional draughtsmen are choosy individuals and like to be at one with their drawing units; a true man-machine unit. They would prefer the machine to be built around *them*, but this is no bad thing and is in the tradition of ergonomics. The trade has responded by producing a very large range of standard and custom-built units to meet this need, even to the point of supplying the parts separately to enable one to choose components for the final assembly into a customized unit.

Basically, a drawing unit consists of:

1 The drawing board.
2 A horizontal straightedge which can be moved, without tilting, in a vertical direction.
3 Some means for supporting the board, whether standing on the floor, or attached to a desk or a wall.

These basic components are subject to many variations, some of which will now be described.

Drawing boards are often quoted in relation to the DIN A series of metric drawing sheet sizes (see Figure 13.1). Allowing for a suitable all round margin, boards quoted as A0 size are typically about 100 cm × 150 cm. A1 boards are approximately 73 cm × 92 cm. It is possible to obtain boards as large as 120 cm × 230 cm, and these are useful for roll drawings, or where it is desired to work with two A0 sheets side by side (for reference purposes). At the other end of the size range, drawing boards come in a range of smaller sizes, even down to portable A4 units.

Straight edges are available in various materials, including aluminium and acrylic, with selected scales. The horizontal straightedges can be moved in a vertical direction, using an arrangement which is known as 'parallel motion' to ensure that no unwanted twisting occurs when the straightedge is moved up and down. This objective is achieved using guidewires, running over pulleys and counterweighted to prevent the straightedge from sliding down the board under its own weight. A more accurate arrangement uses steel reinforced, polyurethane toothed belts. Both methods allow the straightedge to be lifted clear of the drawing during movement, to avoid smudging.

The draughtsman who cannot find a stand to suit his drawing board must be very hard to please. There is something to suit everyone. For example, boards can be attached to floor-standing frames, benches, tabletops, the wall, and even the kitchen sink if necessary!

The simplest models, suitable for students, are fixed frames, but the professionals need fully adjustable stands. Boards can be tilted, and height-adjusted. They are very heavy, so when buying stands the purchaser must look for (a) some counterbalancing of the weight, and (b) quick locking of the board once it is in position. Angle adjustment can be achieved by releasing the locking with a hand-operated control, but stands are made which can be released and locked by a foot-operated mechanism. Counterbalancing can be by counterweight or springs. Height adjustment can be counterbalanced by gas-filled struts.

Looking at fine detail and lines on a white sheet of paper all day can be very wearing on the draughtsman's eyes, so the board must be very well illuminated. Obviously good overhead, shadowless lighting, and individual lamps can be used, but an alternative method is illumination from within the board. Translucent acrylic boards are illuminated from the back by individually adjustable lamps. More sophisticated, but expensive models have integral lighting, known as light boxes. The board-boxes contain cold-light tubes, and often have adjustable intensities. A variation is to have a light-box on one side, with a normal drawing surface on the other, providing two boards in one.

Individual lighting units can be clipped to the drawing board and can be in the form of individual lamps or strip lighting. The individual lamps are either tungsten bulbs of about 75 W, or the more efficient short tube fluorescent lamps (typically 15 W). The main advantage of these individual lamps is that they are easily adjustable to any position over the drawing board.

Large fluorescent strip lighting can be fixed to the top of the board to illuminate the whole area of the surface with shadowless light. These lamps are adjustable and counterbalanced, in some cases using a gas strut and locked by a friction brake. An example is shown in Figure 12.1. Typical lights accommodate board widths of 130 cm to 220 cm.

Drafting machines

Drafting machines are a more advanced alternative to the parallel motion straightedge. After parallel motion they are sheer luxury. Units with parallel motion are essential for drawing parallel lines, but angled or vertical lines must be constructed by placing set-squares with different angles on the straightedge. In the drafting machine the parallel motion straightedge is replaced by a drafting head which is capable of revolving through up to 360° and carries two straightedges set at right-angles to each other. The straightedges can easily be interchanged, allowing a range of scales to be used. A protractor is incorporated in the head (sometimes with a vernier scale for greater precision) which allows the straight edges to be rotated very accurately to the angle required. Vertical and horizontal rails are fitted to the drawing board to allow the respective linear movements of the head.

In the more expensive models, an electronic digital protractor scale replaces the manual vernier head. The advantage of a digital readout is that no optical interpretation of the scale is necessary, and hence it is not subject to

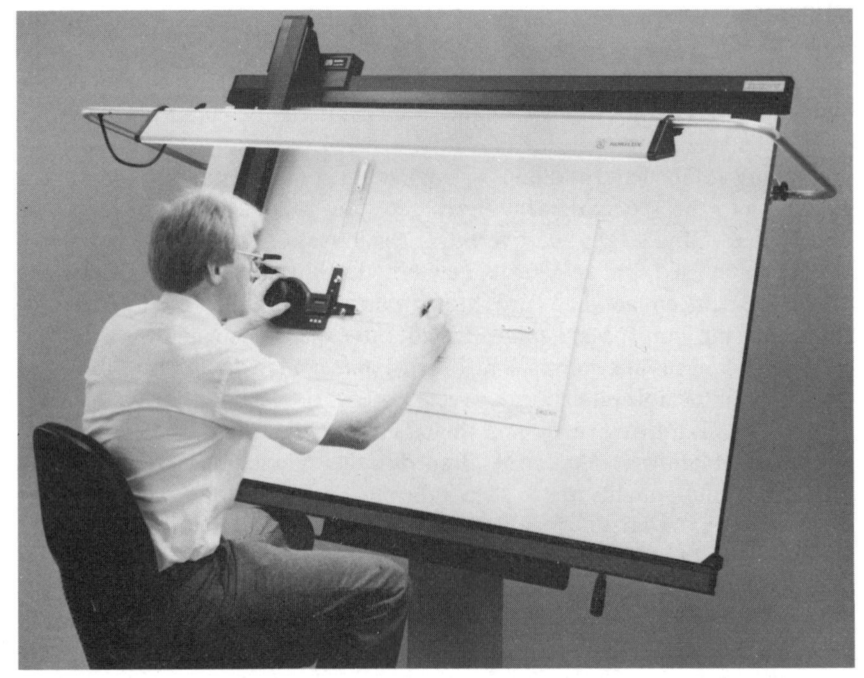

Figure 12.1 *A fluoresccent lighting unit attached to a board with a drafting machine* (Korikux lamp, courtesy of Kings Legend Ltd, Leiston, Suffolk)

misreadings. Typically the accuracy obtainable is five minutes of angle. These heads are fully computerized, incorporating microelectronics with light-emitting diode (LED) or similar displays. An example of a digital head in use can be seen in Figure 12.1, and Figure 12.2 shows a head in more detail.

The precision to which angular scales can be read is called the resolution. Many drafting heads are capable of a resolution of as little as five minutes (one twelfth of one degree of angle) in both vernier and digital models.

Ease of operation can be facilitated by the use of magnetically suspended floating counterweights, or magnetic levitation. A top-of-the-range drafting machine, such as the RXG from Kings Legend Ltd, incorporates the following features:

1 Magnetically suspended floating counterweight.
2 Floating vertical track – no foot roller.
3 Double hinge with floating head balance equalizer.
4 Automatic 15° indexing of the head.
5 Easily removable vertical rail for easy cleaning etc.

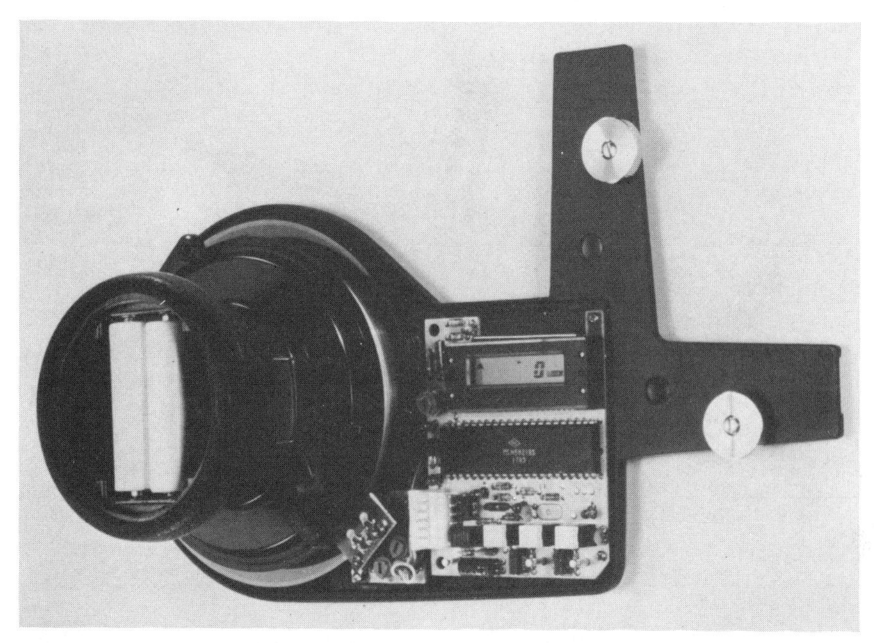

Figure 12.2 *A digital protractor head. The covers have been removed to show the electronics*
(Courtesy of Kings Legend Ltd)

6 Electronic digital angle readout, with a resolution of five minutes.

An innovation

Something which can take a great deal of a draughtsman's time is dimensioning, lettering, text writing, and drawing of symbols. Recently a piece of electronic equipment, known as a Scriber, was introduced to take over this time-consuming work.

The Scriber has a keyboard (either QWERTY or ABC) and a liquid crystal display on which the text or symbols can be viewed before they are transferred to the drawing. For example, dimensions can be typed into the display and, if correct, fed on to the drawing on the board. The unit drives a pen which does the scribing. It can be attached to the drafting head or parallel motion. One such unit, the Stern 100, is illustrated in Figure 12.3.

Latest machines incorporate module cassettes which hold commonly used symbols in such fields as electronics, electrical wiring, hydraulics, pneumatics, architecture, mechanical and others.

Figure 12.3 *An electronic texting machine*
(Courtesy of Stern Limited)

Other equipment and furniture

Desks and chairs

A drawing office will obviously need some desks and chairs but, since there is a plethora of dealers offering new and second-hand items from a wide range of designs and manufacturers, not much need be said here.

The materials used range from steel (not in widespread use) to solid wood (very expensive). By far the most common material is wood veneered chipboard, often combined with plastic mouldings or even hardboard. When buying any furniture constructed from chipboard it is as well to check the method of construction. Some designs rely on screws driven into the chipboard and, while these may give serviceable use with care, the furniture will not stand up to such events as office removals or even layout rearrangements unless special care is taken. A fine-looking, nearly new A0 sized chipboard plan chest has been known to collapse during a removal – it was simply placed very carefully (fully loaded with drawings) on the ground by the removals men. The sides and back fell outwards, and the drawers settled down. A complete write-off. Similar tales can be told about badly designed chipboard desks. Wood veneered chipboard furniture built sensibly around a steel frame is to be preferred.

Modern office chairs can be very comfortable, with many adjustments possible. Some of the cheaper versions are prone to early failure through weld failures, stripped threads, frame fractures or poor castors. The best advice is to choose a reputable brand. The better manufacturers will give a guarantee valid for several years. For draughtsmen, high stools or chairs are preferred, giving them the option to sit or stand and be at the same height as the drawing. Such stools or chairs must, therefore, have suitable height adjustment. Seat height can be altered by means of a column with a screw thread, with a lockable sliding column, or with a lockable sliding column aided by compressed gas (known as a gas lift system). Because of the possible seat height, foot rests are usually built into the frame.

Reference tables

All drawing offices need horizontal plane surfaces on which drawings can be spread out for reference. Reference tables are the obvious answer, and most suppliers can offer a range of tables, with or without drawers, in a range of sizes. If space is short, the tops of plan chests are often used (although these are not ideal). In some arrangements each draughtsman is provided with a reference table behind him, and this may or may not have a drawer for holding drawings up to A0 size. Some managers dislike giving draughtsmen any filing facilities of their own, arguing that this can lead to the unauthorized retention of original drawings which should more properly be held securely in central files.

A useful method for displaying reference drawings is the inclined reference unit (such as the Intercline units available from Ozalid). These units are positioned alongside, and at right angles to the drawing board. They have a reference surface that is angled at 45 or 60°, with a ledge along the lower edge to support drawings. Shelves can be provided below for storage. Figure 12.4 shows how an inclined reference unit can be arranged alongside a row of drawing boards: double-sided units are available for use between two rows of drawing boards. Inclined units save floor space, but the shallower 45° version should normally be chosen in preference to the steeper option because, although it saves less floor space, there is less tendency for drawings to slide off in annoying fashion.

Filing plans

It is one thing to draw plans, but quite another to store them. Many are DIN A0 or DIN A1 sizes, which are 841 mm × 1189 mm, and 594 mm × 841 mm respectively. Almost always they must be stored without folding or rolling.

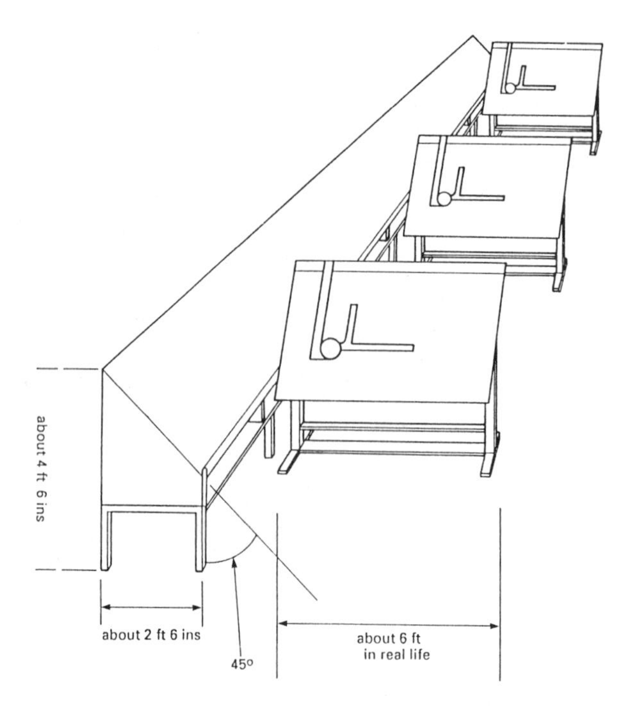

Figure 12.4 *An inclined reference table arrangement*

Thus they must be laid flat in drawers, or hung vertically in cupboards. Please refer to Figure 13.1 for a table of metric drawing sheet sizes.

Shallow drawer cabinets (or plan chests) are made to DIN A sizes, holding several drawers, often with dividers so that A0 can be adapted to store two A1 or four A2 drawings and so on. While filing drawings is quite easy in drawers like these, they have their disadvantages. To find a plan one must search through the stock by lifting the top drawings one at a time to search for the lower ones. This continual ruffling through tends to crease and dog-ear the plans until eventually they lose their corners.

Vertical cabinets (or plan files) have fixing methods at the top. These can be prongs which pass through holes at the edge of the plans. Prepunched adhesive filing strips are stuck to one side of the drawing for the purpose. Drawings are filed or retrieved either by an opening front, hinged at the bottom and/or top of the unit. Examples are the Plandale or Kings Legend models (see Figure 12.5). In another arrangement, the cabinet has a front axial opening, with sliding runners, so that the front half of the cabinet is separated from the back (Graphiteque AC234). In a typical model the cabinet opens partially to enable the operator to select the required compartment. Next the selected compartment is slightly opened but the prongs still overlap. This allows the drawings not wanted to be separated from the others leaving the selected one isolated in the middle of the prongs where they overlap. Retrieval is accomplished by fully opening the prongs and extracting the drawing.

Computer graphics

The science of computer graphics is becoming a very important tool for the designer. With the coming of the relatively cheap computer, available to all, it was inevitable that its power would be harnessed to assist the draughtsman and designer in their craft. The twin techniques of computer aided design/computer aided manufacture (CAD/CAM) were the result of the marriage of designer and computer. Some prefer the term with wider implication which embraces both manufacturing and non-manufacturing sides, computer aided engineering (CAE).

As with any capital project the decision to invest in CAD must rest on both the cost effectiveness of the venture, and other factors such as speed of completion of drawings, and accuracy of the finished result. Advanced equipment can cost hundreds of thousands of pounds, software alone being the most expensive item, running into five or even six figures. This does not mean that CAD is beyond the reach of the small business, because at the lower end of the market quite effective software is available for less than £1000; it all depends on what is needed.

The effective management will research the subject very thoroughly, commissioning a feasibility study to determine the cost effectiveness of its proposed investment. The feasibility study will show whether there is a case for management to choose CAD, or a case for going (literally) back to the drawing board.

An installation project will follow the usual problem solving procedure, which is discussed in Chapter 36. Very briefly, the following steps are involved once it has been decided to proceed. First, all available facts and information concerning the situation must be collected. These can include:

1 The current and forecast future workload in design and drawing.
2 The similarity of work (for example, are standard symbols commonly used which could be stored, with benefit, in the computer memory? Perhaps in electronics, for instance).
3 The complexity of drawings.
4 The need for three dimensional and solid modelling.
5 The speed with which designs must be completed.
6 The amount of finance available for investment in the computerization.
7 The current costs of producing drawings by existing methods, including labour, materials, capital expenditure, depreciation, maintenance charges and all other relevant costs.

Figure 12.5 *A vertical plan file. Filing drawings vertically from suspension rungs aids filing and retrieval, and helps to eliminate damage to the drawing sheets*
(Courtesy of Kings Legend Ltd, Leiston, Suffolk)

The next step is to examine all the collected facts and compile a specification for a system which will satisfy the needs of the organization. The specification then has to be costed, and the proposed total cost of producing designs and drawings must then be compared with the current costs. From these data, a financial project appraisal can be conducted, using payback or other suitable technique (see Chapter 18).

The third step is the presenting of a report to management on the findings of the feasibility study. If this report is accepted, the project moves to step four, which is the development of installation plans for the proposed computer system. A list of furniture, computer hardware and software is prepared, and steps are taken to procure these items. Floor plans and layout diagrams have to be constructed for the new accommodation, including any special services, cabling and environmental control. If the computer is to be left switched on round the clock, some form of automatic fire detection and extinguishing device should be considered, preferably using Halon II (Dupont).

When the accommodation and services have been made available, the final step is to install the equipment and to iron out the inevitable bugs. Periodic reviews will be necessary in case the equipment needs extending, updating or replacing.

Provision must be made in the plans and costing for staff training. The system supplier will probably offer training facilities, either in house or at some other location. Management must realize that it takes several weeks for staff to become fully effective in the use of new computing systems and there will actually be a loss of productivity during the learning period.

Computer systems for CAD/CAE

The basis for computer aided design is the central processing unit of the computer, around which the system is built. The computer's memory holds the CAD software and the input data which forms the drawing. Attached to this are the peripherals, which can include keyboard, digitizing tablet (or bit pad), screen, printer and plotter. Depending on the size of the operation, the computer can be mainframe, mini, or micro. The actual physical size does not affect the principles of operation.

Besides this hardware, one must not forget the essential software without which the computer is just an expensive set of machines capable of nothing. Altogether these items form the CAD system.

Not all computers will be capable of graphics, unless they incorporate certain essential ingredients. When declaring one's requirements, it is important to compare these with the specifications of the computers available. Perhaps the most important thing is to select the software which will do exactly what is needed, and then the hardware which will use the software most effectively. This is why most designers who enter the field of CAD prefer to buy a complete, comprehensive, system.

Specifications should include the following at least:

1 A computer with a memory large enough to hold the CAD software package.
2 A graphics capability (graphics card), with colour if required.
3 A visual display unit (VDU) capable of displaying graphics.

4 Facilities capable of accepting a mouse or digitizing tablet.
5 A backing store (on a floppy or hard disk) large enough to accommodate finished design data.

Some CAD packages require a very large amount of memory space in the computer, to which must be added the space needed for storing input data during design work. Memory is measured in kilobytes (just over 1000 bytes or characters) and megabytes (about one million bytes). Modern microcomputers have memory sizes of at least 500 kilobytes, which can be sufficient for most average programs, but not for the most advanced packages. For example, the RoboSOLID system demands a computer with a memory of at least 640 kilobytes.

Ideally the backing store should be a hard disk drive of about 20 megabytes or, failing this, twin floppy disk drives using double or quadruple density disks. The computer itself should be equipped with a fast processor, preferably with 16- or 32-bit operation. A high resolution graphics card is essential for graphics to be executed. If no card is built in, then the computer should be capable of accepting a chip or card as an add-on accessory.

Surprisingly the IBM PC XT/AT machines are without graphics capability (which can be added later), while much cheaper clones like Amstrad PCs have colour graphics as standard.

Visual display units, or screens, can be equipped with normal raster screens, while other systems can use special screens for the job. These are described in a later section.

The data for actually specifying to the computer the contents of the actual design can be typed in as instructions and coordinates via the keyboard. One problem here is the difficulty of visualization of what one is doing. Most users prefer actually to draw the images on the screen using a mouse, or a digitizing tablet. The mouse is a hand-held device connected to the computer by wires or infra-red signals. Movement about a tabletop or any flat surface causes the cursor on the screen to follow the movements and reproduce them as lines on the screen. Whereas the mouse can be used in conjunction with any surface, another method, the digitizing cursor, needs a special board on which to operate known as a digitizing tablet or bit-pad.

The digitizing tablet can be used with either a cursor box, or a stylus-pen. One of each type is illustrated in Figure 12.6. The tablet itself is a flat surface which accommodates some sort of sensors connected to the computer. In some cases these take the form of a matrix of 1024 wires in each direction at right angles, embedded in the tablet. These sense the contact with a stylus when this is pressed onto the surface.

Another form of tablet relies on sonic signals emitted from the stylus or cursor being located by two linear microphones set into two edges of the board. A light-detector digitizer has rows of tiny light-emitting diodes set

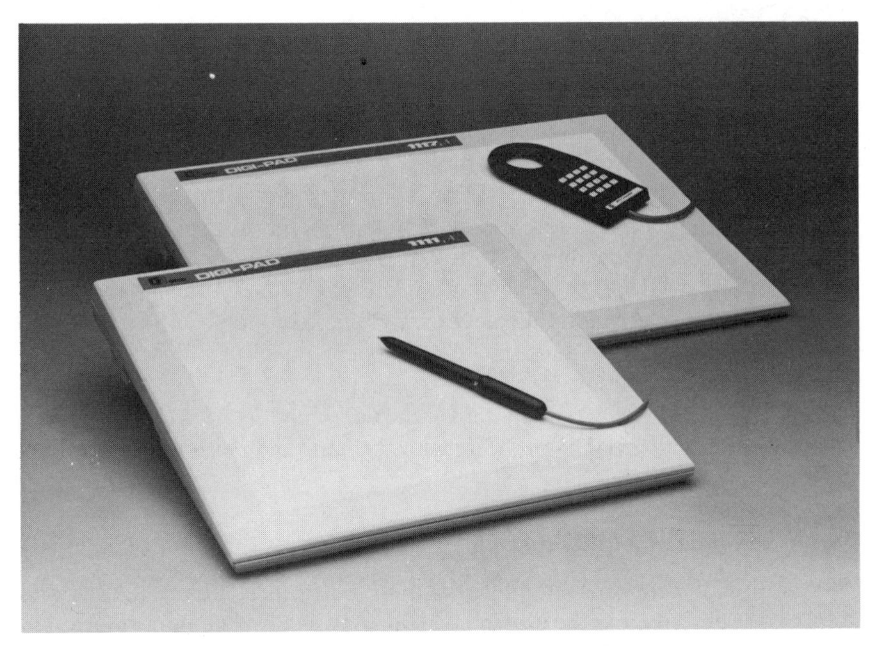

Figure 12.6 *Digitizing tablets with cursor and with pen*
(Courtesy of GTCO Corporation, Columbia, Maryland, USA)

along two adjacent sides of the tablet, and light detecting cells arranged along the opposite sides. These infra-red beams are interrupted when an object such as a pencil is placed in the area of the beams, enabling the computer to fix the position of the pencil. Sets of electromagnetic coils are used in yet other forms.

The cursor is placed on the bit-pad and is operated in a similar way to the mouse. However in this case the computer knows the relative position of the cursor through the sensors in the base of the pad. Not only can drawings be done by hand, moving the cursor around the bit-pad, but drawing can be speeded up by calling upon stored symbols and functions. For example, a straight line is very difficult to draw freehand, so all that is necessary is to mark the start and end of the line using the cursor, and then call upon the computer to join the marks together with a line. Similarly, many figures can be drawn simply by specifying two points. The position of a circle (for example) can be fixed by its centre, and its size can be determined simply by inputting a point on its circumference: the computer will then draw the whole circle. Other regular shapes can be produced in the same way.

A useful addition to this automatic generation of shapes is to have a set of standard symbols stored in the memory, ready for placing on the drawing. These might be electronic symbols (transistors, resistors, capacitors, transformers and so on) or any other set of symbols relevant to a particular engineering discipline. In architecture, windows, doors, brick patterns, planks, trees and other standard features can be stored. Thus sixty identical windows of a building, which with conventional pencil and paper could take hours of laborious drawing, can be created in minutes by picking up the window symbol and positioning it in all sixty places where it is to be drawn. Each time the position is specified an accurately drawn window magically appears. If the image of the building is to be covered in bricks, this can be done in a couple of seconds by selecting this infill and asking the computer to do the filling. All of the drawing, picking up of symbols, specifying coordinates and selecting infill colours and designs is done by positioning the cursor and pressing its buttons. Alternatively an electronic pen can be substituted for the cursor.

The screen is a very important item of the peripheral set, and has been given a separate section in this chapter. Similarly, printers and plotters are described later.

Facilities offered by CAD software

The ability of computer graphics systems to store standard figures and symbols, as described in the previous section, is obviously an important aid to drawing productivity. The addition of standard lettering, in various fonts and variable sizes, is another important aspect, leading to excellent presentation of finished drawings.

Two- and three-dimensional transformations can be made. The more advanced graphics programs allow the image to be rotated or spun so that hidden sides can be turned into view. The image can be viewed from any angle using such software. Panning can be used to see parts of the image which are off the screen, when this is too large to fit the screen.

A useful facility is zooming, so that parts of drawings can be seen in more detail, or, by zooming out, the complete design can be viewed.

Three-dimensional modelling can be wire-frame imagery where solid objects are depicted as if they were transparent so that lines on the 'hidden' side of the object are visible. These lines can be removed to make the object look more solid. A more realistic solid modelling is available which incorporates shading and texture to the surfaces. Other refinements are hidden surface removal, contouring, and piercing of objects, and the sectioning of objects. Certain software will also produce exploded diagrams very quickly.

Purchasing software for CAD/CAE

A wide range of software is available for producing graphics and for CAD/CAE/CAM. Prices vary enormously but, as with most things, one gets what one pays for.

Before buying software, a specification must be carefully prepared. This must take into account both present requirements and foreseeable future needs. This task is easier if the purchaser does not already have the hardware (i.e., the computer) because trying to match new software to an existing computer can restrict the choice. However, much software will be compatible with several models of computer, such as the IBM PC and all its clones.

A typical simple package will produce standard shapes such as straight lines, circles, arcs, boxes, ovals, parallels, tangents, perpendiculars, as well as text and freehand sketching. Boxes, circles and other closed areas can also be filled with colour. Rotate, zoom and mirror-image are also useful facilities.

Micro systems are often purchased complete. Typical start-up prices are quoted later in this chapter. The system in Figure 12.7 is based on a microcomputer (IBM PC) and consists of a high resolution screen, 1k × 1k addressable colour controller and a mouse.

Figure 12.7 *A 1k × 1k addressable colour controller, prism card, 15 inch flat faced monitor and mouse connected to a microcomputer*
(Courtesy of SSI Limited, Pewsey, Wiltshire)

Most usual professional software offers a wide range of facilities and components can be simulated in the solid from the very start of the design process. The image can be solid or wire frame (Figures 12.8 and 12.9) and can be viewed from any angle. The program also calculates physical properties including volume, surface area, mass, centre of gravity, and moments of inertia. This allows accurate estimates of weight, area of surface to be painted, and costings. Selection of modelling tools is by pull-down menus and moveable icons.

Other functions of a typical program are:

- Rotation of solid images.
- Zooming to magnify the image, and shrinking of the view.
- Change of vantage point for viewing the model.
- Extrusion of a profile to make solid.
- Combining two solids to make one.
- Cutting away of part of the solid.
- Finding the solid produced by the intersection of two solids.

Solids can be combined, subtracted from each other and intersected very easily. Systems will work on inexpensive IBM compatible machines but 640 kilobytes of memory are typically needed. This size of memory is no great problem, since many microcomputers (such as the Amstrad PC1640) have sufficient capacity.

Visual displays (monitors)

The most common visual display screen is the face of a cathode ray tube (CRT) which is bombarded by a stream of electrons which cause the phosphor coating on the screen to glow at the point of contact. The screen can be compared with a sheet of graph paper, each square being replaced by a tiny point of glow. Many points of glow make up the final image. Each point of glow is known as a *pixel*.

When choosing a graphics system an important factor is the *resolution*, or degree of definition or precision required in the image. Resolution can be described as low, medium, or high depending on the fineness of the image lines. The resolution depends on the graphics board incorporated in the computer's central processing unit and also on the number of pixels in the screen. A typical high resolution screen measures about 1024 pixels by 780 pixels rising to around 4096 by 3136 pixels for very high resolution (found in storage tubes).

Screens can, of course, be monochrome or colour (the latter having a cluster of three phosphor dots (one each red, blue and green) in place of the single phosphor dot used for monochrome CRTs. Some large graphics systems (such as the Intergraph) use two screens, one black and white for general viewing of the whole drawing, and one colour for zooming and actual working.

One parameter of VDUs is the speed at which they will accept and execute commands, measured in millions of instructions per second (MIPS). Another

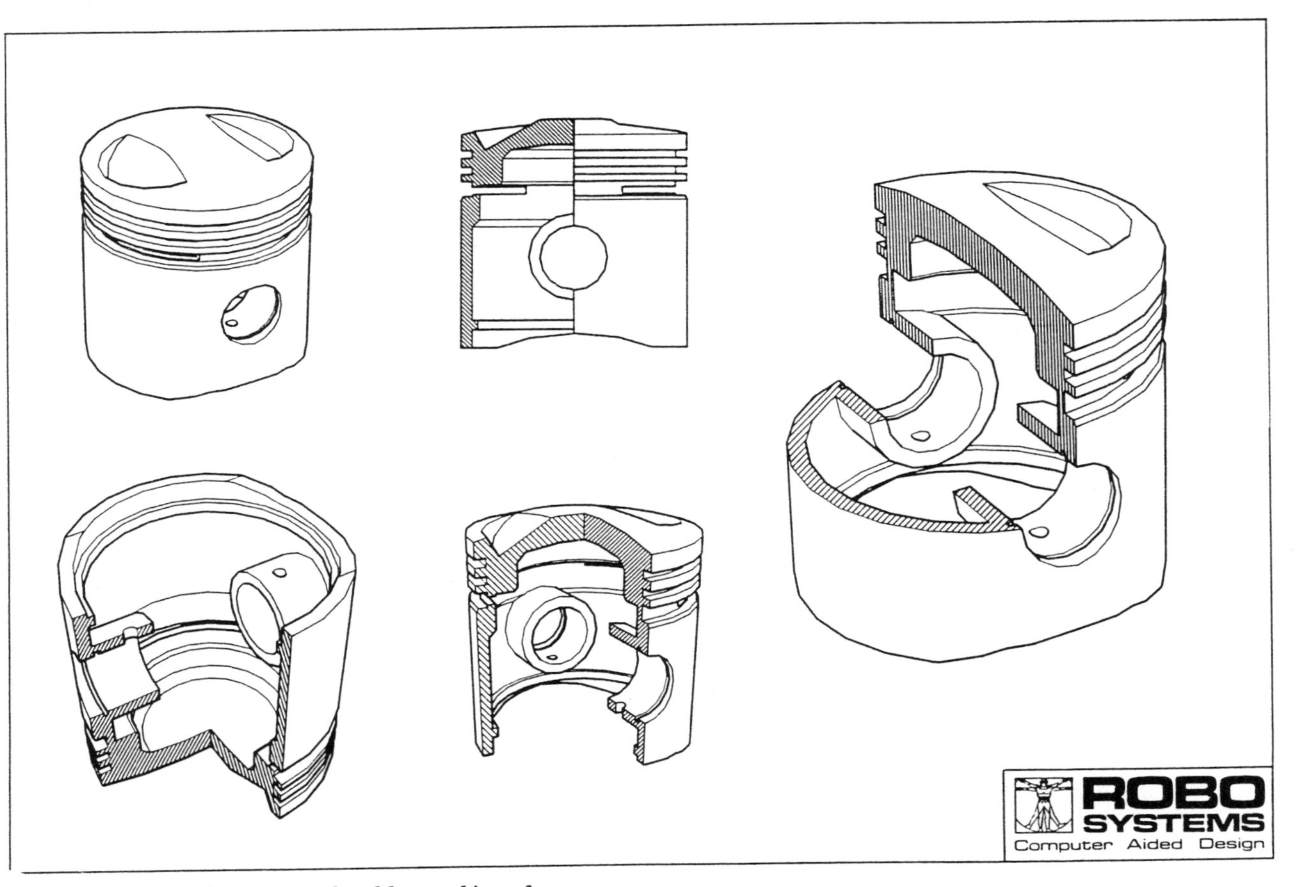

Figure 12.8 *A solid image produced by graphics software*
(Courtesy of Robocom Limited, London)

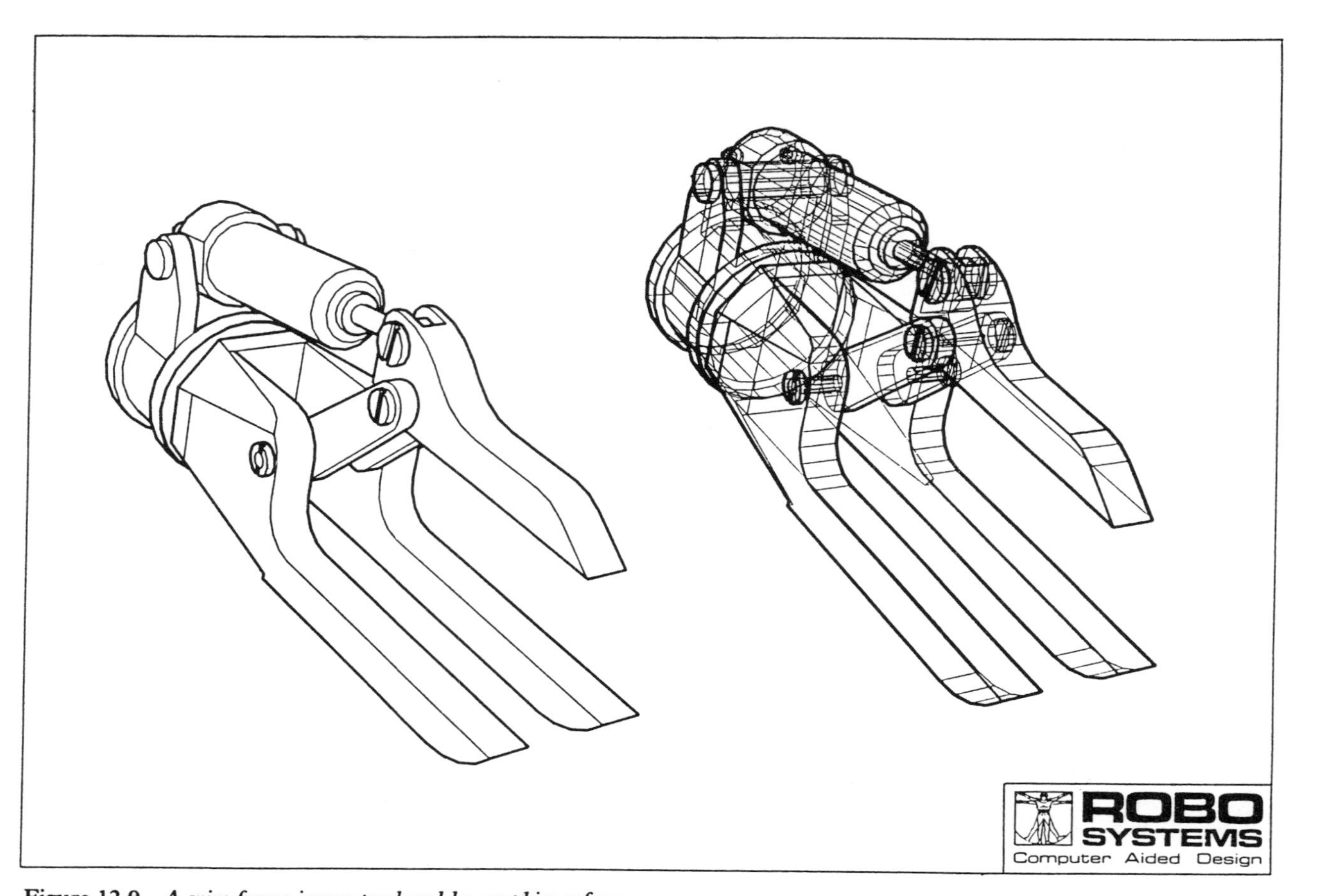

Figure 12.9 *A wire frame image produced by graphics software*
(Courtesy of Robocom Limited, London)

attribute is the speed at which long lines can be drawn (usually expressed in pixels per second).

There are three basic methods for producing images on the CRT screen, and these will now be outlined.

The raster method

The raster system is the familiar process used in television receivers. It is also widely used in computer monitors. It has to be remembered that, in a normal CRT, the phosphor glow at each pixel is very short-lived, lasting perhaps only 100 milliseconds before the glow disappears. In order to maintain a steady picture, all the pixels must be lit many times each second to reinforce or 'refresh' the glow so that it appears to be there all the time. This is achieved by causing the electron beam to move across the screen from left to right (line scan), gradually moving down a line at a time (frame scan) until the whole screen has been covered by a few hundred lines. This line and frame scan forms the complete screen or *raster*.

The picture or image is stored in the computer's memory on a 'bit-map'. The computer scans the bit-map at the same rate as the raster beam scans the CRT's face, switching the beam on and off to produce light where there is an image or dark where there is not, thereby reproducing the memory-mapped image on the screen. When colour is used several bits of memory are needed for each pixel to accommodate the primary CRT colours of red, blue, and green.

Vector refresh

Vector refresh is another method in which a display memory stores the image coordinates. The image is literally drawn on the screen by the electron beam. For example, if a circle is drawn the beam traces out the circle, as one would with compasses. Before the image has a chance to fade, the recircle is redrawn (refreshed), and this is repeated at a sufficiently rapid rate for as long as the display is required.

Storage tubes

The third method of image production on a screen uses the storage tube. This tube, developed by Tektronix, has a fine wire mesh on the inside surface of its viewing face. The electron beam induces a charge on the grid where it hits. The image is maintained by a 'flood-gun' secondary beam, whose electrons are attracted by the positive charge on the grid, thereby creating a steady picture. The image is flicker free. However the image must be completely cleared and redrawn if alterations are made to the picture.

Some commercially available examples

Ordinary computer monochrome monitors costing a few hundred pounds may be used for graphics work, but CAD/CAE work needs high resolution. Many monitors are available for different types of work and resolutions required. The Cambridge Computer Graphics' range of monitors are 20 in (508 mm) high resolution graphics displays, and include a graphics card for the computer. They work with IBM PC/AT and clones. The Cambridge Colour display allows sixteen or 256 colours to be selected from a palette of over a quarter of a million. Cambridge Xcellerator Colour Display also offers sixteen or 256 colours, but out of a staggering 16 million possible colours! The Cambridge Micro 1024 draws at 900 000 pixels per second, while the Xcellerator works at 1.2 million pixels per second.

Plotters and printers

After creating and drawing in the computer's digital memory, hard copies on paper or other sheet material can be produced in several ways. There are three principal methods:

1 Dot matrix printers.
2 Laser printers.
3 Pen plotters.

Dot matrix printers

Dot matrix printers, as their name describes, form images composed of dots arranged in a matrix layout. The print quality depends on the number of needles in the print head, the quality of the ribbon used (and its age) and on the matrix size. Typically the heads hold nine, twelve or eighteen needles in a vertical column, and these are driven electromagnetically to print matrices of dots in arrangements of 7×7, 7×9, and other sizes ranging up to 16×27 and larger.

Both colour and monochrome prints can be produced on dot matrix machines. Another way is to use *memory-mapping*, a system in which each dot has its own location in the computer memory. Images in colour require the printer to make four passes, each over a different part of a four-colour ribbon, to make each dot. This uses a large amount of memory, often around 3 kilobytes for each square centimetre, amounting to 2 or 3 megabytes for an average sized print.

Laser printers

Laser printers use the technology of the plain paper electrostatic office photocopier, in which an electrostatic image is produced on a coated drum. Toner powder adheres to the charged areas of the drum, and is transferred to the print. Heat is applied to fuse the toner on to the print surface. The important feature about laser printers is that the electrostatic image charge is produced by a laser beam, driven from digital information stored in the computer (unlike the more usual office copiers, in which the light is generated by a lamp, and focused by a normal optical system of lenses and mirrors). Extremely good reproduction can be obtained from laser printers.

Digital plotters

A much more aesthetic and professional result is obtained if the image is reproduced on a digital plotter. This literally draws the image using coloured pens in the same way as a human draughtsman. A plotter accepts data from the computer in digital form. The program drives the plotter in two dimensions. The correct colour pen is grasped by a robot arm which moves laterally across the paper, while vertical movement is attained by moving the paper, or the gantry which carries the pen. The combination of movements in vertical and lateral directions produce any two- or three-dimensional shape.

Because plotters actually draw the images with pens, speeds of output are measured in millimetres of drawn length per second. Plotters also write characters, so a second speed in characters per second is often quoted. A typical model such as Hitachi 672-XD is compatible with most personal computers, including IBM PC/XT/AT and clones. This accepts standard (HP-GL) plotting commands including digitization. It has a speed of 200 mm/s (280 mm/s diagonally), and 3 characters per second for lettering. Facilities for four pens are provided, the colours being selected by the operator.

Another characteristic which is important is the maximum size of the plotting produced. Maximum sizes are normally quoted. The number of pens which the machine can accommodate is important too, typical numbers being four or eight but some take many more than this, while others only hold one pen.

When plotting small, detailed drawings the fineness of the lines, or *resolution*, is of great importance. Resolution on some machines is of the order of 0.125 mm, but often it is as fine as 0.025 mm.

Because of the nature of the action of plotters, i.e., drawing as opposed to printing logically row by row, the pen must return often to parts of the plot already drawn, to add dimensions, captions and shading. This requires a high degree of accuracy to prevent overlap or superimposition. Accuracy is something which must be taken into account when specifying the kind of plotter required.

One very large plotter from SSI Ltd, the LP 4000 model, has a maximum plot size of 91 cm × 203 cm, but can go down to 'business-card' size (Figure 12.10). It has an axial speed of 500 mm/s (700 mm/s diagonally). Resolution is 0.0625 mm. The plotter has a capacity of up to 20 pens.

Another type of plotter, used mainly for printed circuit board (PCB) designing uses photographic technology. A typical model has a photo head with a halogen light source, which projects the plotting beam on to the photosensitive surface. This may be film, or a light sensitive glass plate. The

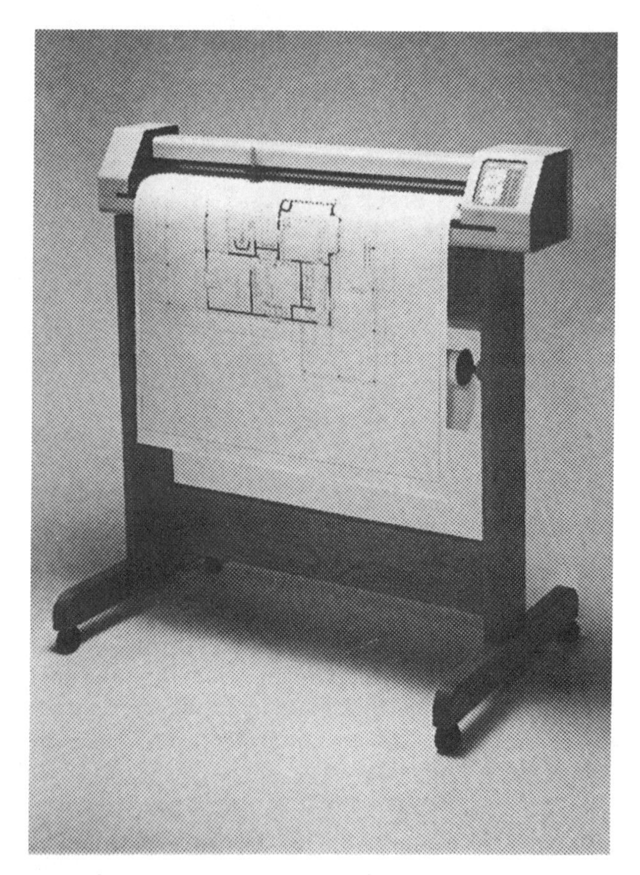

Figure 12.10 *A plotter suitable for drawings of up to 203 cm × 91 cm* (Courtesy of SSI Ltd, Pewsey, Wiltshire)

photo head is equipped with different apertures to produce various thickness of lines. The Glaser Precision Photoplotter, for example, has twenty-four apertures on a disk, allowing up to nineteen different line widths.

Costs of drawing office equipment

Stools suitable for draughtsmen can be bought for around £50 while adjustable chairs with backs are just over £100. Desks can be obtained in steel at under £500, and in wood-veneer at just over half this. Reference tables in steel are around £400 to £500, and in wood £150 to £250.

Costs of filing cabinets vary considerably, depending on the construction and material used. A steel planchest of ten shallow drawers costs around £1000, a five-drawer version being about half this price. Vertical planfiles which take DIN A0 size cost £600 to £700. Vertical planfiles of the Graphitheque type are sold at over £800.

Prices of stands for drawing boards range from the simple tubular frames at less than £100 to the sophisticated at over £700. A very serviceable and sturdy stand with good facilities of which most professional draughtsmen would be proud, can be had for around £400. Stands in the upper range are for very large boards (say, 120 cm × 230 cm) and equipped with more elaborate counterbalance such as oleopneumatic springs.

Drawing boards range from £20 for A1 size to £50 for A0 extended. Typical prices for parallel motions range from around £50 for DIN A1 size transparent blade to £90 to fit a 120 cm × 200 cm size board. Metal blades often are slightly more expensive. Parallel motions are also available with a cutting edge blade for between £150 and £200 depending on size.

Medium-sized light boxes are priced at £700, rising to nearly £1000 for 100 cm × 170 cm size. Thus a light box complete with a top-range stand would cost around £1500, plus parallel motion or drafting machine.

The average drafting machine is about £500, give or take £150. Very simple drafting machines can be purchased for about £200, but the extra money is well spent.

From these prices, shopping-lists can be compiled as follows:

	Average costs	
	Lower end	*Higher end*
	£	£
Drawing board	20	60
Stand	100	740
Parallel motion	50	
Drafting machine (including scales)		825
Light box	150	990
Total for board	170	1625
Total for light box	300	2555

CAD/CAE

Ordinary computer monochrome monitors can be bought for a few hundred pounds, but CAD/CAE work needs high resolution. A typical colour display with sixteen colours costs about £4000, or with 256 colours about £4300. At the bottom end of the price range, a high resolution monochrome screen is around £1800.

Software prices range from a few hundred pounds for a simple package to five-figures for sophisticated software.

Professional software to run on mini- and microcomputers costs around £5000, (1988) but clearly the range is great. A good two-dimensional drafting system from Robocom (RoboCAD/2) is sold at about £2000. The three-dimensional solid modelling version is also £2000.

Complete computer packages have typical start-up prices of about £4000 to £6000 for computer, CAD software, and plotter, with Aplicon Bravo in the top range at a quarter of a million pounds.

Dot matrix printers for making hard-copy prints are available from a few hundred pounds to around £2000.

The cost of the type of plotter shown in Figure 12.10 ranges from £4500 to £5500, depending on facilities.

13

Reprographics and microfilm

Dennis Lock

This chapter describes some of the techniques and equipment available for the reproduction of drawings and engineering documents. The emphasis in this respect is on drawings rather than on office photocopiers and printing equipment. A considerable part of the chapter is devoted to an explanation of microfilm applications in engineering.

Plan printing

Diazo

The most popular system for reproducing drawings has, for many years, been the diazo printer. These machines use a considerable amount of electrical energy owing to the intensity of output needed from their ultraviolet light sources, and ducted ventilation is essential for the larger and faster machines in order to remove the excess heat and surplus ammonia developer fumes. Apart from heat and fumes, they usually also produce a lot of noise.

The drawings to be copied must be made on transparent or translucent material (such as tracing paper or drafting film), through which the ultraviolet light is passed in the machine. The copies are made on coated materials, which can be various weights of paper or polyester film.

The quality of first generation prints from good master drawings is generally very good and, if dimensionally stable film is used, the copies (because they are made by direct contact printing with the originals) can be scaled. Paper prints are cheap.

Prints made on film or translucent paper are known variously as submasters, sepias or reproducibles, and these themselves are capable of being used as secondary masters in diazo machines because their base material is transparent to ultraviolet light.

Second generation prints, made from submasters, are often less satisfactory, especially where paper (rather than film) submasters are used. Dark background areas on the submasters have the unfortunate property of being of a colour that is particularly opaque to ultraviolet light. Unless the submasters are backlit, they may be difficult to microfilm successfully.

Image areas on the submasters can be erased chemically for the purpose of modification. This means that design variants can be created for use on new products or projects. On film difficulties may be experienced in redrawing on the erased areas because the chemical erasure process can destroy the matt surface coating.

For low volume work the sensitized materials are purchased in sheet form, but it is more economical to have a machine capable of taking rolls in two or three appropriate widths. This avoids the need to carry a variety of sheet sizes in stock. The prints are cut to length as the material is taken from the roll by the machine feed, either manually with a knife or drawstring, or by using a printer which is fitted with an automatic shearing knife. Although the materials are not highly sensitive to ordinary levels of office lighting, they will deteriorate when left unwrapped.

The larger machines are very fast, easily accept A0 sized originals (see Figure 13.1 for a table of metric sheet sizes) and they can be fitted with automatic folding devices. Océ (incorporating Ozalid) are major suppliers of diazo machines and materials in the UK.

	DIN A series			*DIN B series (less common)*	
	Length	*Width*		*Length*	*Width*
A0	1189	841	B0	1414	1000
A1	841	594	B1	1000	707
A2	594	420	B2	707	500
A3	420	297	B3	500	353
A4	297	210	B4	353	250
A5	210	148	B5	250	177
A6	148	105	B6	177	125

All dimensions are in millimetres.

Figure 13.1 *Standard metric sheet sizes*

Although diazo machines are still being manufactured and installed, and will undoubtedly continue to be used for many years, they are likely to be displaced eventually by the new generation of xerographic plain paper printers.

Plain paper printers

Until recent years, plain paper copiers using the xerographic (electrostatic) process were confined to office machines, with a maximum possible print size of A4 or A3. That situation has changed, and there are machines becoming available which will take A0 originals. The Océ model 7500, for example, can produce same size copies from originals up to A1 size, and A0 drawings can be accepted to produce reduced size (A1) prints. Efficient folding and collating features can be added. One problem found by contracting companies working on large projects is the difficulty of keeping records of prints made for different clients: in many cases this is necessary because the conditions of contract allow that all prints made can be charged to the client. The Océ 7500 can be fitted with a copy administrator that is capable, among other things, of metering prints made against job or customer code numbers.

Xerox Design Technology also offer a range of advanced plain paper printers, the smallest of which is their low cost, medium speed table top model 2510. This can make A0 or continuous roll prints on paper or other materials. The Xerox 3080 high volume (production) printer is at the top of the range, and is capable of producing prints up to A0 size, or continuous rolls of any required length. Prints can be made on polyester film, resulting in very high quality secondary masters: this feature also allows regeneration of old or damaged drawings, the results often being as good as new.

Plain paper printers offer many advantages over diazo equipment. They are quiet, emit no fumes and use less power. Although they are usually referred to as plain paper printers, they can in fact print on to a wide variety of materials. One of the biggest advantages is seen in the versatility of the input medium possible. Drawings no longer have to be made on translucent or transparent material.

Modifications to secondary masters can be more satisfactory using plain paper printers than with diazo. The xerographic image is normally fixed to the copy material by a heat process which fuses the toner powder on to the surface. If the fusing process is omitted, the image can be removed easily with an ordinary eraser. The erased areas can then be redrawn, before the print is passed through an 'off line' fuser to fix the remaining toner. Engineers will appreciate that the modification of secondary masters in this fashion is a convenient way of incorporating 'retained engineering' in new designs. The off line fusing feature is only found on a limited number of printers (such as the Xerox 3080).

Because plain paper printers use a system of optical lenses (unlike the contact printing method used in diazo machines) it is possible with some models to produce enlarged or reduced prints. Magnification and reduction ratios may be stepped or, in some cases, continuously variable with a zoom lens. The use of this feature to produce A1 prints from A0 originals has also been mentioned (for the Océ 7500) but it also has scaling applications. Caution has to be exercised when scaling such prints in case of optical distortion, but high quality lenses are usually used and Xerox Design Technology (for example) state that their 3080 printer is suitable for accurate scaling up or down.

Developments

Until recently, it was possible to consider the subject of reprographics in terms of separate machines, each of which was able to produce a more or less faithful copy of an original engineering document. As more and more equipment is developed around microelectronics, with the capability of handling digitized information, it has become more meaningful to think in terms of drawing systems rather than of individual stand alone printers and copiers.

Undoubtedly the development for which the industry has been waiting is the capability of production A0 plain paper printers to accept digital input from a computer aided design (CAD) system. The process, using a laser beam to write on the xerographic drum, will be much faster than the current generation of pen plotters. It is already possible to create 35 mm microfilm aperture cards in this way (as described later in this chapter) and technological progress is such that machines capable of printing full-sized drawings from digital input should be available by the time this handbook is published.

Microfilm in the engineering department

Microfilm was originally considered simply as a medium for converting documents into a smaller format, the primary objective being to save space. Microfilm used to be thought of as a system principally for storing archives. But the quality of microfilm possible and the speed with which documents can be found and printed with modern equipment means that microfilm is seen by many as part of their everyday document handling system. It is appropriate to describe the various microforms available and some of their engineering applications. Please note that the term microform is used by the cognoscenti to mean any type, size or shape of microfilm.

16 mm rollfilm

Correspondence and documents (including drawings) up to A3 size can be filmed using 30 m (100 ft) rolls of 16 mm black and white or (less commonly) colour film. One roll can typically hold between 1000 and 1500 documents, depending on whether the images are arranged in comic or cine attitude (Figure 13.2).

Archive film can be stored on simple reels but film for everyday use should be mounted in cartridges. Cartridge mounting speeds up loading and unloading of film in viewers and printers, and protects the film to a large extent against wear, tear and accidental damage during handling. Cartridge systems also provide better indexing options than simple reels.

The choice of cartridge will depend on the viewing and printing equipment to be used. There is no guarantee of compatibility between the various brands and the choice of equipment manufacturer is therefore something of a long-term commitment since any subsequent change might mean remounting and reindexing the entire collection of existing cartridges – which could be a monumental task.

Because it is necessary to spool through any reel to find an individual document, careful attention must be given to indexing. If a film contains, for example, only purchase orders, filmed in sequence of purchase order numbers, and if every purchase order is a single sheet (producing only one frame on the film), then it should be relatively easy to find any purchase order by its serial number without having to provide intermediate index points in the reel. In the nature of engineering files, however, each reel of microfilm is likely to contain batches of different documents, and even if all of them were to be purchase orders, or all of them letters, some might comprise one or two pages while others could run into many pages (some purchase orders might have long specifications attached).

Indexing can be in the form of simple notes that equate document positions with the readings of a footage counter on the reader/printer equipment. This indexing would typically be used to locate batches of documents on a film, such as letters between two dates, or purchase orders within a specified range of order numbers. This effectively narrows down the search for individual pages, which can be found by winding up to the relevant index point at high speed, and then viewing the film by running it through very slowly until the right page comes into view. Obviously the batch sizes chosen for indexing should not be too large; the aim is to strike a sensible balance between attempting the impossible task of indexing every single one of the thousand or so images on each film or going to the other extreme and simply writing on the cartridge that it contains letters or purchase orders. Clearly it is necessary to serial number all the cartridges.

The microfilm index could be written in a notebook. But as the documents are likely to be filmed in mixed batches, chosen for their age and eligibility for filming rather than for their type or subject, a chronological register of filming is unlikely to be of great use in finding individual documents, or even particular batches of documents, as the microfilm files grow larger.

A better system would be to use a flexible indexing system, allowing the

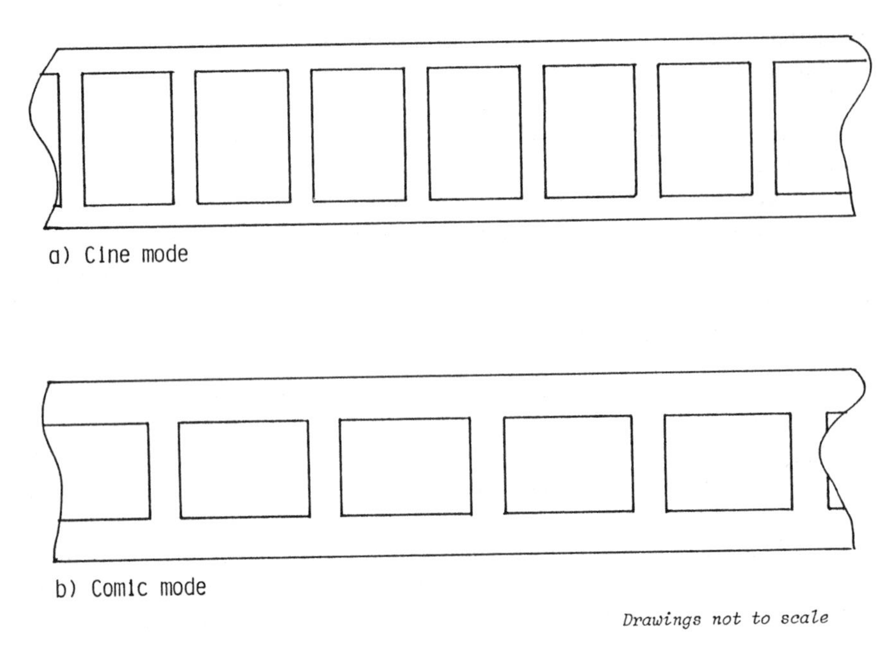

Figure 13.2 *Comic and cine modes on microfilm. Using cine mode, about 1500 A4 documents can be condensed on to a 30 m roll of 16 mm film. In comic mode the figure is around 1000 documents*

insertion of new file additions in some logical sequence. A visible-edge card index can be used for this purpose, arranged alphabetically according to some agreed subject breakdown. The index data would then list the cartridge identifying number and its index points for each identifiable batch of documents.

A microcomputer system is better still, allowing greater flexibility, and giving the engineering staff the ability to search for documents according to one or more identifying characteristics (subject, date, serial number, name of addressee, document type, keywords etc.).

More sophisticated systems exist that rely on edge notching or other markings on the film to provide digital indexing. These systems need more attention during the filming and mounting operation, but they allow rapid location of any document thereafter.

Although it is possible to edit 16 mm film, or to splice in short new sections, this microform is really only suitable for the filming of documents in batches (for example, files which have been closed at the end of a project or at the end of a year). In any case, there is a limit to the amount of film which can be spliced in, determined by the capacity of the reel or cartridge. If the records are active, and are expected to require editing and updating, then jacket fiche is more appropriate.

Microfiche

Microfiche is the name given to sheets of film on to which several images have been filmed in a grid pattern (Figure 13.3). Engineering uses apply to any book, catalogue, list, specification or other document that is made available in microfiche form by its publisher. The system can also be used for internally produced sets of pages, forms or tables which are needed for general reference and which are unlikely to require frequent revision (for example complex parts lists that have been debugged after successful manufacture).

Indexing is achieved by defining the intersections of numbered columns and rows (an alphanumeric grid index). It is generally easier and far quicker to locate an individual page on fiche than with rollfilm. It cannot, however, be edited or updated at all, and any change involves replacing the affected fiche.

Jacket fiche

Jacket fiche is a system which allows strips of film to be inserted into a number of parallel channels in a transparent plastic envelope (in the same fashion as photographic negatives are sometimes returned from a photographic processing laboratory after developing and printing). The jackets are A6, and versions are available for 16 mm or 35 mm film. Documents can be filmed in small groups, or even singly. The processed film is cut and inserted into the fiche jackets, using a special guide tool to make this fiddly job easier.

Jacket fiche has the advantage, compared with other types of fiche, that it can be edited, rearranged or augmented at will. Updating and editing is carried out simply by removing unwanted frames from the jacket and substituting or adding new frames. Where copies of the fiche exist in more than one file, it is only necessary to carry out surgery or transplants on the original fiche, after which it is easy to make quick diazo contact copies of the whole sheet for redistribution (copying is dealt with in a later part of this section).

A typical application for jacket fiche using 16 mm film would be for personnel records. One fiche would be started for each new member of staff, and all documents relating to the progress of his/her employment would be added as various events took place (personal data, salary reviews, pension details, promotions, appraisal reports, and so on).

Viewers and printers for 16 mm film and microfiche

There is a vast range of equipment available for viewing, projecting and printing from 16 mm film and microfiche. Some equipment is limited to film or fiche only, while other units can adapt to take several kinds of input.

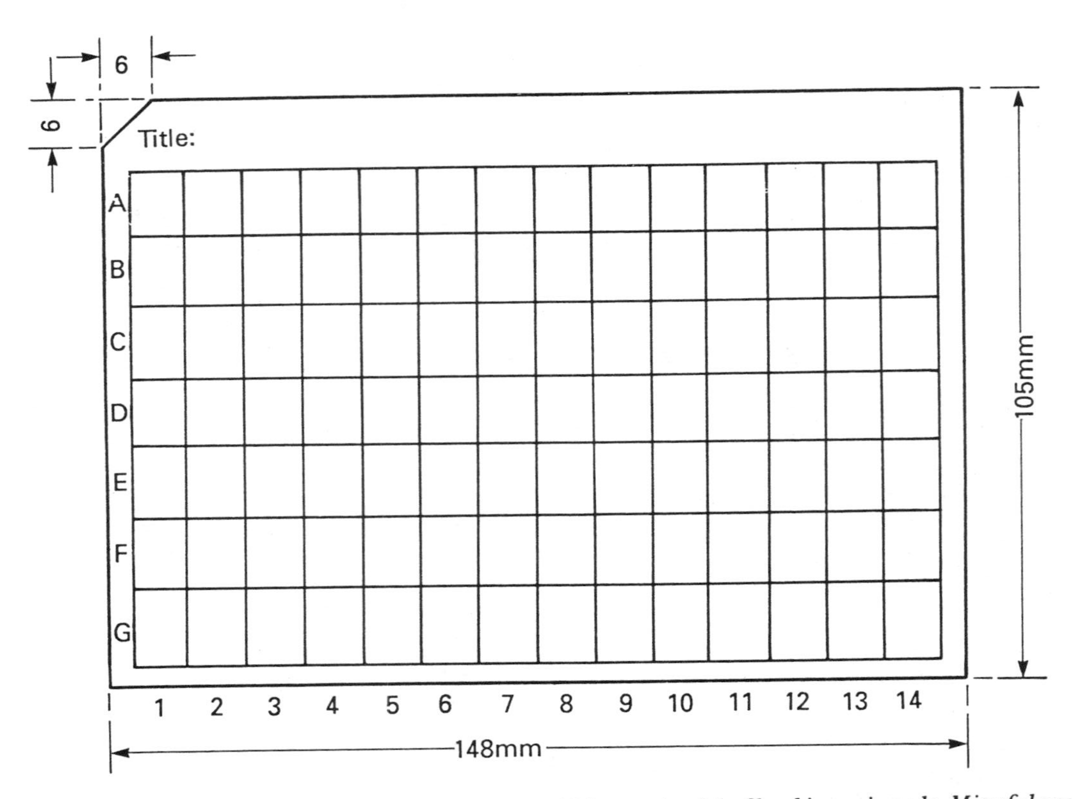

Figure 13.3 *Standard arrangement of an A6 microfiche. This shows a 98 frame microfiche filmed in comic mode. Microfiche can also be arranged in cine mode, in which case the rows would be numbered 1 to 7 from top to bottom, and the columns would be lettered A to I from left to right. Other microfiche reductions allow 60, 208, 270, 325 and 420 frames on one fiche*

Illustrated in Figure 13.4 is the 3M model MFB1100 reader printer, which can take film in several types of cassette or fiche and make prints on plain paper.

35 mm film

35 mm is the most popular size for holding records of drawings on microfilm, being suitable for drawings up to A0 size. Roll drawings larger than A0 size can be filmed in sections. 70 mm film has also been used, but only very rarely by comparison with 35 mm.

When batches of drawings are filmed as a precaution against loss, or just before their destruction well after they have ceased to be of topical use, it is sufficient to film them in numerical sequence on a continuous roll of film which, after careful checking, is listed in a register or index and consigned to a secure place for safekeeping.

If the film is to be used for reference, or as a source of production prints, then it is more convenient to mount single frames in aperture cards. Sometimes it is sensible to consider a dual system, with a rollfilm put into secure storage in case there is a disastrous fire, with aperture cards used for normal reference and printing.

35 mm aperture cards

Standard 80 column punch cards are used as mounts for single 35 mm frames. One frame (and, therefore, one aperture card) is used for each drawing of A0 or A1 size. Smaller drawings, and A3 or A4 sheets of schedules and parts lists can be filmed more than one to a frame. There are three mounting methods:

1. Aperture cards can be purchased which have a transparent envelope fitted over the aperture, allowing insertion of the single 35 mm frame
2. Cards can be obtained with self-adhesive transparent edging tape around the periphery of the aperture. Each film frame (cut as accurately as possible) is dropped into the aperture and pressed on to the adhesive tape to fit flush with the card surface
3. Cards are manufactured with unexposed film already mounted in the apertures. These can be diazo (used only for making copies from existing microfilm) or the more light sensitive silver halide emulsion (used for original filming). The ready mounted silver cards come supplied in light-tight cartridges for use in special automatic cameras (such as the A0 processor cameras manufactured by IMTEC and the 3M company) that deliver a fully developed and processed card about half a minute after each exposure.

After filming and mounting it is necessary to type or print the drawing identification details (number, title, revision number) along the top edge of each card. The layout of this information must be designed with care, and

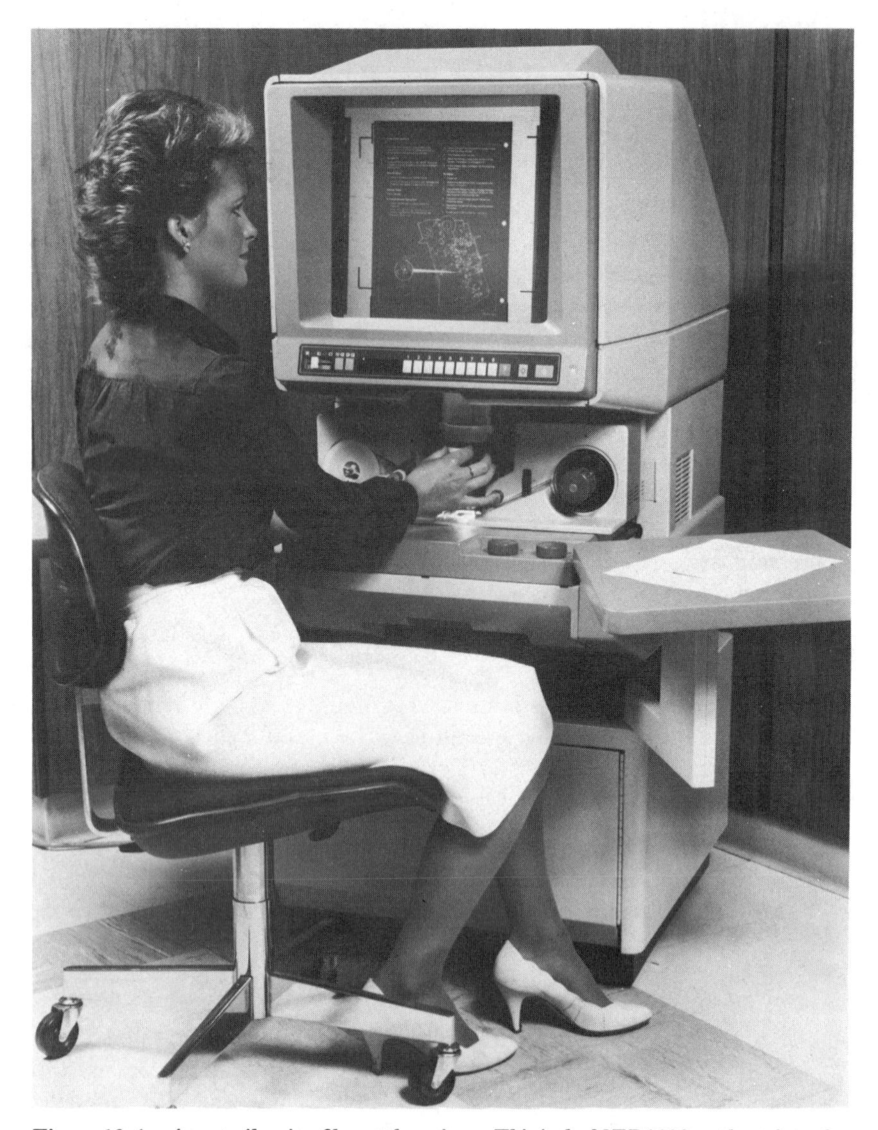

Figure 13.4 *A versatile microfilm reader printer. This is the MFB1100 reader printer from 3M, which can make A4 positive prints on plain paper from 35 mm rollfilm, 16 mm film on reels or in cartridges and all types of fiche. Advanced document search facilities are included* (Courtesy of 3M United Kingdom plc)

then adhered to very strictly. If any variation were to be allowed from one card to another it would be very difficult to locate individual cards from their decks. The process of writing the information along the top edge must be combined with a check, using a viewer to ensure that the correct drawing is actually mounted on the card. This whole operation is sometimes called interpreting the card.

Since the background material is a standard 80 column punch card, there is the obvious opportunity (taken by some companies) actually to combine the printed information along the top card edge with hole punching that allows automatic sorting or card selection. Proprietary equipment can be purchased. It goes without saying that the card punch must be programmed to avoid punching over the picture frame area.

A more recent development, which is preferable (and safer) than punching the cards for identification and subsequent sorting and retrieval, is to use bar codes. Figure 13.5 shows such a card. Optical character recognition OCR is also used.

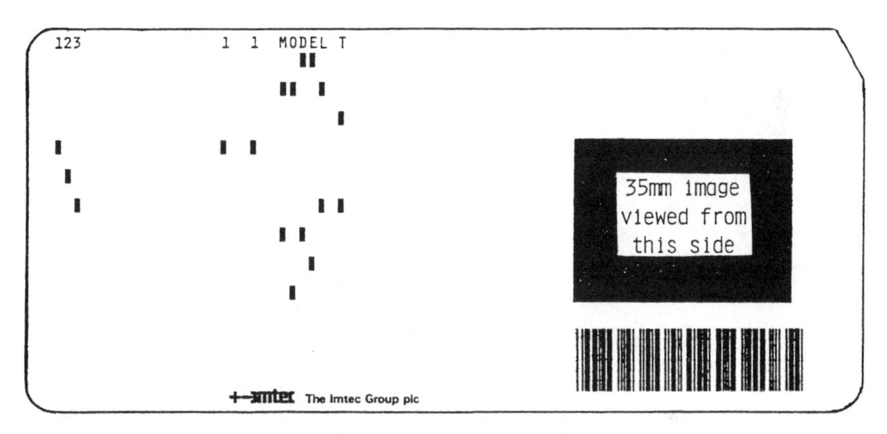

Figure 13.5 *A 35 mm microfilm aperture card. This example shows how drawing information can be coded by means of Hollerith punching or by a bar code. Optical character recognition can also be used*

It is possible to purchase aperture cards which are preprinted with the company logo, or any other required information. Thought should be given to ordering cards in distinctive colours to denote their particular use. For example, a company might choose to use the following colours:

Buff (natural) cards – silver master film of in-house drawings.
Green cards – diazo duplicate cards of in-house drawings (used for reference files).
Yellow cards – silver cards of vendors' and subcontractors' drawings.
Orange cards – silver cards of clients' drawings.

Aperture cards are filed in numerical sequence, and should need no indexing other than that already provided by the company's drawing registers.

Finding suitable filing equipment is never a problem for aperture cards, since they can be kept in cabinets designed for 80 column punched cards.

Companies can choose whether to have all the operations of filming, mounting, checking and interpreting carried out by a professional external bureau, conduct the whole operation in-house, or adopt some shared arrangement which lies between these two extremes.

35 mm microfilm aperture card printers

Printing from aperture cards is a convenient process. A deck of cards can be stacked in the printer, which is then programmed to produce the required number of prints. The 3M machine illustrated in Figure 13.6 can produce sets of prints in programmed quantities up to A1 size.

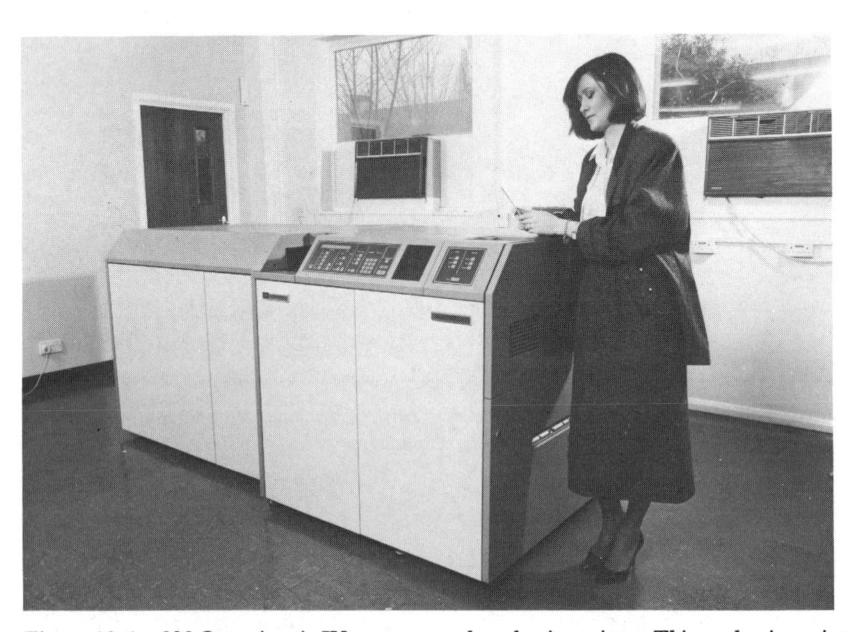

Figure 13.6 *3M Quantimatic IV aperture card production printer. This production printer produces plain paper prints, vellum or offset masters from 35 mm aperture cards* (Courtesy of 3M United Kingdom plc)

35 mm microfilm cameras

The current generation of microfilm cameras has little in common with some of their twenty year old ancestors (many of which are still in use). They are microprocessor controlled to simplify operation and they provide features scarcely considered in the past. They typically offer the user a choice of card output speed per hour. The most advanced, looking dramatically different from their predecessors, integrate a VDU and full alphanumeric keyboard into the control panel (Figure 13.7).

Figure 13.7 *Imtec A0 processor camera. With a keyboard and VDU attachment, this camera not only delivers fully processed 35 mm aperture cards from drawings up to A0 size, but also allows the operator to add drawing information to the cards as the drawings are filmed. Productivity can be further increased by linking the camera to the design office computer, allowing on-line control of duplication, indexing and collation*
(Courtesy of the Imtec Group plc)

There is the option of a built-in computer interface capability to the design office computer and on-line duplication, indexing and collation modules. In this way the engineering microfilm camera can become a highly productive one person workstation, an interactive communications channel and computer peripheral.

With such systems the operator has only to position a drawing on the copyboard and key in the drawing number from the camera keyboard to the design office computer. The computer automatically sets the camera to the correct reduction ratio and the multicopy preselector to the number of silver master cards required. The camera operator just presses the exposure switch to activate the camera.

Please also refer to the section on COM, which follows.

COM

The initials COM are the commonly used abbreviation for computer output on microfilm. This is the procedure where the microfilm is exposed directly from data held in a computer file, with no paper document involved. At one time this process was always achieved by filming from the face of a cathode ray tube, and the system was used for high volume commercial applications (for example, credit card transactions). With the advance in technology, the production of various microforms from digital data is becoming more widespread, using a laser beam to write directly on the unexposed film from digital data held in any computing or word processing system.

Returning to drawings generated from a CAD system, it is now possible to use COM for producing 35 mm aperture cards directly from the digital data. In the past, a full size plot had typically to be produced using a pen plotter and this was then microfilmed using a camera in the traditional fashion. Now this time-wasting process has been made obsolete. Working on-line or off-line to CAD systems, and with in-line camera processing, finished aperture cards can be produced from a typical A1 sized original in only four minutes. Inline modules for duplicating, indexing and collating can all be added to the microfilm plotter, controlled either from the CAD system or by an interfaced computer running drawing register software.

The 3M 2900 CAD film plotter is one example of a machine capable of producing aperture cards directly from digital data. Another, manufactured by Imtec, is illustrated in Figure 13.8.

Microfilm quality

It is essential that all stages of filming and processing are carried out under strict quality control supervision. Otherwise there is a risk that the finished microfilm will be deficient in one or more important respects. Some control of

Figure 13.8 *A digital aperture card plotter. The Imtec Microplotter produces 35 mm microfilm aperture cards direct from a CAD system, and can be fitted with modules for duplicating, indexing and collating. These can be controlled from the CAD system or from an interfaced computer running drawing office register software*
(Courtesy of The Imtec Group plc)

quality can be obtained by the purchase of automatic equipment for filming and also for processing.

Where the expense of automatic processing equipment cannot be justified, unless a company is prepared to employ properly trained staff and set up a well equipped microfilm laboratory, it is best to entrust filming, and certainly processing, to a specialist bureau with a well established reputation.

The first criterion in obtaining good quality is always to buy film of reputable manufacture, and to use it before the expiry of its stated shelf life.

We will assume that the filming equipment is capable of producing good optical results, and that it is well maintained. (Sometimes criticism of poor microfilm prints will arise if the internal lenses and mirrors of equipment are allowed to become misty or dirty, even though the microfilm itself can be of high quality).

Correct lighting is important in order to secure even exposure over the document area. There must be no glare or unwanted reflections. Backlighting is sometimes necessary in order to obtain adequate contrast (this is the technique of spreading the document to be photographed on a glass table which is illuminated evenly from underneath). Backlighting is particularly useful when filming old sepia submasters.

The correct combination of exposure time and developing method, plus quality checks with a densitometer, will ensure that all film produced lies within an acceptable range of image densities (usually between 0.9 and 1.1). Very faint or very dark images will cause information to be lost, while variation from one image density to another can result in unacceptable output from some automatically fed printing equipment.

A quality feature which is more difficult for the nonprofessional operator to identify is the keeping quality of the processed film. This is a function of the developing, fixing and washing processes, and is one of the hidden advantages gained by paying the charges made by a reputable bureau.

Silver or diazo copies?

Whatever the shape and size of the film, all black and white images are available in two different photochemical options.

The usual silver-based photographic emulsion film is used to produce the original microfilm image, which is usually a monochrome negative (black and white reversed). Since microfilm is easily damaged, it is customary to keep the original film in safe storage. For everyday use, the original can be copied or another original can be made (by filming the documents twice).

Copies of the film can be obtained in several ways. These include:

1. Filming the original document several times to obtain several silver images
2. Copying the first film, reel-to-reel or fiche-to-fiche, using silver-based film for the copies, possibly with image reversal to retain a negative image
3. Copying the first film, reel-to-reel, fiche-to-fiche or aperture card-to-aperture card, using the diazo process (ultraviolet light exposure with ammonia developer) which will produce a negative image from a negative original.

Diazo copies are very much cheaper than silver copies, and are generally acceptable for everyday use, with keeping qualities extending over many years. However, provided that the processing has been properly carried out, silver film has superior keeping qualities, especially when very long storage periods are being considered. Silver should therefore be chosen if the film is to be archived for several decades, with the cheaper diazo copies considered as a reliable and cheaper alternative for most other uses.

Copies are made as contact prints. Provided that the automatic equipment is properly maintained, the copy definition should be excellent. However, it is always preferable to archive original microphotographs rather than copies, since this will ensure that the stored material has the best possible image definition.

Note that, since copies are made by contact printing from the original microphotographs, the images will be laterally reversed. This means that the image on a copy must be viewed from the opposite side of the film compared to the original. This means in turn that, when placed in viewing equipment, the focusing distance from the lens will vary by the thickness of the film when a copy is substituted for the original film. This may not sound important, but it will give rise to inconvenience when a deck of aperture cards or microfiche contains a mixture of copies and originals, needing frequent refocusing. This point becomes even more significant when the viewer has an automatic feeding and printing attachment – either the prints from the originals or those from the copies will be sharp, but not both (the degree of blurring depending on the focal length and quality of the lens).

Microfilm satellite files

A practice followed by large engineering companies, especially where several engineering offices are separated by great distances, is to distribute 35 mm aperture cards of layout drawings for inclusion in a satellite reference file in every office. Microfilm is ideal for this purpose, being easy to mail and taking up very little filing space. Satellite file cards must usually only be regarded as reference information for the purposes of assisting in subsequent engineering projects. They are usually not subject to the same degree of discipline in updating and file maintenance as files within the central registries, and the revision status may not always be correct. For this reason, the aperture card material should be chosen with a distinctive colour, and they should be printed with the words 'for reference only: not to be used for production', or something similar.

British Standards Specifications

The following British Standards Specifications are likely to be of interest to engineers using microfilm.

BS 1153: 1975, *Recommendations for the processing and storage of silver-gelatin type microfilm.*
BS 1371: 1973, *Specification for 35 mm and 16 mm microfilms, spools and reels.*
BS 1487, Part 1: 1981, *Microfiche. Formats of 60 and 98 frames.*
BS 1487, Part 3: 1978, *Microfiche. Formats of 208, 270, 325 and 420 frames (except COM).*
BS 4210, *Specification for 35 mm microcopying of engineering drawings.* This specification is in three parts, as follows:
Part 1: 1977, Operating procedures (size and quality of originals, reduction and enlargement ratios, methods of filming large originals in sections).
Part 2: 1977, *Photographic requirements for silver film* (image resolution and densities etc.).
Part 3: 1977, *Unitized microfilm carriers (aperture cards).*
BS 5444: 1977, *Recommendation for preparation of copy for microfilming* (refers to microfiche).
BS 5632: 1978, *Specification for microfilm jackets, A6* (for jacket fiche).
BS 5699: 1979 (Parts 1 and 2) *Processed photographic film for archival records.*
BS 6054, *Glossary of terms in micrographics*, in three parts:
Part 1: 1981 *General terms.*
Part 2: 1983 *Image positions and methods of recording.*
Part 3: 1984 *Film processing.*

PART FOUR

Managing Information

14

Numbering, classification and coding

W. Jack

For the engineering industry there is no system based on the consideration of numbers alone that can survive the test of time. All systems must be built around a classification that has been carefully thought out and based on sound principles. It is no accident, for example, that accountants refer to a code of accounts. They do not talk simply of a 'numbering system for accounts'. A code of accounts is a system for giving numbers to specific accounts which have already been classified.

It is necessary to appreciate the difference between pure numbering and coding. A code is one or more symbols to which an arbitrarily assigned meaning and/or arrangement has been given which, when deciphered, communicates specific information or intelligence. Non-codes are the result of attempts to use numbers to control drawings, materials, components and products, initially by one department and which finish up by each department having more than one 'numbering system' of its very own. It is essential for control and longevity that *classification* is followed by *coding*, and that coding is preceded by classification (Figure 14.1).

Some of the problems

Identification based on numbers alone

Numbering is so important in today's computerized world that questions which need to be asked and answered in considering a database and database standardization are:

1 What is the purpose of classification?
2 What are the principles of classification?
3 What are the principles of coding?

In the absence of known answers to these three questions, companies have found that each department invented its own numbering system for the purpose of identification. It is a fact that most companies have several

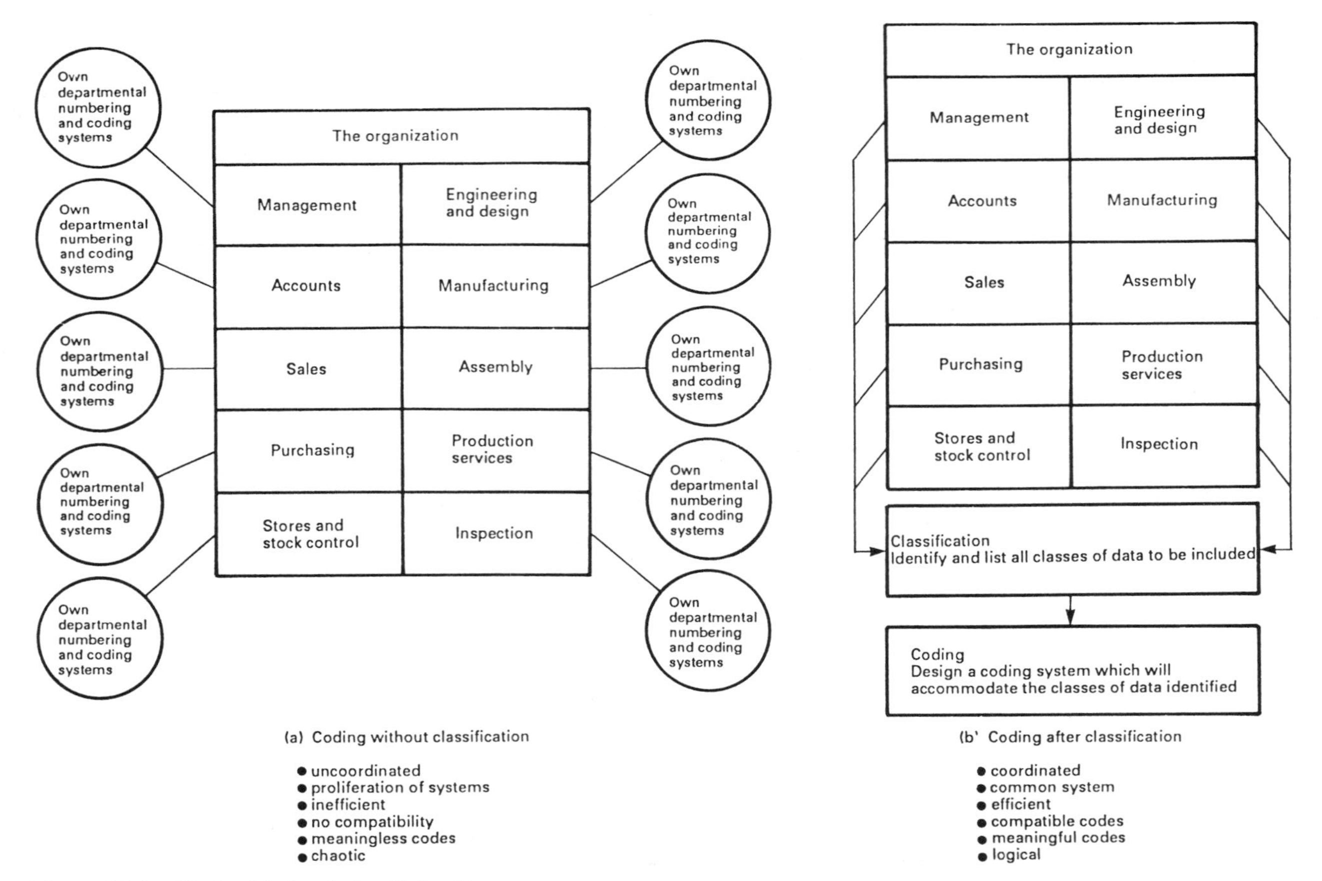

Figure 14.1 *Unclassified and classified coding systems*

'systems', perhaps covering a range of six or more. One company, when investigated, was found to have no less than twenty-seven different 'numbering systems'.

The reason for so many varieties is simply that the 'numbering systems' are not systems at all. Different departments have different needs and are forced to design some sort of identifier because there is no company system. The so-called numbering system of each department soon breaks down, and is modified or replaced with something which is equally invalid and short-lived.

A typical engineering company could have the following minimum requirement:

1 Drawing number.
2 Part number.
3 Material specification number.
4 Pattern number.
5 Casting number.
6 Forging number.
7 Raw material number (bar, tube, plate).
8 Production engineering number.
9 Jig and tool number.
10 Cost office number.
11 Buying office number.
12 Receiving inspection number.
13 Stores bin number.
14 Spares number.
15 Subassembly number.
16 Product number.

When the above needs are considered, it is obvious that the preference should be for one system which would give commonality throughout. In other words, a company system and not a departmental hotch-potch.

A company database used in an integrated data processing system means that all departments would be using the same information, with the same identifier for each item wherever used.

Identification by codes

The actual result of numbering systems designed individually by departments is that companies finish up with identifications using non-codes which are of little value. The following are examples.

Drawings

A frequently encountered system of identification is that which utilizes the size of paper on which the part is drawn. When an engineering change is to be made, it is necessary to know the paper size if the part drawings are filed by size. The drawing size cannot be satisfactorily incorporated as part of the identifier.

Configuration level

The same comments can be said for part identifiers that show configuration level as an integral part of the identifier. There are similar change configuration problems with assemblies and models. Configuration changes affect model designation just as they do parts, materials and components.

Part numbers that reflect the mark level or change level are not codes any more than drawing sizes, random numbers or 'where used systems'.

Part numbers

Often these TAG numbers are referred to as codes because they are arbitrarily assigned to represent something (i.e. one TAG number for one thing). Numbers are assigned sequentially in random order, for example 5320704 could be a rotating assembly, while 5320703 could be a screw. In no way does the number describe any attribute of the part it represents.

Where used

Frequently, part identifier systems relate the part to the model on which it is used. It is important to know on what assemblies and models a part is used. However, if a part is used on one model and bears that model's identifier within its part number, what happens when the same item is used on other models? In neither case does the part number identify any of the attributes of the part itself. 'Where used' is essential information but is a non-permanent characteristic. A bill of material processing system provides the basis for consolidating such requirements in a horizontal search – in other words, the system, not the part number, should produce the desired 'where used' information. Therefore, 'where used' is not a code in this context.

The part creation cycle

Work sampling studies of a number of firms by several investigating agencies has revealed that up to 15 per cent of total available design and draughting hours is spent trying to research information which is known to exist but which, because it is not in logical order, is not easily retrievable. The result is, of course, to create another data item – a new part drawing with a new part number. The following sequence will then occur:

1 The new part concept will be researched to find the state of the art for similar applications.

2 The part will then be physically designed, engineered and placed pictorially on some medium, paper or visual display unit.
3 A prototype of the part or assembly may be built.
4 The part is then signed off, with or without a design/manufacturing agreement.
5 A manufacturing engineer will determine the method of manufacture and the tools and gauges necessary.
6 A tool designer will design tools as required.
7 A toolmaker will make the specified tools.
8 Work study engineers develop the working methods necessary to operate the machines, and establish time standards.
9 A production controller adds the record, and sets aside capacity and storage space.
10 The cost accountant determines the cost of materials, setting up and production labour, adds overheads and creates the record for the appropriate cost system.
11 The relevant data processing personnel add the part number and all relevant associated data to the computer file.

The cost of performing all these tasks will obviously vary from company to company, but it is bound to be significant.

Classification and coding

Fox (1986) states that:

> The processes of classification and coding are the first steps to creating order and meaning in an information system. All the information that management receive will be subject to some form of classification and, in order to record, store and retrieve classifications, a code will be required. Effective management will depend upon the way in which data is classified and coded. For example, poor sales of a particular product are not likely to be recognized by management if an organization only classifies its sales by area. Also, if an unsuitable code is used there may be considerable problems in handling the information however classified. An organization, therefore, cannot expect to be able to take effective decisions or carry out effective planning and control unless it has reviewed very carefully its data requirements (i.e., classifications) and the way such data is to be recorded (i.e., coded).

When the ramifications of numbering requirements in an engineering organization are considered, one wonders how companies can undertake to develop computer aided design and computer aided manufacture without first of all reviewing very carefully their data requirements, and the data to be recorded. Proper classification and coding of drawings, materials, components, subassemblies, products, tools and services are prerequisites to the introduction of computer integrated manufacture. Using integrated codes, one common database system is possible.

The purpose of classification

The user of a classification must decide the purpose for which the classification is to be developed: that is, who in the organization is to be the prime beneficiary of the end product.

If the purpose is to produce one sound database system providing a company service, the beneficiaries should be each and every department. Classification must be seen to provide an economical return on the investment. Consequently, one must start with a design orientated system and, when this has been established, progress to develop the requirements of production and other departments.

Quoting Fox (1986) again:

> . . . accounting does not have a scheme of classification that is universally accepted. The accountant has to create a classification of information that is suited to the particular organization that he serves and even to the particular decisions that management take.

Likewise, for all identification and numbering in engineering, the classification must be designed to suit the user. There is no short cut: no off-the-shelf package and certainly no 'do-it-yourself' job; at least, if the results are to be a return on investment and the creation of order out of chaos.

Principles of classification

In 1947 Edward G. Brisch, a mechanical engineer and designer, adopted the hierarchical classification of the taxonomist and evolved his own classification for engineering applications based on permanent and non-changing characteristics of the items to be classified. The four principles which emerged are:

1 All embracing: a classification must embrace all existing items and be able to accept necessary new items into the defined population of items.
2 Mutually exclusive: a classification must be mutually exclusive, which means that it must include similar things to bring them together while excluding unlike things, using clearly defined parameters.
3 Based on permanent characteristics: a classification must be based on visible attributes or easily confirmed permanent and unchanging characteristics.
4 User's viewpoint: a classification should be developed from the user's point of view – not the point of view of the classifier.

Classification, however, although useful and essential, only becomes efficient when it is married to a coding system.

Some classification systems have been evolved to facilitate computerization.

These are accessed by keywords to track a family of data. They have no code, so they have no use departmentally throughout the organization as a tool of management giving a specific return on investment. For example, simplification and standardization are not elements established for automatic review in such systems.

Classifying for simplification, rationalization and standardization

Design cost avoidance is now easier to achieve than it was thirty or forty years ago. Given a suitable system for the identification of items, much can be done in this area.

If the system chosen for a company is design orientated, with data made visible and arranged in some systematic order, then simplification can lead to rationalization and standardization. Anyone wishing to see the variety in a family can do so by feeding in the classified characteristics of the data to find the family number. Data visibility via a computer is especially useful in curtailing unnecessary designs.

The elimination of unnecessary variety through variety control deserves high priority: any classification system which does not achieve this objective must be regarded as unsatisfactory.

If the classification system selected is orientated towards satisfying production aspects it cannot also satisfy the objectives of variety control. The system chosen must serve all its users, including:

- Design engineers.
- Standards engineers.
- Purchasing controllers.
- Methods engineers.
- Tool designers.
- Production controllers.
- Cost and management accountants.
- Parts and service managers.

The system should give all of these people immediate and direct access to a systematically organized source of information. It should cater for the 'maximum common denominator' of their needs, and it should facilitate the development of ancillary coding to satisfy the particular needs of each specialist. A classification should neither aim to satisfy all users at all times for all purposes nor be biased in favour of any one of them to the detriment of others (see Figure 14.2).

Principles of coding

Classification without coding is like breakfast cereal without milk – cumbersome and inefficient.

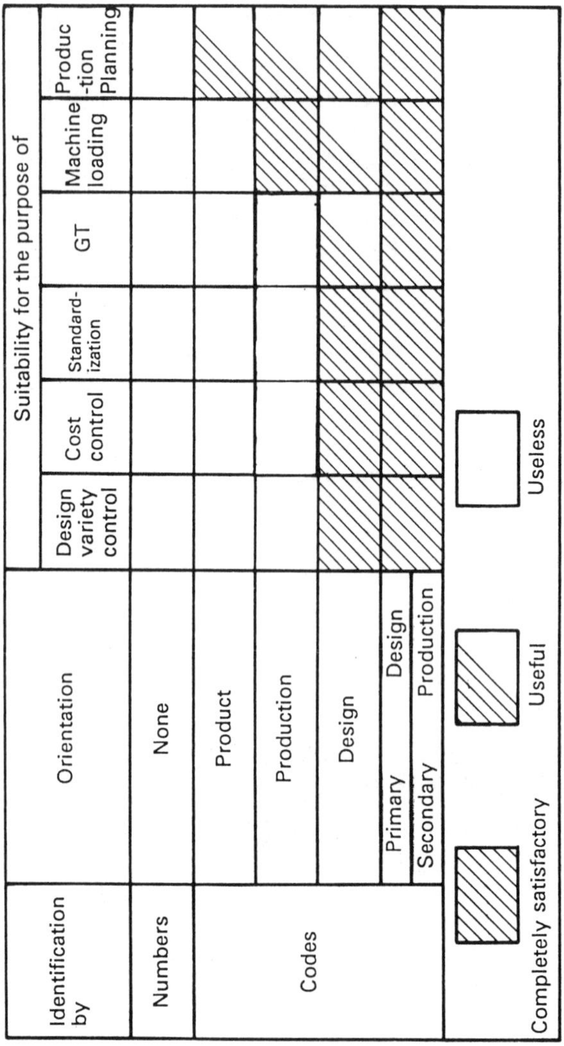

Identification by	Orientation	Suitability for the purpose of: Design variety control	Cost control	Standard-ization	GT	Machine loading	Produc-tion Planning
Numbers	None						
Codes	Product						Useful
	Production					Completely satisfactory	Useful
	Design	Completely satisfactory	Completely satisfactory	Completely satisfactory	Useful	Useful	Useful
	Primary: Design; Secondary: Production	Completely satisfactory	Completely satisfactory	Completely satisfactory	Completely satisfactory	Completely satisfactory	Completely satisfactory

Key: Completely satisfactory — Useful — Useless

Figure 14.2 *Classification methods and their usefulness for different purposes*

As with classification, there is a need for some principles. Experienced practitioners postulate four:

1 No code should exceed five characters without a break in the string.
2 Codes should be of fixed length and pattern.
3 All-numeric codes produce fewest errors.
4 Alphanumeric combination codes are acceptable if the alpha field is fixed and used to break a string of numbers.

Integrated and non-integrated codes

There are basically two methods used in industry to code data. These are:

1 Codes which are integrated with the hierarchical classification. These are called monocodes. A monocode has been defined as 'an integrated code of fewest characters which distributes a population of data in a reasonably balanced, logical and systematic order where each character qualifies the succeeding code'.
2 Codes which are not so integrated, and which are called polycodes. Although polycodes are not integrated with the hierarchy of the data classification, they may relate to a fixed field type of classification.

Proprietary systems

Universal classifications

A universal classification seeks to satisfy the needs of all users. The idea of a universal classification of scientific information stems from the concept of the universality of science. The appeal of such an idea can be readily understood, but there are clearly some practical difficulties. It is impossible to predict which facet of scientific information will be of value to any one potential user at any particular time or place. One may postulate a universal classification that would meet any one of the needs of any one user, but this ideal is unattainable.

The technique actually adopted is to classify information by some of its facets. Multifacet classifications may either be based upon the premise that all the facets are of equal value, or it may be based upon main concepts further qualified by additional facets. Personnel classification is an example of the former, while the UDC and Dewey systems are examples of the latter.

Universal classifications are used by librarians and documentalists, but many special systems have been developed to cover particular industrial or technical subjects. Instances are S and B building classification and the ASM metallurgical literature classification.

Decimal type classification

Melvil Dewey devised the original decimal type of universal classification, since improved in the form of the UDC. The decimal classifications were conceived for use in libraries and for application to various kinds of publication (e.g. books, articles, abstracts). If applied in industry the principle is best explained by means of this example:

Category	*Number*
Metallurgy	669
Steel	669.1
Alloy steel	669.15
Bending strength of	669.15.539.413

Definition is obtained by adding further digits, or by using qualifying code symbols linked by colons or other auxiliaries. Thus the codes vary in format and the number of digits required to identify specific items such as construction materials or an engineering component is so great as to be unmanageable.

Opitz

Professor H. Opitz, in conjunction with some German machine tool manufacturers, in the 1960s designed a system relating to machine tool parts. The Opitz code is a feature code which brings together like items for production purposes, thereby facilitating planning, machine loading, plant layout and group technology. The code is not an identification number so a separate part numbering system must be used.

Multiclass

In the 1970s the Netherlands Metal Institute (TNO) developed a classification and coding system called MICLASS. The objective was to provide a scientific basis for selecting machine tools, based on a consideration of parts to be manufactured.

The system was later developed into MULTICLASS by the Organisation for Industrial Research and applied to a wider range of engineering design and manufacturing situations.

Brisch

In 1948 E. G. Brisch formed a limited company to sell the Brisch system of classification and coding to industry but particularly to engineering companies. This application of classification derives from the works of Linnaeus in the eighteenth century.

The Linnaean hierarchial and systematical taxonomy has a classification structure which is a useful means for branching and dividing populations into defined categories. It is used to solve logic relationships and to display the parameters of commonality (or exclusivity) which hallmark a sound classification (see Figure 14.3).

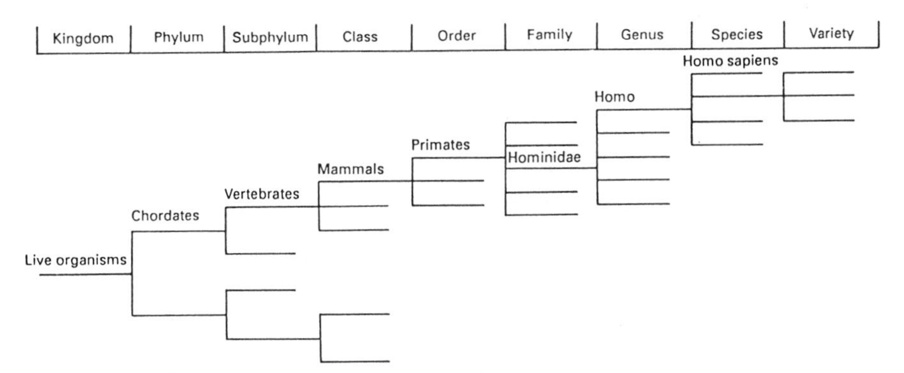

Figure 14.3 *Linnaean taxonomy of human beings: an example of the tree of Prophyrios*

Brisch developed a structure that is not fixed and rigid like the taxonomic structure but which is completely flexible so as to be able to cope with population distribution at the several levels of the classification (see Figure 14.4). Brisch classification is based on the four principles listed earlier in this chapter (all-embracing; mutually exclusive; based on permanent characteristics; user's point of view).

Brisch used integrated codes (monocodes) and non-integrated codes (polycodes). The ideal system would use monocodes for designed parts, operating with a polycode for manufacturing information requirements. Figure 14.5 shows such a system applied to a component drawing.

Group technology and computer aided processes

There is ample evidence that attempts to introduce group technology and computer aided processes have failed because in the rush to get on to the 'bandwagon', company managers did not examine fundamentals deeply enough. One fundamental which was overlooked was identification of data, primary materials, commercially available items, design parts, subassemblies and even products. Result – poor retrieval and CAD becoming known as computer aided disaster!

It is, therefore, of importance to include here some information and views which will convince the reader to examine further the statement that:

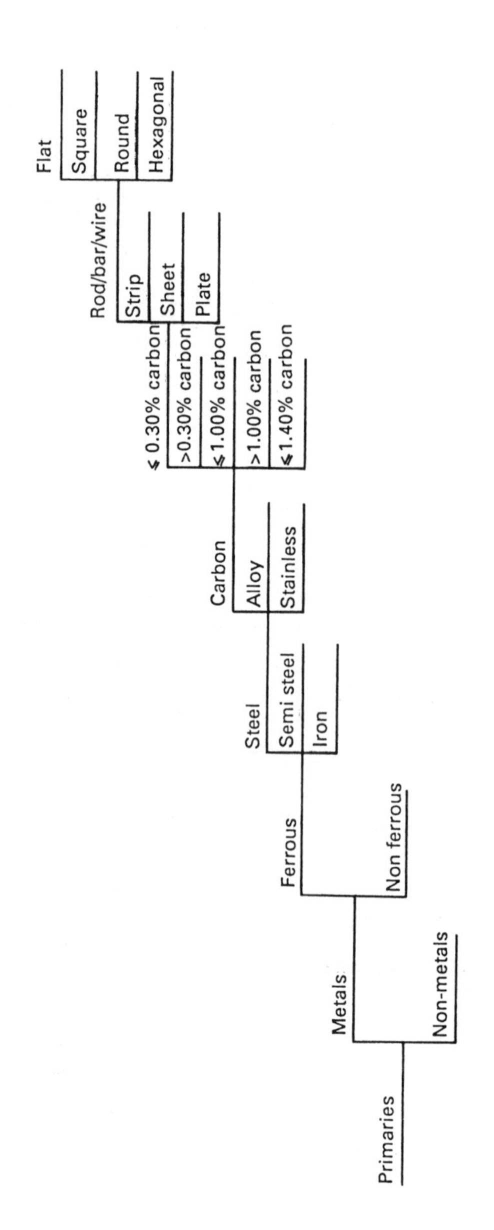

Figure 14.4 *Taxonomic structure for the classification of primary materials*

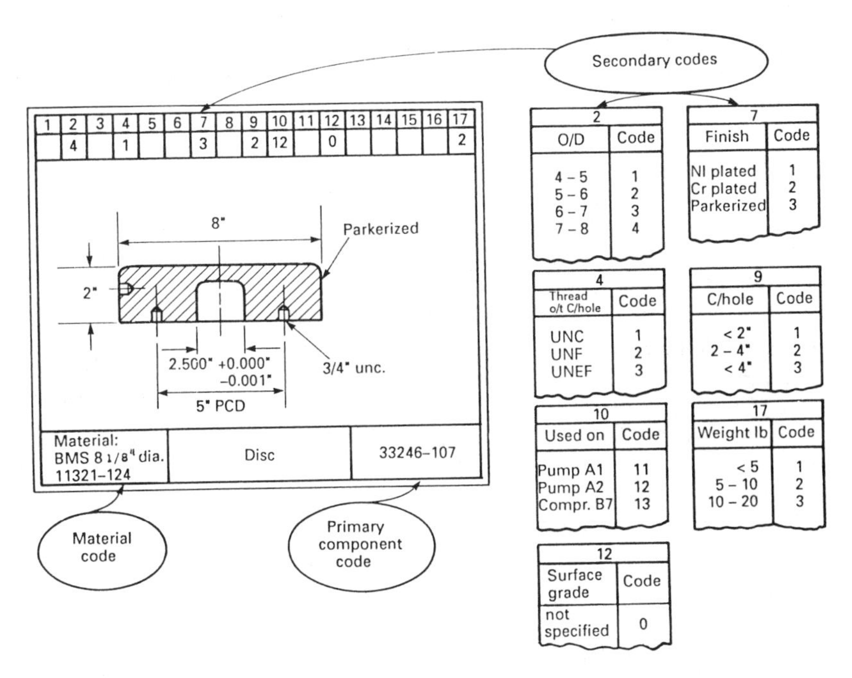

Figure 14.5 *Primary and secondary coding for a component. The left-hand side of this figure is a component drawing which carries the primary code that identifies the component. Specific features of the component are given secondary codes, taken from a code manual (illustrated in the right-hand side of the figure)*

Successful applications in group technology and computer aided processes are not achievable without the prerequisite of a logical classification and coding system.

Group technology

Group technology (GT) in the 1960s and 1970s, as known in the UK, evolved through the availability of a system for the easier retrieval of design based information. It was pioneered by the late Joseph Gombinski who involved a number of companies and academics in the initial development.

Gombinski defined GT as:

> A production method that involves the machining of parts in 'families'. Family Type A consisting of parts similar in shape, having all or most machining operations in common. Family Type B consisting of parts which are apparently dissimilar but are related in that they have one or more machining operations in common.

Finding parts for GT is, of course, a task for classification. Hence the need to mention GT briefly in this chapter. Classification is a prerequisite for GT as it is for all computer aided processes.

Computer aided processes

After years of experience it is becoming evident from press and journal articles that as for GT, logical classification is a prerequisite for successful applications in computer aided design (CAD), computer aided manufacture (CAM), computer aided engineering (CAE), computer integrated manufacture (CIM) and computer aided process planning (CAPP). Blore (1984) stated:

> There are five main parts to any computer aided process planning system, as follows:
> 1 Coding and classification.
> 2 Data base creation and maintenance.
> 3 Logic processor.
> 4 Documentation production.
> 5 File maintenance.
>
> *Coding and classification*
> The use of computers within a planning system forces companies to be logical and to think about standardizing their approach to manufacture, this is because the computer has to act upon certain logical rules. It is therefore necesssary to have some means of identification of individual parts, similar parts, individual routings and similar routings within the numbering system that identifies these features.
>
> Many companies have their own numbering systems. Some are sequential, some alphanumeric, but most suffer from the problem that when they were designed they were not structured for use within a computer. They rely in the most part upon the ability of people to remember what has gone before. This had resulted in a great proliferation of parts within companies and also a greater than necessary proliferation of methods.

It is obviously a prerequisite for a successful approach to GT CAD, CAM, CAPP, CIM for each and every company to ask whether it has a company numbering system which does not cause proliferation of parts, materials, tools and methods. Elimination of the unnecessary and establishment of a tool with which to control variety for the future is a must, otherwise the result will be the computerization of existing chaos and of the unknown.

Classification and coding is right at the heart of control of engineering documentation. The following statement shows that which is required to be defined and installed before examining the requirements for computer aided design and computer aided manufacture.

1 Engineering policies.

2 Engineering standards.
3 Engineering procedures.
4 Drawing content definition
5 Piece part drawings.
6 Revision level.
7 Company classification and coding system for:
   (a) Designed parts.
   (b) Primary materials.
   (c) Commercially available components.
   (d) Products
   (e) Tools.
8 Rules of interchangeability.
9 Product structures.
10 Bill of materials system.
11 Configuration management.
12 Engineering change control.
13 Design retrieval.
14 CAD requirements.
15 CAM requirements.
16 Tool design with CAD/CAM.

System design

Examination of the fundamental concepts underlying various systems leads to the recognition of the following basic models:

1 *Product orientated* whereby the components are grouped by products for which they were originally designed.
2 *Function orientated* whereby the components are grouped by their names, indicative of their function, e.g. impellers, spur gears.
3 *Design orientated* whereby the components are grouped into families by the similarity of design manifest in their overall shape.
4 *Production orientated* where the components are grouped into families requiring closely similar or identical technological processes.
5 *Design and production orientated* (D/PO) which aim at satisfying the requirements of designers and production engineering by a single system.

System 1 is considered obsolete and of limited value. Systems 2, 3 and 4 are single purpose, while system 5 is dual purpose.

The primary object of a design orientated (DO) system is the creation of families of similar design or function and the efficiency of their retrieval for management purposes. In a production orientated (PO) system, just the reverse is true. Similarity of component designs may occur in some of the families but this is not a required condition. A PO system is satisfactory when it is capable of producing families requiring identical, or similar technological processes. Similarity of processes in a family created by a DO system and similarity of designs in a PO system are incidental. They are not the criteria by which each system should be evaluated. Not so for dual purpose D/PO systems which claim to satisfy equally the requirements of design and production. Such systems must be judged on the degree of satisfaction they provide on both counts.

In a DO system one finds in a component family of similar shape, components manufactured by different processes. In a historical file this is acceptable because it permits comparisons between alternative processes. But this is not acceptable in a PO system where a family must consist of components with similar technology.

Classification is a service and as such must provide an economical return. In a DO system this is achieved primarily through the retrieval of existing designs and their incorporation in new products, and also by providing the basis for standardization. Whenever comparisons need to be made with a view to elimination of past errors and anomalies, the reduction of unnecessary variety, and the prevention of its accrual – or in other words, when simplification and standardization are to be promoted – it is necessary to retrieve not an isolated component but a family of components of similar design.

In a DO system the similarity is that of component shape taken 'as a whole', as an entity. There may be variations between members of a family but the components retrieved, to be of interest to a designer, or to a standards engineer, must be all of the same shape, allowing only for such variations of shape features as do not affect the overall shape. Major shape families are subdivided using shape features either singularly, or in combinations and arranging them in order of their relative importance or incidence. Such classifications have therefore a hierarchial structure. The finite shape of a component and its selected features are integrated in a finite code – a MONOCODE – of fixed and uniform notation in which each digit qualifies the preceding one and is qualified by the successive digit. In this way, a true shape code represents a component 'as a whole' and provides its unique identification.

The tussle between production based systems and design based systems, together with the need for retrieval to meet the needs of CAD/CAM highlights the old argument of shape versus function. The user needs both. An overall classification will have a shape basis which will be more than useful to production as a whole. It will have other items based on function that will be useful both to production and the purchasing department. In fact, it is not uncommon to find classifications devoted more to function than to shape. What matters, of course, is the logic behind the classification.

Further reading and references

Blore, David, 'Computer Aided Process Planning', *The Chartered Mechanical Engineer*, May 1984, Institution of Mechanical Engineers, pp. 31–4.

Fox, Roland P., *The Classification and Coding of Accounting Information*, May, 1986 The Chartered Institute of Management Accountants.

Hyde, William F., *Classification, Coding and Database Standardization*, Marcel Dekker, 1981.

15

Distribution, storage and retrieval of engineering documents

Dennis Lock

The information amassed by any engineering company deserves to be treated as a valuable company asset. It has to be protected against loss, distributed according to the organization's requirements and stored in such a way that any given document can be identified and retrieved whenever it is needed. This chapter describes some of the established office routines which engineering companies use to handle drawings, correspondence and other engineering related documents.

Other aspects of this complex subject are dealt with elsewhere in this book. The microfilm section in Chapter 13 is particularly relevant to the efficient filing, storage and retrieval of engineering documents, and should be read in conjunction with this chapter.

Classification

The classification of engineering information into specific categories is necessary in order to be able to establish and implement formal document handling procedures. This classification can be attempted in several ways. One obvious approach is to identify all the different document types (so that engineering drawings and letter correspondence would form two of the many possible separate categories). Within each category there are usually several subdivisions: letters, for example, can be classified as incoming or outgoing, and drawings can be distinguished according to whether they are in-house, vendors' drawings and clients' drawings. Most information can also be classified according to the job (or project or contract), which probably will have a unique identifying number in addition to a name.

Information for projects of any significant size will almost certainly have to be classified further into subprojects or into technical areas of the work, with yet more breakdown necessary to split out the various engineering disciplines (such as civil, structural, mechanical, electrical, piping and process control engineering).

On top of all these ways of classifying information there is another important division. This distinguishes between information needed for work in progress and the information remaining on file after all work has been completed.

While work is in progress, whether in the design or in the fulfilment stage, the engineering documents have great intrinsic value, gauged against the cost and time that would be needed to replace them should they be lost or destroyed. There is also a probability that such loss of information would bring about serious consequential financial loss, owing to the delays and upsets in production departments, and through contractual liabilities with customers.

Many engineering documents will have considerable residual value long after the production or construction work has been finished. All engineers are familiar with the savings that can be achieved on a new design project if a suitable previous design can be identified and re-used or adapted. It may be prudent – indeed obligatory – to retain certain documents in order to be able to provide continuing after-sales service to a customer, to be in a position to modify or extend an earlier project for a client, or to provide evidence against possible claims arising from accidents, breakdowns or contractual disputes.

Naturally some documents are more valuable than others. For example, a file of telexes might include information of only transient value, such as details of flight and hotel reservations, as well as technical and commercial data of more lasting significance. Original drawings, contract documents, test certificates, and the like have obvious claim for safe retention for at least several years. Defining the classes of documents, establishing policies for their distribution, filing, safe keeping and ultimate disposal, and then ensuring that the correct procedures are implemented is one of the most difficult, least glamorous, least liked, but most necessary of all management chores. The most important classes of documents are summarized in Figure 15.1.

Registry organization

Most engineers would accept the wisdom of keeping in-house drawings in a central, secure registry. By storing the original drawings in fireproof cabinets, or by keeping microfilmed copies in some other safe location, there is reasonable protection against loss. The discipline that can be exercised by competent registry staff helps to ensure that new drawings will be registered and numbered according to the appropriate procedures. A good registry also enables any drawings to be found quickly in the files.

It is only when the discussion shifts from in-house drawings to all the other classes of documents that controversy arises, and then there are usually fierce

Document type

Project authorization
Works orders
Project specifications
Product specifications
Drawing schedules
In-house drawings
Customers's drawings
Vendors' drawings
Calculations
General design specifications
Engineering standards
General company procedures
Special project procedures
Purchase schedules
Bills of material/parts lists
Material take-offs
Vendors' literature
Purchase specifications
Purchase enquiries
Purchase requisitions
Purchase orders
Inspection and expediting reports
Test or release certificates
Vendors' invoices
Operating and maintenance
  instructions
Letters
Document transmittal letters
Cables and telexes
Facsimile messages
Memoranda
Progress reports
Cost reports
Bar charts and networks
Cost estimates and budgets
Technical and feasibility reports
Contracts and subcontracts
Variation orders
Engineering change notices and
  modifications
Project closure notices

Document status
(especially applicable to drawings and specifications)

Proposal
Preliminary and unchecked
Issued with information missing
  (released with holds)
Issued (released for construction)
Revised and re-issued
Final version (as-built)

Original or copy?

Subject classification by job or physical item

Main job or project
Important division of job or project
Major components
Other

Classification by specialist work interest

Civil engineering
Structural engineering
Chemical engineering
Mechanical engineering
Piping and fluids engineering
Electrical engineering
Instrumentation and process control
Planning
Cost control
Cost estimating
Personnel administration
Accounting
Purchasing
Materials handling
Production
Construction
Inspection
Testing
Customer services
Maintenance
Security
etc.

Classification by origin

In-house (departments or individuals)
Installation or construction sites
Subcontractors
Head office (if remote)
Associate companies
Joint venture or other partners
Clients or customers
Vendors
Agents
Insurers and brokers
Trade associations
Professional and legal advisers
Finance houses and banks
Government departments
etc.

Classification according to addressee

In-house (departments or individuals)
Installation or construction sites
Subcontractors
Head office (if remote)
Associate companies
Joint venture or other partners
Clients or customers
Vendors
Agents
Insurers and brokers
Trade associations
Professional and legal advisers
Finance houses and banks
Government departments
etc.

Main file locations

Central drawings registry
Central correspondence registry
Library
On-site archives
Off-site archives
Department files
  (eg. personnel, accounts)

Secondary file locations
(Duplicating main files)

Individual
Departmental
Off-site security back-up

Document format

Hard copy (on sheet material such as paper, sometimes subclassified according to sheet size or roll width)
Digital:
- Magnetic tape
- Floppy disk
- Hard disk
- Optical disk

Microform:
- Fiche
- 16 mm jacket fiche
- 16 mm roll
- 16 mm cartridge
- 35 mm roll
- 35 mm aperture card
- 70 mm roll
- 70 mm aperture card

Classification by age

Current relevance
Recent interest
Within or outside legal retention limit
Archive worthy
Obsolete

Security classification

Top secret
Secret
Confidential
Restricted
Staff/personnel department confidential
Private and personal
Company proprietary information
Unclassified

Figure 15.1 *Document classification. This chart shows some of the many ways in which documents can be classified for the purposes of managing engineering information files*

arguments over the merits or otherwise of keeping correspondence (for example) in a central registry. And what about suppliers' catalogues: should these be kept in one central library, or should each engineer be allowed to maintain his/her own mini-library (possibly occupying at least one large bookcase)?

There is probably no universal solution to these problems, since they depend to some extent on the way in which a company is organized, and on the geographical arrangement of its departments (it would be unreasonable and inefficient to expect engineers to use a central facility that is too remote for convenient access).

One solution which can work for long-term projects is to allow engineering departments to hold their own local files, which duplicate those held securely in a central registry. These duplicate files must be properly maintained by a clerk or (if persuadable) a secretary. The obvious disadvantage in this system is that extra space is needed, and additional filing work is generated. The possible advantages are:

1 Time saved for engineering staff.
2 Reduced risk of filing errors, because the technical staff are better able to direct documents into their correct subject files than non-technical central registry clerks.
3 More complete files, since not all documents circulating internally will reach the central registry.

Developing this idea of duplicate filing one stage further, it makes sense to arrange that registry staff file documents, as far as possible, by serial numbers or dates. These are unambiguous and need no subjective decision. Technical staff, on the other hand, can arrange their local files according to a predetermined subject breakdown, which sometimes will demand a degree of technical judgement in allocating documents to specific files. When the job or project in hand is finished, the duplicate files can be sent to the registry to complement their sequential files. Both sets can be microfilmed and archived. This allows documents to be found subsequently according to serial number or date or, if these details are not known, by subject in the second file set. Each file set must be separately and clearly indexed, both as hard copy and after microfilming.

Any concession to allow a project department to keep duplicate files must be coupled with efficient filing discipline and clear filing procedures. Project files must not be retained in the engineering department beyond the end of a project if the provisions of Lock's law are to be avoided (Lock's law states that no engineering company can have infinite life because it must eventually choke to death on its own ever expanding files).

Figure 15.2 sets out some classes of documents likely to be encountered by engineers and engineering managers, and gives some suggestions for their filing and subsequent disposal. These are the sort of recommendations which could be built into company or project procedures. They are given for illustrative purposes only, and each company would obviously have to develop its own rules according to its management policies, organization and the nature of its work.

Registry staffing

For every decision to set up a registry there must be an acceptance that staff have to be provided. This usually means at least one person in each registry. When the need for providing cover during holidays and other periods of absence is taken into account, then it may be found that more than one person per registry is needed, even for quite low volumes of document handling. Although registry staff need not be highly qualified, they are usually classed as indirect workers, which means that their time cannot be charged out to customers, and must be allocated to overhead costs. We have already said that the proper registration and storage of documents is one of the less exciting and least liked management tasks. Now, to rub salt in the wound, it has to be acknowledged that it can also be expensive.

It would be wrong to take the easy option and deny the need to provide competent registry staff in the right numbers and of the right calibre. Apparent cost savings through such short-sighted neglect only lead to increased costs elsewhere, in the time spent by professional staff in searching (perhaps in vain) for vital information and in reduced ability to find appropriate previous design information for re-use on similar new work. On the other hand, it is necessary to strive for efficiency, and to keep indirect registry staff numbers to the essential minimum. This can sometimes be achieved by locating various registries together, so that (for example) the drawings clerk can double for correspondence clerk, and vice versa. This may call for a radical rethinking of entrenched roles, but it has been proved to work in practice. Indeed, if the photocopying and plan printing facilities, post room and messenger base can all be integrated in one central unit with the various registries, staff can be rotated between the various jobs to achieve maximum flexibility and emergency cover without the need to employ excess staff (Figure 15.3).

Registry clerks typically need no academic qualification other than a reasonable level of general education, but they do need to be methodical, fairly intelligent, and able to write words and figures clearly and accurately. They may also be called upon to exhibit firmness with tact, when faced with the need to refuse requests from senior staff that would contravene company procedures, or in calling back documents long overdue for return. Increasingly, registry clerks will be expected to work with computer terminals. It is sometimes thought that older men and women (perhaps between the ages of 40

	Internal documents		**Vendors and manufacturers (for specific enquiries and orders)**		**External purchasing department and purchasing agents**
1	Project authorization	7	Incoming letters	7	Incoming letters
1	Works orders	7	Outgoing letters	7	Outgoing letters
2	Project specifications	13	Incoming DT letters	13	Incoming DT letters
3	Drawing schedules*	13	Outgoing DT letters	13	Outgoing DT letters
4	Drawings*	7	Incoming telexes	7	Incoming telexes
5	Calculations	7	Outgoing telexes	7	Outgoing telexes
6	General design specs	14	Drawings	7	Purchase orders
6	General company procedures	14	Test or release certificates	7	Inspection and expediting reports
2	Special project procedures	15	Invoices		
3	Purchase schedules	16	Operating and maintenance instructions		
4	Bills of materials/ part lists				
7	Purchase specifications				
7	Purchase enquiries				
7	Purchase requisitions				
8	Internal memoranda				
9	Progress reports				
9	Cost reports				
8	Bar charts and networks				
2	Cost estimates and budgets				
5	Technical reports				
3	Engineering change notices				
1	Project closure notice (completion or cancellation)				

Clients and customers

10	Contracts
10	Variation orders
11	Incoming letters
12	Outgoing letters
13	Incoming DT letters†
13	Outgoing DT letters
12	Incoming telexes‡
12	Outgoing telexes‡
3	Drawings

Contractors and subcontractors

10	Contracts
10	Variation orders
11	Incoming letters
12	Outgoing letters
13	Incoming DT letters
13	Outgoing DT letters
12	Incoming telexes
12	Outgoing telexes
15	Invoices

Remote site office

12	Incoming telexes
12	Outgoing telexes
13	Incoming DT letters
13	Outgoing DT letters

Partners, agents

11	Incoming letters
12	Outgoing letters
13	Incoming DT letters
13	Outgoing DT letters
12	Incoming telexes
12	Outgoing telexes

Notes
Recommendations apply equally to first issues and all subsequent amendments

Special arrangements may have to be made for security classified documents and for those which deal with staff matters (which often occur in correspondence with remote site management). When a document cannot be included in the main central files for these reasons, a dummy document can be substituted stating where the actual document is held. This aids subsequent retrieval, and avoids spurious indications of a missing document in sequentially numbered correspondence files.

*The numbering system for drawings and other technical documents should be designed to identify the main job or project, the subproject and the specialist engineering discipline. Filing by document serial number sequence should then produce logical grouping. Drawing sizes should be standardized as far as possible to facilitate a common filing arrangement.

†DT is a frequently used abbreviation for document transmittal letter

‡Electronic mail and facsimile transmissions can often be treated in the same way as telexes

Class	*Official file location (with copies used for daily work in company departments)*	*Filing sequence*	*How long in active file after job/ project completion?*	*Any special safekeeping arrangements?*	*Disposal after active phase*
1	Contracts department	Project or job number	1 year		Microfilm batches and archive. Destroy originals.
2	Originating department	Proposal or job number	1 year		Microfilm and archive. Destroy originals.
3	Drawings registry	Serial number	1 year		Microfilm for archive and reference.
4	Drawings registry	Serial number	5 years	Film new drawings in frequent batches and keep rollfilm in firesafe or offsite	Microfilm and archive as 35mm aperture cards, with extra aperture cards of layouts in engineering offices as retained engineering design reference satellite files
5	Central files	Serial number	1 year		Microfilm and archive. Destroy originals.
6	Originating department	Volume number			Destroy and replace as revised.
7	Purchasing department, if in same building: otherwise in central files.	Serial number, so that a complete dossier builds up for each purchase order	2 years		Microfilm and archive with index
8	Originator	Individual preference	6 months		Review and clear out ruthlessly and frequently. Rarely worth archiving.
9	Project manager	Date	1 month		Microfilm and archive the final report only. Destroy the remainder.

Class	Official file location (with copies used for daily work in company departments)	Filing sequence	How long in active file after job/project completion?	Any special safekeeping arrangements?	Disposal after active phase
10	Contracts or legal department	Firm date	6 years	Firesafe	Contracts manager to decide in each case.
11	Central file (copy to addressee)	Firm/serial number or project number/date	2 years		Microfilm and archive. Destroy originals.
12	Copy in central files	Firm/serial number or project number/date	2 years		Microfilm and archive. Destroy originals.
13	Central files	Firm/serial number	3 months		Destroy
14	Drawings registry (usually as prints)	Allocate new serial numbers within each order or requisition number	1 year		35 mm aperture cards in archives with index
15	Accounts department	Firm/purchase order	1 year		Film and archive unless unpaid/disputed.
16	Central files or drawings registry	Requisition or order number	5 years		Destroy (it might be advisable to let the client or customer know first)

Figure 15.2 *Document filing guidelines. The tables on pp 294–5 list some of the more important types of document likely to be encountered by engineers and engineering managers. The table on p 296 and this page sets out possible ground rules for their filing and subsequent disposal. Although every company will have to make its own rules and issue its own procedures, the formal approach shown in these tables is one universal way of approaching this complex problem*

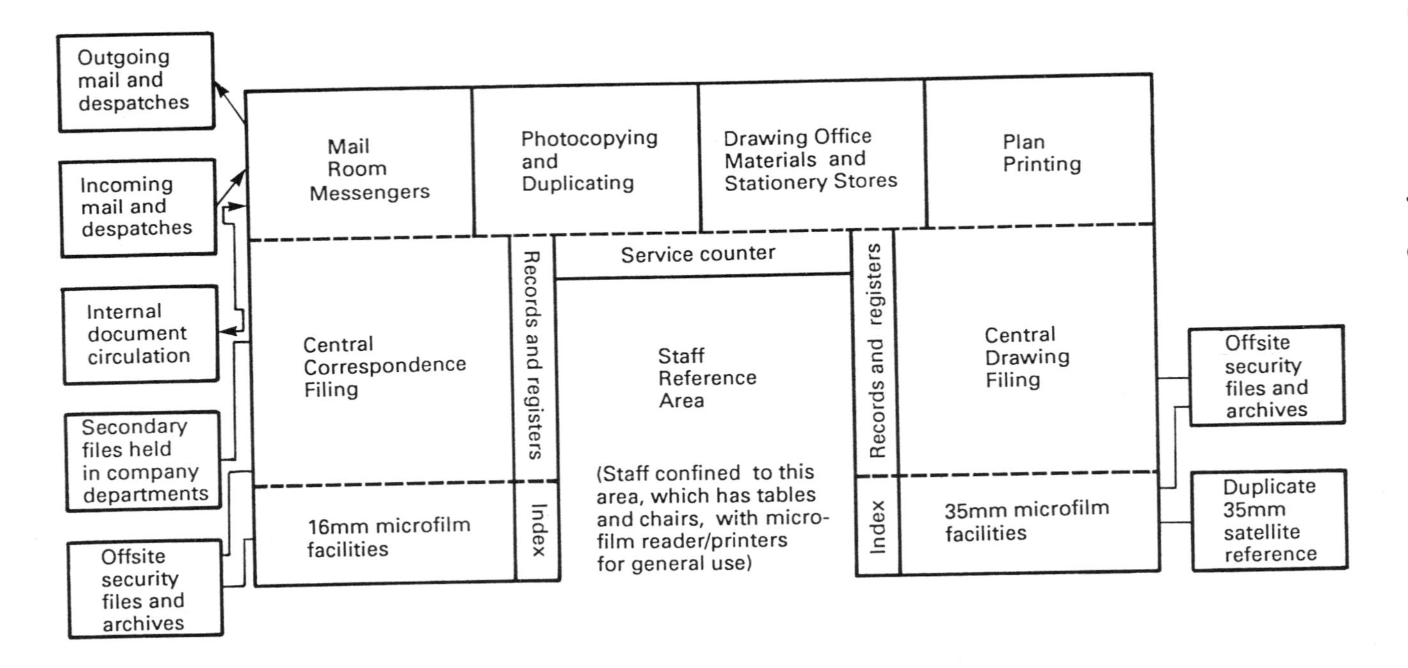

Figure 15.3 *Grouped central office services. If it is possible to group a number of office functions together within one enclosure, and under one supervisor or manager, then increased efficiency can be achieved in a number of ways. There is no need for drawings to leave the security of the unit for printing. Printing and photocopying facilities are on hand where they are most needed, and where they can be kept under control. Mail and despatches received via airfreight or courier can be handed directly to the registrar from the mail room, so that (where appropriate) the originals can be filed without their having to leave the secure area (copies are distributed). It should be possible to train the clerks and printers to cover each other's jobs, to a large extent. It is not then necessary to engage temporary staff on every occasion when someone falls ill or is on holiday.*

and 60) can bring a more responsible and conscientious approach to registry work than younger people (who may have more drive and ambition) but there are exceptions to every rule and there are extreme exceptions to this one at both ends of the working age spectrum.

Clerks with long company service build up valuable experience which enables them to remember and track down obscure items of information from the distant past. Such people are often regarded as difficult to replace. Surprisingly, given good training and clear registry procedures, a suitable new registry recruit can become reasonably effective in a few weeks. However, one aim in appointing new registry clerks should be to look for evidence of stability in employment. Longer-term employees are more valuable through their acquired knowledge, whereas a high staff turnover will produce annoyance and frustration in those experienced registry staff who have the task of training a succession of new recruits, especially when some of these prove themselves to be less than dedicated, leaving a trail of errors behind them that require tedious checking and correction.

Employment conditions and levels of payment should be kept under review, and maintained at standards sufficiently high to satisfy and retain competent registry staff. It also does no harm, and costs nothing, to praise good routine work from time to time. Junior staff will appreciate and respond to management encouragement, rather than having to regard management contact as something that only happens on those unpleasant occasions when criticism is necessary.

Registry practice

Whatever documents are being handled, there is an established set of rules that should govern the supervision and day-to-day conduct of any document registry. These are summarized below, and treated with greater detail for the various classes of documents in later sections of this chapter.

1 Management policy must declare which documents qualify for safekeeping in any registry. For engineering purposes, the most likely relevant documents are indicated in Figure 15.1. It should be noted that this policy must cover documents from every possible source, whether in-house or external. For clarity in this chapter, all documents deemed to require registry care will be referred to as relevant documents.
2 Management should ideally declare the company's policy on the retention period for every class of document, and describe the method of disposal when the initial retention period has elapsed. Such instructions could include, for example, shredding, throw away as normal industrial waste, microfilm and then destroy the originals, archive the originals, or some other combination of these options. Companies engaged in engineering

projects can approach this problem by a formal project close-down procedure, which makes provision for reviewing all project documents and specifying their retention of disposal.

3 Procedures must be established to ensure that all relevant documents are passed as soon as possible to the appropriate registry. This rule should be familiar to all engineers for original drawings (which must usually be taken to the drawings registry as soon as they have been checked and kept there unless required for printing or for modification). The rule applies equally, however, to any relevant document. Thus, for example, it may be necessary to pass certain incoming original letters direct to a correspondence registry, and to send the addressee a photocopy.

4 The receipt of certain documents must be recorded in a register, and the documents serial numbered.

5 Documents must be filed in suitable equipment, with due regard to security from gradual deterioration, or from loss by fire, water, theft or unauthorized borrowing.

6 Document loans must be controlled by a suitable procedure which obtains a signature from the borrower, and which signals any overdue loan to the clerks. If possible, photocopies or plan prints should be issued, rather than lending the original documents. Consideration can be given to setting aside a small reference area in each registry, library-style, where the documents can be perused in comfort at a desk or table, avoiding the need for them to be taken away.

7 Wherever possible, registry documents and files should be clearly identified as such, and with the serial numbers displayed. It may be necessary to use a rubber stamp with red ink on some documents, so that they are clearly identifiable as master documents and not (black and white) copies.

8 All files should be indexed to allow subsequent document retrieval. The index itself must be secured against loss.

9 The registry must be housed within a lockable enclosure, and non-registry staff should be excluded, being served over a counter or in a reception or reference area.

Filing

Provided that clear instructions exist (preferably in the shape of formal procedures), and that documents bear subject names and serial or job numbers, the question of where to file any particular item should be cut and dried. This should certainly be true with engineering drawings, for example, whose serial numbers provide both a unique identity and a logical filing sequence.

Filing decisions may be more difficult with other documents, such as reports, correspondence, contracts, and so on. Wherever ambiguity exists, it may not be reasonable to expect the non-technical registry clerks to allocate such documents to particular files. In these cases, it is sometimes best to arrange that a suitable engineer or project administrator writes the appropriate filing code on each document. For engineering project work it is customary to define a set of filing codes within the specific project procedures, and the person charged with responsibility for allocating documents to the various files would obviously have to follow the list laid down.

Once filed, it has to be possible to find the documents again. Although this may seem to be an obvious statement, it may not be so easy in practice to achieve for documents other than serially numbered drawings.

One idea is to split up the total filing area according to projects, jobs, subjects or whatever criterion is adopted to determine filing. That approach is fine and simple when the system is first set up, but the difficulties come when new work is introduced, and the shelves have to be rearranged as older documents are removed and replaced by the new. Unless the company is prepared to arrange time-consuming and unfunded large scale rearrangements of its files every few months, the original concept of physical grouping will become lost, since the new jobs will not correspond in filing bulk with the old and the system will be too inflexible to accommodate the changing requirements.

Another, more flexible, approach is to regard the location and retrieval of files in the same way as a company would deal with any other stored materials. The registry is treated as a works stores, and the registers become the stock lists. Each filing location is given an address, which defines the filing cabinet or rack, and the shelf level. Alphanumeric codes are most convenient for this application, and an example might be that racks are designated A, B, C, D, etc., with the shelves numbered 1, 2, 3, 4, etc., from the ground upwards (not top down, because some cabinets will have more shelves than others). Now, provided that a suitable index exists which links the various documents with their filed locations, we know where to look. For example, location E4 indicates clearly that the document needed is on the fourth shelf up in the cabinet labelled 'E'. Although there may be 10,000 documents on that shelf, at least the search has been narrowed down. The following section takes this subject further, showing how careful indexing can help to pinpoint the exact location.

Indexing

Bearing in mind that the usual objective of careful document registration and filing is to keep documents ready for possible use at a later date it is prudent to list some of the means by which such individual documents might be located in the registers or index. This problem has to be approached from the viewpoint of the future enquirer, who might have only a vague idea of how to identify the documents and may indeed not know if a document even exists.

One or more of the following factors might be the basis for a document search:

1 The type of document (was it a letter, telex, report, etc?).
2 A unique serial number (particularly for drawings, requisitions, specifications and for some project letters, telexes and document transmittal letters).
3 A job or project number (often built into the serial number).
4 A document title (perhaps for a report).
5 A job or project title.
6 A further title, identifying a part of a job or a subproject.
7 Some non-specific idea of the document's subject (for example, correspondence with a supplier (name forgotten) for flameproof DC motors used on some past job – or on several past jobs.
8 The source of the document (the name of a customer, supplier, in-house department, individual originator, or other.
9 The date (which might be an exact calendar date or some vague idea of a period covering two or three years).

In the past, any hope of tracing documents from some of these hazy ideas would only have been possible through the fortuitous recollections of long-established staff. Some attempts at keeping subject registers were extremely helpful, but tedious when it came to searching. Visible-edge index cards have proved useful, but these are inflexible to change and very difficult to photocopy for the purpose of providing a security copy (there is no point in having security copies of thousands of documents if a security copy of their index is not kept with them). Here is one case where a sensibly designed computer database can provide most of the answers. For this application a library system (such as CARDBOX) can be used, chosen according to the size of the database envisaged, the number of search criteria needed, and the computer hardware available.

The method requires that new files in a registry ('acquisitions' is the approved jargon word) are entered in the database, which replaces the old handwritten registers. Existing files must also be transferred into the new system and, since this is likely to involve considerable time and effort, the job is started from the most recent files, working backwards into history until it is deemed futile to proceed further.

For each document ('record') as much information as possible is entered. This should contain enough information to satisfy the nine factors listed above, together with the file location code. Also to be included are certain 'keywords' which occur in the title or subject of the document. For example, flameproof, DC and motor might be three of the keywords. Only keywords from a pre-established list (the keyword menu) may be used although, subject to the software limitations, new keywords can be added to the menu as the need arises.

The register has now been transformed into a powerful index. Suppose that a letter is required, and the subject, source, and approximate date are known. The computer is fed with the appropriate subject keywords and instructed to print or display all letter records which carry the relevant keywords, and which originated within the specified range of dates.

The system has to be designed carefully, and it must be able to cope when files are moved, microfilmed, archived or otherwise changed. Thus it must be possible to search the database and enter the new location codes for complete batches of records in one operation. For example, if a file of outgoing project letters is to be archived, it should be possible to instruct the computer to find all the relevant records and change their location code in one operation (by searching the database for all outgoing letters bearing project number ABC1234, with old location code F6, and with their dates of origin lying in the range 1 January 1984 to 31 December 1986).

The usual discipline for computer data would apply to ensure that a copy of the index was always kept in a secure place (by storing a regularly updated back-up disk in a suitable fire safe).

Registry equipment and facilities

Perhaps a convenient starting point for considering the facilities needed for a documents registry should be the working conditions and environment for its inmates. Each clerk should have a comfortable chair with sufficient desk area, plus lockable storage for personal clothing and belongings. Ventilation is often a problem in registries with airflow restricted by filing equipment. Lighting should also be adequate and free from glare, not only at desks and the service counter, but also in the aisles between files. Apart from having to satisfy the minimum provisions of the Offices, Shops and Railway Premises Act, staff welfare considerations are not entirely altruistic. Clerical accuracy and good working relationships with professional staff are important, and both will suffer if the registry staff are subjected to avoidable environmental stress.

Security considerations dictate that non-registry staff should be excluded from the registry, and all access to files must be under registry supervision. The most sensible and constructive way in which to arrange this is to provide a reference area where staff can browse through borrowed documents or microfilm in library fashion. This area should be equipped with tables and chairs, and with suitable microfilm reader/printers (most likely to be 16 mm and fiche for the central filing registry and 35 mm for the drawings registry).

Economy can be achieved by combining registries so that one reference area serves both (Figure 15.3).

In a busy engineering company, the central registry will receive many requests throughout a working day for the allocation of serial numbers for various documents. These requests, together with many general enquiries, imply heavy telephone usage. So the registries must be provided with an adequate number of telephone extensions.

Filing equipment for general document files should be chosen with a view to economy of floorspace. Lateral filing equipment is often ideal for this purpose, in tall cabinets. Remember, though, to include suitable short stepladders or other safe means for reaching the upper levels. Provided that the capital expenditure is considered justified, even better economy of space can be achieved using motorized rotary lateral files. These operate on the principle of a paternoster elevator, and can be extended into a ceiling void. The most well known brand is the Lektriever equipment supplied by Kardex Systems (UK) Ltd. Apart from the floorspace saved, these have the additional advantage of allowing the operating clerk to sit at a comfortable bench, to which level any shelf can be brought by pushing the appropriate selector button. Smaller equipment of a similar type can be used for large volumes of microfilmed files (all types of microfilm).

Much of the space in a filing registry is usually taken up not by files, but by the aisles between files. This problem can be overcome (expensively) by buying mobile files – filing equipment which is mounted on rails. The cabinets are closed up together with only one aisle, shared between them. By trundling the cabinets about on their rails, the aisle can be opened up between any two rows of cabinets. This arrangement is a great space saver, but floor loading is considerable, and must be checked first against the building's specification.

File indexes can be arranged on visible edge index cards in trays but more flexibility will be achieved if a small personal computer system is installed.

Microfilm viewing, printing and filing equipment is essential in modern efficient registries. Apart from the enormous space saving possibilities, there are modern systems which allow for very precise indexing, so that individual documents can be located from large files in a matter of a few seconds.

For the purposes of microfilming documents, most engineering offices are within reach of at least one competent microfilming bureau. Consideration can also be given to setting up filming facilities in-house (within the registries) to avoid having to send documents away, and to achieve a faster turnround. In-house equipment might include only the cameras (with films sent away for processing) or a complete installation including automatic film processing. Care must be taken to choose suitable camera and processing equipment for the particular application, and special attention must be given to quality control of lighting, exposure and processing.

Filing cabinets for drawings come in a variety of shapes and sizes. It is probably wise to avoid those with horizontal drawers, since it is not easy to extract or to refile individual drawings when the drawers are fairly full, and precious originals are likely to become worn, torn or damaged. Vertical suspension files are to be preferred generally, and it is possible to have drawing sheets edge-punched (and fitted with self-adhesive reinforcing edging tape where necessary) suitable for suspension from the cabinet bars. With such a system, individual drawings can be found, withdrawn and replaced with comparative ease. It is also easier to mix sheet sizes in vertical drawing plan chests, which may be necessary to allow filing in an unbroken number series.

Drawings registries typically need large flat tables or benches on which to lay out drawings awaiting registration or other attention.

Photocopying or plan printing equipment should be available within the registry confines. This will allow the clerks to issue reference copies of letters or prints from original drawings, rather than having to release original documents against loan requests. Local copying and printing facilities also improve efficiency, cutting down the time taken in issuing prints for production or construction and, again, avoiding the need for original documents to leave the secure area of the registry.

Security of registry and archived documents

The first prerequisite for document security is to exclude unauthorized staff from the registry filing areas. Keen engineers, anxious to pursue their work, tend to regard document procedures as secondary in importance to immediate job needs, and there has always been a tendency for engineers, given half a chance, to borrow drawings or files without the knowledge of registry staff, especially during overtime working or at other times when the clerks cannot be found. Registries should be enclosed within high partitions, with the doors kept locked when no clerk is present.

One file can look very much like another, so it makes sense to make precious central registry files conspicuous as such. This is easily and cheaply achieved by sticking brightly printed labels on all registry file covers, bearing a legend in big red letters 'CENTRAL REGISTRY FILE'. The registry clerks stand a greater chance of recovering lost files if they stand out like sore thumbs on desk tops.

Some files may be sufficiently sensitive to be kept in individually locked cabinets. Where loss from fire or flooding would be disastrous, one sensible precaution is to microfilm the documents as soon as they are received into custody, and file the microfilm off site (on the assumption that any fatal blow will not fall in two places simultaneously).

Specialist companies exist which own secure vaults for the storage of archives, often in tunnels or caves. These places can be considered for safe storage of archives provided that the retrieval delays are acceptable. There is a danger when using such services that the rental payable on space occupied will increase steadily as archives moulder, forgotten, on the shelves long after they have ceased to be of any practical value or significance.

Wherever possible, requests from staff to borrow central files should be deflected into asking the people concerned to stay and view the files within the reference area provided, taking away photocopies of any relevant documents.

If files or drawings are borrowed, a loan ticket system should be operated, and the clerks must follow up overdue loans. It is sometimes possible to use 35 mm aperture cards as loan tickets for drawings. If an original drawing is borrowed for modification, and the registry holds a microfilm drawings file, then the microfilm is also withdrawn, signed by the borrower, and placed in a 'drawings on loan' tray. This system is especially valuable when production prints are issued from the microfilm: it prevents prints being issued for any drawing which has been removed for modification.

Drawings and other documents are possibly at greatest risk when they are sent to external bureaux for microfilming. Strict listing and checking of documents sent away for filming is essential, and both the returned microfilm and the returned documents must be checked for damage or missing information. Some companies take a security print from every drawing before sending the original to an external bureau. No document sent for filming should ever be destroyed until the processed microfilm has been received and verified as being complete and of acceptable quality.

Document distribution

When any engineering design work starts, it is sensible to give special attention to the distribution of the various documents that will be generated. In a small manufacturing company, where the project is aimed at the development of a new product, this may not be a difficult problem. In all probability there will be established procedures that dictate who gets what, and how many copies, and this arrangement will probably not vary from one job to the next. For projects which involve complex engineering teams, external organizations and major communication distances the problem may have to be solved afresh for each new project. It is then expedient to define the agreed distribution pattern as part of a written set of special project procedures. Since these complex projects offer the most difficult problems, much of the remainder of this chapter will deal with that type of project situation.

Distribution matrix

Anyone who has ever had to deal with document distribution systems for a large engineering project will know that one of the difficulties is knowing how to write down the agreed procedures in a way that can be interpreted easily by senders and recipients alike. A very useful tool in this respect is a document distribution matrix, a real-life example of which is shown in Figure 15.4. This version lists the various documents along the top of the form, grouped under their relevant project functions. Potential recipients are listed down the left-hand side. The number of copies of each item to be sent to each recipient is written in the appropriate intersection square.

Provided that the matrix is drawn sufficiently big, it is possible to include more information by coding in each square the form in which the documents are to be sent. Here are some suggestions, which can be used in combination:

2 Two ordinary photocopies (or paper prints of a drawing).
PR One paper reproducible print of a drawing (a paper submaster or translucent print may be more familiar terminology to some readers).
FR One polyester film reproducible print of a drawing (a more expensive but far more durable option to the paper reproducible print).
DA Microfilm (one diazo copy aperture card).
SA Microfilm (one silver original aperture card).
O The original document (for example, drawing to the drawings registry).

Some people like to see their names publicized by inclusion in distribution lists (it gives them a feeling of belonging and status). It is very easy to build up a monumental number of copies in order to satisfy their demands. The principle to be followed in such cases is to test each request by asking the pertinent question 'does he/she *need to know* or is it merely a case of *wants to know*?' It may be necessary to disappoint some people by turning down their requests (politely if possible). Special attention to restricting distribution will obviously be necessary in projects for defence and all other contracts which are similarly governed by the Official Secrets Act or other legislation.

So far we have assumed that every document will be in some tangible form but, if project communications include electronic data links, it may be possible for recipients to receive documents by interrogating their VDU terminals. Some degree of decision might still be necessary, since, quite apart from the provisions of the Official Secrets Act, it is likely that some documents (financial, commercial and staff especially) would be too confidential for all eyes, so that a system of passwords and access levels would have to be determined.

Despatch methods

Having decided who should get what, it may be necessary to give serious thought to the next question, which is 'how?' In other words, how is the external carriage of documents (outgoing and incoming) to be arranged. This

DOCUMENT DISTRIBUTION MATRIX

PROJECT Nº

Key

- * Originator
- O Original
- 1, 2 Etc. No. of copies
- R Reproducible
- A If applicable

General

		Project manual	Client correspondence	Contracts / agreements	Invoices, accounts	Variation order reg	Contract variations
Client	Project Manager						
Home office	Director of Project Engineering						
	Project Manager						
	Project Co-ordinator						
	Project Engineer						
	Commercial Manager						
	Chief Accountant						
	Mining						
	Minerals processing						
	Metallurgical engineering						
	Mechanical						
	Civil / structural						
	Fluids						
	Electrical						
	Process control						
	Planning						
	Estimating						
	Material take-off						
	Cost control						
	Correspondence registry						
	Drawings registry						
	Information						
Procurement	Purchasing						
	Inspection						
Site	Construction Manager						
	Field Cost Engineer						

Engineering

		Process flowsheets	Engineering flowsheets	Plant general arrgts	Project general specs	Hazard analyses	Calculations	Technical reports	Client drawings	Vendor drawings	Site drawings	Equipment / material lists	Spares lists	Operating manuals
Client	Project Manager													
Home office	Director of Project Engineering													
	Project Manager													
	Project Co-ordinator													
	Project Engineer													
	Commercial Manager													
	Chief Accountant													
	Mining													
	Minerals processing													
	Metallurgical engineering													
	Mechanical													
	Civil / structural													
	Fluids													
	Electrical													
	Process control													
	Planning													
	Estimating													
	Material take-off													
	Cost control													
	Correspondence registry													
	Drawings registry													
	Information													
Procurement	Purchasing													
	Inspection													
Site	Construction Manager													
	Field Cost Engineer													

Project control

		Monthly progress reports	Project schedules	Eng progress reports	Drg / spec status reports	Project estimates	Cost control reports	Bills of quantity	Equipment list
Client	Project Manager								
Home office	Director of Project Engineering								
	Project Manager								
	Project Co-ordinator								
	Project Engineer								
	Commercial Manager								
	Chief Accountant								
	Mining								
	Minerals processing								
	Metallurgical engineering								
	Mechanical								
	Civil / structural								
	Fluids								
	Electrical								
	Process control								
	Planning								
	Estimating								
	Material take-off								
	Cost control								
	Correspondence registry								
	Drawings registry								
	Information								
Procurement	Purchasing								
	Inspection								
Site	Construction Manager								
	Field Cost Engineer								

Procurement

		Enquiries	Quotations	Bid evaluations	Requisitions	Orders	Vendor correspondence	Requisition register	Order register	Expediting reports	Inspection reports	Vendor info register
Client	Project Manager											
Home office	Director of Project Engineering											
	Project Manager											
	Project Co-ordinator											
	Project Engineer											
	Commercial Manager											
	Chief Accountant											
	Mining											
	Minerals processing											
	Metallurgical engineering											
	Mechanical											
	Civil / structural											
	Fluids											
	Electrical											
	Process control											
	Planning											
	Estimating											
	Material take-off											
	Cost control											
	Correspondence registry											
	Drawings registry											
	Information											
Procurement	Purchasing											
	Inspection											
Site	Construction Manager											
	Field Cost Engineer											

Construction

		Site correspondence	Site purchase orders	Bulk Matl. inventory	Work order register	Equip. op/maint. reports	Warehouse reports	Ave / shortage reports	Site progress reports	Site variations	Contract reconciliations	Completion / discharge certs
Client	Project Manager											
Home office	Director of Project Engineering											
	Project Manager											
	Project Co-ordinator											
	Project Engineer											
	Commercial Manager											
	Chief Accountant											
	Mining											
	Minerals processing											
	Metallurgical engineering											
	Mechanical											
	Civil / structural											
	Fluids											
	Electrical											
	Process control											
	Planning											
	Estimating											
	Material take-off											
	Cost control											
	Correspondence registry											
	Drawings registry											
	Information											
Procurement	Purchasing											
	Inspection											
Site	Construction Manager											
	Field Cost Engineer											

Figure 15.4 *A document distribution matrix*

problem may be particularly acute in the case of projects which spread across international boundaries. A summary of the principal options now follows:

Private facilities
The methods listed below cover the most widely available methods for sending and receiving documents, but companies with sufficient resources and with regularly used communication routes will consider providing their own independent systems. These can range from local cars or vans to serve one or more locations in the home country to private international data or telephone circuits and private aircraft.

Mail or airmail
Post Office ordinary mail services are usually adequate for handling routine correspondence and packages of documents between outside organizations (such as vendors) where there is occasional traffic, and where extreme speed of communication is not essential. One of the main disadvantages in using ordinary mail or airmail for large projects lies in time taken from door to door, and the difficulty in keeping track of consignments once they have been despatched – once lost they may be impossible to expedite or trace.

Special mail services
The Post Office, in conjunction with many overseas postal organizations, operates a Datapost service which offers guaranteed delivery times. This is one of the options to be considered whenever there is an external project office or site giving rise to regular consignments of documents. The head engineering office will be asked by the Post Office to specify collection times (which are flexible), and the arrangement can be two-way. Special Datapost bags ensure that consignments are easily recognizable.

Airfreight
Consignments of drawings, installation manuals, and other technical documents to construction sites or to overseas clients can prove to be quite heavy (sometimes exceeding 50 or even 100 k) and airfreight may be considered as an economical method for such despatches. Any company using such methods would be unwise, unless it possessed special expertise, to use airfreight without the employment of a local agent or airfreight courier. Agents have good knowledge of available flights, and they operate through their overseas offices or agents to maintain telexed control of each consignment through all stages of the journey, including customs clearance.

Air courier
A sophisticated version of the airfreight option is provided by specialist air

courier companies. Some of these actually send a representative on each flight (hand carried service) who is responsible for clearing the documents through customs and seeing that they are passed on to local services (which may be a car or motorcycle) for the final delivery. Other specialist air couriers exist who operate in remote areas of the world, or in countries considered to be at political risk. Senders should not be surprised in these difficult cases to find that the charges appear to be expensive but, if the documents are urgent and the project is valuable, high costs can be justifiable. This kind of service usually needs no general contract, and consignments can be arranged at short notice, for speedy collection and delivery as required.

Facsimile

Facsimile can be considered wherever the received documents need not be larger than A4 size (A3 documents can be reduced to A4) and where the remote location has compatible facsimile equipment. It is the obvious choice for sending small sketches, drawings and tabulations, in addition to text. There may be serious loss of definition over some long distance overseas routes.

Documents comprising many pages will occupy considerable transmission time (a minute or two, rather than seconds, per page) during which the machine will be unable to receive any incoming message. Other methods should generally be used for non-urgent multiple paged consignments in order to avoid unnecessary expense and to prevent machine bottlenecks that could delay the transmission or receipt of other really urgent documents.

Telex

A telex link is the obvious method for handling technical queries arising between remote project participants, for confirming in writing data or agreements already passed verbally by telephone, and for all manner of urgent communications concerning administrative arrangements (travel, accommodation, progress of shipments, and so on). The system depends on the existence of a telex machine at the other end which, even in these technological times, may not be the case in some remote parts of the world.

Telex used to have the major disadvantage that every document had to be typed into the telex keyboard (duplicating the efforts of the typist if the original draft message was typewritten, and introducing the possibility of keyboard errors). In modern systems the telex system can be linked directly to a word processor or personal computer, or the telex machine can be fed with a prepunched paper tape cut from a number of electronic office machine options, all of which allow the message to be printed, prechecked and, where necessary, approved or authorized for transmission. All of these direct input methods speed up preparation time and reduce the risk of errors.

Despatch controls

In a perfect world every document put into a mail or courier bag would arrive promptly at the correct address and be delivered to the responsible person without delay. In practice things can go horribly wrong even before the documents are consigned, with incorrect or incomplete address labels, transposed contents, missing enclosures and inadequate packaging among the many possibilities. Even supposing that a consignment survives the hazards of a long journey and is delivered safely to the correct address, there is still a chance that it might go astray, especially if it is delivered to a rough and tumble site, or to the administration department of some vast industrial complex in the third world, staffed by clerks without the benefit of sufficient experience or motivation to perform efficiently (I write from experience, not from prejudice).

A sensible engineering company will take steps to prevent mistakes within its own organization and during transit, as far as possible, and will introduce fail safe checks to highlight the failure of any consignment to arrive. In order to keep the argument simple, all the following suggestions will refer to consignments sent to an overseas client. They would, however, apply equally to any other remote project location, including a construction site, purchasing agent, joint venture partner, associated company, consultant, or a major subcontractor.

The measures range from simple commonsense steps to more elaborate procedures that seem to go well overboard in the direction of unnecessary effort. All, however, are based on experience and some were requested by clients (who were prepared to reimburse the extra costs involved).

Correspondence serial numbering

If an arrangement is made that every letter sent to the client is given a serial number, than a gap in the serial numbers received tells the client that he is a letter short. If he receives letters numbered 1134, 1135, and 1137 he will ask 'what has happened to 1136?' The missing letter can then either be traced and delivered or a copy can be sent.

In practice the allocation of correspondence serial numbers can be arranged as follows.

The engineering company must agree with the client (or other external organization) on a reciprocal arrangement of serial numbering. Suppose that the engineering company is based in London, England, and that the client's address is in Zambia. It might be agreed then that letters from London to Zambia would be numbered in a series LZ1, LZ2, LZ3, LZ4, etc. Letters in the reverse direction could then be numbered (by the client ZL1, ZL2, ZL3, ZL4, etc. Different identifiers would be agreed with other addressees (including other clients). Serial numbers for outgoing letters would be allocated from registers by the correspondence registry clerks, on request from the typists or secretaries.

Document transmittal letters

It may be considered too tedious to write a formal letter on every occasion when a batch of drawings or related documents is sent to the client or elsewhere, and a useful substitute is a preprinted letter that acts as a delivery note for the documents. Document transmittal letters allow space for listing the number, revision number, title and quantity of each document. The document transmittal letter will itself be serial numbered, in the same way as a formal letter. If a consignment then goes astray, the client need only say that document transmittal (usually shortened to dt) number so and so has not been received. The sender is then able, through his retained file copy, to identify all the documents in the lost batch, and assemble a new consignment.

Clerical checks

One precaution is to appoint one or more of the correspondence registry or mail room clerks as being specially responsible for checking the contents of every outgoing letter (which might be a document transmittal letter accompanying hundreds of drawings). The clerk must check that every enclosure is present as listed, before sealing the package.

Other checks depend upon the amount of information recorded in the registers from which serial numbers are allocated. If this data includes the name of the signatory, the date when the serial number was allocated and the date when the letter was actually sent, the clerks will be able to question any delay in sending out a letter (sometimes letters are cancelled or subjected to agonizing reappraisal by the signatories). If a numbered letter is to be delayed or cancelled, a telex should be sent to the appropriate addressee to forestall the inevitable missing document alarm.

For large projects, the appointment of a project coordinator may be justified, and that person might be asked to include these correspondence checks among his/her duties for the relevant project.

Now, getting into the realms of overkill, there are circumstances when overseas clients ask for a regular monthly telex, which lists the serial numbers of all letters, document transmittal letters and telex message numbers sent during the month just ended. This might not show a high level of confidence by the client in the ability of his own staff to detect missing document numbers, and the work involved in compiling such telexes can be tedious and time consuming. However, if the client demands such a service, and is prepared to pay for it, then the service must be provided.

Receipt of incoming documents

Arrangements to ensure that incoming documents are properly handled and directed are just as important as the procedures described in the previous section for outgoing despatches. The risks may be less significant in a company with only a handful of staff but, in the larger engineering organizations, it is very easy for important incoming documents to be misdirected or even lost inside the company itself. For simplicity the term mail will be used throughout this section to mean documents received by mail, courier or airfreight.

Address

The first important factor is to ensure that the correct address is advised to all senders. This may seem easy and obvious but beware (for example) giving a Post Office box number as the address when the sender is going to use airfreight (the street address is the only address that means anything to a delivery van driver). Also ensure that individual members of staff do not give senders the addresses of local overspill or branch offices where (for control purposes) all documents should be delivered in the first instance to the central mail room.

Action on receipt

Whatever the method of despatch, all hard copy documents (letters and parcels) should be directed to the mail room. This should be located as near as possible to the other administrative clerical services, including the correspondence registry and, if possible, the drawings registry. It is a further benefit if the telex and facsimile machines can be located nearby.

Unless mail is specifically marked private or personal, it should all be regarded as company property, to be opened and examined on receipt in the mail room. A responsible supervisor should be present when mail is opened, and he must direct each item to the appropriate individual for action, who may not be the addressee named by the sender.

Miscellaneous sales literature, catalogues, individually addressed journals, and similar correspondence of a general nature can usually be taken direct to the addressee by the messengers. In order to ensure that letters with a direct bearing on projects or contracts are placed in central files, these can be copied at this stage, with the originals sent directly to the correspondence registry for filing, and with the copies sent to the addressee (together with any enclosures).

Any clients' or vendors' drawings received should be registered in the drawings registry before they are passed on to the relevant engineer. If this is not done, the busy engineering staff cannot be relied upon to take the drawings along to the registry themselves for entry in the company's registers (and, in many cases, for microfilming).

Many companies operate a rule that all project correspondence (as originals or copies according to their practice) should first be taken to the relevant project manager, no matter whose name appears as the addressee. The project manager can then direct each letter to the individual who is required to act on the information that it contains. The project manager can also arrange for a secondary distribution of the letter into project files, and to other staff or managers who need to know its contents.

It is unfortunate that measures necessary to safeguard the correct distribution and filing of incoming documents appear to create additional work and double handling. Worse, they can introduce delays. The messenger and document handling services must operate efficiently so that the photocopying of incoming documents, or the routing of them through the drawings registry, does not lead to unacceptable delays in their reaching the addressee. Certainly such delays should not exceed two hours or three hours. If the drawings registry has such a pile of drawings that it cannot cope with their registration in a short time, at least the registry clerks must inform the addressee that the drawings have been received.

The flow chart in Figure 15.5 summarizes the routines described.

Receipt and registration of clients' and vendors' drawings

Drawings received from clients (meaning the clients' own drawings and not in-house drawings returned after approval by clients) need to be recorded and controlled in case they become part of a subsequent contractual argument, and to prevent embarrassing losses.

Vendors' drawings, together with associated technical documents (such as operating and maintenance manuals, test certificates, and the like) need to be registered and preserved with care. These documents form part of the total project information which any engineering company needs in order to provide its clients with long-term back up during the operational life of each completed project. There are several reasons why this information may be required again in the future, ranging from operating problems, plant replacement decisions, proposals for modifications and upgrading, or simply that the client has lost his maintenance manual. The engineering company may not be able to rely on every original vendor to be able to supply such information in the future (especially if the vendor has destroyed his records, reorganized or disappeared altogether from the commercial scene).

Provision of adequate vendor documentation must be a condition of each purchase order, with the details usually specified in the purchase specification. The engineering company should specify enough copies of these documents for despatch to the client, and for its own back up registry files. A

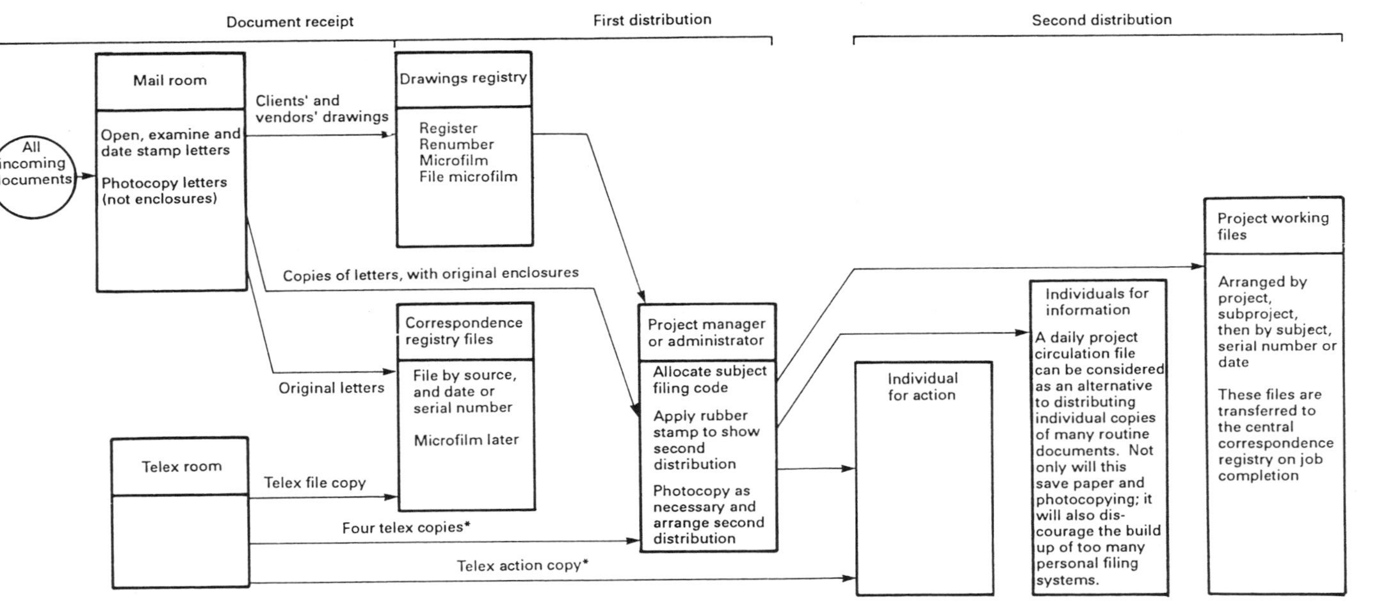

* If several copies of each telex are sent to the project manager, the second distribution can be speeded up, because he or she can send or take the action copy to the appropriate individual at once without the need to carry out more photocopying. If this procedure is still considered to be too slow for urgent telexes, the remedy if for the telex room to send an additional copy to the individual for action – provided that the telex room know who this is. The formal second distribution must still take place, under the project manager's direction, in order that the information is correctly disseminated and taken into project files.

Figure 15.5 *Incoming distribution for engineering project documents in a large organization. This chart shows the principle of two stages of document handling and distribution. Although details will vary greatly from one company to another, the intention is that the first (or primary) stage ensures that every vital document is captured for company files for subsequent availability. The secondary distribution is discretionary, and depends on the needs of each project. The actual secondary distribution must be controlled by the project manager or his delegate. An efficient messenger service is essential in order to prevent internal transit delays.*

vendor's final invoices should not be approved for payment while receipt of any significant document remains outstanding. Progressing the receipt of vendors' documents can be undertaken by the purchasing department as part of their expediting function, by a project coordinator or progress chaser, or it can be dealt with by experienced registry clerks. More senior project staff should only need to be involved when serious default by a vendor calls for tough measures.

Vendors' drawings may be received in the form of prints, reproducible prints or microfilm. 35 mm aperture cards are a convenient method for storing this information in the long term, and some engineering companies microfilm vendor drawings on receipt and send copies of the microfilm to their clients.

The incoming drawings will bear numbers that relate to as many different numbering systems as there are vendors. But, for indexing, filing and retrieval purposes, it is best (and usually necessary) to use a suitable unified numbering system. Here are two solutions to this problem which suggest themselves:

1 Try telling all the vendors to use a common project numbering system.
2 Allocate a new number to every vendor print or drawing upon receipt.

The first of these options is really not an option at all since, although the idea is neat, it would be difficult to get vendors to comply.

The second option, renumbering, is tedious but effective. An index card (either hard copy or in a computer) is set up as each drawing is received, and the numbering can be serial within each requisition or purchase order (e.g. drawings received for purchase order 5432 would be numbered 5432–1, 5432–2, 5432–3, and so on). A large adjustable rubber stamp is a good method for applying these numbers, especially since the ability of registry clerks to write numbers neatly and legibly (to the standard required for microfilming) cannot be relied upon. Care has to be taken to provide a cross reference to the original vendor drawing numbers, since these must be quoted in any future correspondence with the vendors. This cross referencing is also necessary to avoid the mistake of giving a drawing a new number when, in reality, it is merely a duplicate or a revised issue of a drawing already received and renumbered.

Before leaving the subject of dual numbering, the case should be mentioned when the engineering company, as contractor for an engineering project, can itself be regarded as a vendor, with the client as purchaser. If the client is a large organization, with plenty of muscle, it might be able to insist upon being supplied with the complete set of project drawings, all numbered according to its own numbering system (and very likely drawn on its own drawing sheets too). A wise contractor will comply. Here is a case where every drawing might have to be given two numbers when it is drawn, one for the client's benefit and the other for internal filing and subsequent retrieval (with both cross referenced in the drawing registers).

16

Information and library services

Elaine Bimpson

This chapter deals with the services which can be expected from an information unit or library, and illustrates some points of management which occur when the library is considered in its context within a larger organization. The standard functions of a library are the acquisition, storage and subsequent retrieval of documents, together with the maintenance or records needed for these activities. In other words, the library is concerned with the physical management of relevant material, with information retrieval, and the dissemination of that information to others who need it.

The management context

Nothing can develop in an organizational vacuum. The first decision which may have to be made is whether or not an engineering company should have a library at all. Whichever way this decision goes, it has to be recognized that the organization is still going to be involved in certain continuous costs in respect of gathering, storing and retrieving information.

All engineers need specialist information sources to hand in order for them to do their jobs effectively. A cost element relating to books and periodicals will appear in every detailed budgetary analysis. This represents expenditure against which a company should seek to obtain maximum value. Purchasing through the centralized function of a library can ensure control, make information available to a wider selection of company staff, and provide proper physical management of the information materials themselves. The drawing together of expenditure under one budget will enable the company not only to see the full extent of its spending in this field, but it will also minimize potential duplication and wasted costs.

To function effectively within the wider organization, the terms of reference for the library should be worked out as thoroughly as possible and agreed at management level so that the procedures will carry authority and provide a clear working framework. A short code of practice should be agreed and issued to staff to ensure that material is handled in a uniform way. This applies where material might be received elsewhere than by the library and should ensure that it reaches the library to be formally recorded. Even if the library is not to be the final home for a particular document, then at least the library should see it for formal record so that its final location is known for later retrieval. Awareness of what is available is vital to the library, which will act as a point of continuity in what may be a fluctuating organizational environment.

Changes in the responsibility of the library or the setting-up of a new library from scratch are likely to cut across established document flow patterns. The conflict which may become apparent here can be minimized by early agreement on what the library is for, who it is to service, how it is to accomplish this and under what budgetary constraints. Its success will depend on the efficiency with which this is carried out together with adequate management support. However, these terms of reference should not be considered to be sacrosanct but should be reviewed on a regular basis, thus ensuring a clear and workable framework for operations.

Housing the collection

There are several factors to be considered in connection with accommodating the collection, all of which will have a bearing on the amount of space required and, therefore, on the cost of the library or information unit. Principal factors include:

1. The initial size of the collection.
2. Space available for expansion (if any).
3. The envisaged final size within some reasonable foreseeable future.
4. Space needed for staff.
5. Space needed for equipment.
6. Floor loadings permissable.
7. Space for users to work within the library.
8. The varying formats of the material itself.

Every library must justify the amount of space needed for its operations as concretely as possible, as space is often at a premium. In terms of expense per square foot, and also in terms of competition for that space from other expanding departments, office managers look for a return on space: information unit managers must be prepared to try and put a value on the area for which they are asking. The established library must not be afraid to look hard at its material to see if it is earning its keep, while the new library (which may grow rapidly in its initial stages) must be able to provide a reasonable forecast of requirements.

All this will be reflected by the availability of local external library services. A library situated near to a major institutional or university library (where back-up material may be available quickly) will have differing stock requirements or retention periods from one in a more isolated situation (dependent upon an inter-library loans delivery service or the post). Time and project deadlines equal money. In any case, restrictions may require that an orderly weeding/disposal programme be carried out at regular intervals.

A large organization, housing many different departments under one roof or faced with multisite operations, may have found that it is inappropriate for all the material to be housed in one location, which may be too far from the immediate user.

Separate libraries or information units may have been set up to fulfil certain specialist needs for particular types of information. These may be quite automonous reporting specifically to, and funded by, their own local management. What is of importance here is that these units and their holdings be known to each other, with full liaison. This broadens the information base available without incurring some of the costs of going outside. It also helps to keep the handling of sensitive queries in-house. Costs can also be minimized by non-duplication of stock, for example by the reduction of retention periods (and therefore shelving costs) in one area which may be able to use another as back-up. The establishment of company-wide contracts with suppliers may mean that better terms can be negotiated than those which would be offered to a small unit on its own.

However, there are other problems with information units in large organizations. The establishment of separate cost and profit centres may mean that all charges incurred must be recovered as far as possible. The separation of activity into different business streams may mean that the corporate HQ can only pay for itself by passing its charges back to its corporate users, who may then be free to take its services only if they feel that value for money is given or that the charges are affordable. A head office library may be faced with having to pass on charges which include expensive overheads because of its location, a factor that is totally outside the control of the library itself. The introduction of charges is one that should be very carefully looked at as it is likely to change radically existing working relationships.

The stock

An initial review of the established procedures will identify the types of material that the library will be asked to manage. These might include:

Books	Hardback Softback	
Pamphlets		Varying pages sizes and bindings
Journals	Bound volumes Unbound	
Standards		
Reports		
Government reports, laws etc.		
Newspapers		
Audio visual material	Audio tapes Overhead slides Slides Film Fiche etc.	
Maps		
Trade literature	Various formats	

It will be seen from this by no means exhaustive list that not only does the type of material vary but so does its size and format. Shelving widths will have to accommodate oversize material, with boxes provided for the protection of floppy material such as pamphlets (the page sizes of which can also be very variable). The storage methods chosen must reflect this and the frequency with which the material is used (and therefore the wear on the storage medium itself). Should it be decided that newly received books, journals and newspapers will be put on display before being permanently shelved, then extra space must be allowed for display racks. Boxes or lateral filing systems are available for overhead slides, while 35 mm slides can be accommodated in boxes or pocketed sheets. Plastic boxes are suitable for unbound journals. The question of whether or not to bind journals will be affected by cost, heaviness of use and maybe the need for preserving the volumes for posterity.

Where floor loading is not a problem, moveable racks may be considered although the problem of access (only one row at a time) may arise with material that is heavily used.

The lower costs of off-site storage for infrequently used material must be set against problems of security control and access. In some cases it may be feasible to put material on to microfiche or microfilm. The quality, and readability both of the film and any reproduction from that film depend upon the quality of the original. Quality control and storage of masters must be gone into thoroughly before such a project is undertaken.

Acquisition

Control

Material may be acquired by the library in a number of ways and from a number of different sources such as direct purchase, exchange, gift or donation together with internally generated material. Levels of authority to place orders and pass invoices should be established with management and with the appropriate accounts department. Procedures for passing the charges back to departments or to other cost centres must be agreed and incorporated into a code of practice. To minimize costs and administration, purchasing agreements can be made with book and journal supply agents who may be able to provide discounts. In any case, the incorporation of large multititle journal orders into the provision of one annual invoice submitted by an agent represents greater efficiency. However, the library must maintain flexibility of supply in order to respond to requests at speed, so that facilities for quickly raising cheques or access to petty cash should still be available to the librarian.

It is advisable for the library to keep track of its own accounts in order to monitor expenditure throughout the year to ensure budgetary control. This is also necessary to act as a check against the figures produced by the accounts department. Difficulties will be minimized where procedures for handling invoices and allocation to authorized budgets are agreed by all parties. Again for checking purposes, it may be useful for the different cost elements of the library to be itemized individually by the accounts department as well as within the library's own records. This may be necessary where the library forms part of a larger organization in order that expenditure can more easily be monitored by higher levels of management. In any case referring to previous detailed records will help in future budgetary planning.

Sources

Material may be obtainable from a variety of sources, including, for example:

Externally generated – widely available
- Commercially published material: books, periodicals, pamphlets.
- Government publications.
- Standards, specifications, codes of practice.
- Patents.
- Trade literature.
- News media: newspapers, press releases, transcripts.
- Audio visual material.

Externally generated – limited availability
- Material not formally published but available on a need-to-know basis.

- Consultants' reports – client private or subscriber only, specially commissioned.

Internally generated
- Internal reports, manuals, specifications.
- Project files.
- Laboratory notebooks.
- Expertise register.
- Audio visual material.
- Other records with non-ephemeral information.

Awareness that a particular item is available, followed by a decision to acquire it, is often not initiated by the librarian alone, but results from discussion and interaction between library staff and library users. This interaction should be encouraged at all levels. The library staff develop a knowledge of particular publications and information sources needed to answer the questions put to them. The engineer user, at the forefront of his developing branch of technology, will come with his own special information needs. Interaction between the two can only enhance the awareness of each. Constant awareness is the keynote, from the availability of material to the changing needs of the user.

Unsolicited announcements from publishers come everyone's way, but the library should keep track of:

1 Specialist publishers' mailing lists.
2 Book agents' specialized or customized lists.
3 Advertisements and reviews.
4 Book exhibitions and fairs.

The above information from publishers and book agents will come free of charge, often *ad nauseam* and in duplicate. But there are also a number of important secondary sources which, however, have to be paid for. These are the published indexes and abstracting services relevant to a particular speciality and more and more are now available on-line from computer databases. Use of these for searching the literature on a given topic can be combined with regular search profiles to keep abreast of the latest developments.

Provision of up-to-date standards, specifications, codes of practice, patents and suchlike material is of primary importance to technical staff, having implications for safety and legal obligations for example. These, too, can have a historical value relating to an operation or design of a specific project, so that care should be taken in the weeding of such material. Care should also be taken in the acquisition, storage and ultimate weeding or disposal of material which may have a security classification because it was obtained in association with a government contract or because it has been the subject of secrecy agreements between companies, or between companies and other organizations such as universities or research institutes.

In all cases of acquisition and destruction it will be helpful to have established clear lines of authority in order that these procedures may be carried out quickly and efficiently.

It should be remembered that where an organization publishes its own material in the form of a journal or monograph series then it may be possible to minimize costs of obtaining other, similar material by exchange: in effect swapping one's own publication in order to eliminate ordering and invoicing procedures. But care must be taken to establish that this represents value for money. Where an abstracts journal is produced, more and more organizations are marketing the information, not only as hard copy but also through the various on-line bibliographic database lists. Various deals are available not only to the marketer but also to the eventual user including, for example, special rates to members of the abstracting organization or to existing purchasers of hard copy, together with special rates linked to a guaranteed minimum usage. Smaller departmental libraries may therefore benefit from a contract negotiated on the basis of company wide use.

Organization of acquisitions

It can be seen that substantial costs will have been incurred in obtaining material and for its proper housing and storage. To get a return on this the material must earn its keep – that is it must be used to maximum advantage. In order to do this standard housekeeping routines must be applied, demanding consistency in application and attention to detail. Information must be correctly recorded, designated locations known and whereabouts tracked. Maximization of use in order to get value for money demands accurate and efficient retrieval. This falls into two parts: information concerning the physical document itself and information concerning the contents of that document, be it book, trade catalogue, microfilm or other audio-visual material.

Dealing with the physical entity first, one is looking at various elements of the document which are standard, though they may only apply selectively:

- Author(s), editor(s): individual/corporate.
- Title.
- Conference title.
- Publisher, organizer.
- Place of publication, conference venue.
- Date, date received.
- Volume, part number, page numbers, text figures.
- Edition.
- ISBN/ISSN.

- Classification/shelf mark/location.
- Status: in order/approval/shelf/loan.
- Price.

Some idea of the clerical effort involved in the management of a relatively modest list of 100 serial titles ranging from annuals to dailies can be seen below:

Number of titles	*Frequency*	*Number of issues received*
20	Annual	20
20	Monthly	240
20	Weekly	1040
10	Daily (5 working days per week)	2600
10	Irregular (say, 5 per year)	50
Plus		
10	Monthly duplicates	120
10	Weekly duplicates	520
Total number issues received in one year		4590

Add on to this claims for missing items, confirming renewal both internally and with the supplier, checking and passing invoices, together with maintaining constantly changing circulation lists, and an idea can be gained of the amount of work involved in a task that may be expressed in a job description merely as 'journals management'.

All the various types of document which have been dealt with in the physical sense must also be dealt with from the point of view of the information which they contain – not always clearly expressed in the title of the document itself. Where information retrieval is concerned it is the concepts expressed in the text that are of paramount importance – can they answer the questions of what, how, why, where, when and by whom, in terms of equipment, process, method, product, designer, etc. It is here that the understanding of the science and the industry in which the work is based is of great importance. The interpretation of a classification system and allocation of keywords may vary in detail but gross errors should not occur where there is a sound understanding of the scientific principles, involved backed by professional expertise in librarianship and information science.

Staffing

Library users may be specialists in their own field but, generally speaking, are not specialists in information sources and resources and their efficient exploitation. This is the function of the library and other information staff, who must act as the interface between the information and the users. Their qualifications, both personal and professional, must reflect the levels at which the staff are expected to operate. It is to be expected that all operations will be run professionally and efficiently and that staff will have the necessary professional qualifications, training and experience to do this at every level of operation. Further training should be encouraged in order to maintain efficiency and keep up to date: short courses in special topics and new developments are available. Library staff will also be dealing with all levels of user, both in terms of area of specialization and managerial seniority. Professional qualifications for library assistants through a degree in information science and librarianship, or a post-degree qualification are indicated levels for manning an information unit. The level of background knowledge required in relation to the particular industry or specialization must also be considered, in which case a first degree followed by an information science qualification may be most appropriate. In order to index and catalogue correctly a certain knowledge of the subject matter is necessary. Dealing with high level technical queries will be quicker and get a better response where there is an understanding of the subject matter. The members of staff required and the level of operation will depend upon the flow and volume of work at different intellectual levels together with the levels of responsibility designated to each function.

Service

In-house function

Much mention has been made of the 'subject matter', 'speciality', 'technology', 'science', without being very specific. This is to some extent deliberate as the mechanics of librarianship and information retrieval cut across boundaries, and professional standards apply commonly no matter what the subject matter. Even when considering the various specialities which fall under the heading of engineering there are certain types of publication which form a useful background to any specialist collection. Core publications, the dictionaries, handbooks, manuals, standard texts and journals specific to each discipline would be prime acquisitions. However, although each information unit operates within a particular technical and economic context in a particular location, it may well be that the library is asked to provide a wider service than that which covers just the relevant technical aspects.

The function of the library is to provide for the acquisition, storage and retrieval of material which is central to the activities of the organization it serves, while also possibly providing back-up to other managerial and administrative functions. To that end material in the collection needs to be well publicized. This can be done quite simply by means of acquisition lists, abstracts journal or current awareness bulletin depending on the size of the organization, the number of people it must reach and, last but not least, the availability of staff to produce such in-house publications. Where these publications are to be made available to external users, it is important that they be as up to date as possible.

External services – the wider context

The library network

Limited by the constraints of space, time and money, the in-house library cannot hope to cover the world's literature in one speciality let alone cover those which, though normally are not relevant, from time to time will contain vital information. The interface between disciplines has often been as fruitful a field for new developments as advances within the core of one in particular. Therefore, there will be a requirement for extra material from external sources together with a dependency upon those sources. An awareness of external sources and resources available to the librarian is vital to the provision of information to the library user. Time should be taken, and allowed for, to build up contacts locally, nationally and internationally.

Speed is often of the essence in answering a query, so that what is available locally may provide the first fall-back. Sources which should be cultivated locally will include not only the local public library network but also the local technical college or university. Care should be taken to establish contact with learned societies, institutions, institutes and research associations nationally where specialities are relevant. These will be prime sources of information although their use may well be dependent upon either corporate or individual membership. A provision for membership fees may well be appropriate within the library budget. It will also be useful for the library to know which members of staff belong to external organizations. While these exist primarily to serve their own members they may be willing to help outsiders – a service for which there may be a charge. It is expected that any postal or photocopy charges would be paid by the requesting library together with costs of returning borrowed material.

External databases and their access

Here the librarian is concerned principally with secondary sources of information, i.e., the use of published indexes, catalogues and abstracts. Providing access to the world's literature has become a growth industry in recent years and ranges from the wide coverage of chemical abstracts to the more specialized databases. At one time only published in hard copy, these were always bulky to store and laborious to search using their own annual published indexes. Publications of this sort were always, and are still, expensive both to produce, to purchase and to shelve.

Advances in telecommunications and in computing have radically changed the way in which information on the world's literature is now available to us. The abstract journal is still seen as a useful way of quickly scouring the current literature to spot new developments of particular interest. This can be combined with a regular computer print-out, based upon an established user profile which can be modified as necessary to reflect the changing needs of the user as he progresses from project to project.

With the advent of powerful computers not only can all the standard bibliographical details be recorded, abstract and full-text stored but also information within a book or journal article can be covered in detail by the keywords assigned. Keywords may be thesaurus-controlled or free-text, or a combination of both can be used in order to cope with the non-uniformity of natural language and its evolution. There are many steps between the original writer and the final user, even after publication:

1 Publication.
2 Receipt of publication by abstracter.
3 Preparation of abstract.
4 (a) Publication of abstract journal.
  (b) Updating of computer database.
5 (a) Receipt of abstract journal by user.
  (b) Access to database by user.

The time delay in getting recent publications into the abstract journal may well be outside the abstractor's control (such as delays in the post or material sent by sea-mail). Fresh developments in electronic transfer may well reduce these delays, but the librarian should be aware of the normal delay period of the particular database being used (i.e., the time between receipt of publication and its appearance in the database itself).

Purchase of hard copy is a direct transaction between the purchaser and publisher. The on-line database is more usually handled by a database host, who may have many databases available with common routines combined with the individual structures of each separate database. However, a contract giving access to the database host may not always give access to each separate database on the system, some of which may have to be subject to a special user contract.

Access is dependent upon the sophistication of equipment and telecommunications available locally. A typical route would be:

User – terminal – modem – local node – international network – host – database.

Substitution of, say, a microcomputer for the terminal would enable material to be spooled off, to be worked on later off-line.

Costs for all this will include elements for:

1 Telecommunication time.
2 On-line time for the searches.
3 On or off-line charge per citation printed.
4 Postal charges.

Database hosts as well as database producers all run training courses in order to enable the searcher to use his time with maximum efficiency.

Internally there is also the question of the possibility of networks within the organization itself, so that separate sites may be able to access each other's databases and/or operate systems in common.

The technological revolution

The application of new technology has indeed resulted in a revolution within the library and information science world. This applies to the management of material as well as to the management of the information contained within that material. There has been a progression, for example, from the standard card catalogue, through optical coincidence cards and punch-cards for computer input, to direct input into the computer. Hardware has increased in power and capability enormously so that sophisticated systems can now be run on microcomputers; a factor that has brought the cost within the range of quite limited budgets.

Before any commitment is undertaken, however, a careful analysis should be done in order to establish which system is right for the job in hand. A listing of elements and functions as they exist, together with extra desired elements and functions, should be carefully matched against available systems in terms of software and the hardware on which the system is to run. The sort of coverage that might be looked at would include:

1 Catalogue.
2 Suppliers.
3 Borrowers.
4 Loans.
5 Circulation list record.
6 Journal holdings.
7 Journal check-in.
8 Reservations.

Packages range from those designed to cope with vast collections, to those which may only cover one particular aspect. A recent development has been that of the total library management system, available on microcomputer with all functions interrelational. It is for the buyer to decide whether a system should be completely tailored to his detailed needs or whether a preformatted system is acceptable. Time and system support are prime factors here. A small, one-man band unit may find it difficult to cope with the detailed planning involved as well as running activities on a day-to-day basis. It may be easier in this case to accept a pre-formatted system ignoring those elements which do not apply. Large, multisite organizations may wish to consider the possibility of all sites sharing one system and how far this is practical in terms of standardization of input, multi-user access for input and retrieval, and apportioning of costs internally.

The degree to which any system may be termed 'user friendly', especially on the retrieval side will affect the use of the system. Where the library user is to be allowed to search the catalogue himself then the system must be simple; a menu can be devised or bought-in to facilitate this for example. Each element within a computer-based system must be carefully controlled so that procedures are not altered by unauthorized users. The interrelational nature of the new LMS (library management systems) can be extremely useful and time saving; the linking of the catalogue to the loans/borrowers/reservations parts of the database results in instant updating whenever changes are made. Linking of suppliers to the catalogue which includes all ordering data can result in the ability to pull out budgetary information. Detailed statistics on the usage of material and tracking of journals is also possible.

Indexing of material has also become much easier in one sense, although there are problems where vocabulary control is necessary. In operating conventional classification systems there was always the problem of dealing with the latest development, application or concept – the problem of new terms. In computerized systems keywords can be accepted freely. Depending on the size of the system and degree of control required it may be possible to buy in a thesaurus where one exists covering a particular speciality. The relationship between terms then being already set up:

BT – Broad term
NT – Narrow term
RT – Related term
SA – See also

Retrieval uses standard Boolean logic 'and', 'or', 'and not'. Detailed, complicated search statements may be constructed in order to get at the publication which contains the information required. The initial input of course must be done by someone with an understanding of the subject matter.

Storage of information on computer where only one entry is needed per item cuts down on time and eliminates the need for the boring task of multi-entry as in a card-based system. Material can be pulled out to produce general acquisition lists or for other purposes.

Where records are being computerized for the first time then the question of whether or not to computerize the entire collection arises – dependent again upon time, funds and staffing. Standard procedures for safeguarding the information must be instituted; personal identity codes and passwords strictly controlled; back-up procedures put on a regular basis.

Costs involved include hardware, software licence, maintenance and support and staff training. The whole field is one which is developing rapidly, in terms of power and capability of the hardware together with a growing range of software programmes. Care is needed to match requirements with availability at an affordable price.

Legal aspects

Data Protection Act (UK)

Storage of information on authors, borrowers and reservations, together with special listing of staff (perhaps allied to areas of expertise for example) means that the information stored in this way by the library comes within the scope of the Data Protection Act. Care must be taken to register and comply with the requirements of the Act.

Restrictions on the use of external databases

There may be restrictions preventing the supply of data obtained from particular databases to third parties. This can apply where the third party in question is a subsidiary or associate company, and where the company holding the database contract does not own a specified proportion of the subsidiary or associate company. This is a point that must be checked when the database user's contract is entered into.

Copyright

All company staff should be aware of legislation affecting copyright. In the library this is particularly applicable to the photocopying of published material. This subject is dealt with in more detail in the following chapter.

Useful organizations

Alan Armstrong and Associates, 76 Park Road, London NW1 4SH. Training courses.

ASLIB, Information House, 26–27 Boswell Street, London WC1N 3JZ. Promotes the interests of special libraries through research, training courses, publications, etc.

British Library Document Supply Centre, Boston Spa, Wetherby, West Yorkshire.

Engineering Information Inc., 345 East 47th Street, New York, NY 10017, USA. Information service providing abstracts and indexes, engineering literature and conferences worldwide.

Institute of Information Scientists, 44 Museum Street, London WC1A 1LY. Professional body concerned with information science.

Library Association, 7 Ridgmount Gardens, London WC1E 7AE. Professional body concerned with librarianship and education for the profession.

Library Technology Centre, Polytechnic of Central London, 308 Regent Street, London W1. Advice and demonstration of available systems.

Further reading

Anthony, L. J. (ed.), *Handbook of Special Librarianship*, Aslib, 1982.

Burton, P. F. and Petrie, J. H., *The Librarian's Guide to Microcomputers for Information Management*, Van Nostrand Reinhold, 1986.

CD-ROM Directory 1988, TFPL Publishing, London, 1987.

Cook, M., *The Management of Information from Archives*, Gower, 1986.

Dyer, H. and Brookes, A., *A Directory of Library and Information Retrieval Software for Microcomputers*, 2nd ed, Gower, 1986.

Roberts, S. A. (ed.), *Costing and the Economics of Library and Information Services*, Aslib, 1984.

Rowley, J., *The Basis of Information Technology*. Clive Bingley, 1988.

Turpie, G., *Going Online 1987: An Aslib Online Guide*. Aslib, 1986.

Webb, S., *Creating an Information Service*. Aslib 1985.

17

Intellectual property and information protection for engineers

Iain C. Baillie

Business people, particularly engineers, may tend to regard intellectual property as essentially an ancillary aspect of operating a business to be attended to by specialists who will occasionally advise them on those occasions when litigation may be involved. Although the identification of inventions may be significant depending on the industry, the value of a technology is rarely seen in terms of its intellectual property content. Yet as is evidenced every year, both in terms of licensing and litigation to say nothing of lost opportunities, inadequate attention to intellectual property can result in considerable loss, both of technology and of the commercial value to the company.

The purpose of this chapter is to review intellectual property in the context of company management; to indicate how intellectual property can be identified, its signficance and its correlation with management. The issue is when and what decisions must be made to ensure good management. No attempt has been made to outline the precise procedures which must be followed for protection except in so far as they are significant to general management. A reading list is provided for some further information on detailed procedures but the aim has been to set out the type of questions which must be asked by management rather than to provide specific answers, many of which would require considerable background and training for their preparation. To provide satisfactory advice in intellectual property for a particular situation requires a highly skilled professional who has several years of training and who is qualified by membership of the appropriate professional bodies; there is provided later a short discussion for those engineering managements who have no immediate adviser as to where to seek and how to seek such legal advice.

At the time of writing this chapter there is a Bill before the UK Parliament which will effect significant changes in the law of copyright and protection for industrial designs. One can only predict the final form and try and give guidelines for UK managers to follow by the time the Bill is enacted.

Nature of intellectual property

It should first be borne in mind in regard to intellectual property that it is essentially exclusionary in nature. Unlike other branches of the law which must be considered by a manager, for instance product safety, it does not in itself limit the operations of the technology or the management decisions that must be made. To some extent counselling on technology transfer will involve issues of competition law. However, given that one has the technology, no protection is necessary for use of that technology. What is important is that technology or reputation be kept exclusive to an owner; because of that very exclusivity it will have a market value. The maintenance of that market value which can be attributed to exclusivity is the business of intellectual property. Without intellectual property there would be no exclusivity since there would be freedom for others to use the information or reputation and therefore one could not have any added value from that exclusivity; for that reason therefore intellectual property has a direct definable value measurable by the exclusivity it protects.

There can be many reasons for market exclusivity which have nothing to do with intellectual property. A company may make and be the only manufacturer of a product simply because it has an established marketplace which, although reasonably profitable and although the product is easily copyable, does not justify a competitor entering into the investment necessary for that size of market. This is true perhaps of many companies who work in what might be called last stage markets, where other companies have moved on to growing markets. It is important therefore that, as a first stage in managing intellectual property, one identifies the significance of exclusivity in a given marketplace and in relation to particular product line. The product line which does depend on some exclusivity but where that intellectual property protection is diminishing because of limited life (e.g. in patents) may be one in which management decisions must be made as to additional research to enhance the technical content, its exclusivity and therefore its protection. In so far as information is concerned, therefore, constant awareness of the value of that information must be maintained. The same is true of reputation. First there is created an identification feature which creates to the customer a unique image of a particular operating entity or product; anything that weakens that uniqueness or exclusivity and opens it up to others weakens the investment which has been entered into to create that identification.

Classification of intellectual property

As will already have been seen from the introduction, intellectual property divides itself into two main areas, the first being protection of information and the second being protection of reputation, although these overlap slightly in the area of so-called unfair competition. Moreover, one can be used to assist and enhance the other.

There are other ways of classifying intellectual property. The different types of right have varying advantages and disadvantages and apply to different aspects of a business. Analysis is important to *create* a balanced package of strengths, and *maintain* that package all in the most economic way possible. Some rights are basically statutory, therefore in order for them to be defined one must examine the specific statute under which they are created. These are patents, registered designs, the new 'design right', copyright and registered trade marks. Others are created primarily by reason of common law in those countries which follow that concept or under general codes of commercial behaviour. This usually means that they are much more general, less easy to define and very often primarily dependent on contract or other legal relationships between two parties as distinct from a statute defining a relationship of the owner with the world in general. These rights are primarily trade secrets or confidential information on the one hand and the general concept of goodwill or protection against unfair competition on the other in the field of reputation.

Another definition or distinction can be drawn between those rights which are objective (i.e. in which a boundary is drawn) and it is irrelevant how a third party comes to employ the subject of the right and those that are subjective, where it must be demonstrated that an 'infringer' *obtained* the concept from the original owner. For an objective right the mere comparison between what a third party is doing and the defined right is sufficient to determine an 'infringement'. Such rights are again usually statutory. Such 'objective rights' are, for example, patents, registered designs or registered trade marks. Not all statutory rights are however objective. Copyright is defined solely by statute but it is necessary to demonstrate that a third party 'copied' the owner's property before any legal remedy is available. Similarly the non-statutory right of trade secrets requires a relationship be proved between owner and alleged illegitimate user. A chain of title must be shown for trade secrets and copyright for there to be any form of action. In the case of trade secrets there must be a breach of a confidentiality which arose between the acquirer of the information and the person who originally possessed it. To some extent goodwill falls within this class although it is possible to infringe upon a goodwill without being fully aware of it.

Types of intellectual property

An itemization approach to intellectual property is probably the most useful practical way of specifying its nature. 'Broad' definitions are somewhat confused but probably a statement which is a paraphrase of the definition in the International Convention is useful, namely: intellectual property is the result of intellectual effort in which an owner can claim exclusive rights. While there may be some question as to whether exclusivity is a consequence of a right or an aspect of it, probably the truth is in practice that is a bit of both. However, this gives us the main categories, namely:

1 *Information*: Patents, registered designs, design rights, copyright, confidential information (trade secrets).
2 *Reputation*: Trade marks (registered or unregistered), trading names, trade identifiers and goodwill.

There is a suggestion that the law should develop a so-called innovation warrant, the concept of which is that an innovator who is first on the market with a product would be entitled to 'exclusivity' until there had been a reasonable return on investment. This idea is attractive in avoiding some of the problems of patents in that it would deal only with commercially marketed projects, but analysis indicates the result would be some form of modified commercial patent and many of the present problems in patents would still be found. Protection for 'commercialized' concepts would leave the usual problem – is it justifiable to exclude competitors from what is essentially not a 'real' advance in technology? It is this type of problem which makes one doubt whether the innovation warrant would have sufficient additional value to merit its creation.

The importance of a listing at this stage of the various intellectual property rights is that, in considering each of the individual rights, it must always be borne in mind that they are correlated and interrelated and that a choice for or against one will depend on the extent to which one can rely on one of the other rights. The question is never 'Do I want a patent?' but 'Do I need protection and, if so, what is the best protection?'. The test in the end is one of cost effectiveness in that there are many patentable inventions which are totally commercially useless – a fact probably not always appreciated by the private inventor. This chapter will therefore be broken down into consideration of the different types of intellectual property; there will then be a consideration in one place of the question of ownership and finally a brief analysis of exploitation particularly by limited transfer (sometimes popularly known as licensing).

Requirement for clearance

The importance to the manager of intellectual property is not just in its existence but in its function in management. Since it is an exclusionary right there is a possibility that a third party will be able to prevent entry into its own protected area. The first question therefore that the average engineering manager will ask is not 'What right can I obtain?' but 'What rights exist to prevent me continuing with a new project or indeed continuing with an existing project?'. It is important, therefore, that in the following discussions there is always consideration of the question of clearance as much as the question of creation. This is certainly very clearly seen in the case of trade marks but failure adequately to take into account potential patent rights can be fatal. Patents can be a devastating protection as proved by the Kodak against Polaroid litigation.

The national aspect

It must also be borne in mind that intellectual property, as for most branches of the law, represents a series of national rights. Patents exist in terms of national laws both for creation and enforcement. There is no such thing as an international patent or an international trade mark. Decision making must therefore be on an individual country basis both when considering protection and when considering enforcement and transfer. It is incumbent on the manager therefore in reviewing intellectual property always to bear in mind this international aspect. This is somewhat inconsistent with the fact that the underlying technology or reputation will be international and can be frustrating for the manager who has no reason to draw national boundaries over a technology map. However, advice given solely on the basis of one law can be both misleading and indeed erroneous. No study this length can take into account all the complexities of international law but it is incumbent on the manager to make sure that he asks and receives the answers to questions on an international basis and consider these issues in other than a purely local context. For that reason this discussion is directed primarily to issues and answers from an international viewpoint although stress is laid on UK law.

Nature of treaties

As a background to the international aspect there are two types of treaty involved. The first and more universal are those treaties which coordinate international systems so as to ensure that each country gives to foreigners the same rights as it does to its own nationals and provide certain limitations on the freedom of each country as to the laws it may enact. These are the so-called governing conventions; the most important of these for patents and trade marks is the International or Paris Convention originally signed in 1885. In copyright there are the Berne and UCC Conventions.

Administrative conventions set out actual procedures by which intellectual property rights can be secured, for example, The European Patent Convention and the Patent Cooperation Treaty, both of which will be discussed later.

Summary of types of right

As an assistance in bearing in mind all of these different rights they can be very crudely analysed as follows:

Information (technology)

1 *Patents*: limited life (twenty years); protect concepts; must define protection sought early on; elaborate procedures to obtain and maintain protection.
2 *Designs*: protects appearance; limited life; must define protection sought early on; legal procedure to obtain and maintain.
3 *Copyright*: long but limited life; comes into existence automatically; need not define protection claimed; protects expression only not concept; generally no formal procedure.
4 *Design rights*: a new type of protection outlined in the pending UK Copyright, Patents etc. Bill; limited life; no formal procedure; protects appearance.
5 *Confidential information*: e.g. trade secrets, know-how; no defined life; no definition of protection but must prove breach of confidentiality to enforce.

Reputation

1 *Registered trade mark*: indefinite life, registration procedure with definitions circumscribing protection.
2 *Goodwill*: protects reputation but no defined boundaries; must prove confusion; protected by a legal remedy – passing off or unfair competition action.

Patents

Patents are perhaps the best known type of protection for technology and therefore the term is sometimes applied to other types, for example, trade marks or copyright where it is totally inapplicable. The term 'patent' in English comes from Letters Patent, a specific state document defining and giving protection for ideas, but is now applicable only to technical ideas. A patent therefore will always be a defined right obtained under a Statute for a limited number of years for a technical concept. A procedure always has to be undertaken to secure rights in a given country and moreover it is usual to pay fees regularly for the maintenance of the right.

Although the legal requirements differ from country to country they can be generalized under three headings:

- Subject matter.

- Novelty.
- Inventiveness.

Patentable subject matter

The first test as to whether a development is patentable is whether there is appropriate subject matter for a patent. There are ideas which although undoubtedly new and original are not considered appropriate for patents. This can lead to problems if the idea is self-teaching once it becomes publicly available, since it will not be possible to create some form of reasonable exclusivity. This is particularly true, for example, of business plans. This issue is important in considering whether or not an investment is justified in development of a concept. While each country has its own list of subject matters not considered to be patentable, those which are listed in the European Patent Convention are a fairly reasonable international guide. The statement of patentability is that the subject matter must be 'industrially applicable'. The definition of industry in most countries now includes agriculture but it is as well to remember there are countries which do not regard purely agricultural inventions as patentable: for example, a method of killing weeds.

Exclusions and special situations

With the broad definition of patentable subject matter given above there are certain clear exceptions:

1 *Mere discoveries*: this can apply in the situation where the reason that a certain method works is discovered. If as a result one does not change the method there will not be an invention in the mere discovery. However, if the method can be controlled because of the discovery so as to operate more effectively there could be invention even though in many cases the method will in fact operate just as it did before. This is a mixed question of patentable subject matter and novelty but it may be possible, in some circumstances, to define this controlling step as the invention in a new sequence.
2 *Business plans*: this problem will often arise in regard to operating systems for offices in which more efficient use is made of conventional materials, the distinguishing feature being purely intellectual.
3 *Immoral, anti-social or illegal*: there can be difficult questions if 'illegal' means breaking a rule which is applicable only in one country: for example, food treatments. Usually the invention will be patentable but its use, of course, may be restrained by other laws.
4 *Computer programs*: generally computer programs are not protectable as

such but protection has been given to total systems of operation including programs and to hardware incorporating programs.

5 *Chemical products*: apart from areas which tend to be recognized by most countries there are special rules which depend on the individual countries. For example, at one stage many countries would not permit protection of chemical products and this is still true, for example, in Spain; the same is true of pharmaceutical products. This is a decision based on the law of an individual country rather than on any general principle. Most major countries have now changed their law to protect pharmaceutical products, chemical products and concepts of this type but it is not wise to assume that this is generally true.

6 *Treatment of the human body*: only a few countries give protection for this concept which would include, for example, treatment with drugs. There are special rules as to whether one can protect a known product for a new medical use or a second medical use. Generally these rules do not cover mechanical devices used in treating the body. If the device is old and can only be defined in relation to a particular context in treating the body, this may be a problem on protection: for example, a known filter which proves very efficient in a known treatment system which operates directly on the blood flow.

It must be emphasized that giving the bare bones of these exclusions does not indicate the precise extent to which protection may or may not in fact be obtainable in each case. For example, a method of laying out information in printed form is not usually patentable as it is a mere presentation of information but, if the printing in its disposition on the paper has functional aspects (for instance, a ticket on which information was available whichever way it was torn) then it might be patentable. The importance is that, if a particular project appears to involve one of these excluded areas then, as will be seen on other areas of this field of law, one should:

1 Secure an evaluation of the likelihood of patentability.
2 Make sure that that evaluation applies to all possible markets.
3 Then decide, assuming it is not patentable, as to whether the commercial success of entering into the project for the given time before it can be picked up by others justifies the investment. As has been stressed, the mere fact that something is not patentable does not mean that it cannot be commercially developed.

Novelty

A requirement for a patent is that, before an application is filed, the subject matter is not known to the world. Again this is a test which varies from country to country. The requirement of many countries is that before the day on which the first application is filed there must not have been any disclosure. Accordingly therefore any disclosure of any type anywhere in the world will prevent patenting in, for example, Europe and most major countries. Here again, however, one should be aware of important exceptions. One of the most important of these is the US where the disclosure, though 'anywhere in the world', must be in some 'printed' form (although the definition of this tends to be very liberal), or there must have been actual public use of the invention *in the US*. The further proviso is that the publication must be before the inventor made the invention, but this proviso primarily applies to inventors in the USA. It is roughly true to say that publication by the inventor within one year of the application date in the USA will not prevent securing of a patent in the USA. This is an exception which must always be borne in mind for any product which has a future market in the USA.

Other countries will have different rules. For example, some at the moment require any disclosure to be within the country. The safest rule, however, to bear in mind is that any publication is going to be dangerous and should be restrained if patent protection is desired. It is only where publication has taken place and one still wishes protection that one will look to these exceptions. Therefore control of information is vital in ensuring that patent protection is securable. A lecture by a research scientist, an unnecessary statement of future products by a sales person, can equally be fatal. Control of information about a new project must always be strict.

Obviousness

It is not enough that a concept be new for a patent to be secured. Although defined in different ways, most countries require that an invention not be 'obvious'. This is a term which gives both laymen and lawyers great difficulty. Clearly one is always looking back. If the situation ever reaches a court the test will be 'At the time the invention was made was it almost inevitable that someone working in the field would have gone in that direction?'. Another form of the test is 'Was the change any more than an ordinary workshop improvement?'. If one adopts a favourite approach of the European Patent Office, one first defines that there was a problem in the industry or art, one defines the improvement as the solution to that problem and then determines how likely that solution was in the light of what was already known. Clearly all of these tests are subjective. Many of the major inventions appear blindingly obvious once made. One can use subsidiary tests. Did the development or invention have instant commercial success? Was there a long-felt need which was not met until the invention was made? Was indeed the realization that the problem existed an invention?

In practical terms, this lack of certainty about obviousness can be advantage and a disadvantage. As a disadvantage, the patentee may never be completely sure that the patent will stand up in court since clearly there will be always an argument between two experts as to the state of knowledge at the time the invention was made and the likelihood of conclusions being drawn from that knowledge. As an advantage, those who propose to infringe will equally never be sure that they can safely ignore the patent. In the context of licensing a 'weak' patent can nevertheless be a strong feature of a total programme involving associated know-how and confidential information. The weaker a patent, however, the likelier that it will be narrow and therefore the more likely that it will be possible to design round it.

Disclosure

A patent requires that one discloses the invention. Some countries such as the US require that there be a disclosure of the best mode of carrying out the invention. Since the patent application will be published about eighteen months after the initial application date, the patentee has constantly to be aware of the risk of premature publication which may take place before there is certainty that a patent will finally be granted.

Duty of disclosure

Particularly for the US but also to a lesser degree in other countries, a potential patentee must be prepared to disclose everything that it knew about the development prior to submission of the application and to disclose any potential inadequacies in the invention. Knowledge about prior disclosures is a particularly difficult item to handle. It is important therefore, when a patent application is being considered, that those involved provide a full statement of all that they know which might be pertinent to deciding the issues of novelty and obviousness and, if advantages are to be argued, the extent to which the invention, particularly in its broadest form, does provide these advantages for every embodiment of the invention.

UK procedure

A UK patent, being a registered right, must be applied for, usually examined by a government department prior to grant and secured and maintained in accordance with the statute of the relevant country. Normally the procedure would be as follows:

Pre-application

The invention will be identified together with inventors and the ownership (for a discussion on ownership see later). Sufficient information for the description should be gathered initially together with the relevant commercial information for assessment of the justification of patent protection. This information will be submitted to the professional adviser (in Britain usually a Chartered Patent Agent). If necessary certain additional searches may be carried out to determine what information previously existed. With the technical and commercial assessment a decision will then be made as to whether to proceed with the patent application.

Initial patent application

In the UK, a first patent application will usually be submitted in a relatively informal manner. This gives the first application date in the UK Patent Office which is the date which will determine such issues as novelty (this date is the 'priority date'). One then has a year within which the commercial value of the invention can be assessed. A search can be requested from the Patent Office or from an international searching group to assist in further evaluation.

Completion, initiation of international applications

Towards the end of the year a decision must be made as to whether it is appropriate to proceed with the application and also whether protection is desired in other countries. If the decision is made to proceed in Great Britain either directly or through one of the international procedures, usually the application is resubmitted since the life of the patent depends on the date of this resubmission and therefore resubmission gives one an extra year's life. Additional material may be added at this stage and the final scope presented in its broadest form. At this stage one must request a formal search from the appropriate organization.

Publication

At the eighteenth month after the initial application date, the application will be published and will be available throughout the world.

Examination

Under the British system one first requests a search and then, depending on the results of that search, one can request formal examination. Government fees are payable at both stages. This final stage of examination is usually requested about two years after the initial application date of the first application, i.e. one year after the resubmitted application. The application in the Patent Office is submitted to a technically trained person, a Patent Office Examiner, who will examine the results of the search and determine whether the content of the application justifies a patent. Depending on the country, the tests as to inventive content can be relatively lenient or quite stringent.

Grant

Assuming the patent application is considered acceptable, it will then be granted. The life will then be twenty years from the date of application, which in the case of a resubmitted application within the first year will be from the date of resubmission of the application. This twenty year term is becoming a universal one but there are exceptions, for example, the US where it is seventeen years from the actual date of grant. Sometimes an extension can be obtained, for example, in the US for pharmaceutical inventions where there is a great delay in reaching the market. Generally speaking annual fees have to be paid to maintain the patent (see maintenance of intellectual property later).

Procedure in countries other than the UK

The above procedure is essentially one for the UK. If one were to wish protection in other countries throughout the world the procedure would have to be repeated in every single country. Because of the publication at eighteen months, foreign applications should be submitted before any publication (preferably during the first year from the priority date). In non-English speaking countries translations would have to be prepared. The resulting combination of translation costs, government fees and professional charges for each of the countries can make patent applications throughout the world extremely expensive. There are two procedures by which, although the expenses may not be minimized, they can be somewhat alleviated.

European patent office

As far as Europe is concerned an alternative procedure to applying in the British Patent Office, particularly for the so-called resubmitted application, is to take the application first submitted in Great Britain and apply in the European Patent Office (EPO). This organization will then proceed substantially similarly to the procedure mentioned above but for a single application covering up to twelve countries (Austria, Belgium, France, Germany, Great Britain, Greece, Italy, Luxembourg, The Netherlands, Spain, Sweden, Switzerland). It will be noted that this list does not cover certain EEC countries such as Ireland and Denmark or certain non-EEC countries such as Norway. At the end of the submissions to the European Patent Office a patent will be granted which is essentially a bundle of patents which will then have to be construed in accordance with the national laws of each of the countries, although certain general guidelines are in existence. Also at the date of grant it will be necessary to prepare translations into the languages of the individual countries with the exception of Germany. The procedure is more efficient in that one has only one application to guide through a patent office and one can postpone the cost of the translations which would otherwise be necessary at the time of application. With these various factors it is possible therefore to postpone making a decision as whether one wants all of the twelve countries until one reaches the actual final stage. If one took three countries (for example, Britain, France and Germany) and followed the European procedure, the costs would probably be about 10 per cent less than if one had applied individually in the three countries. The more countries one submits the greater the saving, up to a probable saving about 20 to 25 per cent comparing national as against EPC procedure costs. On the other hand it must be said that the European Patent Office has somewhat higher standards for patentability than the British Patent Office and certainly as compared with some of the other countries so there is a greater risk of not getting any patent protection at all. However, the European procedure is now generally favoured for all companies who want protection in three or more of the relevant countries.

Patent cooperation treaty

The other procedure which is available for a group of countries is the so-called Patent Cooperation Treaty (PCT) procedure. In this procedure a single application is submitted, for example, through the British Patent Office in which one can designate a substantial number of countries, including the European Patent Convention countries as a group (which would include Britain) together with countries such as the US, Japan, Russia, the Scandinavian countries, Australia. Countries are constantly being added to this procedure. One pays a reasonably substantial fee and the application would be submitted towards the end of the first year from the priority date. One receives an international search report and about twenty months after that initial application date or priority date a decision has to be made as to whether to proceed in the various countries. This decision can again be postponed by using what is known as Part II Procedure in which you request an initial decision as to patentability or preliminary examination which would allow postponement of up to thirty months from the priority date before a final decision as to the national procedure has to be made. Once one reaches the end of these postponements one will have the benefit of international search and the preliminary examination before a decision must be made as to whether national costs should be incurred.

However, at that point one has, in contrast to the European system, to continue the application in the various individual patent offices, with various national costs and procedures including translation (although, in the case of the election of the European Patent Office as a unit, one could for these European countries still further postpone actual national costs and only pay European Patent Convention costs). Probably the main advantage of the Patent Cooperation Treaty is that it allows one to postpone major costs including translations and national fees for up to thirty months, although in the end, if the decision is made to go ahead in all of the countries, it is likely that the final total costs will be greater than would have been incurred if one had initially applied in each of the individual countries. Of course, there are countries (such as South Africa and presently Canada) in which neither of these procedures is available and therefore a decision must be made very early on as to whether the protection is desired in these countries. The position can therefore be summarized as shown in Figure 17.1.

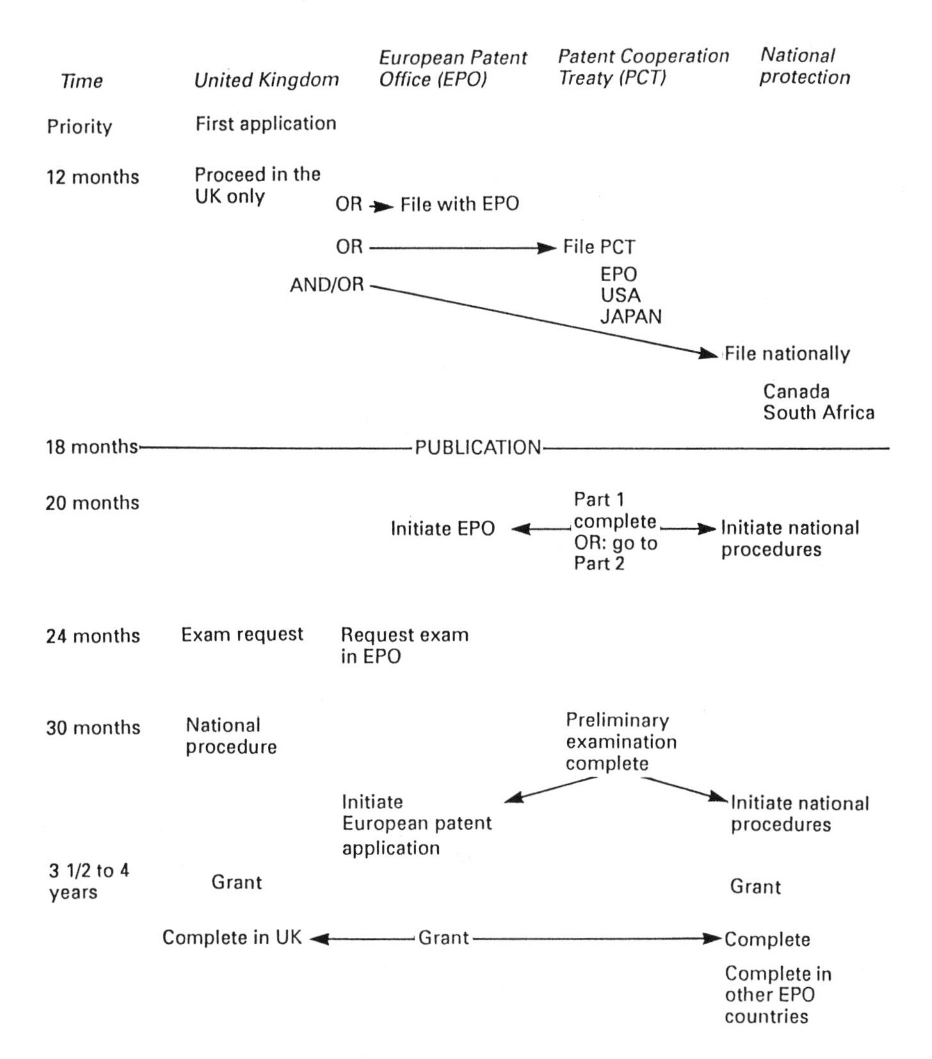

Figure 17.1 *Summary of alternative international patent procedures*

Costs

This issue will be discussed again when we are looking at the total analysis of management of intellectual property. Basically, even a simple invention for protection in Great Britain alone will probably cost initially, if one takes into account professional services whether by way of fees of outside consultant or costs of an in-house patent department, in the region of £500 upwards which could easily be £2000 to £3000 for a complex electronic invention with a total cost after going through all the procedures of £1000 upwards to £5000 to £10 000. For applications outside Britain it is probable that the initial cost of a European Patent *application* would be in the region of £2000 to £3000 and total cost per final country £1000 upwards. Going by the PCT route would initially cost from £2500 to £3000 but of course one will have eventually all of the national costs. At the point where one reaches national costs in individual countries, it is likely that an English-speaking country will cost in the region of £500 to £1000 and a translation country cost will be from £1000 to £2000. However, these are at best very broad estimates of cost and are merely intended to emphasize the likely expense of patent protection. For any individual invention one will have to consult a professional adviser for more defined costs both as to the individual countries and the various stages.

It is also important to remember that with the renewal or maintenance of a patent programme, fees will be payable at regular intervals annually which can mount up to £1000 to £1500 per country in the later stages of the life of the patent.

Maintenance of a patent – use requirements

There is no compulsion actually to use an invention but on the other hand a patent can be attacked if the subject matter is not used. This does not usually result in invalidity of the patent but rather in granting of a so-called compulsory licence, by which the licensee is entitled to use the invention but only the invention as expressed in the patent, not the associated secret information. This procedure is only rarely used, since when the situation arises, normally the parties will enter into a voluntary licence. The attack can be brought, not only for non-use of an invention in a country but also because, although there is use, the demand is satisfied primarily by import into the country concerned rather than local working. Depending on the country, attacks can be brought also on the basis that a patent is being used to hold up the development of another significant invention. If therefore a patent programme is used to try to minimize competition in countries in which the owner does not have an operation or a licensee, the risk must be assessed that a compulsory licence may be grantable at some time and, depending on the country, this could even give a right to export.

Invalidation of patents – monitoring other people's patents

Patents, of course, can be attacked by third parties and a manager must also consider whether he/she should attack the patents of others. In this connection it is important to be kept aware of patent protection sought by other parties using the various search systems (see below in the discussion on patents as information). The assessment of the risks of litigation lie outside the scope of this article. In most countries litigation of this type is in the courts although in some, as in Germany, any attack on the validity of a patent must be brought in the Patent Office while the issue of infringement will be tried in the courts. In some countries, such as Britain, there is an optional procedure by which one can go into the Patent Office first of all to determine whether one infringes and secondly to attack a patent which is believed not to be valid.

Improvements

Very often there will be an animated discussion as to whether some development is merely an improvement. Strictly speaking, to the patent lawyer the word 'improvement' is meaningless. A simple issue for any technical development is whether it is patentable as compared to what was already known. What is important, however, is the realization that once a patent application has moved beyond the initial year mentioned above it is impossible to enlarge it and to add new concepts. If therefore an 'improvement' is made after the patent has 'solidified' then the only protection available under most circumstances is submission of a further patent application which will have to be tested not only against the knowledge in the industry but against the disclosure in the earlier patent. Premature patent applications, i.e. where development is still being assessed, can therefore be counterproductive in damaging the possibility of patent protection for what eventually proves to be the real commercial development. It is for this reason that it must be emphasized that there must be a most careful assessment of everything to do with the development before a patent application is presented.

Practical aspects in the preparation of an application

If a new technical development is made and the possibility of patent protection is to be considered, a great deal can be done to minimize unnecessary costs in securing patent protection. Apart from appropriate identification of the inventors, they should be encouraged to develop a formal written statement of the invention. A useful model is to follow the type of consideration which would be argued by those in the European Patent Office as being the best means of assessing an invention, although there would be disagreement on this point in other countries.

Thus, the inventor should be encouraged first of all to state what was known at the time the invention was made: what problem existed and how that problem was solved. With this background of information one can then turn to the preparing of a patent application. The drafting of a patent application is a task for a skilled professional. Nevertheless a technologist who has been involved in the preparing of patent applications will often develop considerable knowledge of the requirements and some of the techniques at least for the preparation of the description of the invention. Much of the cost of patent application is time related and governed by the effort of the intellectual property adviser in converting a technological statement into a legal description and definition. The more material provided either in reasonably final form or in easily convertible form, the quicker the task of preparing a patent application and the more the saving in professional charges. It is useful therefore to study some typical patents in the relevant technology to secure some hints as to the presentation of the relevant information.

Drafting of a patent application falls into three stages:

1 The initial assembly of information characterizing the nub of the invention and its essential elements and description.
2 The enlarging of this concept to ensure that all reasonable alternatives are encompassed. The aim here is not merely to protect that which is to be manufactured by the patentee but also those equivalents or alternatives which a competitor might choose to try.
3 In the light of the broader statements produced out of the second stage, to ensure that there is adequate basis for any speculation on these elements in terms of the available technology to carry them out.

In enlarging the concept, the danger is that the patentee will reach the point at which the definition of the invention encompasses that which was already known or which was obvious and we would refer to the early discussion of these concepts. Another danger is that one would include in the statements possible alternatives which cannot be made to operate or can be made to operate only by exercise of a further inventive skill. As already emphasized, a patent does not prevent future patents on inventions made by way of improvement; nor does failure of operability close to the edge of the boundary of the protection necessarily invalidate the patent, but it is wiser for the draftsman not to speculate too far.

A patent application is not an encyclopedia so the patentee need not include descriptions of conventional operations or items of an apparatus which are irrelevant to the inventive context. It is also desirable that the invention actually proves to work and not be a mere speculative concept. A skilled engineer may, for a mechanical invention or even an electronic invention, be able on the basis of engineering or electronic knowledge alone to predict operability and the results. Inventions involving chemistry tend to be much more unpredictable and it is a brave inventor who proceeds with a chemical or pharmaceutical or biotechnology case without at least some real experimental work.

A draftsman will usually start off with a relatively specific embodiment. For a complex system one starts off by eliminating those parts which do not bear on or have an effect on inventive areas. For example, the base of a machine need not usually be described unless it is important to stress that the base must have a certain rigidity or some other characteristic. Starting materials in a chemical process should only be described if there is not likely to be a description in the art sufficient for the obtaining of such materials. If a particular test is necessary to determine success of the invention and that test proves not to be a common one, then a description should be included. Each draftsman has his/her own style and approach to drafting; for the lay draftsman perhaps it is best to take the one or more specific embodiments and provide the fullest possible information on these. In a case of mechanical system modifications, the engineering drawings can be used omitting any essential features like nuts and bolts. Each significant part should be given a numeral and then described in writing with a name for the part and a description of its relationship and function. In going through the description, possible equivalents should be envisaged. To obtain a smoothly running narrative it is probably best to describe the machine, a system in action and then list the alternatives and developments that might be possible. The draftsman should then try to develop the broadest possible statement of concept involved. This definition of the breadth of the invention is set out in one or more statements called 'claims'. The drafting of these represents the most difficult aspect of the arcane art of the patent agent. After the patent application is prepared, those responsible for its technical development must be ready to give sufficient time to assist the professional, i.e. patent attorney or patent agent, in answering the technical queries which will be raised from time to time by the different patent offices. Indeed, this cost of time must be assessed in considering the potential value of the patent application to the company.

Patent protection can be a considerable drain on technical resources. For example, in litigation the potentiality of tying up valuable technological time should never be underestimated. Of course, this can also be a weapon used against others.

Patents as information

As already mentioned, patents must contain a reasonably complete disclosure of the relevant invention. Patents are classified under an international system which analyses subject matter in considerable detail and therefore represent, particularly in certain fields of engineering, one of the best sources of classified technical information available.

It is important therefore, as far as possible, that patents be used as a source of information in assessing the potential protectability of new developments. In addition, because so many countries publish patent applications after eighteen months from the initial application date, they can be an invaluable source of information about competitors and developments in the industry. There are various journals available by which one can monitor duly published applications; this can also be arranged through professional representatives who maintain watching systems. In addition, certain journals for specialist engineering fields and other specific interests publish details of patent applications relevant to a specific industry. By developing a long-term monitoring strategy it is sometimes possible to predict the direction in which an industry will go in the future.

Clearance of projects

This topic will be discussed again under management of information and development, but clearly patents represent one of the most significant 'barriers' to potential development of technology. It is therefore vital that any projects be 'cleared' before they are entered into and that, during the course of development, clearing continues. Failure to clear a project can be disastrously expensive if it is found that products placed on the market are not free of patents. In addition to potential liability for infringement of patents there may be issues of liability to customers and warranties as to freedom from infringement. Under UK sales law there is an implied warranty that the product is free for sale or use by the purchaser, which includes freedom from claims under patents and other intellectual property rights. One problem is that, because of the multitude of patent systems, one for each country, a clearance in one country may not be sufficient evidence of freedom to use in other countries. However, at a certain point after searches have been carried out, one can be reasonably assured in most cases that, if that project has been cleared for some of the major industrial countries, it is unlikely that problems will exist in other countries, although the risk assessment will have to be undertaken.

Other types of 'patent' protection (petty patent)

In Britain there is only one type of patent protection. Other countries have a secondary form of patent protection sometimes called a petty patent or utility model. Depending on the country, this can be for a shorter time or have a narrow scope of protection. Sometimes it is not available for protection of other than mechanical systems, although usually this type of protection has now been extended to cover electronic circuits. It is not usually available for processes. In some cases, if a patent application runs into problems it can be converted over into this somewhat lesser type of protection. It is useful to bear in mind the possibility of this type of protection if one is assessing a large international programme and would like to take every advantage of different systems for narrower aspects of the programme which might not otherwise be worth protecting by patent. One should also always inquire as to the existence of this type of protection if one would like protection in a given country for a rather narrow concept and consideration should therefore be given to this type of protection in countries such as Italy and, particularly, Japan.

Scope of protection

Geographic

Scope can be considered from two points of view, first of all geographic. As already emphasized, a patent is usually only effective in one jurisdiction. Thus a British patent is only effective in the UK, Great Britain and Northern Ireland and the Isle of Man and does not encompass the Channel Islands which have their own separate patent system. There can be certain eccentricities in regard to the question of protection of patents in colonies and dependencies. For example, many former British territories still provide for patent protection by way of registration of the British patent with certain time limits being available within which such patents must be registered. This is one exception to having to make a decision as to patent protection at the time of the initial year or so from date of application. In considering an international programme, therefore, if for particular territory it might be significant that it is a former colony or a dependency, then the precise patent status of that territory should be determined to ensure that it falls within the scope of the various patents which are being sought.

British patents, for example, now extend over the North Sea oil fields, but for many years did not do so.

Scope – extent of technical protection

The other meaning of 'scope' is the technological meaning: i.e., how far does the protection of a patent extend over the technology? It must be emphasized, and this is frequently a source of confusion to laymen, that the exclusionary significance of a patent means that it is possible to have multiple patents one within the other: i.e., a broad concept and then an improvement within the concept are separately patentable. It must always be borne in mind that there is a clear distinction between the scope of a patent in terms of its exclusionary power and the effect of a patent as a piece of prior information. For instance, a broad patent on a new brake system would be infringed by a narrow development on the material from which the brake disc is made but clearly would not anticipate or prevent issuance of a patent on such a brake disc. In trying to determine whether or not infringement exists for a patent, issues will arise as to whether the wording of the patent is 'infringed' by reason of an equivalent. This is a very difficult area of the law and demonstrates the importance of a most careful definition of the wording of the protection, i.e. the statement of claim. Does the word vertical mean vertical or almost vertical? Does the presence in the statement of the combination as including a base mean that the device sold without a base does not infringe? It is the aim of the patent attorney using all of the many skills he/she has developed over many years of training to ensure that the technological concept is converted into this legal definition to provide as broad a protection as possible but clearly he/she must be guided by the original technologist/inventor. As we have indicated in the discussion of preparing a patent application, each word used in the claim must be analysed to make sure that it is not unduly limiting.

Infringement – real and contributory

In Britain (and this is true of many countries) there are two types of infringement. The first is the actual making or selling of something which is defined in the claim in all of its features. The second is the selling of something which is not fully within the claim (in the sense of the device without the base just mentioned) but which is an essential feature or means for carrying out the invention. This is known as a contributory infringement. Whether or not the act of a contributory infringer is infringement will depend very often on the relationship of the party concerned to the party who will finally carry out the actual infringement. For example a statement of claim, i.e. invention, for a combination of five different components will *not* be infringed by kit of four of the components if that kit is to be sold overseas, particularly to a country where no patent protection exists.

There are other technical aspects of protection which can also be significant from a managerial point of view. For example if, as already mentioned, it is not proposed to manufacture a given machine in a country in which patent protection is proposed then it might be desirable to ensure that the patent protection contains statements of protection for other aspects of the technology, e.g. use of the machine. One can then provide licences to the local customers for such use and argue that at least the invention 'use aspect' is being 'operated' in the country, so avoiding attacks of non-use. It is therefore commercially as well as legally important, in drafting a patent application, to make sure that every approach to the technology is considered in terms of definition in one or more patent applications. This can also have significance in anti-trust law. For example, the European Guidelines on Patent Licensing allow distinction between technically different uses or applications which can be licensed separately. If a patent, however, does not make that distinction then it may be difficult to justify when it comes to licensing.

Design protection

Registered designs

The patent covers the concept behind a machine and can cover such things as operations and processes. Design protection protects the appearance of the article, particularly its external appearance. The word 'design' is a confusing one in engineering since it can extend from legal meaning of 'design' (appearance) into areas such as choice of materials which a lawyer would not recognize as being design in the technical sense of the word. Usually design encompasses an aesthetic concept although strictly speaking a protectable design is one which is novel to the eye irrespective of its attractiveness. The concept dealt with here is that of registered design protection. Some other aspects of design protection more akin to copyright will be considered after the discussion on copyright.

UK procedure

Registered design protection is similar to patent protection in that it has to be sought on a country-by-country basis by registration in each of the individual countries. This involves a procedure substantially similar to that of patent protection (i.e., preparation of a submission to the Design Registry, which may be just copies of the design or may include a short description and definition of the design features to be protected). The procedure involves payment of statutory fees, investigation of the design by the Design Registry for differentiation from known designs and a decision of registration followed by a limited life during which renewal fees may be payable. The period of protection for designs is usually shorter than for patents, being fifteen years in the UK (an initial period of five years renewable for two further periods of five years). The proposed Bill amending copyright, patent and design protection may extend this protection for a period up to twenty-five years. As in patents, disclosure of the design (e.g. sale or publication before the application date) can be fatal. In the case of Britain (unlike patents) the disclosure must take place in Britain, but this is not true of the law in other countries. One can protect sets of articles when the same design is repeated perhaps in slightly different form on different articles of the set.

Usually for the protection of the design one will prepare various views of the design which can be presented as either photographs or drawings. In some countries drawings are compulsory. Where only part of an article is to be protected, for example a new handle for china, one could have a drawing of, say, a cup and then circle either the handle (indicating that this is the novel feature) or the cup (not included in the design). The registration thus will protect only that part which is indicated as novel. This issue has to be carefully considered since it means that one has to characterize in an article those features which one believes are of novel appearance as distinct from the rest of the article; an erroneous decision at this point could create problems at a later stage if it is in fact found that other features of the article are worthy of protection. This type of procedure is not necessarily available in other countries. Some difficult decisions may have to be made as to how far one can protect novel parts of an article having a distinctive appearance, where the rest of the article is conventional and could be of various shapes.

In preparing the design protection, many of the steps that one goes through in patents will have to be taken although in perhaps lessor rigorous degree, since one is looking essentially at appearance. If the articles can have different appearances when, say, opened or closed, then views open and closed may have to be provided.

Procedure for countries other than the UK

Internationally, applications would have to be submitted in each of the desired countries. The procedures mentioned with regard to patents (such as those of the European Patent Office etc.) are unfortunately not available, so a decision has to be made quite quickly (in the case of designs it is within six months of the first application) as to whether foreign protection is to be sought. One cannot 're-apply' in Britain, as one can for patents. There is an international arrangement for designs but this is not available for British companies. The cost of protecting designs throughout the world is much less than that for patents because of the briefer nature of the documents to be prepared, but it can still be quite significant (£500 to £1000 per country) with, as already mentioned, renewal fees payable every year or, in most cases for designs, every five years.

Design and functionality

One major issue in design protection is the issue of 'functionality'. A novel feature of a design appearance which is governed solely by its function (the so-called key and lock problem) may not be separately protectable, although here again there is some indication of change in this rule in the new law presently proposed.

Design protection has the advantage over copyright (which we shall review shortly) in that it is an objective protection, i.e. one merely has to look at the certificate eventually issued, compare the alleged infringement with it and this will allow one to determine whether or not infringement exists. One does not have to prove copying. Also it does allow protection in other countries by reason of initial protection in Great Britain which permits a claim of priority in the manner just indicated. If, therefore, an article is being developed which has a distinctive appearance, design protection should certainly be considered whether or not other types of protection for appearance exist in the forms that we shall presently analyse. Depending on the distinctiveness of a design and the likelihood that it will create a major impact on the market, the cost of protection as against the likelihood that others can create 'competitive designs' will give some guidance as to the appropriateness of protection.

Copyright

For many years copyright was considered to be a type of protection more appropriate to artists and authors and to have little to do with commerce and industry except in those fields specifically directed to the exploitation of art, for example publishing, dealing in art works, films, music etc. However, protection by copyright is the only protection available for many important aspects of commerce, business and industry. Essentially copyright is the protection of the expression of a work as distinct from the concept or theory behind it. Copyright is an immense topic and cannot be adequately covered in all of its aspects in a chapter of this nature. From a managerial point of view, the main requirement is to make sure that if copyright is going to be important in the future one has taken the necessary precautions to be able to demonstrate existence and ownership.

Nature of copyright

Basically, copyright protects against *copying* any type of drawing, whether an artistic work or a non-artistic work such as a technical or an engineering drawing, and any literary work, by which is meant any work using information. Leaving aside the issue of performing rights, however, it does only protect those works which are in permanent form. For the engineering manager, therefore, copyright will protect technical and architectural drawings regardless of 'aesthetic' content and assemblies of information such as manuals, tabular correlations of information instructions, directions and any form of assemblage of information in which there has been a distinct effort in the compilation of the information. Thus, for example, timetables have been protectable by copyright. It must be emphasized that this scope of protection by copyright is not shared universally by other countries who may attempt to apply an obligation of some aesthetic content. Three-dimensional articles may also be protectable and this will be examined slightly more closely in the discussion of industrial design protection; usually three-dimensional articles are protectable only as works of artistic craftsmanship. Thus, for example, a suite of furniture would have to have aesthetic content in addition to its original appearance.

Therefore almost any 'visual' work and almost any work involving a compilation of information is protectable, at least in the UK, by copyright. This protection by copyright of assemblages of information is very important in the management of information and its control. Although it must be fixed in permanent form to be protected, this is an advantage in that it can be identified (as distinct from oral information or know-how) and can therefore be more easily controlled by contract whether in terms of employer/employee relationships or in terms of technology transfer and licensing. Copyright owners will try, in the sale of an article, to control separately the rights from ownership in copyright which exist in the article. It must be emphasized that the courts are quick to find implied licence of right to use information when that is inherent in the article being sold. Copyright is always distinct from a particular article. Thus sale of a particular device would not convey the copyright in that device, so reproduction by the purchaser, if such copyright existed, would be prevented.

Establishing protection

Copyright rarely requires registration. For example, in the US, where registration was once necessary, this requirement was abolished and registration is only now necessary when initiating litigation. On the other hand, if it is likely that there will be problems in certain countries, for example Taiwan, then the possibility of protecting copyright information in that country by registration should be considered. There are a few countries where copyright registration is possible and such registration is frequently desirable even when not mandatory in that it will expedite any possible suit for protection and can give certain additional rights such as extra damages. On the other hand, for the vast bulk of copyright work no specific procedure for protection is necessary.

Because of this lack of procedural protection, however, there can be slack internal control. To evidence copyright when one wishes to enforce it one has to demonstrate:

1 Ownership.
2 That there has been copying (this implies existence of the original).
3 In some instances, that copyright has been alleged (marking of a work).

Ownership will be discussed later. Non-existence of the original copyright can often be a problem: for example, in engineering drawings which are amended over the years it may be difficult to trace the original drawing which is alleged to be copied and it may be difficult to convince a court that a copy allegedly made of a drawing made five years ago can be proven in relation to a drawing which shows amendments dated up to within a few weeks of the court case.

It is important, therefore, for materials which might be valuable and which others might wish to copy or use, to maintain copies of all of the relevant documentation in its original form, not amended, and, where there is any element of confidentiality or non-publication, to be able to demonstrate that the alleged other party could have had access. A document or other copyrightable material need not be published or otherwise released to the public to be the subject of copyright. The copyright comes into existence with the creation of the work. But copying (i.e., access) must be demonstrated. Thus if a work had never been published, one would have to show that it could have reached the hands of the alleged infringer and the manner in which such access could have taken place. It is always useful to place on a work, whether it is to be released to the public or not, some form of copyright notice. The well known notice:

© Year, Name of copyright owner

is only really appropriate for published works. In other forms (e.g. drawings) where it is desired they be maintained confidential, wording such as 'copyright reserved' should be used. Every work should carry a date and the name of the alleged copyright owner. It is most important, as will be seen under the discussion of ownership, that adequate record exists as to the original 'author'.

Marking is compulsory to maintain existence of copyright in countries such as the US.

Clearance of copyright

It is clearly easier to determine whether a work is free of copyright of others than it is to clear for patents or trade marks. To ensure that one is free of a patent or design or trade mark one has to carry out a search to determine what patents exist and whether an activity is within the scope of those found. From a copyright point of view, all that one has to do is ensure that those who are creating the work do so originally and without reproducing the work of others. On the other hand, it is rare for the creation of any work to be completely free of a surrounding body of information. Works of others will be the inspiration for either the technologist or designer. Copyright is a narrow protection and therefore only protects against a *substantial* taking of the previous information or artistic work. The test of whether there is a substantial similarity between an earlier work and a later one is essentially a practical one rather than a test of law. How far do the two resemble each other? One can look to similar works to give guidance as to the comparisons but one does not use the tests of patent law (i.e. originality or obviousness) in this connection. What designers and writers of manuals etc. must guard against is the taking of blocks of material from other works, actually physically copying in any way and, particularly, using features which might be specifically recognized as characteristic of another's work in style and concept. For example, the designer of a particular container should be provided not with samples of other containers made by competitors, but rather with a specification for what is to be achieved. At least in theory, defining what is to be done in verbal terms (including numerical definitions) will avoid any risk of infringing copyright in the article – although there is some case law to the contrary. The worst situation is where the designer is found by reason of court order sitting in front of the competitor's articles busily reproducing their features in his own designs.

Copyright has a very lengthy life of up to fifty years from the death of the author or artist (i.e. designer). Particularly with the current popularity of 'nostalgic' designs, one must be careful that reproduction of earlier works does not still involve copyright infringement.

One other area where the manager must be very careful in regard to copyright is in advertising. The temptation to develop a campaign which makes use of some current popular feature has to be resisted. For example, 'slogans' which are characteristic of a service should not be used, even in pastiche form, to promote sales of other goods. For instance, American Express will tend to take unkindly (and to the extent of law suit) the use of their slogan 'That will do nicely, thank you'. Advertising campaigns based on popular films or other works can also run into the same problem. The advice that 'a lawyer has reviewed the situation and there is not likely to be a problem' is enough in itself to indicate that a problem exists, since why otherwise would there need to be advice? It is probably safer to approach the owner of the film or TV series and work out a proper reputation or character merchandising licence even where the connection appears to be somewhat remote.

Rights from copyright protection

Copyright protects against manufacture, i.e. reproduction of the work in which copyright subsists or its importation. Generally speaking, once a work has entered the marketplace there can be no further copyright infringement in merely selling on the work in question nor would a retailer usually be guilty of copyright infringement. Very often one must know that there has been copyright infringement to be guilty of it and therefore the owner of copyright must seek out the original person responsible for the reproduction or manufacture or first import. Thus the acts restricted by copyright are copying, issuing copies to the public in the case of artistic works, performing, playing or showing in public or broadcasting or making an adaptation of the work (e.g., a translation). While there can be copyright infringement in importation, sale and possession in the course of trade or business, this will usually require knowledge of the existence of the copyright and that there is a dealing in an infringing work.

Photocopying

One important question is of course reproduction of copyright works by photocopying, a subject which is going to be dealt with in the new Act. There is no general right of fair use which automatically permits limited copying.

One has to look to the statute for any exemptions from the rule that *any* copying is infringement. At this stage the Bill contains a provision that reproduction of a work (for example, a journal article) for 'commercial research' will not be exempt from copyright infringement and may require some form of elaborate licensing scheme. The Government is apparently reconsidering this issue. The restrictions on reproduction of copyright works are customarily ignored in many libraries but it is wise to remember that wholesale reproduction of articles or other material is copyright reproduction and even if the new Bill is modified one can never suggest free use of the photocopying machine. This can also be important in the management of internal proprietary information. If material is to be controlled partly as a trade secret and partly by copyright then it should be marked that copying is not permitted except under authorized circumstances or by authorized parties. Numbering of such controlled works can be important so that if they are reproduced one can identify the source of the original. Particularly in security controlled situations it may be desirable to have physical control over copying machines so as to limit unauthorized copying.

Unauthorized copying can result in having to deliver up infringing articles. Under the present law it can also result in so-called conversion damages being awarded (i.e., a damage which is equivalent to the price of the article when sold by the infringer as distinct from loss of profits to the original owner). This would obviously be quite a drastic remedy. It is likely that this particular remedy will be abolished by the new law but the risk should be borne in mind until such time as the Act is finally passed.

Unregistered design protection

Design copyright

There has already been a discussion of protection by way of registered designs. Copyright can provide protection for industrial designs and there has already been a short discussion of the protection for three-dimensional 'works of artistic craftsmanship'. Under the current UK law (which is reflected in the law of some of the other so-called Common Law countries which stems from British law), if there is copyright in the original drawing (e.g., an engineering drawing from which the form of the final work can be visualized) then, if an article made from that drawing is further copied (e.g., by reverse engineering), that final article is an infringement of the copyright in the original drawing. This is true even though the article may be extremely functional. The courts have expressed some concern with this particular concept and although it was extremely effective in certain areas (e.g., spare parts for motor cars) at least one judge tends not to favour a claim of this nature where the article finally produced is extremely functional.

Nevertheless, at the present time this type of protection, sometimes called design copyright, is an important weapon for the protection of most articles made from drawings. If such an article is, however, a spare part, the House of Lords has held that there is a right to obtain spare parts for an article which has been purchased and that consequent upon that right is a right to make and supply such spare parts. The exact scope of the term 'spare part' is obscure. Nevertheless, at the present time it must be appreciated that, until the law is changed, an article made from a drawing is protected by the copyright in that drawing and reproduction of that article is copyright infringement and, as already mentioned, is subject to the penalty of conversion damages. Both in terms of an enforceable right and in terms of a business risk, this type of situation must be carefully reviewed by any manager. Producing exact copies of the works of others in the engineering field is as risky as reproducing artistic works. This right may not extend to reproduction of works originally produced as prototypes. If a prototype is produced, converted into a drawing, an article is manufactured and the final article is copied by another, the situation is not clear; some recent cases indicate that this could be copyright infringement.

The distinction between this type of 'design copyright' protection and registered protection is that, unlike registered protection, one does not have to define what is considered to be the protectable aspect of the appearance of the article nor need any registration procedure be undertaken. One waits until there has been apparently some infringement and then proceeds to determine from comparison of the 'infringement' with the original article whether copying can have taken place; if so, one can then enforce one's rights.

It is likely that this particular aspect of British law will be swept away by the Bill which has been mentioned several times in this chapter. Essentially, the law will provide that it will not be infringement of the copyright in a drawing or other artistic work to reproduce an article which would not have a copyrightable content in its own right. Thus, a three-dimensional very functional engineering article would not be protected. The exact scope of this amendment of the law is still, however, obscure.

Unregistered design right

In place of design copyright there will probably be enacted a design right. The 'design' which can be protected in this manner is the aspect of shape or configuration, whether internal or external, of an article, although not its principle of construction. Excluded will be features of an article which permit it to be interlocked or connected with another article, or a feature which is determined solely by the total appearance of an object of which the particular article is to form a part. The purpose of these two exclusions is to avoid protection in what is called the 'lock and key' situation, i.e., protection for the lock extending to the key, and to avoid so-called body panel protection, i.e., protection for one part of an automobile, the shape of one part being determined by the shape of the whole. Designs will not be protectable if they are 'commonplace' in the design field in question at the time of creation. Such protection will be available only to UK citizens or to citizens of the Common Market countries or certain other countries with similar law. Computer aided designs will be protectable. The design must be permanently recorded in some form before it is protected. Unlike design copyright, a drawing is not needed. Prototypes will be sufficient to create protection. Reproduction of articles made to the design, or importation, will then be an infringement, with impoundment of infringing designs. At the present time the Bill proposes a life of ten years from first marketing or fifteen years from the end of the year in which the design was first 'recorded' or made, whichever ends earlier, the last five years being subject to compulsory licences.

Clearly until this Bill is finally enacted one cannot predict the full scope of this particular right. From a managerial point of view, whether this right comes into existence or whether one continues with a present law there will still be the necessity, as mentioned under copyright, of having comprehensive records of works which are liable to be copied by others and, as will be discussed shortly, clear records as to ownership.

Confidential information

Perhaps the most important aspect of any protection of information within a company is the protection of its confidential information, whether as trade secrets, know-how, show-how or promotional information (e.g., customer lists). Curiously enough, despite the value of this aspect of intellectual property, it is perhaps one in which there has been least statutory enactment. Indeed strictly speaking 'information' is not a property right. Some countries do have codes of law relating to trade secrets and confidential information but this is not true, for example, of Great Britain. Indeed to some extent it is true to say that, because there is no law of privacy in Great Britain, there is in fact very little law on the maintaining of secrets. Generally speaking, a company is free to do what it can to find out the secrets of others providing it does not commit an illegal act such as telephone tapping nor attempt to interfere with contractual relationships. For example, flight over a competitor's plant would probably not be illegal, although it might be so in certain states of the US, if the object of the flight was to view a secret part of the plant.

In so far as Britain is concerned, one must find an appropriate relationship between two parties before the disclosure of information from one to the other can be regarded as being in breach of confidence. An obvious case would be the employer/employee relationship. This subject is of course integrally related to the question of control of security over information within a business. Once commercial information has been 'released' it cannot normally, unlike the situation with national security, ever be regarded as having a continued element of confidentiality. Moreover, a party who acquires confidential information initially without knowledge that it is confidential cannot be compelled to maintain it confidential or be penalized for using that information.

The problem therefore usually breaks down into two areas:

1 Maintenance of control over employee/employer relationships;
2 Prevention of leakage of information outside the business.

Employer/employee

A relationship exists between employer/employee, irrespective of express contract, which compels the employee to maintain confidential the information acquired about the employer's business. During employment, providing the employees understand what information should be maintained confidential, there will not usually be a major problem. The problem arises when the employee leaves the employment; what obligations can be imposed on a departed employee? Here the Courts will be much less strict. There is no precise classification, but a useful one is that the courts will not limit disclosure of information with a trivial confidential content, for example information already available from other sources (e.g., information in published patents) or 'trivial in nature' (e.g., the location of the proverbial key to the executive washroom). At the other extreme there is certain information which is self-evidently confidential; this would include the type of information subject to copyright, for example, drawings, manuals etc., secret formulae, specifications of materials. Intermediate between these two, however, there is a body of information which the courts may not regard as confidential particularly in the absence of any specific contract. Thus, customer lists have been indicated as non-confidential where the employee had the information memorized.

Restrictive covenants

Employers may seek to try to expand protection by means of non-competition covenants. While it is possible to enforce such a covenant while the employee remains an employee, again the courts are reluctant to enforce such covenant if it is seen to restrict the employee's freedom to work after departure. Undertakings not to compete must be reasonable in time and geography. For example, not permitting an employee to work in any other company in the same industry anywhere in Britain would be extremely unlikely to be upheld in the absence of very specific and good reason: the covenant is too broad in geography, irrespective of the length of the time involved. Indeed, such a clause in Germany requires the employer to continue to pay the employee recompense for the restrictions imposed. Restrictions as to specific companies who are competitors may be held acceptable if the employee is otherwise still able to secure employment.

It is best, therefore, although one should always explore the possibility of non-competition covenants, to consider making sure that the specification of what is considered confidential clearly includes material which might not be automatically protected by a court in the absence of express statement by the employer. Specification in manual form of information which is indicated as confidential can be very helpful, thus:

List of customers
Lists of materials and suppliers
Specifications of materials

should all be indicated as confidential.

Know-how

Know-how can be a problem where one is concerned particularly with operational skills rather than information which can be presented in some form of fixed statement. Nevertheless if such information can be presented precisely and in manual form (marked confidential), the greater the likelihood of protection. This can be particularly significant, for example, in franchising.

Establishing confidential information for employees

As far as an employee is concerned it is important:

1 To ensure that there is some form of agreement setting out the duties (see Figure 17.2, for example).
2 To characterize operations and skills which will be taught to the employee in the business and are not part of his normal skill as a tradesman or skilled professional and to establish a record in the form of a manual.

Any information which is circulated should be controlled and marked as confidential.

Training sessions should clearly indicate what information that is being transferred is considered as confidential and some form of record should be kept if at all possible (e.g., by video).

Employees should be obligated to return all materials provided to them in the course of employment or produced by them during the course of employment at the end of such employment.

AGREEMENT TO MAINTAIN SECRECY AND ASSIGN INVENTIONS

THIS AGREEMENT MADE this day of 19 , between a limited company organised and existing under the laws of the United Kingdom, hereinafter called "Employer", and an Employee engaged in sensitive work, that is, work in which the Employee is exposed to confidential information, hereinafter called "Employee".

WITNESSETH:

1 In consideration of the employment of the Employee by the Employer, the Employee agrees that he will not now or hereafter use for his own benefit, disclose to, or use for the benefit of any other person, firm or company, any secret or confidential information which he may or has obtained during his term of employment, whether prior to or after the date hereof, relating to the business or affairs of the Employer, and affiliated or associated companies, which shall include any parent company, including but not limited to secret or confidential information relating to any apparatus, process, formula, product, design, system, or computer program and copyright information.

2 The Employee also agrees that any and all new or improved apparatuses, processes, formulae, products, designs, systems or computer programs conceived, discovered, authored or invented (whether patentable or otherwise protectable by registration or not) by him during the term of his employment by the Employer and related to any business or activity in which the Employer is engaged at the time of such conception, discovery or invention, shall fully, freely, immediately and confidentially be communicated by the Employee to the Employer and to no other person, firm or corporation whatsoever.

3 The Employee agrees that the duties for which he is engaged would normally be such that inventions and registrable or protectable designs might reasonably be expected to result from the carrying out of said duties and shall

Figure 17.2 *Example of an employee agreement*
(Courtesy Ladas & Parry)

Third parties

In regard to outside parties, information should always be indicated as confidential if that is the intention. For example, in tendering or licensing programmes there should be an analysis of what information will have to be released and when. One should always try to release only results, not the means of achieving the results. Information should be categorized and assigned degrees of confidence. At the earliest possible date a confidentiality agreement should be executed.

As with employees, the more that information can be specifically characterized and not transmitted merely by oral discussion, the better the possibility of control.

Staff should be alerted to the consequences of disclosure (note also the relevance to protecting patentable concepts).

belong to the Employer. The Employee also agrees that he will, if and as desired by the Employer, execute all documents and do all such acts and thing at the cost of the Employer as may be necessary or desirable to obtain or join with the Employer in obtaining Letters Patent or other registration protection in respect of any such invention or improvement or design and for other similar protection in any part of the world and to vest the same in the Employer for its exclusive benefit together with all rights which may belong or accrue thereto.

4 The Employee also agrees, if the normal duties for which he is engaged cease to be such that an invention or design would normally be expected to result therefrom but duties are specifically assigned to him which would be such that an invention or design would be expected to result therefrom, or, if the nature of his duties and particular responsibilities arising therefrom create a special obligation to further the interest of the Employer, and an invention or design is made in the course of said duties, that said invention or design belongs to the Employer and the Employee will comply with the obligations of paragraph 3.

5 The Employee also agrees that in respect of any invention or design made by the Employee which is made otherwise than in accordance with paragraphs 3 and 4 and if said invention or design belongs to the Employee then if such invention or design is related to any business or activity in which the Employer or affiliated or associated company of England is engaged at the time of conception, discovery or invention, then the Employee shall offer the invention or design for licence or assignment first to the Employer before disclosing to any other party. The Employee understands that any decision made by the Employer as to whether or not to take an assignment or licence under such invention or design will not derogate from any duty of confidentiality owed to the Employer by the Employee by virtue of this agreement or any rule of law.

Figure 17.2 (*Continued*)

If a confidentiality agreement is entered into it will usually have certain exemptions for the receiver. These are principally:

1 Information which is already in the public domain other than by breach of the agreement.
2 Information already in receiver's possession.
3 Information acquired from an independent third party.

A fuller discussion of these exemptions will be found in works on licensing (see Baillie, *Licensing – a Practical Guide for the Businessman*, 1987).

Computers and computer programs

The protection of computer programs has been the subject of considerable anxiety because of, admittedly, some uncertainty in the law. However,

6 At such time that this employment is terminated, the Employee will turn over to the Employer the original and all copies of papers, notes, books or other documents belonging to the Employer relating to its business or containing a record of work done or experiments conducted or designs made by Employee.

7 In the event of this employment being terminated for any reason Employee undertakes that, for a period of two years after date of termination, Employee shall neither directly nor in the course of other employment undertake any design or development work for any company, individual or other entity which, during the term of employment, was a customer of Employer or a major commercial recipient of Employer's products or was, at the date of termination, reasonably likely to become such customer or major commercial recipient.

IN WITNESS HEREOF, the said parties have severally executed this agreement the day and years first above written.

____

By: ____________________________

Managing Director

____

Employee

In the presence of: ____________________________

Note: Two copies of this Agreement are completed. The top copy is kept by Employer and the other copy is for the employee's retention.
COPYRIGHT RESERVED: LADAS & PARRY
Ref ICB/1978/88(L9259)

Figure 17.2 (*Concluded*)

generally speaking, most countries now recognize that computer programs are at least protected by reason of copyright. This has been true for some time in Britain and is expressly stated in the proposed Bill on copyright. On the other hand, as already mentioned, computer programs are not usually protectable by patents unless they can be associated either with hardware or with some other form of technological progression. For example, a process controlled by a program can sometimes be protected in terms of defining the basic process and then reciting within that process the particular computer program control. In some countries it is possible to claim a computer programmed in a given manner even though the computer is otherwise conventional. However, it is important that a computer program be indicated as protected by copyright (preferably the necessary statement should be encoded as part of the program). As already mentioned, if it is also to be maintained as a trade secret then any marking should indicate 'copyright reserved' but not that it is a published copyright.

Computer programs can be registered, for example, in the US but this is usually only done for what might be described as mass distributed programs where speed of enforcement is necessary to avoid piracy and counterfeiting. As with confidential information, the problem is usually with the reproduction in an unauthorized manner and strict controls over the use of computer programs are important. It is also important that it be recognized that unfortunately copyright law may not yet be fully adequate to deal with such problems as use of a single program on a multitude of units. Accordingly, the computer program contract, whether or not dealing with computer programs as confidential information, is extremely important in defining the permissible use of a program by the 'purchaser'. However, for the purpose of this work it is probably sufficient to state that on the whole computer programs are protected by copyright but if there is to be any sale on of a program to another party with problems regarding the use of that program then consultation with a professional skilled in the designing of computer program contracts is essential.

Ownership

Patents

The subject of ownership has been touched upon a few times and in this connection the statutory British law is very important in relation to patents. Basically as far as patent law is concerned the inventor must be identified and named in the patent and this is indeed also compulsory in the US where a patent application must be filed in the name of the inventor. The right to ownership is defined by the Patents Act 1977. Somewhat simplifying the situation, an invention will be owned by the employer if the invention results from the employee's normal duties such duties being likely to result in inventions or, depending on the status of the employee, if he has been given special duties from which inventions are likely to result or he/she is in a position of responsibility. All other inventions belong to the employee. The significant difference between these two situations is that, when an invention belongs initially to the employer, the employee may be entitled to recompense if the invention is a very significant invention of considerable value to the employer. If the invention belong to the employee and has been assigned or licensed to the employer then the employee is entitled to additional compensation if, all things considered, the return to the employee from the payment under the original contract is less than the employee might reasonably expect having regard to the benefit to the employer. In either case the employee can apply to the courts or the comptroller of patents for an award. Clearly in the case of an employee invention the likelihood of being able to claim further compensation is much greater than in the case of an invention which initially belongs to employer, where an additional reward would only be likely in the case of a very significant and important invention. In either case the reward to the employee will have to take into account the contribution of other parties (the employer in developing the invention etc.).

Most of the problems which have arisen to date have centred on the issue of the duties of the employee. For example, is a sales engineer expected to make inventions although he may normally solve certain types of problems? For the purpose of patents, therefore, it is important that employer, as far as possible and as permitted by the Statute, establish that any invention made by an employee does belong to the employer. The duties of the employee should be established and any duties likely to lead to inventions characterized and this should be updated regularly. A typical employment contract for such an employee is provided as figure 17.2. When an invention is made within this category the employee should sign *immediately* an acknowledgement that the invention did arise within circumstances which made it an employer's invention.

It will be noted that, in the UK, but not in other countries such as Germany, the reward is related only to patented inventions so that an employer is perfectly free therefore to decide whether or not to patent a particular invention and in which countries it should be protected.

If the invention does not fall within the category which belongs to an employer it does nevertheless appear that the employer has the right to at least those inventions which relate to the employer's business, although this has been questioned by some commentators. Certainly an employer has a right to control disclosure of any invention which incorporates confidential information belonging to the employer; indeed, the employer has the right to prevent the employee preparing a patent application which includes such confidential information so that an impasse could be reached in which, although the employer does not own the invention, nevertheless it incorporates confidential information and cannot be patented without the consent of the employer. It appears appropriate therefore that all employees should sign an undertaking, if they make any invention which relates to the business of the employer, that the employer be at least given the first opportunity of taking rights under the invention by way of assignment or licence subject to the appropriate reimbursement of the employee.

It will be noted that this statutory arrangement relates only to employer/employee relationships. Insofar as outside persons are concerned, common law applies which is that, in the absence of contract, the devisor of an invention owns it. Therefore, if a consultant is retained to develop a concept, a question could arise as to the ownership of any invention made; the same could apply to, say, a subcontractor given a new piece of equipment to manufacture or develop. It is desirable therefore that in any relationship with a consultant or outside manufacturer who has discretion as to the work to be done there be a clear understanding as to the ownership of inventions and developments.

Designs

In the case of designs there is no specific statutory regulation as in the case of patents but it may be useful to use the same language for both concepts. Probably in the case of designs the relationship to the employer's business will be much clearer as will the duties of the employee.

Copyright

In copyright the author of a copyright (i.e., the artist, writer or person who creates a work) is the owner. In the case of a computer the exact owner of the work generated by a computer may be somewhat obscure as between the person owning rights in the program and the person owning rights in the data to be operated upon by the program. The proposed definition under the new Bill would appear to make the latter the owner of the copyright work. There are some eccentricities in ownership: for example, the current law would make the owner of stock on which a photograph is taken owner of the copyright as distinct from the photographer. These will probably be rationalized by the new law so that the photographer, as distinct from the owner of the stock, will become the author.

The author is the first owner except in the case of a work made by an employee in the course of employment where the employer is first owner. Under present law the commissioner of a portrait is the owner of the copyright in that portrait but that right will be abolished under the new Bill. There are some special provisions applicable to journalists which could apply to those providing material for a company journal or newspaper.

Employer/employee

This issue of employer/employee relationship in copyright demands a finding of that relationship (i.e., that there is such a relationship between the employer and employee) as to be defined as employer/employee under the Statute. This is technically known to lawyers as a provision of a contract 'of service' rather than 'for services', which would apply to a consultant. There has been some question under these provisions in the existing law, as to whether the provision that the employer owns the copyright would apply in the case of a team working on a computer program, where one member of the team is not an employee. Since it is possible to have ownership in the future work still to be created it is probably desirable that any contract make it clear that, in the case of an employee, any works created become the property of the employer as soon as created in the course of employment and that care be taken that any consultant should have an express contract setting out the disposition of any ownership of rights created by such consultancy.

It would be noted that in copyright that the work must be created 'in the course of employment'. Presumably a work created outside normal employment conditions (e.g., at home) and not directly and closely related to the employment would be outside the course of employment. Again there could be doubt on this issue which should be resolved by the appropriate employment contract.

Design rights

The new 'design right' has something of the same language on the proposed statute as copyright protection. It is important to note, however, that, unlike patent law, the law is not as rigid as to the possibility of bringing into the ownership of the employer works which might be questionable as being within the course of employment. For instance, a computer programmer could be forced to transfer to the employer all programs written whether or not written within office hours; the same might be true of the design rights. How far one wishes to go in this context of course depends on the relationship of employer to employee but careful phrasing of an employment contract is desirable in this connection.

Confidential information

Confidential information by definition is information which belongs to the employer. Here this can be tied to the general conditions of employment, i.e. the responsibility of the employee to apply all of his/her working effort to the affairs of the employer as distinct from having 'outside interests' which might compete with employer's rights to the working time of the employee.

This area fringes upon employment law but it is important in drafting any employment contract that adequate advice be taken from a specialist on intellectual property law.

It should be noted that in the area of patents the transfer of ownership of an invention to an associated company will not cut off the employee's rights, but this would not be true of ownership of design rights, copyright etc. Special problems could arise in relation to employees who are not normally located in Great Britain.

Control of information

General principles

At various points there has been mentioned the question of control of information. Control of course can involve many other issues, including those of national security, particularly for companies involved in defence and the general protection of a company's integrity.

In general terms it is first desirable that all employees be made specifically aware of their responsibility to maintain information within the company confidential. If necessary differentiation should be made between classes of employees in terms of their access to certain types of company information. This may occur naturally but in more sensitive situations express statements to this effect would be desirable. As already mentioned under confidential information, as far as possible items which are confidential should be spelled out. In more sensitive positions, an employee should be specifically briefed as to the confidential nature of the information which he/she will receive. Documents circulated within the company and not intended for distribution outside the company should all have a marking both as to copyright and as to company property. It is probably desirable not to mark every document in the business confidential, which might lead to a dilution of the meaning of that word. One has seen even items so innocuous as the works newspaper marked 'company confidential', which renders the whole system somewhat farcical. A useful distinction, therefore, is between 'company property' and 'confidential'.

The control of information within the company can be examined in two areas:

1 Access.
2 Copying.

Where appropriate, premises can be divided into two or more sections with controlled access and staff classified in regard to freedom of movement. Distribution can be determined as to material distributable to the different classes of employee on a need-to-know basis.

Employees should be discouraged from excessive copying whether by making copies of existing documents or making excessive numbers of copies of documents where the copies of existing documents are materials being circulated by way of memo etc. Creation of copying facilities should always be carefully analysed as to the desirability or otherwise in relation to this possible problem. Controlled circulation must always be considered. It is worth bearing in mind that the concept that a document will be circulated and then destroyed usually fails in practice unless there is a specific means established for receiving back the document and undertaking its destruction. Any document therefore which is likely to be significantly confidential should be controlled in some manner both in terms of a register of issue of the number of copies and the likely disposition. Where it is desirable that it be destroyed there should be control to ensure that the document has been recovered and destroyed and entry should be kept in a specific register. For that reason use of electronic mail is to be discouraged for any document which is regarded as reasonably confidential, simply because, unfortunately, it is proven that electronic mail is on the whole rarely adequately controlled both in terms of access and in terms of disposition of material by the recipient if a hard copy is made.

In the drawing office clearly confidentiality can be controlled, but again reproduction of drawings may require establishment of control as to the number of copies made and their disposition, particularly if they are transmitted to outside entities.

Establishing a programme of information control

A programme of control over confidential information should start with an assessment of various types of information in the company which require control. Clearly information which requires control because of reasons of national security will require a stricter regime than those which are merely to be controlled for commercial reasons and may indeed carry their own security system imposed by the appropriate governmental authority.

At the same time as there is an inventory or assessment of information there should be a parallel assessment of those persons requiring access to such information.

With that assessment of the areas in which control is required there can then be established the appropriate security provision. In the case of computerized information, appropriate coding may have to be inserted into the computerization to permit only limited access. Here the possibility of access to the computer by breach of these codes must be taken into account, particularly if the computer system is accessible by outside means (for instance, telephone lines) or one has a range of terminals in different establishments. In connection with the computer one should also have regulations as to the preparation of hard copy from computer stored information and how that hard copy can be controlled (if possible, electronically) so as only to permit copying by specified individuals. Persons within the company can be ranked as to access to the confidentiality of electronically stored information.

In assessing confidential information care must be taken to regulate not only technical information but also financial information which may be of sensitive nature.

Those concerned with the commercial side of the company may have greater familiarity with such means of control. The problems that arise may be with technical personnel who are not normally accustomed to dealing with information on a confidential basis. Those in accountancy and personnel sections are accustomed to the concepts of certain information being confidential, and to controlling access.

For the average company it will usually be sufficient for those concerned to know that certain types of information are confidential, that they should not be discussed outside the company or, even within the company, should only be discussed with limited groups of persons, and that copies should not be made unnecessarily and, if made, should be stored under reasonably confidential conditions. Excessive security can be counterproductive.

A clean desk routine should be established whereby papers are not left on desks overnight so as to be accessible by cleaning staff. Drawings and similar material should be kept in locked filing or storage cabinets.

Only in more unusual circumstances is it necessary to go as far as to prepare registers of documents in which original documents are given specific numeral systems and copies are in turn numbered so that circulation and storage can be strictly controlled. However, if it is necessary to go that far, then a person in the staff must be appointed with responsibility for the control of such information and the registers must be kept up to date.

It is always important that staff appreciate the necessity of nondisclosure of information outside the company, particularly to customers and colleagues in other companies. Any development department must be made aware of the dangers of excessive enthusiasm in discussing developments until they had been cleared by the patent department and the head of technology development as to potential publication.

A significant problem can be the sales section: sales staff have to be warned as to the dangers of promoting to customers products which are still in development, if there is any likelihood that the information has not been protected legally. In the description of any system the aim should be to disclose only results rather than means by which the system operates.

Control of access is quite important from the points of view not only of security and safety but also of control of information. New developments should not be exposed in parts of an establishment which are likely to be visited by third parties (for example, sales personnel from suppliers etc.). Any person allowed access to a plant should give an undertaking, even in brief formal form, that anything viewed in the establishment is kept confidential. It is surprising how many establishments have yards or peripheral areas where access by outside persons is not adequately controlled.

Managing intellectual property

From the discussion of the individual kinds of property and from areas such as security control, many of the initial management problems can be seen.

Inventory of property

If one can assume an establishment in which there is no present 'systematic' management of intellectual property, the first step would be to determine what types of information in the establishment are the subject of intellectual property. Patents and registered rights should be listed, and designs, confidential information or potential developments identified. Thus existing rights should be systemized particularly in the area of copyright which, because of its rather tenuous nature, can sometimes be difficult to identify. Proper recording systems should be established for maintenance of records of intellectual property. On the personnel side, appropriate contracts should be drawn up; both general contracts for all staff and specific contracts for any staff who have the potentiality of being involved in development.

An important aspect of any management of intellectual property is the identification of those developments which may require some form of registration. For that purpose it is desirable, if a company does not have an in-house professional staff, that somebody be appointed who can gain some acquaintance with intellectual property and its needs and who has the responsibility of constantly monitoring the developments in the company so that they can be identified at an early stage before any risks are run as to possible inadvertent disclosure.

Similarly at that point areas needing clearance must be identified. To what extent should identification be made of potential interference with the freedom to exploit, in other words, patent searches? No project should ever be allowed to develop too far in terms of financial investment before some form of clearance has been effected as to its freedom from the conflict with the rights of others, at least in the fields of patents and designs. As already indicated, there can be less problem with copyright or, for obvious reasons, confidential information providing that the creative staff are fully aware of the risks of undertaking development based on study of other people's products.

Assessment of the need for registration protection

At the time a development is identified which requires some form of protection by registration, the first assessment will have to be of the best type of protection. Clearly a development which would lead to a product which, on sale to the public, immediately disclosed its nature (for example, a new mechanical device) requires some form of protection by patent or registered design or perhaps the new design right if copying is to be resisted. The opposite extreme is some form of method involving confidential information.

The measurement of the value of intellectual property is the exclusivity which it gives to a development. If that development is likely to reach its maximum commercial value in a very short period of time then the cost and effect registration protection will normally not be justified. On the other hand if concepts are involved with a life which can be several years, then a decision as to registration protection must be taken.

One must assess the degree, after considering professional advice, to which protection by way of registration (e.g. a patent or design) will provide an actual exclusivity for a significant period of time. How soon will it take a competitor to 'design' around the best patent protection that can be secured? It is that competitive engineering time which is a considerable measure of the value of a patent to a development company. Psychologically, does the existence of patent protection afford some competitive edge while the product achieves a market segment so that, even if it is possible to design around the protection, the exclusivity gained by protection will have a marketing advantage whatever the long-term legal advantage may be?

In assessing protection one of course has to evaluate the possible cost of international protection and one has to assess the use of such systems as the European Patent Office and the International Patent Cooperation Treaty as a means of deferring cost while the commercial value and technical value of a development can be assessed. In selecting countries for protection one often finds that the selection is made according to countries in which the owner of the technology already has factories. This is not always the best solution. Very often the protection should be sought in the location where the competitor is active, so as to cut off competition at source.

Maintenance of rights

After a programme of registered rights has been established, it is important that these be maintained. As already discussed, patents require payment of fees each year and in addition areas such as use made of the patent (i.e., working) should be monitored. In a larger company this will probably be established by the appropriate professional department. Sometimes smaller companies will establish parallel control/record departments but they should also use the facilities provided by their normal professional advisers to maintain such records rather than rely solely, in the first instance, on their own records. Laws change, and a professional adviser will have a system of updating requirements as to renewal fees and other requirements.

The important point is that, upon generation in the professional system, whether in-house or by a professional adviser, of the notification that a renewal fee or maintenance fee is due, an appropriate system should be established in the company for attending to this at a reasonably high executive level of the company. One should not rely on loose systems of being reminded on the second notice from the professional adviser.

Probably the best system to employ is to ask the professional adviser to provide an annual inventory and to make a systematic approach to reviewing this once a year and instructing the maintenance of such patents as are to be continued. If a notice comes in during the year because of intermediate grant etc. this could be then dealt with on an individual basis. Sloppy handling of renewal procedures for patents result in a surprising number of lapses each year, particularly in smaller companies. The problem is usually identified as having an insufficiently defined system which goes out of control with alterations in staff.

In terms of the management decision again one should determine the continuance of a patent with the same rigour as would apply to the initiation of the costs for securing patent protection. Very often it will be found that the question of whether a patent should be maintained is answered by stating that the product is still in manufacture. That is not necessarily the correct answer to the question. The issue is, does the maintenance of a patent in practice prevent competition with a product? For example, competitors may have already invested in competitive products so that, even if the patentee's product became free to copy, there would be no commercial reason for a competitor to initiate such reproduction in view of potential costs of re-tooling etc. The patent may have run its effective course in maintaining exclusivity even though it may have legally some years still to run before it expires.

Assessments of the 'package' of rights is also desirable in examining the continued life of commercial product. If, for a valuable product, patent protection is drawing to the end of its life then assessing existing the protection may lead to an appreciation of the possibility of further development so as to upgrade the product and patent protection on it so as to maintain a competitive standing.

Similar consideration would apply to registered design rights. The assessment of the value of investing in registering designs as against relying on unregistered rights (copyright or the new 'design right') must always be made.

Staff responsibility

As already mentioned, it is desirable to have a person in the company who is responsible for monitoring intellectual property protection and its management and exploring the need for some further protection. Such a person can acquire over the years considerable knowledge of intellectual property. For the smaller company it can be very helpful not to regard their outside professional adviser as merely a person to be called upon either in the case of problem or when a determination has been made that there is an invention, but rather as a general consultant in regular touch with the development of the company.

The question is never 'Is a particular development patentable?' but 'What protection is required having regard to the economic status of the development and its future?' There are many developments which are legally protectable but commercially worthless.

Protection of reputation

The engineer manager may be more concerned with the protection of technology or confidential information. Nevertheless the reputation of a company can be as significant as its technical development. A company which depends for the particular product on its high quality of production rather than a specific technical development will sell as much on its reputation as on the 'innovative' aspect of the business. Basically, reputation is protected by trade marks and by general goodwill and indeed trade marks are a specific form of what the lawyers describe as goodwill.

Goodwill

Goodwill has a number of aspects. There is 'location goodwill' which refers to a certain type of business at a certain place which will receive customers. This is the so-called corner-shop type of goodwill. More commonly, goodwill is an aspect of a company which draws customers to it. It can be represented by the name or by other features which create a recognizable entity in the customer's mind: for example, in franchising, the particular style of an establishment, the appearance of premises or vehicles and the appearance of certain of its products. In the UK this type of goodwill is protected against encroachment by others by the so-called action of passing-off.

The specific aspect of goodwill known as a trade mark will probably be the one which will have to be most actively considered by most managers.

What is protectable

Many aspects of a company are protectable either by the common law rights of goodwill and passing-off or by specific registration. The significant difference between a trade mark registration and protection of goodwill is that a registered right which is a trade mark is protected against use by others on the basis of registration. The rights are defined by such registration. To determine whether use by another of the somewhat similar concept is an infringement will depend on that registration rather than having to prove, as would be a necessity with passing-off, that there was confusion in the public mind.

Thus if a company develops a characteristic business in, say, London and the name or other aspects of the company are copied by a company in Scotland, to enforce rights of passing-off one would have to demonstrate that the customers in Scotland were aware of the London company and were confused. For obvious reasons this can be an expensive and difficult matter to evidence in the courts. A trade mark once registered is much more easy to demonstrate in terms of its enforceability and it is for that reason that one rarely finds cases in the courts involving infringement of registered trade marks. When problems arise they are usually quickly settled simply because of the clearness of the rights.

Registration of marks

Generally speaking, for registration a mark must be distinctive of a product or service. Marks are often classified either as *trade marks* which relate to products or as *service marks* which relate to services. The distinction is somewhat blurred and the present statutory system in Great Britain is perhaps undesirable in drawing such a clear distinction between the two types of mark, unlike other countries. Perhaps the distinction can be defined most clearly by saying that a trade mark applies to the product irrespective of who actually supplied it to the customer. A service mark relates to the relationship between supplier and customer.

Whether or not one is concerned with trade marks or service marks it is necessary that the mark be distinctive. There are certain features of business which could be protectable by goodwill but may not be protectable as registered protection simply because the registrar will not permit them to be 'taken out' of the common usage by the device of registration. Such marks might be names of common large towns or countries or common laudatory words such as 'best'. However, the greatest class of this type of non-distinctive mark is that which describes the product or service.

A mark must therefore not be:

1 Lacking in distinction.
2 Misleading (i.e., indicating something that the product or service is not).
3 Property of another party.

Choosing a mark

The manager developing a new product, when it comes to identifying that product, will first wish to give it some clear-cut distinctive aspect which preferably is one which can be registered. It should be emphasized that marks need not be words or even so-called logos or designs in the commonly understood sense of these words. They can be features of a product, e.g. particular striping or a coloured thread on a hose or even the colour of certain products. Generally this type of identification builds up by reason of usage but a company should always be alert to realize that a particular feature has become distinctive of its product, even if not intentionally designed to be so, and be ready to consider the possibility of protection for such features. Registrability requirements differ from country to country and therefore one should look to the country of the market in reviewing this issue.

In selecting a mark one will look to something which is commercially sound in that it conveys to the customer the image desired by the company but, as mentioned, is not descriptive. It can be said therefore that good marks evoke some identification of the product or service but do not describe it. Too often products are named without adequate consideration of these trade mark considerations. It is desirable that a professional be consulted early on in the identification of a product. For the same reasons as apply to the technology, it is also important that use be cleared ahead of time. Searches should be carried out of the proposed identifier in the various registers, which should include not only the official register of the Trade Mark Office but also the Company Registry and various directories and other listings of names.

Company names

At this point is perhaps worth while to mention company names. Registration of the name of a company does not afford any protection for the name. It merely prevents formation of a company with an identical name. The Registrar of Companies will allow two names which are very similar to each other where there is no real likelihood of confusion in the industries. Equally, forming a company to do business in a certain area which involves use in the company name of the trade mark of another party could result in infringement action, and the fact that the name has been registered as a company name will not afford any defence.

It is always wise, therefore, in choosing the name of a company, to make sure that appropriate trade mark investigations are made before registration. This is not done by the Registrar of Companies nor is it done, unfortunately, by many of the so-called company formation agents. If a so-called ready-to-use company or off-the-shelf company is purchased one should consult a trade mark professional to make sure that the name is usable if it is intended to retain rather than change the name of the company. It must be emphasized that this sort of advice should be sought from professional advisers in the field of trade marks and not from accountants who, excellently qualified though they may be, are not qualified to advise on the legal problems of names.

Procedure for registration

Under the system applicable in Great Britain, marks are classified into some forty-two classes of goods and services. In considering whether a mark is free for use, therefore, discovery that the mark is already registered is not a bar if it is registered for some business which is remote and not under consideration. Care, however, must be taken to ensure that the owner of the other registration is equally convinced as to the remoteness. Very often the sequence of choosing a mark will consist of the company first discussing with professional advisers some of the aspects of mark protection and, after preparing a short list in the light of these guidelines, clearing such short list by appropriate searches in one or more countries. On a final short list resulting from this screening process the commercial determination can be made as to which mark is best for the purpose.

Where a mark may be used internationally it should be examined for linguistic problems: for example, meanings in foreign languages which could be singularly unfortunate in relation to the product.

The procedure in protecting a trade mark is somewhat similar to that of patents, i.e., an application is prepared setting out the mark and the specification of goods and is submitted to the Registrar of Trade Marks. In Britain, if the mark is to be used on a number of different goods, these may fall in different classes and a separate application must be filed for each class of goods or services which is to be protected.

The mark is investigated by the Registrar and, if distinctive and not already the property of another party, it will eventually be published. If there are conflicts these may be settled by securing appropriate consents if there is a clear distinction between the rights. Once a possibility of registration has been decided by the Registrar, the mark will be published and oppositions can be brought within thirty days by those who believe that the mark should not be registered (usually because of potential conflict with existing marks in use, many of which have not been registered).

As in the case of patents, worldwide trade mark protection must be sought in each individual country, but for UK companies there is no equivalent system to the European Patent Office at the present time. The so-called International Trade Mark System is available only to nationals of participatory countries, which do not include the UK.

The cost can vary widely. In Britain it will probably cost for the average trade mark between £500 and £600 to secure registration and will take approximately two years. For a mark which is not clearly distinctive but requires evidence of the distinctiveness acquired by use of the mark, then there might be greater costs in collecting evidence of such distinctiveness. Outside Britain the cost of securing registration can vary from a few hundred pounds in an English speaking country up to several thousand pounds in a country in which there is any difficulty in demonstrating distinctiveness. The importance therefore of early searching of a mark to minimize the cost of securing protection is clear. Generally those marks which meet with greatest difficulty are marks which have been selected without any great thought as to distinctiveness and possible conflicts but which have acquired such goodwill that the owner cannot afford to give them up.

Once secured, as in patents, renewal fees will have to be paid on a regular basis. Because of the long period between such renewals, which can be from five to ten years, care must be taken that adequate maintenance systems are operated. As in patents, trade marks which are not used become subject to attack. In deciding whether to go worldwide therefore one has to balance the chance of a good name being pirated against the risk which can be run if the name is not used in a certain country.

Exploitation of intellectual property

Intellectual property will generally be used to claim protection of the mark of the owner and to defend it against encroachment by others. In some instances it may even be used as the means of advertising the company by indicating that a product is patented thus giving the status of something which is being indicated by the Government as being inventive.

Licensing

Most commonly the alternative to use of intellectual property as a defence is exploitation by licensing. This is a subject in itself but for this purpose the manager assessing intellectual property will have to realize that licensing, although primarily concerned with the sale onwards of reputation or technology, has to proceed from something which is exclusive. A licensee will not purchase a technology, however advantageous, if there is no way in which the purchaser can control that purchased material against encroachment by others. Therefore, as in deciding upon patenting, decisions as to licensing will depend on the degree to which exclusivity can be maintained by patents, trade secrets or similar concepts. Accordingly, if a licensing programme is to be undertaken, one must assess the development against the rules applicable to intellectual property rather than the value of the technology to the present owner. The value of a licence will be determined by the exclusivity factor.

Although it is frequently said that licensing is primarily concerned with confidential information, this is not always true, particularly in engineering and in certain aspects of chemical products.

A licence will be affected by intellectual property apart from the recitation of the intellectual property itself and the type of rights granted, which will vary from property to property. There will be the warranties given by the owner. For example, an owner may warrant title to the various types of intellectual property but must be sure that it owns title. What if the property was originally secured from a third party consultant? Is that title clear, particularly in relation to copyright? It is probably unwise to warrant validity of the right because of the difficulty of ensuring, from searching, all of the information which might be relevant to validity being determined. Similarly, it is probably unwise for a licensor to warrant that it will undertake to protect the rights completely, since the licensor's investment by way of royalty can be considerably less than the value accruing to the licensee by virtue of sales.

A license should contain clauses which relate to the upkeep of the rights, registration of users in the case of trade marks and also in the case of trade marks stringent quality control which is mandatory in trade mark licensing. A licensee should never be permitted to acquire rights in the original intellectual property, although in the case of improvements competition law does not permit the licensor to take over any development of the licensee but merely to secure some form of licence under them. Under European law a licensee will always be free to challenge the rights, but this is not true of agreements in other countries.

Much of licensing is governed by anti-trust law. It is therefore desirable for a company which has not had much experience of licensing to ensure that its potential programme is discussed with a professional adviser who is expert in licensing of technology and reputation and knowledgeable as to the effects of anti-trust and competition law in the EEC, and their effects on licensing. There are many firms of solicitors and international lawyers who have this expertise but, since it is not necessarily found in every firm of solicitors, one should ensure that one's normal legal adviser does have expertise in this area. If franchising is to be undertaken, only a relatively small body of English solicitors have the necessary background of experience to be able to advise competently in this field.

Professional advisers

For most aspects of intellectual property law one tends to rely on the advice given by a chartered patent agent, who will usually also be qualified as a European patent attorney practising before the European Patent Office in Munich. A list of patent agents can be obtained from The Chartered Institute of Patent Agents, Staple Inn Buildings, High Holborn, London WC1V 7PZ.

For trade mark advice and certain aspects of copyright, one can rely on a member of the Institute of Trade Mark Agents, the address for which is The Institute of Trade Mark Agents, 4th Floor, Canterbury House, 2–6 Sydenham Road, Croydon, Surrey CR0 9XE.

In fields of pure copyright law and technical information many solicitors have considerable experience. This would also apply in licensing. But here, again, many chartered patent agents have considerable background in industrial copyright and licensing skills.

Internationally there are some firms of lawyers in London who practice in international intellectual property and licensing law. For example, details of US law firms with offices in London can be obtained from The British-American Chamber of Commerce, 19 Stratford Place, London W1N 9AF, or The American Chamber of Commerce, 75 Brook Street, London W1Y 2EB.

Also helpful in matters of American law in this field are The American Embassy, 24–31 Grosvenor Square, London W1A 1AE.

In licensing, the following organizations can offer help The Licensing Executives Society, c/o Dr R. C. Cass, Secretary, The Licensing Executives

Society of Great Britain and Ireland, Borax Research Limited, Cox Lane, Chessington, Surrey KT9 1SH and The Institute of International Licensing Practitioners, Suite 78, Kent House, 87 Regent Street, London W1R 7HF.

Further reading

Intellectual property law in general

Baillie, Iain C., *Practical Business Management of Intellectual Property*, Longman, 1987 (a Crown Eagle commissioned report).
Blanco-White, T. A. and Jacob, Robin, *Patents, Trade Marks, Copyright and Industrial Designs*, Concise College Text, 3rd edn, Sweet & Maxwell, 1986.
Cornish, W. R., *Intellectual Property, Patents, Copyright and Allied Rights*, Sweet & Maxwell, 1981.

UK patents

Blanco-White, T. A. *et al.*, *Encyclopedia of United Kingdom and European Patent Law*, Sweet & Maxwell, 1987 (a loose-leaf volume with supplements up to 1987).
CIPA *Guide to the Patent Act*, 2nd edn, Sweet & Maxwell, 1977 (plus supplements to date).
Terrell on the Law of Patents, 13th edn, Sweet & Maxwell, 1900.

European patent law

Chartered Institute of Patent Agents, *European Patent Handbook* Oyez-Bender, 1900 (a loose-leaf volume with supplements to date).

US patents

Rosenthal, Peter T., *Patent Law Fundamentals*, 3rd edn, Clarke Boardman, 1900.

Registered designs

Russell-Clark (Michael Fysh), *Copyright in Industrial Designs*, 5th edn, Sweet & Maxwell, 1900.
Johnston, Dan, *Design Protection*, 2nd edn, The Design Council, 1986.
Morris, Ian and Quest, Barry, *Design – The Modern Law and Practice*, Butterworth, 1987.

Copyright

Copinger and Skone-James on Copyright, 12th edn, Sweet & Maxwell, 1900.
Laddie, Prescott and Victoria, *The Modern Law of Copyright*, Butterworth, 1980.
Whale, R. F. and Phillips, Jeremy J., *Whale on Copyright*, ESC Publishing Limited, 1983.

Confidential information

Mehitan, S. and Griffiths, D., *Restraint of Trade and Business Secrets – Law and Practice*, Longman, 1986.
Milgrim, *Trade Secrets*, Matthew Bender, 1900 (a loose-leaf volume with supplements giving a survey of US law).

Ownership of invention

Hodkinson, Keith, *Employee Inventions and Designs – Law and Practical Management*, Longman, 1986 (in the Longman Intelligence Reports series).

Computer law

Brett, Hugh and Perry, Lawrence, *The Legal Protection of Computer Software*, ESC Publishing Limited, 1981.
Carr, Henry, *Computer Software – Legal Protection in the United Kingdom*, ESC Publishing Limited, 1987.
Hanneman, Henry W., *Patentability of Computer Software*, Kluwer, 1985.

Trade marks

Blanco-White, T. A. *et al.*, *Kerly's Law of Trade Marks and Trade Names*, 12th edn, Sweet & Maxwell, 1900.
Michaels, Amanda, *A Practical Guide to Trade Marks*, ESC Publishing Limited, 1982.
Reid, Brian C., *A Practical Introduction to Trade Marks*, Waterlow, 1986.

Licensing

Baillie, Iain C., *Licensing – A Practical Guide for the Businessman*, Longman, 1987 (a Crown Eagle commissioned report).
Christou, Richard, *International Agency Distribution and Licensing Agreements*, Longman Commercial Series, 1900.
Eckstrom, *Licensing in Foreign and Domestic Operations*, Clarke Boardman, 1900 (a US work on licensing, published in three loose-leaf volumes with updates).

Reputation licensing

Adams, J., *Merchandising Intellectual Property*, Butterworth, 1987.

Franchising

Adams, J. and Pritchard-Jones, K., *Franchising*, 2nd edn, Butterworth, 1987.
Mendelsohn, Martin (ed.), *International Franchising – an Overview*, International Bar Association (North Holland), 1984.
Mendelsohn, M. and Acheson, D., *How to Franchise Your Business*, Franchise World, 1987.

Educational courses

Courses on intellectual property licensing and the managing of intellectual property are given by various commercial seminar organizations including:

Crown Eagle Seminars, Hawksmere Limited, 12–18 Grosvenor Gardens, Belgravia, London SW1W 0BD. Tel. 01 730–1902.
European Study Conferences Limited, Douglas House, Queen's Square, Corby, Northants NN17 1PL. Tel. 0536 204–224.
International Business Communications Limited, IBC House, Canada Road, Byfleet, Surrey KT14 7JL.

Note Since this chapter was written, the law, particularly in relation to copyright and designs, has been changed by enactment of the Copyright, Designs and Patents Act, 1988 of the bill mentioned in the discussion. Although enactment was toward the end of 1988 most of the provisions will probably not come into force until towards the end of 1989. Generally, however, the Act reflects the forecasts in the text as to the effect of the provisions.

PART FIVE

Commercial and Financial Management

18

Financial project appraisal

Dennis Lock

The financing of engineering projects is of vital importance to engineers and engineering managers. All engineering projects (and the employment and associated benefits that they bring) must depend on the willingness of investors, who will usually want to make sure that their money is well spent. This chapter is concerned with some of the techniques used by potential investors to weigh up the financial pros and cons of new project proposals before committing their funds. These arguments are addressed principally to the financial aspects of project appraisal, it being assumed that engineers involved in appraisal decisions will be equipped through training and experience to assess the technical and forecast performance factors relating to new projects. Our case examples will range from those that are simple, offering obvious investment solutions, to others that are less obvious, and which can involve a degree of risk.

Expenditure recovery options for projects

There are several ways in which the costs of an engineering project can be funded. The particular method used may influence the appraisal calculations. The principal options are:

1 Client funded, for an agreed fixed price or according to an arrangement that reimburses the engineering company for all expenditure incurred on some time and materials basis.
2 Funding from overheads. The project costs are recovered by amortizing them over overheads built into the cost accounting and pricing structure for sales of the company's range of products and services.
3 Funding from capital. The project expenditure is covered by spending cash reserved from profits accumulated from past sales (and therefore not

distributed to shareholders as dividends), from money raised as a capital loan, from a new issue of shares, or from the sale of a capital asset.

The engineer as purchaser and seller

Every engineering manager (and many less senior engineers) have to become involved with investment appraisal decisions. This involvement can arise in a number of ways, and the proposed expenditure might be forthcoming from a customer, or it might be for an in-house project, funded from within or from money raised by the company for the purpose. Here are a few of the many possibilities:

1 Projects initiated in-house for the development of new saleable products.
2 Proposals for the purchase and installation of new equipment (for example, a computer aided design facility).
3 Professional advice given by the engineer, in the role of consultant, to a client wishing to invest in new plant or a major industrial project. This case includes selling activity by sales engineers, who must use well reasoned and convincing arguments to persuade potential customers to opt for their proposed solution, rather than for any of those put up by competitors.
4 Projects in which the engineering work is to be funded by a client as part of an industrial or construction project.
5 Many situations arising within projects in which alternative purchasing options have to be considered.

It is apparent that an engineer has to see the problem of investment appraisal from two sides. On the one hand, he is dependent upon the investment decisions of others for funding the work upon which he is engaged. In this capacity, the engineer is selling his own services, and has a responsibility to the investor to produce the required results as efficiently as possible. He will probably be required to keep the client informed of progress and expenditure on a regular, formal basis. Knowledge of the criteria used by clients to make investment decisions, particularly the amounts and timings of payments (cash outflows) and any resulting financial benefits (cash inflows), must help the engineer to construct his reports in the format best suited to the client's requirements for monitoring and control.

Now consider the engineer on the other side of the investment fence, in the role of investor or purchaser (or at least as one giving advice or recommendation to his own management or to a client on a proposed investment). In this case the engineer is identified actively with the investment appraisal and must be able not only to appreciate the financial techniques involved, but should be able to apply them himself. He can then support his recommendations for spending (or for not spending) with properly reasoned and quantified arguments.

Payback methods

Simple payback

Consider the following proposal to commit expenditure (based on an actual case).

A 35 mm microfilm viewer was in regular use in a busy drawing office. It was of excellent quality, and gave good service. The device had an inbuilt facility which allowed any user to push a button and get an A2 sized print from the microfilm. The only snag lay in the cost of the print materials, the specially prepared paper having a silver based coating which was very expensive. The company spent £20 000 each year in the purchase of paper and chemicals for this printing application. This amount (chargeable to engineering overheads) was increasing year by year.

The engineering manager was approached by a sales representative from a well-known supplier of office printing and copying equipment, who proposed that the company should install a plain paper microfilm printer to reduce greatly the cost of materials. Since this printer had no viewing capability, the proposal was that the company should keep the original viewer printer but (by depriving it of supplies after the exhaustion of existing stocks) prevent its continued use for printing.

The additional machine could be installed for an initial installation payment of £100, plus an annual rental of £3 600, payable in monthly instalments of £300. The agreement was to run for a minimum period of one year, after which there was an optional and reasonable cancellation clause. The rental was to include all maintenance, including parts, and all other consumables except the plain paper itself. At the current volume of printing, the paper costs were expected to be only £2000, as against the existing level of £20 000.

Now this is how the figures looked:

Annual saving (cash inflow)		
Sensitized paper purchases saved		£20 000
Offsetting costs (cash outflow)		
Initial payment	100	
New electrical socket (spur)	80	
Cost of floor space to accommodate the new machine.		
One square metre at £150 per metre all in cost	150	
Estimated annual electrical power consumption cost .	120	
New machine rental and maintenance	3600	

All risks insurance (obligatory under the terms of contract	50	
Plain paper purchases	2000	6100
Net annual saving (net cash inflow)		£13 900

It is obvious from these figures that the company needed no time or complex techniques in order to reach a decision. The machine was installed, and the company began to enjoy the expected substantial savings straight away. This is a case where the investment was seen to provide an immediate and assured payback. Notice, however, that the company was careful to consider the secondary costs of the new machine, such as its power consumption and the notional cost of its accommodation. On too many occasions such factors are ignored, sometimes with awful results.

Payback period

Consider now the situation facing the engineering manager in the previous case example if the salesman, instead of offering the new machine on rental terms, was only prepared to sell it outright. However, because of some special promotional deal, the company could purchase the machine at a substantial discount for a total price of only £9000, provided that the order was placed during the limited period of the offer (two weeks). This makes the investment decision more difficult, especially when the further complication is taken into account that for the purchased machine a separate maintenance contract would be needed, costing £1000 annually in advance, to cover spares and servicing. This complication is compounded by the influence of any initial guarantee period, that might have the effect of delaying the start of the maintenance agreement.

The biggest argument in favour of purchasing the machine outright is that, although the first year costs will be high, in subsequent years there will be no rental payable. Consider the figures in the first year, assuming (for simplicity) that the transaction was to take place at the start of the company's financial year. The first annual maintenance contract charge becomes payable when the six months guarantee expires. Here are the figures:

Annual saving (cash inflow)		
Sensitized paper purchases saved		£20 000
Offsetting costs (cash outflow)		
Purchase price	9000	
Initial installation charge	100	
New electrical socket	80	
Accommodation, one square metre	150	
Electrical power consumption	120	

All risks insurance (optional)	50	
Maintenance contract	1000	
Plain paper purchases	2000	12 500
Nett first year saving (net cash inflow)		£7500

Once again there is an obvious case for installing the new machine, with an assured saving of £7500 in the first year, and much larger savings in subsequent years. The machine can be said to have paid for itself within the year or, in other words, the payback period is less than one year. But what is the indicated payback period? This is certainly not obvious from the above figures. In order to discover the break even date it is necessary to tabulate the cash inflows and outflows in spreadsheet fashion, taking careful account of when they fall due for payment. This has been done in Figure 18.1. It is seen that the cumulative monthly result becomes a net inflow during the seventh month, so that the payback period is just under seven months (seven months for practical purposes).

This example has been simplified by ignoring the delays that occur in practice between invoice receipts and actual payments and there is also no consideration of the effects produced by taxation. In this case these omissions are unimportant. In commonsense terms, if any project investment produces a payback forecast of around one or two years or less, there is no need to test the wisdom of the proposals with more sophisticated techniques. The obvious recommendation is to get on with it as soon as possible and start to reap the benefits.

Total cost importance – a cautionary tale

The two cases examined so far had obvious solutions. Before going to more difficult cases, here is another case with an apparently obvious solution. This one is again taken from real life, combining events from two different projects known to this author. Names and figures have been changed, but the scale of the disaster has not been exaggerated.

It is well known that computing managers are often involved in investment decisions. Their very existence depends on their employers investing substantial sums in hardware and its associated software. John Jones was such a person, and he was in the difficult position of having no computer at all, so that he and his team were entirely dependent upon external bureaux or on data links with computers in associated companies. He decided that it was time to persuade his superiors to invest in an in-house computer, and he decided to go about it by recommending the purchase of a small computer for £50 000, which would be used in the first instance as a central word processing facility for all the company's secretaries. The initial concept was that six printers

	MONTH												Year
	1	2	3	4	5	6	7	8	9	10	11	12	total
Cash outflows	£	£	£	£	£	£	£	£	£	£	£	£	£
Purchase machine	9000	–	–	–	–	–	–	–	–	–	–	–	9000
Maintenance contract	–	–	–	–	–	–	1000	–	–	–	–	–	1000
Install power supply	80	–	–	–	–	–	–	–	–	–	–	–	80
Install machine	100	–	–	–	–	–	–	–	–	–	–	–	100
Insurance premium	50	–	–	–	–	–	–	–	–	–	–	–	50
Notional rent for space	37	–	–	38	–	–	37	–	–	38	–	–	150
Electrical power usage	–	–	30	–	–	30	–	–	30	–	–	30	120
Plain paper purchases	166	167	167	166	167	167	166	167	167	166	167	167	2000
	9433	167	197	204	167	197	1203	167	197	204	167	197	12 500
Cash inflows													
Saving on sensitized paper purchases	1666	1667	1667	1666	1667	1667	1666	1667	1667	1666	1667	1667	20 000
Net monthly cashflows													
Outflows	7767												
Inflows		1500	1470	1462	1500	1470	463	1500	1470	1462	1500	1470	
Net cumulative cash flow													
Outflow	7767	6267	4797	3335	1835	365							
Inflow							98	1598	3068	4530	6030	7500	7500

Figure 18.1 *A simple cash flow schedule. This example shows that the intended project will begin to produce net cash inflows from the second month and that the cumulative result achieves a net saving or inflow during the seventh month. In other words, this projected cash flow schedule shows that the proposed project will have a payback period of approximately seven months*

would be shared among about twenty-five secretaries, and that some VDU terminals would also be shared. Mr Jones's financial argument went like this:

Proposed expenditure	
Initial purchase of computer with word processing software, printers and VDU terminals (to include initial staff training):	£50 000
Expected annual savings	
Two secretaries fewer required, saving (salaries and benefits):	20 000
Notional saving in engineers' time through having specifications and similar documents available on file, equivalent to an annual saving of:	10 000
Total annual savings expected	£30 000

Here was a case with an apparent payback period of less than two years, offering additional benefits of increased efficiency. This is what happened in practice.

Incredibly, no one had thought about accommodation for the computer. When a suitable room was found, this had to be fitted out and air conditioned at a cost approaching £10 000. Automatic fire protection equipment had to be installed (which in fact proved itself to be necessary on one memorable night).

A 'clean' 60 A power supply had to be run from the switch room, at a cost exceeding £600. Operation of the computer and its air conditioning plant round the clock added about £5000 per annum to the company's electricity bills (increasing the peak maximum demand by 20 kVA).

Fireproof document safes had to be purchased for the storage of back-up disks and tapes. The printers (although costing over £1000 each) proved to be unreliable after heavy use, and their output was less pleasing on the eye than the typewriters that they were supposed to replace. The noise which they created led to more expense in the purchase of acoustic hoods.

Training of secretaries took longer than anticipated and there was some resistance at first to using the new system. There was no immediate reduction in the number of secretaries (although this was achieved in time) but additional computing staff were required to support the system (at higher salaries than those paid to the secretaries).

Over the next couple of years, which was supposed to be the payback period, new printers and VDUs had to be purchased, and the computer itself had to be modified and provided with more disk drives. Secretaries were faced with delays, because the system was overloaded and too slow, and they not infrequently lost several hours' work – right across the company – either through being unable to work when the computer was 'down' or as a result of data lost through a variety of causes.

The increased efficiency in engineers' time was achieved, but took over a year because of the work involved in entering the various specifications into the system.

Maintenance costs climbed to and beyond £10 000 per annum. There were increased stationery costs, since some of the company's forms had to be redesigned to suit the new printer formats. The company's insurance premiums were increased and the all-risks schedule became very complicated and difficult to administer.

The upshot of all this was that the true payback time was never. The system never stood a chance of recovering its investment. The problem with the investment appraisal calculation was that it was over optimistic, both in the technical sense, and by not including all the ancillary expenses needed to achieve the performance objectives.

In any cost estimating calculation, no matter how simple it may seem to be, the concept of total cost consideration must be satisfied. Any engineer or manager who is likely to be involved in investment appraisals (whether as a proposer or as an adjudicator) should learn to assemble and apply a few sensible test questions in order to reduce the possibility of forgotten cost items. For this purpose (especially for major capital projects) the development and use of standard checklists is recommended.

The principles of discounting and present value

Suppose that a company has undertaken to carry out a small project for a client, and that the project is expected to cost £100 000. The agreed selling price is £115 000, and the project is expected to be finished and fully billed within one year. In this kind of situation, the financial performance of the company would be adjudged by the return on its investment, and not on a payback calculation. The gross profit here is planned to be £15 000, representing a return on investment (ROI) of 15 per cent before tax. A few more suppositions are necessary before this example can be continued, namely that:

1 This company's corporate objectives are aimed at an overall ROI of not less than 12.5 per cent.
2 The interest rates obtainable at the time of this project by investing funds in a deposit account would be around 10 per cent per annum before tax.
3 The current cost of loan financing, by a bank overdraft or similar method was 12 per cent per annum.

It often happens, unfortunately, that projects overrun their budgets and timescales, sometimes by alarming amounts. Assume that this project, instead of taking one year to complete, took twenty-three months, cost £115 000, and was billed at the agreed fixed price of £115 000 one year late. Engineers working on this project, when learning of these figures, might have been inclined to shrug their shoulders and say 'Well, at least we broke even'. In fact, the company made a loss. This result can be argued in several ways, two of which follow:

1 *Cost of financing*. The £115 000 due from the customer was received one year late, although most of the project costs (assume £100 000) had to be found in the first year. If this company had an overdraft at the bank, or was dependent upon other loan financing, it would have incurred a year's additional interest charges on most of the £115 000 (say, 12 per cent of £100 000, which is a loss of £12 000 attributable to the project).
2 *Opportunity cost*. If the company had not undertaken the project at all, but had decided to invest the equivalent costs in a deposit account, the £100 000 spent in the first year, and the additional £15 000 spent in the second year, would all have been earning compound interest at the rate of 10 per cent per annum. So, without doing the sums in detail, the company could have finished up with something well in excess of £12 000 pretax profit, for no effort and without risk. This can be regarded as the cost of not taking an alternative opportunity.

These arguments illustrate a principle that has been well known to investment, life insurance and pension fund actuaries for many decades. This is that a pound spent or earned today has a higher potential value than a pound spent or earned in the future (even without the additional complication of inflation). This principle becomes very important in industrial or investment project appraisals which have planned durations of more than a year or two. Indeed, if the sums of money contemplated are very large, months become significant.

If a payment of say, £100 is due, but paid one year late, and if prevailing interest rates are 10 per cent, then the value of the £100 when it is actually received will only be equivalent to £90.09 at today's value. The £90.09 is said to be the *present value* or *discounted value* of the £100. The factor by which it is necessary to multiply the original sum to arrive at the present value (0.909 in this case) is called the *discount factor*.

Discount factors are obviously dependent on two parameters, namely the percentage discounting rate chosen and the number of years (or other chosen periods) elapsing between project start and the actual cash flow date. Tables are published which give discount factors, often to five or six places of decimals, and covering a wide range of discounting rates (including fractional percentages). Figure 18.2 gives a range of discount factors that readers should find useful in many cases, and which are adequate for application to the case examples in the remainder of this chapter.

Lease or buy?

One of the dilemmas which often faces managers wishing to invest in a project for the acquisition of plant or equipment is whether to purchase the new item outright as a company asset, or to lease it. The principal options are usually:

1. Purchase, from cash reserves.
2. Purchase, on deferred terms (lease-purchase or with a loan from a bank or other financial institution).
3. Renting, which is a hiring arrangement that is continued period by period, usually with the risk that rentals can be increased as time goes by. There will probably be a minimum rental period, and the shorter this minimum period the higher will be the rentals charged. Rental agreements often include free maintenance or even the promise that replacement equipment will be supplied in the event of breakdown.
4. Leasing is a fixed-term rental option, usually with regular payments in advance that are fixed for the term of the lease. Maintenance is not usually included, but the user will be required to keep the equipment in good repair on behalf of the leasing company (who become the legal owner of the goods) and the user may also be obliged to take out all risks insurance. If the user defaults during the lease term, or seeks to terminate the lease early, he will almost certainly be asked to pay all the remaining rentals in one immediate lump sum (possibly less some of the interest and with an allowance for any residual value of the goods realizable by the leasing company). Lease rentals are likely to be substantially lower than ordinary rentals for the same goods, but it is clear that the hirer accepts a far greater degree of financial commitment when entering into a lease agreement.

Case study: new telephone exchange

This example is another adapted from actual experience, and is concerned with an engineering company that was faced with the need to install a new private telephone exchange. A consultant was employed to recommend the size and make of exchange that would represent the best value. It was left to the engineering company to decide the terms on which to finance the acquisition.

Two options were considered. The first was to purchase the equipment outright, with the intention of disposal at the end of its expected ten-year life. The alternative case was to take the exchange on a seven-year lease (the

Year	1%	3%	5%	6%	7%	8%	9%	10%	11%	12%	13%	14%	15%	16%	18%	20%
0	1.000	1.000	1.000	1.000	1.000	1.000	1.000	1.000	1.000	1.000	1.000	1.000	1.000	1.000	1.000	1.000
1	0.990	0.970	0.952	0.943	0.935	0.926	0.917	0.909	0.901	0.893	0.885	0.877	0.870	0.862	0.847	0.833
2	0.980	0.943	0.907	0.890	0.873	0.857	0.842	0.826	0.812	0.797	0.783	0.769	0.756	0.743	0.718	0.694
3	0.971	0.915	0.864	0.840	0.816	0.794	0.772	0.751	0.731	0.712	0.693	0.675	0.658	0.641	0.609	0.579
4	0.961	0.888	0.823	0.792	0.763	0.735	0.708	0.683	0.659	0.636	0.613	0.592	0.572	0.552	0.516	0.482
5	0.951	0.863	0.784	0.747	0.713	0.681	0.650	0.621	0.593	0.567	0.543	0.519	0.497	0.476	0.437	0.402
6	0.942	0.837	0.746	0.705	0.666	0.630	0.596	0.564	0.535	0.507	0.480	0.456	0.432	0.410	0.370	0.335
7	0.933	0.813	0.711	0.665	0.623	0.583	0.547	0.513	0.482	0.452	0.425	0.400	0.376	0.354	0.314	0.279
8	0.923	0.789	0.677	0.627	0.582	0.540	0.502	0.467	0.434	0.404	0.376	0.351	0.327	0.305	0.266	0.233
9	0.914	0.766	0.645	0.592	0.544	0.500	0.460	0.424	0.391	0.361	0.333	0.308	0.284	0.263	0.225	0.194
10	0.905	0.744	0.614	0.558	0.508	0.463	0.422	0.386	0.352	0.322	0.295	0.270	0.247	0.227	0.191	0.162
11	0.896	0.722	0.585	0.527	0.475	0.429	0.388	0.350	0.317	0.287	0.261	0.237	0.215	0.195	0.162	0.135
12	0.887	0.701	0.557	0.497	0.444	0.397	0.356	0.319	0.286	0.257	0.231	0.208	0.187	0.168	0.137	0.112
13	0.879	0.681	0.530	0.469	0.415	0.368	0.326	0.290	0.278	0.229	0.204	0.182	0.163	0.145	0.116	0.093
14	0.870	0.661	0.505	0.442	0.388	0.340	0.299	0.263	0.232	0.205	0.181	0.160	0.141	0.125	0.099	0.078
15	0.861	0.642	0.481	0.417	0.362	0.315	0.275	0.239	0.209	0.183	0.160	0.140	0.123	0.108	0.084	0.065
16	0.853	0.623	0.458	0.394	0.339	0.292	0.252	0.218	0.188	0.163	0.142	0.123	0.107	0.093	0.071	0.054
17	0.844	0.605	0.436	0.371	0.317	0.270	0.231	0.198	0.170	0.146	0.125	0.108	0.094	0.080	0.060	0.045
18	0.836	0.587	0.416	0.350	0.296	0.250	0.212	0.180	0.153	0.130	0.111	0.095	0.081	0.069	0.051	0.038
19	0.828	0.570	0.396	0.331	0.277	0.232	0.194	0.164	0.138	0.116	0.089	0.083	0.070	0.060	0.043	0.031
20	0.820	0.554	0.377	0.312	0.258	0.215	0.178	0.149	0.124	0.104	0.087	0.073	0.061	0.051	0.037	0.026

Figure 18.2 *A table of discount factors. This table will be found useful in the calculation of present values. For example, the present value of £100 due in year 10, where the percentage discounting rate is 11 per cent, is £(100 × 0.352) or £35.20*

primary term) plus a further three years at greatly reduced rentals (the secondary term). Here are the parameters which applied:

1	Installed cost if purchased	£100 000
2	Planned equipment life	10 years
3	Capital allowances against corporation tax possible on annual decreciation cost based on 25 per cent of the reducing value	
4	Corporation tax rate	35 per cent
5	Estimated resale value after ten years (approximating to the written down value)	£5600
6	Primary lease term	7 years
7	Primary lease rentals (payable annually in advance, and tax deductible)	£20 000
8	Secondary lease term	3 years
9	Secondary lease rentals (payable annually in advance, and tax deductible)	£1700
10	Tax cause and effect lag (may be longer than this in practice)	1 year
11	Maintenance (whether leasing or buying): First year Each subsequent year (payable in advance and tax deductible)	 Free £8000
12	Discounting rate, per annum (advised by this company's accountant)	11 per cent
13	Value added tax, chargeable on all fees, rentals and the prices in this case, but each payment is recoverable in full by the company within the same year, and so this tax cancels out and is not included in any calculation.	

First, calculate the total cost of each option in terms of all the outgoing payments due over the ten-year period, without worrying about discounting or tax complications (this is how the average non-financial engineer might approach this problem). The comparison becomes:

Item	*Purchase* £	*Lease* £
Price	100 000	—
Maintenance (nine years)	72 000	72 000
Primary lease rentals	—	140 000
Secondary lease rentals	—	5100
Estimated resale proceeds	(5600)	—
Totals	166 400	217 100

On this basis, it seems that the option to go out and buy is the clear favourite – to the tune of £50 700. This is what happens when we take tax into account, set out the cash flows, and then discount them at 11 per cent. This has been done in Figures 18.3 and 18.4.

Project:		Condition:			Date:	
NEW TELEPHONE EXCHANGE		PURCHASE			MARCH 1988	

Year	Item	Cash flow at present cost			Discount factor at 11%	Discounted cash flow
		Inflow	Outflow	Net		
				£		£
0	PRICE	—	100 000	(100 000)	1·000	(100 000)
1	MAINTENANCE TAX RELIEF – CAPITAL	8750	8 000	750	0·901	676
2	MAINTENANCE TAX RELIEF – CAPITAL – MAINTENANCE	6563 2800	8 000	1 363	0·812	1107
3	MAINTENANCE TAX RELIEF – CAPITAL – MAINTENANCE	4922 2 800	8 000	(278)	0·731	(203)
4	MAINTENANCE TAX RELIEF – CAPITAL – MAINTENANCE	3691 2 800	8 000	(1509)	0·659	(994)
5	MAINTENANCE TAX RELIEF – CAPITAL – MAINTENANCE	2769 2 800	8 000	(2 431)	0·593	(1 442)
6	MAINTENANCE TAX RELIEF – CAPITAL – MAINTENANCE	2076 2 800	8 000	(3 124)	0·535	(1 671)
7	MAINTENANCE TAX RELIEF – CAPITAL – MAINTENANCE	1557 2 800	8 000	(3 643)	0·482	(1 756)
8	MAINTENANCE TAX RELIEF – CAPITAL – MAINTENANCE	1168 2 800	8 000	(4032)	0·434	(1 750)
9	MAINTENANCE TAX RELIEF – CAPITAL – MAINTENANCE	876 2 800	8 000	(4324)	0·391	(1 691)
10	SALE REVENUE TAX RELIEF – CAPITAL – MAINTENANCE	5 600 657 2 800	—	9 057	0·352	3 188
11		—	—	—	0·317	—
		Project net present value (NPV)				(104 536)

Figure 18.3 *Net present value calculation for a purchase option. Please refer to the text for an explanation of this case study*

Before looking at the results, it might be wise to run through the mechanics of the calculations. The first point to note is that the orderly entry of data has been assisted by the use of a simple form. Discounted cash flow calculations can seem very complicated at first, and the use of a pro forma to slot each entry

Project: NEW TELEPHONE EXCHANGE Condition: LEASE – 7 YEARS Date: MARCH 1988

Year	Item	Cash flow at present cost			Discount factor at 11%	Discounted cash flow
		Inflow	Outflow	Net		
0	LEASE RENTAL		20 000	£ (20 000)	1·000	£ (20 000)
1	LEASE RENTAL MAINTENANCE TAX RELIEF: RENTAL	7 000	20 000 8 000	(21 000)	0·901	(18 921)
2	LEASE RENTAL MAINTENANCE TAX RELIEF: RENTAL – MAINTENANCE	7 000 2 800	20 000 8 000	(18 200)	0·812	(14 778)
3	LEASE RENTAL MAINTENANCE TAX RELIEF – RENTAL – MAINTENANCE	7 000 2 800	20 000 8 000	(18 200)	0·731	(13 304)
4	LEASE RENTAL MAINTENANCE TAX RELIEF - RENTAL – MAINTENANCE	7 000 2 800	20 000 8 000	(18 200)	0·659	(11 994)
5	LEASE RENTAL MAINTENANCE TAX RELIEF – RENTAL – MAINTENANCE	7 000 2 800	20 000 8 000	(18 200)	0·593	(10 793)
6	LEASE RENTAL MAINTENANCE TAX RELIEF – RENTAL – MAINTENANCE	7 000 2 800	20 000 8 000	(18 200)	0·535	(9 737)
7	SECONDARY TERM RENTAL MAINTENANCE TAX RELIEF – RENTAL – MAINTENANCE	7 000 2 800	1 700 8 000	100	0·482	48
8	SECONDARY TERM RENTAL MAINTENANCE TAX RELIEF – RENTAL – MAINTENANCE	595 2 800	1 700 8 000	(6 305)	0·434	(2 736)
9	SECONDARY TERM RENTAL MAINTENANCE TAX RELIEF – RENTAL – MAINTENANCE	595 2 800	1 700 8 000	(6 305)	0·391	(2465)
10	TAX RELIEF – RENTAL – MAINTENANCE	595 2800	—	3395	0·352	1195
11					0·317	
		Project net present value (NPV)				(103 485)

Figure 18.4 *Net present value calculation for a lease option. This is the same project for which the purchase option was shown in Figure 18.3*

into its appropriate year greatly simplifies the exercise. Although a computer can be used, in the simple examples given here it will probably be faster to use an ordinary calculator, unless the user already has the computer, the software, and the necessary practice in using both.

In column 1 the years are numbered chronologically according to when each payment or receipt is expected to happen. The only comment here is that the initial year is always numbered O (O for origin, if you like). The periods do not have to be years, and in some cases with large sums of money involved the periods used might be shorter (in which cases, of course, the discount factors will have to be modified accordingly).

In the second column the various items have been entered in the years in which they are expected to occur. In our examples all payments (outflows) are due annually in advance. Tax reliefs are shown occurring one year after their causal events (for the purpose of these examples the possibility that tax relief inflows can take longer has been ignored).

The third and fourth columns are used to show the expected inflows and outflows, respectively, before discounting. The net resulting inflow or outflow for each year is written in the fifth column. The accountants' notation for negative amounts is used (i.e., outflows are denoted by enclosing them in brackets).

Recapitulating, at this stage we have a year by year schedule of net expected total income or expenditure. This is similar to the arrangement used in Figure 18.1, but made less familiar to engineers and other scientifically trained people because the time axis reads from top to bottom instead of the more familiar left to right.

The discounting factors listed in column 6 have been taken from the table in Figure 18.2, appropriate to an annual rate of 11 per cent. The choice of this discounting rate is obviously very important. It will vary from time to time with the economic climate, prevailing interest rates, the company's expected rate of return on capital employed, and so on. An engineering manager wishing to carry out a project appraisal using discounting techniques would be well advised to consult his financial director or chief accountant for advice on the appropriate rate.

The figures in the final column are calculated simply by multiplying the net cash flow figures in column 5 (the present cost figures) by the relevant discount factor to arrive at the present values. Summing all the present values leads to the net present value (NPV) for the whole project.

Now compare the results with the pretax present costs obtained earlier (before tax considerations and discounting). Here are the relevant figures:

	Buy £	*Lease* £	*Option indicated*	*Advantage* £
Simple pretax cost	166 400	217 000	Buy	50 700
NPV (outflow)	104 536	103 485	Lease	1051

So, there would actually be a small advantage in leasing the new telephone exchange, even though the initial simple calculations gave a clear thumbs-up to the case for buying. Leasing will not always prove to be the best option of course, but on some occasions it will be – and with far more decisive results than the small margin shown in this case.

Return on investment

Suppose that a company was contemplating a major capital investment which, instead of merely generating inflows through savings, was expected to produce revenue related directly or indirectly to the sale of a product or commodity. Some of the characteristics typical of such projects, and which would figure in appraisal calculations, are:

1 An initial capital investment is required, which may be very substantial.
2 There is likely to be a delay of one, two or more years before any return is seen on the investment.
3 In many cases, the amount of product or commodity that can be sold is finite without additional investment, either through eventual design obsolescence or, in the case of a commodity, when natural resources are used up (e.g. coal, oil, natural gas, mineral ores, etc.).
4 There will probably be a significant element of risk in the proposals, since many of the data must be estimated, and economic, environmental and political factors well beyond the control of the project manager can intervene to distort plans and make nonsense of financial targets. All projects are risk enterprises.

The question of risk will be left to the final section of this chapter. First, here is an example of a project appraisal in which the project owner wishes to know the probable return on investment that can be expected from a given set of estimates.

Mining project – rate of return calculation

A mining company plans to carry out a project to extract a mineral ore from a new underground mine. For simplicity, assume that road and rail links exist in the vicinity, and that most of the investment is required for shaft sinking, and for surface and underground equipment. The project parameters have been assessed after a feasibility study. They are:

1 Investment cost £100 million, spread over the development period in the following amounts:

First year	£20 million
Second year	£40 million
Third year	£30 million
Fourth year	£10 million

2 Ore production is estimated to be:

Fourth year	200 000 tonnes
Then at	500 000 tonnes
Until the final tenth year at	300 000 tonnes

(after which the remaining ore is forecast to be unsuitable for futher mining)

3 Operating and maintenance costs for the mine, based on the production levels given above:
Fourth year (project start up year) £3 million
Fifth and subsequent years to close down £5 million

4 Ore would be sold to another division of the company's group (which has nearby processing plant) at a price of£70 per tonne

If the estimated investment costs, operating costs and sales revenues are set out in a discounted cash flow schedule, it is possible to calculate the net present value of the project for a range of annual discounting rates, repeating the scheduling and calculation until a rate is found that gives an NPV of zero. The rate which gives this zero result is the estimated rate of return on the investment, since the net inflows will have recovered all the expenditure. In this example, the following results can be calculated:

Annual rate per cent	*NPV £m*
10	11.8
11	7.7
12	3.8
13	0.2
14	(3.0)
15	(6.2)

So, if all the project estimates and forecasts turn out to be true, the expected rate of return on investment is likely to be about 13 per cent. Figure 18.5 shows the cash flow schedule which gave this result. If it is required to get an NPV of exactly zero, instead of the approximation used here, the final narrowing down can be done by plotting a graph and interpolating (Figure 18.6) or the calculation can be carried out reiteratively (preferably with a computer).

Risk considerations

The expected outcome of a capital project might be at risk from a number of causes. Here are some of them, which may or may not apply according to the type of project:

1 Underestimating the investment costs, through any one of a number of reasons.
2 Unexpected changes in international currency exchange rates.

Project: UNDERGROUND MINE		Condition: 13 PER CENT DISCOUNT			Date: 1988	
Year	Item	Cash flow at present cost			Discount factor at 13%	Discounted cash flow
		Inflow	Outflow	Net		
				£m		£m
0	LAND AND DEVELOPMENT		20	(20)	1·00	(20)
1	DEVELOPMENT		40	(40)	0·885	(35·4)
2	DEVELOPMENT		30	(30)	0·783	(23·49)
3	DEVELOPMENT INITIAL OPERATING COSTS ORE SALES	14	10 3	1	0·693	0·69
4	OPERATING COSTS ORE SALES	35	5	30	0·613	18·39
5	OPERATING COSTS ORE SALES	35	5	30	0·543	16·29
6	OPERATING COSTS ORE SALES	35	5	30	0·480	14·4
7	OPERATING COSTS ORE SALES	35	5	30	0·425	12·75
8	OPERATING COSTS ORE SALES	35	5	30	0·376	11·28
9	OPERATING COSTS ORE SALES	21	5	16	0·333	5·33
10						
11						
		Project net present value (NPV)				0·24

Figure 18.5 *Using discounted cash flow to find the expected rate of return on project investment. This is a cash flow schedule for a mining project described in the text. It is one of a number of such schedules which were discounted on a trial and error basis, using different discounting percentage rates, until a project net present value of approximately zero resulted. The rate which yields an NPV of zero is the expected percentage rate of return on investment*

3 Changes in loan interest rates charged by financial institutions.
4 Default by suppliers, agents or subcontractors.
5 Strikes, lockouts, or other problems associated with the project workforce.
6 Natural disasters, such as floods, earthquakes, lightning damage, storms, etc.
7 Unexpected difficulties at the project site, perhaps with underground

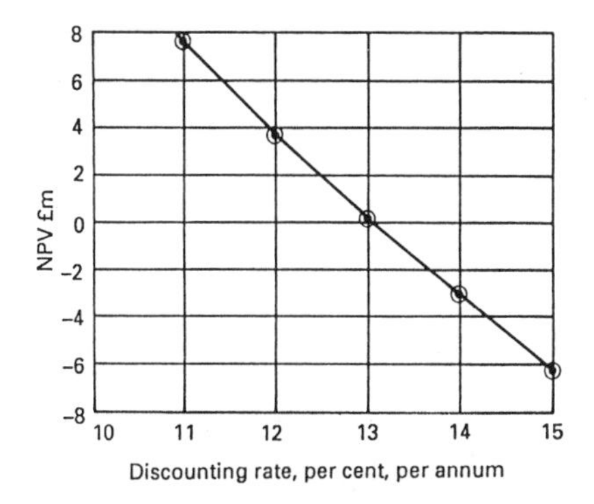

Figure 18.6 *Graphical solution for zero NPV. A graph can be plotted of NPV's found using a number of annual discounting rates in order to interpolate the results and find the rate that would yield an NPV of zero – about 13.2 per cent in this example. This is equivalent to the expected project return on investment. This method avoids tedious reiterative calculations (for which the discounting factors may not be obtainable from published tables)*

water, running sand where rock was expected, or rock where sand was expected.

8 Adverse legislation by governments, or other political factors (even including war).
9 Hostile action by bandits or terrorists.
10 Failure to achieve project completion on time, for one of the above reasons, or for any other reason, so that production start-up is delayed.
11 Technical or design difficulties that lead to increased operating costs of the completed project plant.
12 Flawed market research, so that the anticipated sales do not materialise at the planned price.
13 Early obsolescence of a product.
14 Early exhaustion of natural resource reserves.

This depressing catalogue is not complete, but should be sufficient to make the point that some attempt must be made to consider potential risks before any project funds are committed.

The risk viewpoint between owner and contractor

Industrial projects typically have two principal parties to the contract, namely the purchaser or eventual owner, and the contractor. Either the owner or the contractor could be a complex grouping of companies, organized for the purposes of the project, but it will be sufficient here to consider each as a single entity.

Very broadly speaking, the intentions of the owner are to invest funds in the project in order to achieve a planned return on his investment or (in the case, say, of a government funded public works or defence project) to contain the expenditure within authorized budget limits. The owner will probably try to limit risk on project expenditure by insisting on fixed price quotations from contractors, or by exercising very careful monitoring of project costs and progress.

The contractor's motives are different, since (with the exception of a few rare cases) he will be seeking to earn profits which are related to his expenditure in terms of labour, materials and overheads. The contractor will want to maximize his profits, but in a free market he is obliged to work within limits dictated by the competition.

Such projects should be regarded as partnerships, with the aims of the owner and the contractor being complementary in some respects. It may not be in the owner's interests to see the contractor bound by a price or agreed payment rate that is far too low, since this could itself introduce a serious risk element. In the extreme case heavy losses made by a contractor could lead to bankruptcy, leaving him unable to complete the project, and leaving the owner with an unfinished project and much of the money already paid in progress costs unrecoverable. The contractor, on the other hand, will want to achieve a project result that can be demonstrated and publicized as having been completed to the owner's satisfaction, since this will help to generate future business from the owner and from the world at large.

Some risk countermeasures for contractors

For any contractor, the consideration of risk must start when the project costs are estimated. The project definition must be as detailed and precise as possible, with a specification that is understood and agreed by both contractor and owner to contain the same objectives.

It is obviously important to reduce the risk of errors of omission in the cost estimates. The use of checklists can be invaluable, especially when the company can draw upon the experience of many past projects for their compilation. Given sufficient experience, the probable degree of estimating error should be known. The extent of such errors can be evaluated in terms of the likely range of total project costs and, as far as the competition allows, be used to influence the quoted price.

When the degree of uncertainty warrants it, the estimates can include a contingency allowance to cater for possible unforeseen problems. Other allowances can be made to cover the possible effects of inflation or alternative causes of escalating costs. But the need to quote a competitive price will always cause the evaluation of these amounts to be a matter of fine judgement.

Even in fixed price contracts it is usually possible to include clauses that will allow for price increases related to certain specified events (such as nationally agreed wage increases). The contractor will have to bear in mind the possible effects of general cost inflation or of currency exchange rate fluctuations, and it may be possible for these to be cushioned through the inclusion of suitable contract clauses.

It often happens that there are areas of a project which cannot be estimated accurately at the time of quotation because the extent of work necessary is not known, or the need for a particular piece of equipment is not adequately defined. One method often used to contain these risks is to exclude the costs of these items from the fixed price part of the quotation, and to include them separately as provisional sums (or PC sums). If, for example, a project is being undertaken to build an office block, it may not be known at the outset whether or not the local fire authority will insist on a sprinkler system as part of the fire precautions. The contractor could, in this case, include the estimated cost of installing sprinklers as a maximum possible PC sum, to be invoked at cost plus agreed mark up only if sprinklers should subsequently prove to be necessary. This procedure can work to the benefit of the project owner since, if PC sum costs do not materialize, he will not be asked to pay, even though the project was quoted otherwise at a fixed price.

The word risk automatically conjures up the thought of insurance. Many project risks would be uneconomic or impossible to insure, but other risks should be insured and some cover (such as third party liability risk) may be mandatory within the terms of contract or through legislation. The following list summarizes risks that are typically insurable in projects (in addition to other insurances that might be taken out by an employer for the protection or benefit of employees):

1 Loss or damage to property (including plant, buildings, stores, work in progress and so on) through fire, flood, theft or other accidental causes.
2 Losses or liabilities arising from the transport of project personnel or materials.
3 Liability to third parties, which can be subdivided into several categories, among which the more important are the mandatory employer's liability insurance, more general third party accident insurance, and professional liability insurance.
4 Other financial losses, including legal costs, dishonest actions by employees, and loss through accidental interruption to the business.

There is always a risk that the owner will run out of funds and be unable to meet the contractor's legitimate claims for payment. In fact, both the owner and the contractor will wish to satisfy themselves that there is no undue risk of the other running into financial problems during the course of the contract.

There are agencies which can research these risks and provide confidential reports (e.g., Dun and Bradstreet). Their methods include the examination of published accounts, and confidential approaches to other companies for independent trade references.

Default risks can be higher for overseas contracts, if only because the debt recovery process becomes more difficult. Substantial protection against the risk of non-payment by overseas project clients may be obtainable through a government-supported export credit agency. This facility is available to UK exporters (subject to certain conditions) through HM Government's Export Credit Guarantee Department (ECGD).

While the contractor may not be able to avoid or insure against all the risks caused by factors outside his control, he must obviously set up an effective project management organization that can control all phases of the design and implementation. An important objective of project management is to foresee and counter the many avoidable and unnecessary risks. Otherwise, all the preliminary effort expended in careful project appraisal will have been a dangerous waste of time.

Some aspects of risk from the investor's point of view

In any financial investment decision, it is usually recognized that the higher the potential risk, the higher the rate of return the investor will expect. A deposit account in one of the major clearing banks might be regarded as a very safe investment, but it will not make the investor's fortune. It is well known that there is a whole range of investment options, with government securities, large building societies and similar investments up at the low risk end, with considerably more risk to capital in some stocks and shares, and with very high risk to capital in certain inducements to invest in some business partnerships, certain franchises, and other more shady propositions (Mr X doubled his money in only two years completely free of all UK tax. You can do the same – just send us a cheque for £10 000 – but you must act now!).

An investor in a proposed industrial project faces risks that are often very difficult to assess. It is often argued that, if the risks are high and the forecast return is low, the owner would do better to place his funds with one of the safer investment options, giving an assured rate of return with virtually no risk to capital. Alternatively, the company may wish to place a risk premium on the target rate of return expected from a project, so that the mining company in the previous case example may have decided to rethink its expenditure and revenue plans in order to raise the discounted rate of return from 13 per cent to, say, 18 or 20 per cent.

It may be possible for an investor to gain some idea of the potential risk in a new project by studying reports of previous projects carried out in the same industry, for which published statistics may be available in some cases.

Although it is fine in theory to say that the investor will want to set his rate of return target higher to compensate for increased risk, in practice he may not be able to afford this luxury. Suppose that in the case of the mining project (for example) the same company needed the mine output to feed its existing process plant, replacing the output from another mine which was nearly worked out? The risk of proceeding with this project could be weighed against the certain loss to be suffered if the project did not go ahead.

The project owner may be able to limit his financial risk. For example, he can reduce the risk of overexpenditure during the development phase by the negotiation of contracts with fixed prices, or which are structured to provide incentives for the contractors to keep their costs and prices down. If late project completion would be particularly serious, penalty clauses can be included in the relevant contracts, so that (in theory at least) any contractor not meeting the agreed finish date will be asked to deduct an amount from his price, the total amount of the penalty being assessed in proportion to the extent of the delay. Under these conditions, the owner has to be careful not to start changing the project specification, since modifications can open the flood gates to price increases, and could well render any penalty clause unenforceable.

Another area of risk arises later in the project programme, after it has been built. Either the operating costs or product manufacturing costs could prove to be higher than expected, or the planned selling price (and therefore the estimated sales revenue) cannot be achieved owing to competition or other market conditions.

It may be possible to use statistical methods to determine the probability of some external event. For example if dry weather happened to be specially important over a particular period, meteorological records could be studied for the area in question to determine the rainfall patterns experienced over many previous years. It should then be possible to state that there are x chances in y of rain falling during the period in any future year. Other risk assessments will have to be purely subjective, so that predicting future movements in market prices, the intervention of competitors, and a whole range of other commercial and economic factors may, in spite of extensive market research, boil down to a matter of judgement based on very shaky or non existent data.

When DCF project appraisals are aimed at highlighting the most profitable choice between a number of options, there is a danger that they can tempt the investor to travel along the path with the highest risk. For example, leasing might be indicated as the clear best choice when compared to an alternative buying option, but this result does not consider the financial stability of the leasing company. Is it possible that the leasing company is having its own cash flow problems, with the possibility of bankruptcy in the future?

Potential risk factors can be tested for their possible damage to planned project success by inserting them into the discounted cash flow schedule in order to gauge the effect on the net present value. Note that this technique is not limited purely to financial changes but, since cash flows are set out against time, the effects of any possible delay to project start-up or any other change in the timing of cash flows can be tested.

Summary

Here are some generalizations.

1 In any project appraisal calculation, always try to include the total costs. Failure to consider all the incidental expenses associated with a proposed investment can lead to a dangerously misleading forecast of the project outcome.
2 If a proposed investment is expected to produce very quick returns, carry out a payback calculation. If the payback period is forecast to be less than two years, there is probably no need to use more sophisticated appraisal methods.
3 If discounting methods are to be used, take advice from the company financial director or senior accountant on the discounting rate appropriate to the company's policy and the prevailing economic conditions.
4 In order to evaluate the financial potential of a project, or to indicate the best choice between one or more alternatives, use a discounted cash flow schedule in order to find out the net present value in each case.
5 If the estimated rate of return on investment is to be calculated for any proposed project, this can be found by repeating the net present value calculation for a range of different discounting rates until, by trial and error, a discount rate is found that yields an NPV of zero. (In some exceptional cases with complex cash flows there may be more than one discounting rate which, disconcertingly, produces this result.)
6 Remember that the predictions from any of these financial appraisal techniques may be overturned by risk factors. The methods can, however, be used to test the probable financial effect of predictable risks. Otherwise, the investor must reduce the possibility of avoidable risk through good management, and use sound commercial judgement to weigh up the implications of other hazards.

Further reading

Franks, J. R. and Broyles, J. E., *Modern Managerial Finance*, John Wiley & Sons, 1979. This is a comprehensive text which includes an excellent account of appraisal techniques and the financial treatment of risk. There are many worked examples and exercises plus some useful discounting tables.

Lock, Dennis (ed.), *Project Management Handbook*, Gower, 1987. A general treatment of project management, but with specific chapters on project financing, cost estimating, project definition, contract administration and insurance.

19

Cost accounting principles

Michael Tayles

Accounting has been defined by the American Accounting Association as 'the process of defining, measuring and communicating economic information to permit informed judgements and decisions by users of the information'. Accounting has three branches, which are:

1 Financial accounting.
2 Cost accounting.
3 Management accounting.

The emphasis of this chapter is on cost accounting, since it is this aspect which the average engineer will encounter most often. It is appropriate, however, to start by distinguishing between the three accounting branches.

Introduction – the three branches of accounting

Financial accounting

Financial accounting is the recording of past transactions for the whole organization in order to fulfil legal requirements and to report to external users of information (for example, the shareholders, creditors, lenders and government departments). It is regulated by accepted accounting principles and Statements of Accounting Practice (SSAPs). The accounts of all UK public limited companies are audited.

Cost accounting and management accounting

The distinction between cost accounting and management accounting is not always clear-cut. They both serve internal members of the organization, and often relate to segments of it (such as departments, machine groups, sales regions or individuals). They are future orientated, and not governed by any legal regulations (although there are some generally accepted principles and practices). The best cost and management accounting information is that which is the most useful to the manager using it.

If a distinction can be made it is that:

Cost accounting is concerned with accumulating cost information for the valuation of stock, the calculation of profit and to help with judgements on pricing.

Management accounting has a wider role relating to all information which can help management to make better decisions, extending (for example) to sales revenues, market shares and cash flows.

In practice accounting executives have various job titles and there is no strict guide to their functions and responsibilities. All accounting information should support managers in their functions of decision making, planning, controlling, organizing, communicating and motivating.

Cost objectives

Information will be passed between cooperating accountants and used for various purposes depending on the problem posed and the circumstances. It is important that the way in which cost information is to be used is clear, both to the preparer and to the user: otherwise there is a danger that the information could mislead. For example, the acquisition of material can be regarded, according to the circumstances, as a purchase, held as a stock value, issued to a department, related to a finished article, be compared with revenue to reveal a profit or loss, and feature as a cash flow.

Costs are often related to a cost objective, which is any activity for which a separate measurement of cost is desired. Examples of cost objectives are a product, a service, a department, a machine and a sales territory.

A useful and common framework is to consider cost objectives that relate to the major areas in which accounting information supports management. These are:

1 Stock valuation and the calculation of profit.
2 Decision making.
3 Planning and control.

For example, the cost of operating a machine may be required:

1 To assist in determining a stock value for the profit and loss account; or
2 For a decision concerning the possible replacement of the machine; or
3 To ensure that the costs of running the machine are known and controlled, so that no more than is necessary is spent in running it.

Costs for stock valuation and the measurement of profit

The methods by which cost data are collected, particularly for product costs, are influenced by the production process. An industry which produces discrete items, perhaps to special order, will employ *job costing*. Where, on the other hand, a succession of similar items are produced with no break in the operation, *process costing* is applicable. A common requirement of both systems is the calculation of a product cost.

Product costs

The elements of cost which comprise a product cost are shown in Figure 19.1. The calculation of a product cost requires the identification and measurement of all costs that are directly or indirectly associated with the product. The calculation of a product cost up to the point of manufacture will be considered first, because it is this cost that will be applied to finished stock.

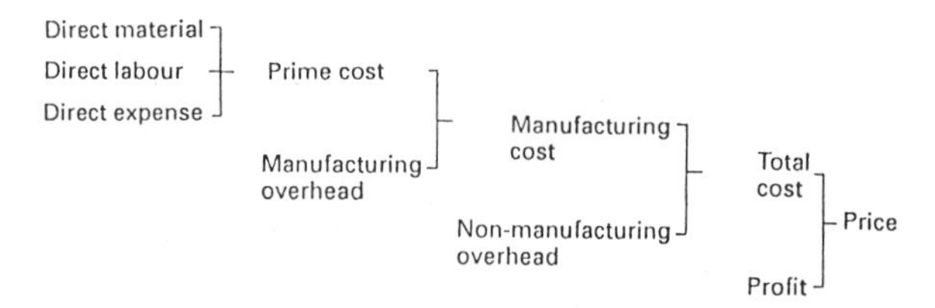

Figure 19.1 *Elements of cost. This shows the build up of costs towards a manufacturing and total product cost*

Direct costs

All product costs are either materials, labour, or other expenses. Some of these costs can be easily and conveniently associated directly with the finished product. These are called *direct costs*, and the word direct is also applied to the materials, labour and expenses that make up the direct costs of a product.

Thus, for example, the costs of sheet steel used in pressing out car body parts, or of the plastic used in an extrusion process are direct costs (because these costs are readily attributable to the product) and the materials are called *direct materials*.

Similarly, *direct labour* is the time and cost of production operatives which can be identified specifically with the manufacturing process.

Direct expenses occur less frequently. An example would be royalty payments to be made for each unit produced under a licensing agreement.

Indirect costs (overheads)

All costs which are not direct are called *indirect costs* although *overheads* is a more usual term. *Indirect materials* might include such things as free issue consumables (cleaning rags, lubricants, welding rods, and so on) which are used generally across a range of products and cannot easily be attributed to a particular job or product.

Indirect labour includes the costs of management and supervision, and can apply also to those operatives working in general service areas such as maintenance, heat treatment plant, paint shops, and so on.

It is relevant to note here that although the term direct workers is frequently used to describe those who are always engaged on direct operations, any time spent by them other than on direct production must be regarded as indirect. This situation arises whenever they are unable to carry out their direct work through sickness or other absence, or it might be due to waiting time, machine failure, scrap or rework. This leads to the condition generally referred to as indirect bookings by direct workers.

Indirect expenses generally are the costs incurred in providing and supporting the general facilities of the company. Rent, rates, water and sewerage, heat, light, insurances, legal fees and catering subsidies are all examples of indirect expenses that contribute to the general overheads and which cannot be attributed directly to any specific product or operation.

Blanket rates for overheads

The objective when dealing with overheads in product costing is to try and apportion them as equitably as possible to products. One principle reason for this is to assess a factory cost for each product that (when used as a basis for its selling price) will recover in sales revenue the direct costs plus a fair and sufficient share of the indirect costs. This process is termed overhead recovery because, if the apportionment has been correctly gauged, the company will recover all of its indirect costs from the product sales.

It would be possible simply to add up all the overheads for a whole factory, and apportion these as a universal average rate, applied possibly in accordance with the number of each type of unit produced, or on the basis of the relative number of production hours consumed by each product. Such a rate would then be incorporated with the direct material, direct labour and any direct expense costs previously identified to form a product cost for each item in the product range. But such a rate, by averaging out the impact of various labour and machine operation cost rates, would significantly reduce the accuracy and usefulness of the information. A blanket overhead rate, as this is called, is only suited to very small factory operations or where only one product is manufactured.

Establishing departmental rates for overheads

A usual method of dealing with overheads starts with the calculation of an overhead rate for each department. Some costs may be indirect to a product but direct to a department (for example, a supervisor's salary). Other indirect costs relate to several or all departments, so that an equitable method for apportioning these has to be established (such as the floor area occupied for rent and rates and the value of plant for plant insurance premiums). As the prime objective is to arrive at a product cost, the costs of service departments need to be reallocated on the basis of the service they provide to production.

Stages in the calculation of departmental overhead rates are:

1 Allocate or apportion all predicted overhead costs for the period under consideration across all productive and service departments.
2 Separate the productive and nonproductive departments listed during stage 1. Then apportion the service departments' costs across all the productive departments.
3 Knowing the planned number of direct manhours to be worked during the period under review, calculate an overhead rate for each productive department. Assume that the company is in the UK and that the currency is sterling. The rate would be worked out by dividing the number of direct hours planned into the total departmental indirect cost, with the overhead rate being expressed in terms of pounds sterling per direct labour hour (in some companies this is expressed in terms of a percentage rate, relating overheads to the direct labour costs instead of to time).

Figure 19.2 shows in principle how an overhead analysis (stages 1, 2 and 3) would be set out. The numbers of items and departments shown in the example have been scaled down for the sake of simplicity.

It is now possible to relate the productive department total overheads to products by means of a suitable index according to the use made of each department by each product.

The overhead absorption process

Most overheads accrue on a time basis, so such a basis is appropriate for charging departmental costs to products. Machine hours are used for a mechanized department, and labour hours for a department which is labour intensive. The rate which is produced is called a departmental *overhead absorption rate*. The application of these rates is illustrated in Figure 19.3.

Predetermining overhead rates

The allocation and absorption process can be undertaken using actual (measured) information or budgets (predictions or estimates). The use of actual costs and hours involves a time delay (waiting for the information to be measured and processed) which would reduce its usefulness. It is preferable to

Expenditure	Basis of apportionment	Total	Production departments			Service departments	
			Pressing	Drilling	Assembly	Stores	Maintenance
		£'000	£'000	£'000	£'000	£'000	£'000
Indirect wages	Direct labour	665	205	210	147	29	74
Rent and rates	Area occupied	100	10	20	30	30	10
Plant insurance	Book value	15	5	8	1	–	1
Plant depreciation	Book value	150	50	80	10	5	5
Supervisors' salaries	Number of employees	105	18	29	32	16	10
Subtotal (stage 1)		1035	288	347	220	80	100
Reapportionment of service department costs:							
Stores	Value of materials		30	40	10	(80)	
Maintenance	Engineers' time		32	48	20		(100)
Subtotal (stage 2)		1035	350	435	250	–	–
Machine hours (000s)			50	100	–		
Labour hours (000s)			–	–	200		
Departmental overhead rate (stage 3)							
Per machine hour			£7	£4.35			
Per labour hour					£1.25		

Figure 19.2 *Manufacturing overhead analysis. The completion of stages 1, 2 and 3 as indicated in the next is shown at the subtotals*

		Rate £	Cost £
Direct material			
0.25 square metres		32.00	8.00
Direct labour			
	Time	*Rate £/hour*	
Press	8 minutes	10	1.33
Drill	25 minutes	8	3.33
Assembly	20 minutes	6	2.00
Direct expense			0.00
Prime cost subtotal			14.66
Manufacturing overhead			
	Machine time	*Rate £/hour*	
Press	10 minutes	7	1.17
Drill	15 minutes	4.35	1.09
	Labour time		
Assembly	20 minutes	1.25	0.42
Total manufacturing cost			£17.34

Figure 19.3 *Manufacturing cost of product 'hypoth'*

use estimates of future costs and workload, and integrate these with the planning process.

Annual overhead rates

Overhead rates can be revised as frequently as desired, but this should be done at least in line with the annual budgeting cycle. Calculating overhead rates at more frequent intervals (monthly or quarterly) can introduce seasonal fluctuations, which will probably be considered undesirable.

Under or over recovery of overheads

One effect of using predetermined overhead rates, whatever the accounting period, is that they are unlikely to agree exactly with the actual costs and activity levels (workload) experienced. It is important to know the extent of these differences, so that action can be taken for the future and to reconcile the periodic financial accounts.

Under recovery of overhead occurs from one or both of the following reasons:

1 The planned product output does not materialize, either through lack of customer orders or from production difficulties of one kind or another. This

would result in fewer direct hours than those planned, with a proportional reduction in the planned overhead recovery.
2 An underestimate of total indirect costs (for example, an unexpectedly high increase in rent).

Over recovery of overhead results from the opposite of either or both of these conditions (increased workload or decreased total indirect costs).

Non-manufacturing overheads
Administration, selling and distribution overheads will not be applied to stock values because these functions do not relate to the manufacture of goods for stock. Where, however, a product cost is to be used to help set a selling price, then these additional overhead items must be incorporated into any cost projection. As there is only a tenuous relationship between these overhead costs and individual products, a sophisticated approach to allocation is impossible. Instead, an average rate per unit, or a percentage of the manufacturing cost might be predetermined from the year's budget, and such a flat rate applied to all products to determine the total cost and examine a proposed selling price. This is demonstrated in Figure 19.4.

	£
Manufacturing cost (established in Figure 19.3)	17.34
Non-manufacturing overhead (based on 30 per cent of the manufacturing cost)	5.20
Total cost	22.54
	£
Projected selling price (determined after market research)	26.00
Margin on sales	3.46
which, expressed as a percentage of the selling price is	13.3 per cent

Notes
The overhead rate of 30 per cent used here would have been established from the company's annual budget.
The selling price would have been decided after management discussion, bearing in mind the total cost and the knowledge of the market.

Figure 19.4 *Total cost of product 'hypoth'*

Costing materials and labour

In many organizations the most significant proportion of costs is associated with materials and labour. It is important, therefore, that appropriate systems and records exist to control and account for the disposition of these resources. Control implies the objective of ensuring that minimum funds are invested in these resources, while ensuring the survival of the firm in the long term. Accurate costing of materials and labour requires that the appropriate job, department or customer is charged with the cost of any resource used.

Materials

Costing of materials involves identifying the use of all stores issues or direct purchases. This is achieved through the medium of a stores or purchase requisition procedure (using either manual methods or an integrated computer system).

Each requisition acts as a control (authorizing the issue or purchase) and as a document raising a charge on the receiver. If the materials are to be used on a specific job, then the cost should be charged to the relevant job account (which will have a unique job number) and, ultimately, to the customer. If material is part of the direct materials needed for a process, it should be included with all other costs for the process for the accounting period.

Materials issues that are not identifiable with a job or process (indirect or overhead materials such as machine oil) should be recorded in an account along with all other overhead expenses. For the purposes of product costing, this overhead would be apportioned over all products to which it has contributed.

The value attached to a material could be its cost as determined by the price paid against the relevant purchase order. This would seem to be logical, and might perhaps be applied on the basis of using the oldest material first (a method known as FIFO, meaning first in, first out). This approach to cost flow aligns with the physical flow of many items.

As many organizations adopt some sytem of predicting costs, it is usual to find that material is stored at standard cost (a technically predetermined cost based on managerial estimates) and charged to users at this rate. Any difference between actual and standard cost would be revealed at the time of purchase (not issue) and reported within the organization to senior management in the purchasing department.

Costs at which material is issued, whether at actual or standard cost, might have to be adjusted to account for stock losses, evaporation or waste. It is customary to estimate a normal or standard allowance for this.

Labour

Costing of labour is based on principles similar to those relating to materials. Labour is a resource which cannot be stored, so that the total charges made to accounts for jobs, processes, customers or overhead should equal the total payroll costs for the same accounting period.

Organization members are paid on time rate, piece rate, a fixed salary, or some combination of these. This will form the basis for the payroll, to which the organization's employment costs (especially National Insurance) must be added.

The total labour cost is chargeable to the profit and loss account, and it is usually analysed into main business functional areas. The accounting records depend for their input on some form of time booking by direct employees to jobs, customers, processes or authorized overhead account (e.g. waiting time). Indirect staff may also be required to book their time to the relevant overhead accounts. A manual or (far more likely) a suitable computer system is used to analyse the time bookings, usually on a weekly basis with weekly and cumulative monthly reporting.

As a general rule, it is desirable to incorporate as many labour charges as possible as direct costs, rather than merge them into overheads where they would lose their separate identity. This is only appropriate where such charges are legitimately attributable to jobs: for example, where a customer has specifically ordered work.

If overtime premium, shift premium, or the costs arising from some incentive scheme payments apply to specific work, then it is correct to charge the extra cost exclusively to that work, as a direct cost. If, on the other hand, such premiums apply regularly as a matter of course, it is possible to inflate the hourly labour rate used to include them. A frequently used alternative is to merge such premiums with other indirect costs (including indirect labour) in overheads.

The treatment of engineering costs

Clients of the engineering function (including development and design) could be, for example, internal departments of the company or external customers to whom the company sells its engineering expertise as a contractor or consultant. Costs relating to engineering activities may be treated within the accounting system in broadly the same manner as costs of other departments. In the context of this handbook, however, brief individual elaboration is justified.

Where engineering labour can conveniently be associated with a project, it is regarded as a direct cost. The exact charge can therefore be made to the project concerned. The specifically identified labour time or cost will also be used as the basis on which overhead costs are attributed to the project by means of an overhead absorption rate (based on a rate per hour or a percentage of the labour cost). Any identifiable expenses (for example, machine hire) would be individually charged, also as a direct expense.

Labour costs, material costs or any other expenses incurred generally for engineering design and administration (and not for the benefit of any particular project) will form part of the engineering overhead. These overhead costs will be applied to all clients of the engineering department through the overhead rate. In the case of an external client, the charge for the job will be based on all direct costs, apportionment of general overheads and a profit margin (see the section on direct costs and overheads).

Process costing

Process costing principles apply when a number of identical products are produced by a continuous operation. In such cases it is impossible to relate any cost to an individual unit. Production of chemicals in a refining process, or food industry products are often process orientated. The product cost is taken to be the average cost applicable to the time period over which the total costs are collected, given by the total cost divided by the total output.

The identification of direct and indirect costs with departments or processes differs little from any other industry. The accountant has to deal with the product flow through a sequence of operations, perhaps involving losses, the need to add more material, and the production of byproducts or joint products at the same time.

For a detailed exposition of the principles involved, see Drury, 1987.

Costs for decision making – marginal costing

The previous section emphasized the calculation of costs for stock value, for the setting of prices, and for periodic profit measurements. Such costs, unadjusted, would not be appropriate for short-term tactical decisions. If the total cost of a product has been established as £22.54 (according to the method shown in Figure 19.4) and a potential customer offered only £20 each for ten such products, how should the company respond? The offer might not be as unreasonable as it first appears, but more information than that currently available is needed.

If the company had spare capacity, then one extra unit of product would not increase the total cost by £22.54. Some costs would remain virtually unaffected (factory rent, salaries, depreciation, and many other administration and selling costs).

If these unchanged costs were removed for the purposes of providing accounting information for this specific problem, the extra cost of producing one unit might be nearer £16 (said to be the cost at the margin or the *marginal cost* of the extra unit). Hence the order might be worth accepting. A tactic such as this must not, however, be allowed to offend regular customers who pay the full price.

Variable and fixed costs

Arrangements of costs for decisions rely on the identification of fixed and variable costs. Variable costs are those costs which change approximately in proportion to activity (for example the materials used in production). Fixed costs are those which remain unchanged despite changes in activity (rent, salaries, depreciation, for example as mentioned above).

This approach must be used with care, because the assumptions only hold good over a certain operating range (within the capability of the current facilities). This range is known as the relevant range. Furthermore, a number of costs do not fall neatly into one or other category, and may be semi-variable. In other words, they will change for different levels of activity, but the change will not be in simple proportion. Figure 19.5 shows the typical cost behaviour patterns in graphical form.

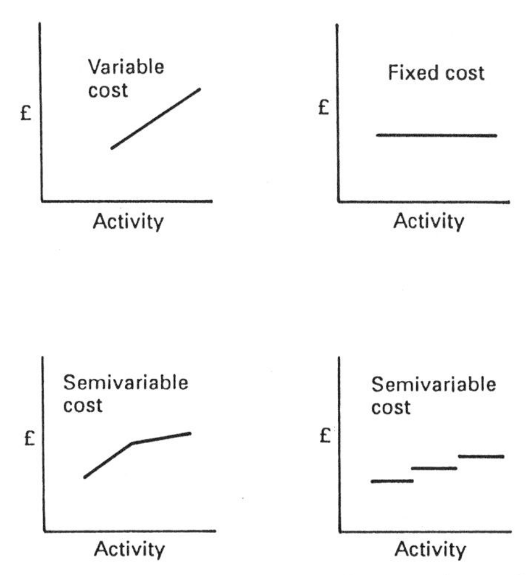

Figure 19.5 *Typical cost behaviour patterns. Each chart shows the level of total cost related to a level of activity within the operating range*

A marginal product cost is acceptable for internal management information and as a basis for making decisions, but it cannot be used for stock values in external, published accounts. Standard accounting practice dictates that stock values should include all costs involved in manufacture, including fixed costs.

Accounting for decision making

There are no hard and fast rules governing accounting for decision making. The marginal costing example just described only introduces the issues. Each decision-making situation requires its own unique arrangement of costs and revenues. An accounting statement drawn up on absorption costing lines, including all costs, might show a product or a department as a loss maker and provoke speculation about closure. When it is realized that absorption costs include some fixed costs which cannot be avoided, the decision is not so clear cut.

The decision situations envisaged above are based on short-run predictions of cost and revenue behaviour. It is inappropriate to make long-term decisions in this manner. In the long run all costs must be covered, or the company will not survive. Different approaches are necessary for long-term pricing or investment decisions.

Costs for planning and control – standard costing

Planning is a most important managerial function, and associated with this is the need to implement plans and monitor them. Accounting can support this planning and control process, using budgeting, standard costing and variance analysis.

Budgets focus on the short-term future of the company, usually over the following accounting year. They state in quantitative and financial terms the future sales, production levels, manpower levels, machine capacities and profits that are expected, analysed over convenient time periods and segments of the company.

Material and labour resources used in product costing will have been acquired as part of an integrated plan. Overhead rates used can be established from budgeted departmental overheads and production levels.

A carefully prepared budget clarifies to all organization members what is expected of them and the cost or spending levels that are organizationally authorized or desirable.

Responsibility accounting

The issue and use of departmental budgets is part of a system of responsibility accounting. Responsibility accounting involves the regular reporting of actual events, which are then compared with the predicted results. Action must be taken to correct any divergencies.

Only those costs over which a manager can exercise significant influence should be included in his/her responsibility report – a fundamental principle of controllability. So not all costs used for product cost ascertainment would be incorporated into a report. A productive department manager might typically control direct material usage, direct labour efficiency, power, and some indirect labour.

Standard costing

The application of responsibility accounting is usually undertaken using the technique of standard costing. Whereas company budgets relate to the whole organization, standard costs are the predetermined costs of one unit of each product. Direct cost budgets can be set based on quantity and price standards. Standards are, therefore, the building blocks of a budgeting system. Overhead rates for standard costs are usually derived from the annual budget.

Standard costing requires careful determination of the quantity of each resource which should be used to make a product, and of the prices which should be paid to acquire the resources.

Standard cost (SC) = standard quantity (SQ) × standard price (SP)

Any deviation or *variance* from standard costs (and hence from standard and budgeted profit) will be due either to non-standard resource usage and/or to the payment of a price other than that specified in the standard.

This approach can be applied to all materials and grades of labour and, somewhat less specifically, to overheads. Once a system of standard costing is in use, all operations, raw materials, work in progress and finished goods are accounted for at standard cost. Variances are reported at the first possible opportunity to alert management to the deviations from expectations.

Extended control reports

The reporting of variances in a system of responsibility accounting should not be limited to costs and revenues. Control information should be in the form most useful and understandable to the appropriate manager.

Financial reports can be embellished and are often accompanied by non-financial performance data, such as:

1 Kilogrammes of material wasted or saved.
2 Hours of labour wasted or saved.
3 Labour turnover.
4 Hours of plant capacity unused.
5 Proportion of rejects to good units.
6 Plant efficiency.
7 Orders taken, in relation to the number of calls by sales staff.

The list is limited only by the ingenuity of the manager and the accountant working together to focus on the key control information required.

The profit and loss account

The profit and loss account is a fundamental accounting statement, usually emanating from the financial accounts department. It presents the manufacturing and trading results of the whole company for a defined period of time usually a month a quarter or a year. A profit or loss is determined by matching the revenue earned in an accounting period with the expenses incurred in earning that revenue.

In a manufacturing company, revenue is largely determined by products sold. Costs incurred up to final manufacture of all products sold are set against this revenue to reveal a manufacturing or gross profit. The costs (both direct and indirect) incurred in manufacturing all products not yet sold are removed from the profit and loss account in the form of a value of closing stock. The closing stock value will be carried into the following accounting periods to be charged when the product is sold. In contrast, all non-manufacturing costs are treated as period costs (not product costs) and charged as expenses in the current accounting period. A simple profit and loss account for a manufacturing company in a vertical style is shown in Figure 19.6. Note how costs relating to stocks of raw materials, work in progress and finished goods are deducted before the remainder, cost of sales, is deducted from sales revenue. This style of layout follows the structure of costs established in Figure 19.1.

Any costs relating to items manufactured for use within the company will not appear in the profit and loss account. They should be recorded separately and then listed as an asset in the balance sheet. It would be appropriate to charge as a cost of this asset all material labour and overhead expenses incurred in its internal manufacture, a process called capitalization.

Conclusion

The chapter has drawn out and explained the main principles governing cost accounting. As engineers are employed in a variety of manufacturing and service industries no emphasis is placed on a specific industry group.

Technology used to process and prepare cost accounting information is undergoing continuous development. The result is that accounting information is available quicker, and with a greater number of alternative arrangements. Cost accounting information can be updated more rapidly and is available from external as well as internal sources. Management can receive (perhaps even read) greater quantities of information. It is vital that the accountant keeps up to date with the potential of new technology but this will influence, to a limited extent only, the principles of cost accounting to which the chapter is addressed.

Cost and management accounting is generally seen as an aid towards maximization of company profit. It is indeed true that in the long run

Hypoth company	*£000*	*£000*	*£000*
Sales			6500
Direct costs			
Material	1200		
Labour	1000		
Expenses	500		
		2700	
Manufacturing overheads			
Indirect wages	900		
Rent and rates	400		
Depreciation	200		
		1500	
Work in progress adjustment			
Add opening stock	70		
Less closing stock	50		
		20	
Manufacturing cost of goods completed		4220	
Finished goods stock adjustment			
Add opening stock	100		
Less closing stock	110		
		(10)	
Manufacturing cost of goods sold			4210
Manufacturing profit			2290
Administration selling and distribution overheads			
Salaries	600		
Printing and stationery	100		
Motor vehicle expenses	100		
			800
Net profit			1490
Corporation tax			500
Profit after tax			990
Dividends			400
Retained profit			590

Figure 19.6 *A simple profit and loss account*

businesses must earn a profit to survive but some activities defy the assignment of objective cost. To this extent engineers and managers must be aware that businesses are not run with a knowledge of cost alone, but it is better to know what something 'costs' than to be without this information.

Further reading

Drury, Colin, *Costing, An Introduction*, Van Nostrand Reinhold, 1987.
Drury, Colin, *Management and Cost Accounting*, Van Nostrand Reinhold, 1988.
Horngren, Charles T., *Cost Accounting: A Managerial Emphasis*, Prentice Hall International, 1987.
Rockley, Lawrence E., *Finance for the Non-Accountant*, Business Books, 1984.
Sizer, John, *An Insight into Management Accounting*, Penguin, 1980.

20

Establishing engineering budgets

Paul Plowman

Much has been written on the subject of budgets. For example, should they be 'top down', where the overall budget level is dictated by senior management and spread to the best possible effect? Or should budgets be 'bottom up', the result of an evaluation by managers of the resources needed to meet their set objectives? Is management commitment really important and, if so, to what degree? What happens if activity changes during the year?

Since the role of engineering is central to most industrial companies, any shortfall in performance can have significant effects, not only on the short-term company performance, but also on long-term profitability. Budgets, therefore, need to be forward looking, and they must help in efficiently meeting departmental targets. This chapter concentrates on the nature of engineering departmental budgets.

The purpose of budgets

Whatever the area of a company, available resources are scarce. So they need to be used as efficiently as possible. Costs must be controlled if target profits are to be met without prejudicing the long-term future of the company. This is particularly true for the engineering department, where adequate controls are needed for individual jobs, for projects and for the department itself.

Budgets set out the responsibilities of individual executives in financial terms relative to the overall company objectives. Continuous comparison is made of actual results against budgets, either to secure objectives or to indicate when the objectives need to be revised. The commitment of individual managers is crucial to this process, and budgets should be developed and agreed with them rather than having them imposed from top management. In other words, the 'bottom up' approach to budgeting is to be preferred.

The overall purpose of a budget is to:

1 Provide top management with a picture of the overall performance of the engineering department.
2 Serve as a guide to managers within the department.
3 Measure departmental performance.
4 Provide a central aid to cost control.

Even though the budgeting process is coordinated by accountants, it should not be regarded as just an accounting exercise since this might result in unnecessary constraints on managing the department.

Company objectives and plans

The technical objectives and costs of the engineering department need to be viewed in the light of overall company objectives. Most good company plans can be reduced to a set of simple objectives. For example:

- Sales must grow at a rate which will produce at least a 20 per cent increase over last year's sales.
- Profit before tax should be not less than £1 million.
- Profit margins on sales should be not less than 10 per cent.
- The return on capital employed should be not less than 25 per cent before tax.
- A range of new products should be developed that will form 20 per cent of sales within two years.

Company objectives will mainly be stated in financial terms and will serve as the background for the development of sales and production plans. These, in turn, will dictate the level of support required from the engineering department.

The budgets so produced provide the framework for evaluating the overall company plan in financial terms. It is only at a fine level of detail within each department that performance can be monitored and any necessary actions be taken.

Overall cost objectives

Engineering, as other departments, should provide an acceptable level of service at the lowest cost. It therefore requires to be managed and cost controlled with the emphasis on controlling future costs, and use of historical discrepancies as a guide to better performance. The point at which costs can be controlled is central to this and effective controls are needed for individual jobs and for projects as well as for the total department.

Operating and capital budgets

Before getting into the detail of operating budget preparation, it is necessary to reflect on the possible sources of funds from which the budget is to be financed. By far the most significant source of funds will be the planned revenue from sales of products or services. As far as engineering design and development is concerned, this activity will either be funded directly from the revenue paid by engineering project clients, or it will be recovered indirectly from the overheads built into the company's pricing structure for the sale of manufactured products and other services. These cost recovery processes were described in Chapter 19. Most of this chapter will be concerned with the determination of operating budgets funded by either of these methods: that is, directly or indirectly from the revenue from sales.

There is, however, another important source of company funds. This is the capital invested in the company by its shareholders or partners, and also the capital that might be provided by various financial institutions, such as the banks. These capital funds include money reserved from profits in earlier years, which arguably might otherwise have been paid as dividends to the shareholders or other funding sources.

All capital funds are required principally for the purpose of buying company assets or for acquiring other businesses. These assets (buildings, plant, machinery, computers, for example), although deemed necessary for the company's operations, are not themselves usually purchased for the purpose of being built into products or projects for profitable resale. In fact, these capital assets (apart from property) usually depreciate, and their depreciation costs have to be allowed for in the operating budget, where they appear among the overhead items.

Different companies have different ideas about what constitutes a company asset chargeable as a capital item, and what is simply a consumable item to be regarded as a charge to overheads. However, most engineering managers will at some time in their working lives expect to want to recommend the purchase of equipment (such as a new computer, a word processor or a plan printing machine) that will unquestionably be regarded by the company's accountant as a capital item.

Since every capital purchase involves the expenditure of funds that were provided by the shareholders or as loan capital, the purchase of capital items requires a separate capital budget forecast, and acquisitions need approval by the board of directors (as being responsible to the shareholders).

In a well run engineering company, therefore, every engineering manager should expect to be asked before the start of each new financial year what he/she expects to need in the way of new capital purchases for the relevant area of responsibility in the coming year. It is a sobering thought that if, for example, the company were to purchase a new computer for £100 000 as a capital item, and the funds for this were to be generated from retained profits, then if the company normally achieved a 10 per cent profit margin on sales, the new computer would require something like £1 million of sales to cover it.

There is often an alternative to the acquisition of a capital item by renting or leasing it instead. This means that the item does not become a company asset (it still belongs to the rental or leasing company) and the rent or lease instalments are charged in the operating budgets and recovered as overhead against sales revenue. At one time there were considerable tax advantages in leasing in the UK: these are now greatly reduced, and the company accountant should always be asked to decide on the best method of acquisition. Leasing may be indicated if the company is short of spare capital: it is a well known method of improving the immediate cash flow position as against purchasing. But leasing can involve accepting a high commitment in the event of early termination or default: again the company accountant must be asked to advise.

Cash flow schedules

On some engineering contracts the engineering company may be asked to purchase large items of capital equipment for clients, the costs of which are to be met by the client (often at cost plus a small handling charge). These costs do not fall within the usual expenses budgets, because they are reimbursable. But the engineering project manager responsible will probably be asked to estimate the timing of such payments and their corresponding repayments, and set these out in a cash flow schedule. This will enable the company's financial management to plan that suitable funds are available.

Sometimes arrangements are made for the client to pay suppliers' invoices directly. Another arrangement is that the client deposits money with the engineering company in the form of a purchasing fund. Yet another possibility is to arrange for progress stage payments, but these may be received after suppliers' invoices fall due for payment, in which case the engineering company has effectively to advance the sums involved for periods of one or more months. If the amounts are large, the cost of such advances might be significant.

Although this type of expenditure does not fall within the scope of preparing an engineering department budget, cash flow scheduling is, nevertheless, a form of budgeting. Reference to the process has to be made here, since the engineering project manager may have responsibility for assisting the company's financial management, and indeed the client, in planning cash flows. This responsibility might well extend beyond the question of major equipment purchases to cover all aspects of project costs.

Cash flow schedules can be for expenditure (cash outflows) or for income (cash inflows). In the case of major equipment purchases, an outflow schedule would be prepared, with the payment timings based on the dates when payments to the equipment suppliers are expected to fall due. This information obviously requires access to detailed project equipment cost estimates, to the equipment deliveries schedule, and to the terms of payment determined by each equipment subcontract or purchase order.

A cash outflow schedule of this type is prepared by listing the major items of expenditure in the left-hand column of a spreadsheet. For every item, the cost is then distributed along its row, with the amounts expected to fall due for payment put in the relevant periods (usually months or quarters). The top part of Figure 20.1 illustrates this process: this is the format usually presented to clients to help them plan their project funding.

For the purposes of the engineering company, if they are going to have to fund project purchases or other expenditure for any length of time before payments are received from the client, it is necessary to take cash flow scheduling two steps further.

The first of these additional steps is the construction of a cash inflow schedule. This has the same cost items as the outflow schedule, but the entries in the monthly columns are timed according to when payments from the client can realistically be expected. This may depend on certain contract conditions, and might be influenced by expected foreign exchange restriction difficulties or other problems.

The next step is to reconcile the inflow and outflow schedules, to produce a net cash flow schedule. This is simply a question of comparing the totals at the foot of the corresponding monthly columns from the inflow and outflow schedules. For each month there will be a net inflow or a net outflow, and these results are entered on one of the schedule forms. This result shows the net funding required each month.

Responsibility for preparing budgets

The concept of budget managers

Budgets should ideally be prepared by, or at least agreed with, the managers who are going to be responsible for managing the operations to which the specified expenditure corresponds. This is the bottom up approach to budget preparation that was mentioned and recommended at the beginning of this chapter.

Before any attempt at preparing company budgets can start, it is necessary to break down the budget into parts that can be assigned to individual managers. For this purpose, each manager can be designated the budget manager for that part of the total company budget. The budgets parts can be for direct or for indirect (overhead) budgets.

As examples, project managers would be the budget managers for their

SCHEDULE A CASH OUTFLOWS

(payments to suppliers and subcontractors)

Cost code	Item	Total cost	1989 Nov	1989 Dec	1990 Jan	1990 Feb	1990 Mar	1990 Apr	1990 May	1990 June	1990 Jul	1990 Aug	1990 Sep	1990 Oct	1990 Nov	1990 Dec	1991 Jan	1991 Feb	1991 Mar	1991 Apr	1991 May	1991 June
123	A	480		48				100			150				134				48			
418	B	55		6					20				23			6						
206	C	150	14			35			38				49				14					
045	D	136	22			13											110			13		
999	E	225											80		100				23			
Monthly totals		1046	36	54	—	48	—	100	58	—	150	—	152	—	234	6	124	—	71	13	—	—

SCHEDULE B CASH INFLOWS

(receipts from the client)

Cost code	Item	Total cost	1989 Nov	1989 Dec	1990 Jan	1990 Feb	1990 Mar	1990 Apr	1990 May	1990 June	1990 Jul	1990 Aug	1990 Sep	1990 Oct	1990 Nov	1990 Dec	1991 Jan	1991 Feb	1991 Mar	1991 Apr	1991 May	1991 June
123	A	480					48			100			150						134			48
418	B	55					6						20			23			6			
206	C	150					14			35			38			49			14			
045	D	136								13									110			13
999	E	225					22									80			100			23
Monthly totals		1046	—	—	—	—	90	—	—	148	—	—	208	—	—	152	—	—	364	—	—	84

SCHEDULE C (B - A) NET CASH FLOW

Inflows or (outflows)	(36)	(54)	–	(48)	90	(100)	(58)	148	(150)	–	56	–	(234)	146	(124)	–	293	(13)	–	84
Cumulative cash flow	(36)	(90)	(90)	(138)	(48)	(148)	(206)	(58)	(208)	(208)	(152)	(152)	(386)	(240)	(364)	(364)	(71)	(84)	(84)	—

All figures are £1 X 1000

Figure 20.1 *A simple example of a project net cash flow schedule. The number of items listed has been kept low for simplicity. The top schedule shows payments expected to be made to suppliers of equipment according to the contract terms. The client will reimburse the contractor, but funds are only going to be approved at quarterly board meetings. Thus the contractor effectively has to fund the purchases for several months and, as this schedule shows, is really lending the client nearly £400 000 at one stage. A sensible contractor will add a handling fee that will cover the notional interest on these funds*

particular projects (direct budgets). The chief engineer or engineering manager could be the budget manager for each new product development job, and also for the budget for time spent on engineering management and supervision. The buildings or office services manager would be responsible for a budget that included all costs associated with the provision, maintenance, heating and lighting of premises. And so on, for every part of the company budget.

Cost coding

The company accountant will operate according to a code of accounts, almost certainly linked to a suite of accounting programs on computers. These will allocate codes to various cost elements in a hierarchical fashion, so that each cost item is assigned a number, and every element of further cost breakdown with each item is given a logical code from which the detail and its parent cost can be identified. This system allows the computer to analyse or group costs according to management requirements.

Budgets for jobs, for projects and for indirect costs can all be coded in accordance with the company code of accounts. Then, when the actual costs of salaries, wages, materials and other expenses are collected and fed into the cost accounting system, they can be given their appropriate cost codes in order that the computer assigns them into their correct places in the job costing or overhead costing structure. The use of such codes should mean that the actual codes are recorded as they arise in a way that allows analysis and direct comparison with their similarly coded budget counterparts.

By arranging that like can be compared with like, there is a basis for providing data for cost control. But the cost collection and computer system must be operated efficiently, so that cost data collected from timesheets and other sources are processed within a few days, rather than several weeks. Otherwise the information provided will be too long delayed after the event for any control action to be applied effectively.

Obviously a job number can be assigned when each job or project is authorized. A budget manager can be assigned to each job, group of jobs, or a project. For indirect budgets, the breakdown could appear like this, for an engineering company:

Code	*Budget*	*Budget manager**
0002	Absence through illness	Personnel manager
0003	Special leave	Personnel manager
0004	Statutory holidays	Chief accountant
0005	Annual leave	Chief accountant
0006	Training	Personnel manager
0010	Company management	Company secretary

0020	Premises	Property manager
0025	Office services	Office manager
0030	Plant maintenance	Plant manager
0040	Production management	Production director
0045	Production services	Production engineer
0050	Estimating	Production engineer
0060	Engineering management	Chief engineer
0065	Project services	Chief cost and planning engineer
0080	Depreciation	Chief accountant
0090	Marketing management	Marketing director
0092	Sales development: Europe	Sales manager A*
0094	Sales development: Americas	Sales manager B*
0096	Sales development: Australasia	Sales manager C*
0098	Sales development: Far East	Sales manager D*

and so on.
*In practice, names rather than job titles would be shown.

Please note that Chapter 14 explains the principle of hierarchical numbering systems, and cost accounting is dealt with in Chapter 19.

Responsibility for approving budgets

Since the engineering budget forms part of the overall budget of the company it will require approval not only by the departmental head, but also by the company board of directors. Because the budget review must match departmental objectives with cost constraints, this approval process can need several attempts, unless the budget has been very carefully developed in the first place.

Estimating costs

Since every budget is composed of items for which many or all of the costs have been estimated, it is appropriate to discuss briefly some estimating methods before going on to examine the detailed process of compiling an engineering department budget.

The Pareto principle (the 80:20 rule)

Any project consists of a multitude of individual parts. It would not be possible or economic to estimate the costs of all these items, many of which will probably only be covered by general arrangement drawings in any case. The emphasis should be put on high value items that contribute most of the overall value. Pareto's principle states that 20 per cent of the items are likely to account for 80 per cent of the total value (see Figure 20.2). If reasonable estimates can be made for the costs of this top 20 per cent, broader estimates can safely be made for the remaining items.

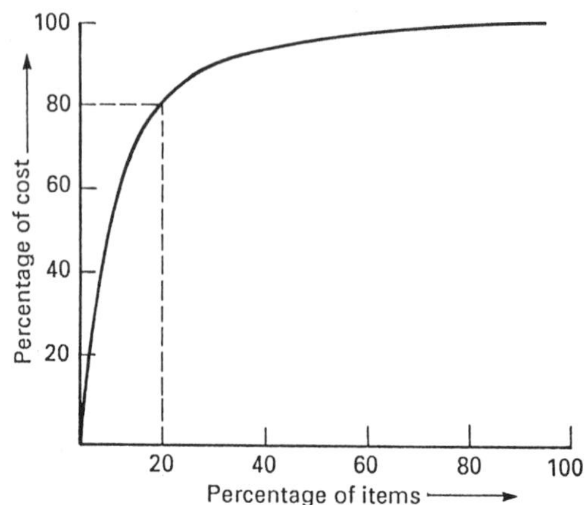

Figure 20.2 *Pareto analysis*

Methods of estimating job costs

There are three basic methods for estimating job costs. These are:

1 Group method.
2 Comparison method.
3 Detailed estimating.

Group method

This is the broadest method of estimating. Representatives of the various departments involved (such as purchasing, engineering, manufacturing and accounting) confer and produce a joint estimate of costs, often from information obtained from past experience. The main advantage is speed and this method allows information to be pooled at minimum cost. It can however suffer from lack of accuracy and a higher level of contingencies needs to be included.

This method is particularly useful when information is vague, for example when drawings and detailed specifications are not available.

It is important to have a clear team leader for these estimating meetings and to record all commitments made.

Comparison method

The comparison method relies on the existence of previous experience and data. Where the new item is similar to work undertaken in the past it may be possible to factor previous actual costs. While some elements will remain fairly constant, others will change owing to differences in size (material usage and scrap rates, for example). Quick estimates can be made by making allowances for costs in relation to a particular parameter, such as a diameter (Figure 20.3).

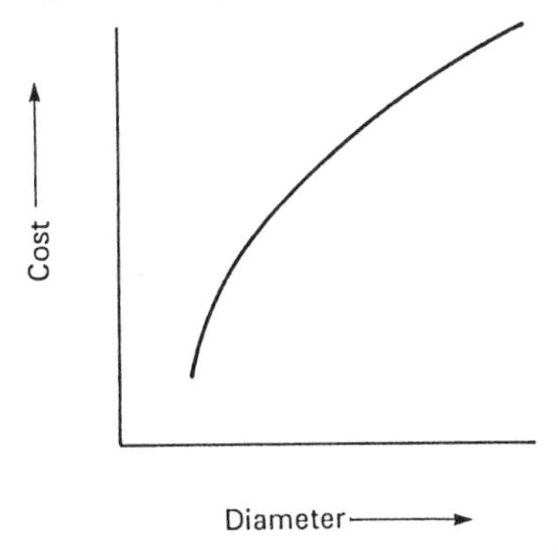

Figure 20.3 *Example of cost variation by parameter*

Detailed estimates

All estimating techniques depend to a large degree on the availability of historical data but this is particularly true for the preparation of detailed estimates. Historical data may be largely unsorted or may be analysed to show changes in cost behaviour with particular parameters or factors.

This method is particularly appropriate for items of high value, and where detailed drawings are available. The estimate has to be produced by following a series of steps:

- List and calculate the usage of materials, including scrap allowances.
- Detail the design time required.
- Detail the sequence of operations for each component and calculate the labour times.
- Detail any subcontract work required.
- List the equipment required.
- Estimate special tooling and test equipment requirements.

Where drawings are available, and if there is sufficient time, it may be possible to obtain two or three quotations for bought out materials and tooling. An example of a detailed estimating form is given in Figure 20.4.

COST ESTIMATE

Part name: Motor components (prototype) Job number: A1300

Part number	Operation	Standard hours	Material	Labour	Tools and equipment	Overheads	Total
	Design	25		125		100	225
	Materials:						
	– housing		80				80
	– stator		40				40
	– rotor		65				65
	Machining:						
	– stator	6		30		60	90
	– rotor	15		75		150	225
	Outside heat treatment		30				30
			215	230		310	755

Figure 20.4 *Example of a detailed cost estimate record for part of a manufacturing project*

Project cost estimates

Estimates are prepared for the major components of a project in the same way as for job cost estimates. At the beginning of a project, particularly if it is a development project, the only detail available may be some general arrangement drawings. On the Pareto principle, further detail should be worked out for the major items. A contingency allowance has to be included to cover potential understatements of the costs for other items.

Many projects can last for periods exceeding one year, which causes particular problems when company financial systems focus on the current year alone. For an eighteenth-month project, most of the effort in the first six months will be in design, and in the procurement of long lead items. As this is the time when there is the greatest influence on cost, it is important to be able to identify any shortcomings in design so that alternative designs and courses of actions can be considered in time. Figure 20.5 shows the pattern of a typical manufacturing project, and Figure 20.6 illustrates how the possibility of influencing project costs reduces as time proceeds.

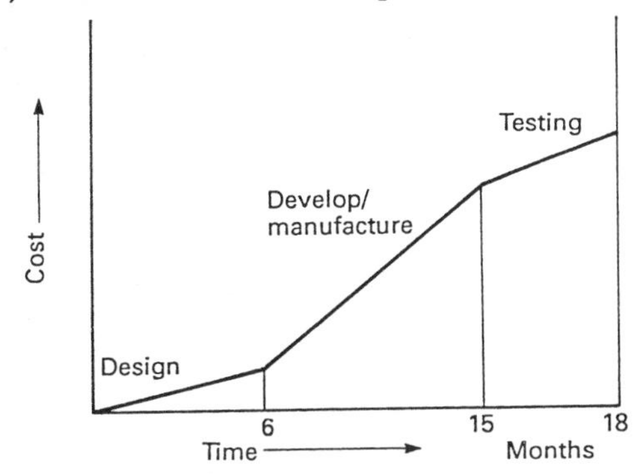

Figure 20.5 *A typical life cycle for a manufacturing project*

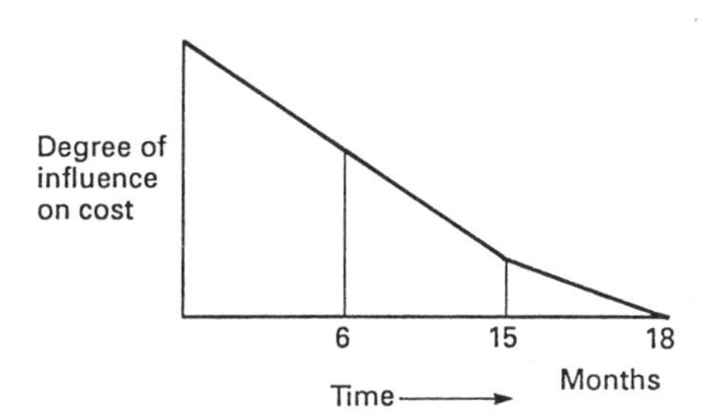

Figure 20.6 *Degree of influence over cost with relation to time. This curve relates to the project time cycle shown in Figure 20.5, and illustrates how the effects of design and other decisions have decreasing effect on total costs the longer they are delayed*

Estimates for large scale projects, although they may include several detail estimates and actual price quotations for equipment, are likely to be made generally on the comparison basis. Figures 20.7 and 20.8 show the documents used by one mining engineering company for the head office labour content of their large engineering projects. Figure 20.7 summarizes all the labour estimates in terms of manhours, categorized according to standard cost grades and the departments. The departmental breakdown is of obvious use when it comes to integrating such project estimates into the annual departmental budgets. This process will be described later in this chapter.

Figure 20.8 is used to collect the estimates from the manhour form (Figure 20.7), translate them into money at standard cost rates, and then add various mark ups for contingency allowances, overheads and profit.

Figure 20.9 shows a companion form, used for estimating the cost of capital equipment to be purchased for the project.

For the purpose of establishing departmental budgets in any particular year, it is very important to know how much of each project's costs are expected to occur during that year (since the total costs of a large project will be spread over several years). For this purpose, the cash flow scheduling technique described earlier in this chapter can easily be adapted, simplified by using years instead of months as the time divisions.

When the cumulative costs for any engineering project are plotted against time, an *S*-shaped curve always results. These curves, and their use in cost control, are described in Chapter 23 (see Figure 23.3). These curves will have the characteristic shape whether they are drawn for total project costs, or for some part of the project (such as equipment purchases). To illustrate this point, the figures used in our very simple cash flow example (Figure 20.1) have been plotted as an *S* curve in Figure 20.10.

Preparing an annual budget for the engineering department

A total company budget comprises several components, each of which may be for the total fulfilment of an individual job or a project, or to define the expected expenditure for a particular department or other group over a period of time. These budgets are obviously interrelated.

When compiling an annual operating budget for an engineering department, it is necessary to examine each possible reason for needing expenditure, and then identify that part of each item which is planned to fall within the relevant financial year.

The reasons for expenditure split three ways:

1 Capital expenditure (which was dealt with at the beginning of this chapter).
2 Direct expenditure (costs that are identifiable against customer or stock orders, and which are going to be charged out to customers or clients).

MANAGEMENT INFORMATION SYSTEM

ESTIMATE OF MANHOURS

SUMMARY

SELTRUST ENGINEERING LIMITED

DPT No.	DEPARTMENT	MANHOURS BY GRADE								TOTAL HOURS	% OF TOTAL	LINE No.
		11	21	22	31	41	51	52	61			
01	Company Management											1
												2
11	SEL Consultants Management											3
12	Geology											4
13	Mining											5
14	Minerals Engineering											6
15	Metallurgical Engineering											7
												8
												9
22	Project Management											10
23	Sec. & Clerical Services for P.E.D.	–	–	–	–	–	–	–				11
24	Estimating											12
25	Cost Control/Planning											13
26	Project Analysis											14
27	M.T.O.											15
29	Purchasing											16
												17
												18
												19
												20
												21
41	Civil								–			22
42	Structural								–			23
43	Mechanical Engineering								–			24
44	Fluids								–			25
45	Electrical								–			26
46	Process Control								–			27
52	Special Projects											28
												29
61	Finance Management											30
62	Accounts											31
63	Commercial											32
												33
71	Administration Mangement											34
74	Computing											35
												36
76	Personnel											37
79	Office Services											38
												39
81	Business Development											40
												41
												42
87	External Services (1)											43
89	Secondments											44
99	Other Allowances											45
TOTAL HOURS BY GRADE											100	
GRADE AS % OF TOTAL HOURS										100		

Notes

(1) For staff external to SEL who submit weekly timesheets

PROJECT				PROJECT No.		
Originator			Project Manager	PLANT		
Date			Date	EST./CV No.	REV.	SHT. OF

Form 171 Rev 11

Figure 20.7 *Estimating form for summarizing manhours on a large mining engineering project*
(Courtesy of Seltrust Engineering Limited)

ESTIMATE OF ENGINEERING COST SELTRUST ENGINEERING LIMITED

Date .... Project No .... Estimator .... Approved ....

Project ....

SECTION	CATEGORY								
	11	21	22	31	41	51	52	61	Total
Project Engineering Hours									
Engineering Discipline Hours									
Services Hours									
M.T.S. Hours									
Total Hours									
Rates (£/hr)									—
Project Engineering Cost									
Engineering Disciplines Cost									
Services Cost									
M.T.S. Cost									
Total Cost									

Allowance for Other Direct Bookings at % of		
Total Salary Direct Costs		
Overheads at %		
Salary Plus Overheads		
Expenses		
Total Inhouse Engineering		
Outside Engineering		
Vendors Licence Fees		
Total Engineering Cost		
Contingency at %		
Total Cost (Inc. Contingency)		
Profit on Inhouse Engineering at % (Inc. Contingency)		
Profit on Outside Engineering at % (Inc. Contingency)		
Total Profit		
Total Engineering Cost Plus Profit		

Remarks

Form 56K Rev 2

Figure 20.8 *Cost summary document for the engineering work on a large mining project. This form takes part of its input from the manhour summary form that was shown in Figure 20.7*

(Courtesy of Seltrust Engineering Limited)

Seltrust Engineering Limited London England ESTIMATE OF CAPITAL COST

Estimate No.

Company
Project No.
Plant
Plant Section

All values are in
• Curr = Currency in which value is to be paid
Exchange Rates
Estimate Base Date

Estimate Identifier
Sheet of sheets
Estimator (s)
Approved by

Rev. 0 | 1 | 2 | 3 | 4 | 5

Description	Discipline / Detail	Detail Extension	Classification	Quantity	Unit	Supply			Freight				I. D. S. T.			Sub-contract			Total cost
						Unit Rate	Cost Fob / For	• Curr	Gross / Ocean Tonnes	Rate	Cost	• Curr	Rate	Cost	• Curr	Rate	Cost	• Curr	
Sheet Totals																			

Form 56 A Rev 2

Figure 20.9 *A form suitable for use in estimating capital equipment costs for an engineering project* (Courtesy of Seltrust Engineering Limited)

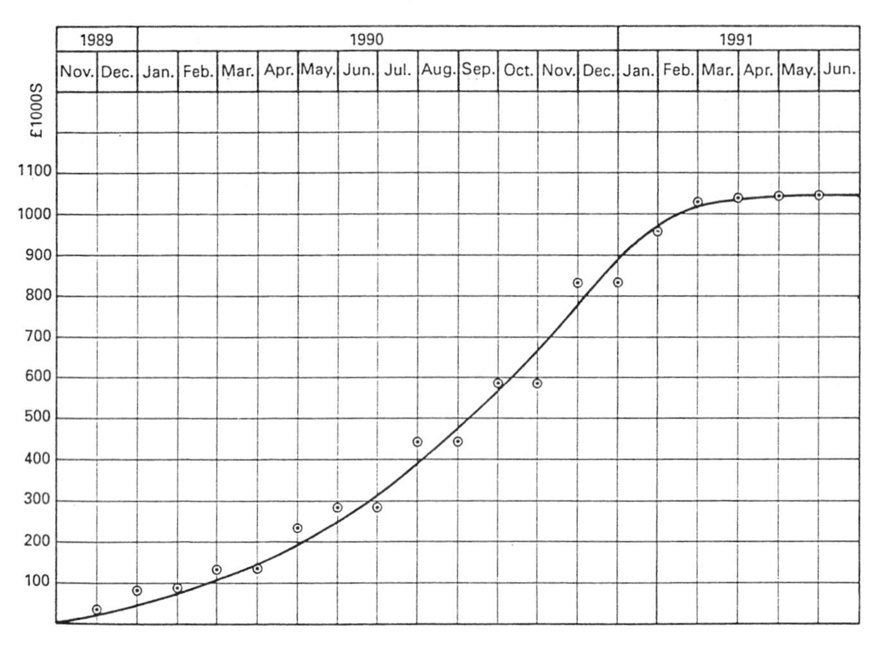

Figure 20.10 *A project S curve. When project costs are plotted against time, an S curve always results. This is especially true for engineering salaries: the curve shown here was obtained by plotting the equipment costs from the cash flow schedule of Figure 20.1, and the start is a little steeper than usual owing to the deposits paid with the issue of the purchase orders as part of the conditions of contract. See Chapter 23 for more information on the use of these curves for project cost control*

3 Indirect expenditure (overhead costs that cannot be identified with specific jobs, and which must be recovered by spreading them as an extra layer of charges to customers on top of the direct costs.).

Both direct and indirect activities can split into labour and other expenses.

In the explanations which follow, several references will be made to the practices that were actually followed by a medium-sized engineering company (about 500 staff) operating in large scale mining and metallurgical projects. Other examples will be given that are more applicable to a manufacturing company.

Budget coordination

Before any attempt at preparing a budget can start, it is necessary to remember that each project and each overhead account will have its own budget manager. Every budget manager will have to assemble his/her own budget for each account for which they are responsible. Each departmental budget will then be assembled by taking the relevant constituents from all the different project and other accounts budgets that are going to require the department's resources during the year.

One difficulty with this arrangement is that the sales department, for example, will probably want to provide for the use of engineers from the engineering department to assist in the preparation of new sales proposals throughout the coming year. Since this activity will eventually be charged on their weekly timesheets by the engineers to the relevant sales cost account, the sales departmental manager (the budget manager for the sales budget) will have to show these engineering hours in the sales budget. Now we have the situation where engineering hours are being allocated for the coming year into a budget which is outside the engineering manager's control. This can happen also in a number of other circumstances (for example in support to manufacturing groups).

If the engineering manager is to compile a complete budget for the department, he/she must have access to all budgets that predict usage of engineering staff, including those compiled by other managers. The total engineering commitment throughout the company, over all jobs, for all support reasons, must be known by the engineering department manager.

Conversely, engineering project budgets will contain estimates for man-hours needed from non-engineering departments. Activities designated by the accountant as indirect (overhead) do not have to be listed in project budgets, because their costs will be recovered through the charging structure as described in Chapter 19. But some direct activities from non-engineering departments will appear in budgets compiled by engineering project managers. In some projects, for example, printing and clerical services are charged out to clients as direct. In a manufacturing company, it is possible that very substantial manufacturing costs will be listed in budgets compiled by the engineering project managers.

Now it is seen that, in any but the smallest engineering company, there will be several managers from different departments compiling job budgets, project budgets and annual overhead account budgets, all of which use resources from outside their own departments. It is obviously vital that this is not done without reference to those other departmental managers. This involves considerable cooperation and cross-communicating, especially at the annual budget preparation time.

In practice, this need not be nearly as complicated an exercise as it sounds. The key to success is to make certain that there is a budget coordinator in the company who has copies of all project and overhead account budgets. In the case example now being described, this function was performed by a central project services group. This group, part of the engineering department, carried out planning, scheduling, progressing, estimating and cost control. Its cost engineers were always available as budget coordinators.

If such a group should not be available in a company, then the accounts department would be the obvious choice for performing the coordinating role. Another option is to send a copy of every account budget to every manager in the company, so that each keeps a reference dossier (not the best method, because it is almost certain that some managers will not update their files correctly, leading to corrupted and conflicting data across the various departments).

The labour budget

Available labour hours

The first thing to decide, before any consideration of forthcoming workload, is how many manhours are going to be available in the budget year from each person employed. In the case of the company in this example, the normal working week comprised 35 hours (09.00 to 17.00 for each of five days, less one hour each day for lunch). Each person had, therefore, to account for 35 hours on his/her timesheet (plus any overtime worked). All staff had to complete timesheets, including the directors, so that every manhour in the company, direct or indirect, was entered in the cost accounting system.

The company accountant knew the duration of each forthcoming financial year, from which the gross number of hours available per person was calculated. Supposing that there were 52 weeks in the budget year, then there were (35 × 52) hours to be accounted for in respect of each person (1820 hours).

Not all of those 1820 hours were available for engineering work. Allowances had to be made for statutory holidays, annual holidays and illness. It is generally acknowledged in companies that these items are outside the control of engineering managers, so that the hours involved should be removed from the hours available for work budgeting. The solution in this case was the one usually adopted, namely to use past experience of statistical illness levels, and to use accurate information on forthcoming annual holidays and statutory holidays to compute the relevant manhours. When this calculation was completed (by the company accountant) the net hours available for allocation to work budgets reduced to 1645 hours per person.

In some companies, holiday entitlements and even the contracted hours of work may vary considerably from one individual to another. In this less fortunate circumstance, the number of hours available will have to be worked out for each person, deducting holiday entitlements according to the individual contracts of employment, and using either a statistically derived allowance for possible illness or an estimate based on knowledge of the individual. An example of a calculation of this type of company is shown in Figure 20.11.

Remember that in the case example company all hours were logged on

Employee	*Total hours available*	*Statutory holiday*	*Annual holiday*	*Training*	*Illness etc.*	*Total available*
Mechanical design						
Harrison	1950	60	112	75	25	1678
Marsh	1950	60	127	–	25	1738
Oliver	1950	60	112	37	25	1716
Electrical design						
Brotherton	1950	60	127	37	50	1676
Electronic design						
Boxall	1950	60	112	37	25	1716

Figure 22.11 *Analysis of hours available for charging to work in a budget year*

weekly timesheets, whether direct or indirect. So time estimated for non-productive absence on training (or for any other indirect reason except illness or holidays) was allocated to budgets by budget managers in exactly the same way as time estimated for projects or other direct engineering work. The accounts computer simply recognized which of these were direct or indirect from the system of cost codes used for the various budgets. (Illness and holidays were in fact allocated to their own budgets, but this was done by the company accountant, since engineering management did not need to be involved in the calculation.)

Salary rates

In the case example, all staff were graded into eight different standard hourly rates. Thus, all project managers (for example) were graded as 21, and the same cost rate was applied to every manhour spent by this group irrespective of their actual individual salaries. The rate for each grade was calculated by the accountant, this being an average over the total number of staff in the group, with salary costs, BUPA subscriptions, company pension contributions and National Health Insurance all added into the equation.

The salary grades used in this engineering company were:

11 Directors and some senior managers.
21 Departmental managers, project managers and chief engineers for the various engineering disciplines.
22 Project engineers.
31 Senior engineers.
41 Engineers.
51 Drawing office group leaders and senior designers.
52 Draughtsmen and tracers.
61 All non-engineering staff below managerial level.

Other codes were added to identify the different engineering disciplines, these being identifiable by the company's computer based management information and costing system.

The use of such standard hourly rates, especially when the number of grades used is only around six or so, greatly simplifies cost estimating, budgeting, pricing and cost control. Some clients for cost-plus projects, however, insist on being charged on the basis of actual salaries paid to the individuals working on their projects: this causes unnecessary complication and probably results in the clients having to pay more in the event because extra overhead work is caused.

If any reader is forced to prepare budgets where standard grade rates are not applied, then every individual in the department must be listed by name in the budget. This is not a sensible approach, because no manager can predict what every named individual in the department will be doing for the next twelve months. Some will have accidents or babies, some will leave, the work situation will change, and so on. The sensible way is to work with numbers within each grade of seniority (so many mechanical engineers, so many mechanical draughtsmen, and so on).

Adopting the principle that managers should only be asked to plan and manage factors within their control, engineering managers should be asked to prepare their budgets in manhours per grade of staff. It is arguable that engineering managers cannot be expected to have complete control over salary costs (as opposed to manhours) because they have no control over cost of living increases, other awards, and changes in pensions and other salary burden elements (some of which might be altered by government legislation).

In the case example, therefore, the company accountant was responsible for converting manhour budgets into cost budgets, applying the standard costs appropriate to each grade. Each rate included an uplift to allow for expected salary increases (the rate of increase used being the expected average percentage as determined by the board of directors and advised to the accountant in confidence).

Balancing manpower requirements

To illustrate how an individual departmental manager might approach the task of preparing an annual budget, consider the manager of an engineering project services department. This example is again based on the engineering company quoted throughout this section, except that the figures used here are fictitious.

The starting point for this manager is the existing level of staffing in the department. Suppose that it comprises the following:

Title	*Number*
The department manager (job title: chief cost and planning engineer)	1
Senior estimating engineer	1
Estimating engineers	4
Cost control engineers	2
Senior planning engineer	1
Planning engineers	3
Tracers	2
Coordinating clerk	1
Secretary	1
Standards engineer	1
Total departmental strength	17

The manager has to decide whether this level is going to be appropriate for the coming year. He therefore collects all known facts about the planned workload from project budgets, the sales budget, and from sales forecasts for new projects. The department manager has an account number 0065 allocated to him for indirect time bookings needed for the management and internal running of his department. All other indirect time (such as support to sales) and direct project work is to be allocated to other account numbers, each with its own separate budget. The resulting calculation might look like that shown in Figure 20.12.

Current department strength			Direct hours needed for projects *d*	Indirect requirements			Remarks *h*
Job title *a*	Present number *b*	Hours available *c*		Sales proposals *e*	Production support *f*	This department *g*	
Chief cost and planning engineer	1	1645	—	—	—	1645	
Senior estimating engineer	1	1645	—	—	—	1645	
Estimating engineers	4	6580	25	2500	1000	3055	
Cost control engineers	2	3290	—	—	—	3290	
Senior planning engineer	1	1645	—	—	—	1645	
Planning engineers	3	4935	175	1000	—	3760	
Tracers	2	3290	820	2000	—	470	
Co-ordinating clerk	1	1645	—	—	—	1645	
Secretary	1	1645	—	—	—	1645	
Standards engineer	1	1645	—	—	—	1645	
Total hours		27965	1020	5500	1000	20445	
Equivalent people	17	17	0·62	3·34	0·61	12·43	

Figure 20.12 *Example of a manpower reconciliation calculation for a department within the project engineering division of a medium-sized engineering company*

It is not necessary to use a standard pro forma for this purpose, and the calculation might well be recorded in a notebook. However, the layout of Figure 20.12, in which the columns are given identifying letters, will help in this detailed explanation.

The department manager starts by listing existing staff on the form, as shown in columns *a* and *b*. The accountant has decreed that there will be 1645 available hours per person in the budget year (equivalent to 47 weeks at 35 hours per week). Multiplying the numbers of staff employed in the department by 1645 gives the results written in column *c*.

The manager now has to consult the budgets for all known forthcoming direct project work, including an allowance for new work that the sales people predict will be ordered. In this engineering services department, not much work can be charged direct to projects, and all management and supervision time is classed as indirect. But that time which has been allowed in project budgets has been recorded in column *d*.

Consultation with the sales manager to discuss his budget requirement shows that the estimating engineers, planning engineers and the tracers will be expected to provide some sales support, with their time to be charged out to the sales budget. The agreed hours have been entered in column *e*. A similar discussion with the production manager has indicated the need for an estimated 1000 hours to be set aside for sundry support to production during the budget year (column *f*).

The departmental manager now has to assess the numbers of hours not allocated, and these remainders are written in column *g*. The magnitude of these figures has to be assessed to see if they appear reasonable: are the hours remaining sufficient for the purposes of internal departmental administration and activities? If the other demands were high, the remainders could be negative, indicating the need for the recruitment of additional permanent or temporary staff.

In the example, the manager (chief cost and planning) has decided that the remainders have worked out reasonably, and has made no application for increased staff. He can now submit his annual budget application for account 0065, project services, using the form already sent to him by the accountant for this purpose (Figure 12.13). In its first, unapproved state, the budget statement is given the revision number 0.

Had recruitment been indicated, the budget request would have been increased accordingly, and steps put in hand to obtain authorization for recruitment through the use of staff requisitions (Chapters 4 and 5).

The company accountant collects the budget statements from all budget managers and has the manhours converted to costs using standard rates plus an escalation factor.

All budget statements will be sent for approval to senior management after the computer has done its initial number crunching exercise to assess the potential level of company overhead in the forthcoming year. The board of directors might not like what they see: there might be a request sent back to the various budget managers to effect economies. If the workload is falling, with no sign of improvement, this is the time when early indication of the need for interdepartmental transfers, retraining or even possible redundancies might be seen.

The timing of manpower requirements

This chapter has so far dealt with labour budgets in simple annual terms without regard to peaks and troughs throughout the year. The resolution of this problem needs the application of planning and scheduling techniques such as those described in Chapter 22. Figure 20.14 shows a time schedule of

ANNUAL BUDGET STATEMENT

Budget: Project services | *Year*: 1991

Budget manager: Joseph Bloggs | *Account no*: 0065

Job title	*Grade*	*Hours per week*	*Hours for year (see) footnote)*	*For accounts use*	
				Rate	*£*
Chief cost and planning engineer	21	35	1645		
Senior estimating engineer	31	35	1645		
Estimating engineers	41	65	3055		
Cost control engineers	41	70	3290		
Senior planning engineer	31	35	1645		
Planning engineers	41	80	3760		
Tracer	51	10	470		
Coordinating clerk	61	35	1645		
Secretary	61	35	1645		
Standards engineer	41	35	1645		
	–	435	20 445	–	

Notes: Available weeks this year are: 47
Escalation used in cost rates: %
Equivalent staff numbers: 12.43

Rev: 0

Figure 20.13 *Example of a budget statement for an indirect account*

manpower usage applicable to a drawing office in a manufacturing company. As in the previous example, additional requirements would be met by recruiting, and other options might be to use overtime or subcontract work to an outside organization.

Budgets for materials and expenses

As with the salaries budgets, expenses can be divided into direct and indirect. The headings under which expenses are likely to occur include the following.

1 Direct expenses. These are costs directly chargeable to customers. They can include such things as:

(a) Expenses incurred against projects. Note that these can include the

Drawing office manpower	1	2	3	4	5	6	7	8	9	10	11	12	*Average*	*Total cost*
Production support Mechanical Electrical Electronic *Development* Mechanical Electrical Electronic														
Total Mechanical Electrical Electronic														
Overtime programme % Mechanical Electrical Electronic														

Figure 20.14 *Manpower and overtime requirements by month for a drawing office*

costs of mail, plan printing, telex, and other items that would normally be considered as indirect, but which the clients have agreed to pay as direct, with specific clauses to this effect in the relevant contracts. This is quite usual in contracts for mining, petrochemical, construction and consultancy work.

(b) Travel, accommodation and subsistence expenses on visits to or for clients.

(c) The purchase of subcontracted services for design or other professional reasons.

(d) Materials required for incorporation into engineering projects. Note that materials required for manufactured goods (as opposed to engineering projects) will appear in manufacturing department budgets, but engineers may have to provide input data to manufacturing management in the form of materials usage estimates.

(e) Patent and licensing fees, where these are specific to a project.

If estimates have been completed for all projects planned during the year it will be possible to sum the total requirements for materials and other direct costs such as professional fees and patents. Figure 20.15 illustrates such a reconciliation for a manufacturing company.

2 Indirect expenses. These might include:

(a) Training fees and expenses.

(b) Travel, hotel and subsistence expenses not reimbursable by customers or clients.

(c) Stationery, printing and other drawing supplies (although in some companies these items are covered in the general stationery and printing budget under the responsibility of a budget manager outside the engineering department).

(d) Insurances, especially such as professional liability insurance.

(e) Purchases of books, engineering standards.

(f) Subscriptions to professional or trade associations.

(g) Service or maintenance contracts for equipment used in the engineering department (such as computers).

(h) Depreciation allowances for capital equipment in the engineering department.

(i) Some professional and patents fees.

Of course there are many more items of indirect expense occurring in an engineering company (see Chapter 19). But the above list includes those that are most likely to be influenced by decisions of engineering managers. These are therefore likely to be included in engineering department budgets.

In the case of the mining engineering company used as a case example in this chapter, each indirect budget account had one budget statement sheet for salaries (see Figure 20.13) and one for the associated expenses. The expenses

	Projects					Total
	RO1	RO2	RO3	RO4	RO5	
Direct project costs						
Materials						
Materials						
Subcontract work						
Tools/equipment						
Other						
Salaries						
Mechanical						
Electrical						
Electronic						
Other costs						
Professional fees/ patents						
General costs						
Supervisions						
General salaries						
Technical publications						
Stationery						
Travel and subsistence						
Other						
					Total	

Figure 20.15 *Reconciliation of total project costs in budget*

sheet would bear the same headings (budget account title, budget manager, account number) but was a very simple document: virtually a 'shopping list'.

General costs

Figure 20.16 is an example of a form used by the engineering department for annual budgeting in a company handling manufacturing projects. It has general relevance to most engineering companies, and the inclusion of a column showing the basis on which each charge has been calculated is very useful, both in compiling the estimates and in retrospect.

For significant costs it is advisable to list, wherever possible, all the major items of expenditure likely to be incurred. For example, the charge for patent fees should reflect not only the maintenance of existing patent cover, as

Cost for previous year	Expense	Basis of charge	Budget for year
	Labour: production support	Manning table	
	Materials	Project analysis	
	Other direct costs	Project analysis	
	Support staff	Listing of personnel	
	Supervision	Listing of personnel	
	Overtime	Reconciliation of project requirements	
	National Insurance Pension	Related to labour costs	
	Depreciation: Equipment Computers CAD	Specific write down rules.	
	Computer costs	Previous year adjusted for volume and economic increases	
	Repairs and maintenance	Previous year plus inflation or specific listing	
	Trade marks/patents	Specific listing	
	Stationery	Estimated usage	
	Training	Specific listing	
	Travelling and subsistence	Listing by individuals or groups	
	Sundry	Various allowances	

Figure 20.16 *Basis for calculating budgeted costs*

appropriate, but also the patenting of new products or the extension of existing patents into new markets.

Where it is not possible to produce detailed listing of expenditure requirements, then costs will have to be estimated in relation to the previous year's charge, after making appropriate allowances for changes in usage as well as allowing for inflation.

Controlling costs against budgets

The traditional method of cost control is to compare actual expenditure for each category of expense against the budgeted cost for the same period. Monthly reports are then produced showing this comparison for the month and for the cumulative position.

The annual budget is a commitment to spend no more than a certain level of cost during the year. If overspends are shown for periods early in the year, there is an implicit assumption that similar levels of overspend could continue through the year. It is therefore important to re-examine regularly the forecast expenditure for the full year, and to report this at the same time. In the event that additional expenditure is required to support a higher level of activity, then this should be capable of being justified by reference to progress against the relevant project budgets.

Figure 20.17 shows a typical format for a monthly departmental cost report. A similar example, for use in helping to control the costs of an individual project, is shown in Figure 23.6.

It is possible to set out some principles and guidelines that apply generally to cost control. These are not listed in any special order of significance or priority. It will be seen from the following that budgeting is only one of a family of steps that must be taken to control costs:

- The company must be organized for effective control, with jobs and responsibilities adequately defined.
- Work must be defined properly in specifications or other documents so that the scope of work relating to any budget item is known without ambiguity.
- All items of work or expense must be broken down into manageable parts, each of which bears a code number allocated from a logical system of cost coding.
- A budget manager must be appointed as being responsible for each of the budget accounts identified from the work breakdown.
- Cost estimates should be made using Pareto's principle, concentrating attention on the most expensive items, with broader estimates permissible for the lower cost items.
- The cost accounting system must provide for collection of all costs,

Engineering Department	*Month*				*Cumulative*				*Full year forecast*			
			Variance				*Variance*				*Variance*	
	Budget	*Actual*	£	%	*Budget*	*Actual*	£	%	*Budget*	*Actual*	£	%
Chargeable Salaries												
Mechanical												
Electrical												
Electronic												
Materials												
Other direct costs												
Support staff												
Supervision												
Overtime												
National Insurance												
Pensions												
Depreciation												
Equipment												
Computers												
CAD												
Computer costs												
Repairs and maintenance												
Trade marks/patents												
Professional charges												
Stationery												
Training												

Figure 20.17 *Departmental analysis of actual expenditure against budget*

analysed according to the relevant cost codes. This information must be presented to budget managers regularly and promptly, because early information provides the basis for early action.

- For the control of labour costs, the most important factor is the control of progress: if work is finished on time using the planned labour force, it is unlikely that the labour budget will be exceeded.
- A big factor in the successful control of direct equipment and materials costs is an efficient purchasing system. Cost control for these items is exercised up to the time the orders are placed. It is too late to attempt control once excessive costs have been committed in bad purchasing contracts (see Chapter 35).
- Engineering changes and modifications must be controlled (this subject is covered in detail in Chapter 25).

21

Engineering contracts

Stephen H. Wearne

An oral contract isn't worth the paper it's written on
Attributed to *Sam Goldwyn*

Definition: **Contract** (noun): A mutual agreement between two or more parties that something shall be done . . ., an agreement enforceable by law. (*Oxford English Dictionary*)

Introduction

Contract relationships in engineering are numerous and vary greatly in type and importance. Employment is contractual. So is the purchase of pencils, the supplying of design services or the ordering of mechanical plant. All these agreements are subject to the law, but the practices followed vary from one type of transaction to another. The most significant differences are in the financial risks involved to one or more of the contracting parties.

In general, the longer lasting a transaction or the larger the financial risks, the wiser it is for the parties to an agreement to negotiate a specific written contract. The contract should define what each party is to do: typically one party to deliver 'the works' and the other to pay for them. Prudently, it should also include agreement on how problems which may arise in the work, or from the behaviour of any individual or organization, are to be solved.

A transaction such as the purchase of pencils is normally agreed quickly on the basis of statements of quality and price. Doubts and disputes are unlikely and, if there are any, these should be settled by custom and practice under the law. But before the purchase of complex plant there are usually discussions of the proposed design and of responsibilities for the engineering and other work involved. A specification of quality and performance is written, prices are negotiated and a contract is drafted that incorporates conditions of contract stating the remedies which are intended to be available in the event of faults, disputes, etc. that might occur after the agreement.

Various model conditions of contract have been evolved for engineering and construction work, particularly in the UK. They have been evolved chiefly by the professional institutions and government departments, largely by a process of trial and revision designed to codify the lessons of experience. This chapter concentrates on contractual practice and the use of these conditions.

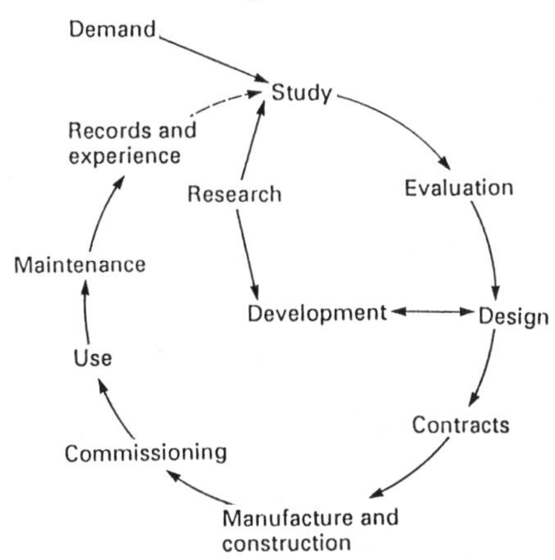

Figure 21.1 *A project cycle – a contract is the link between design and reality*

Types of contract

Several types of contract are in use in engineering. These differ in the ways in which they divide risks between a purchaser and a contractor.

Minimizing risk

Risks can arise from a purchaser's uncertainty at the start in specifying what he wants. For instance, the contract might be for the design and supply of plant that is to be part of a system which will be affected by design discussions during the contract. Another example would be where the initial prediction of the plant's purpose might have to be varied during the work to meet changes in the expected demand for the goods or services that the completed plant is to produce. The choice of contract type should depend on the extent to which such uncertainties may affect the contractor's planning of the development and use of his resources – people, finance, equipment, materials and subcontractors (Figure 21.2).

Commonly a purchaser specifies his requirements before inviting tenders, although not in all detail. But however well detailed these requirements, a contractor's planning depends on his predicting how the work expected may be affected beyond the control of either party (for instance by subcontractors'

		Impact of a risk	
		Low	High
Frequency of the risk	High	Share	Customer bears
	Low	Contractor bears	Avoid

		Specification and drawings	
		Complete and final	Vague or changeable
Resources required	Certain	Lump sum Fixed price	Bill of quantities Schedule of rates
	Uncertain	Reimbursable Convertible	Staged Mixed

Figure 21.2 *Choices in contractual risk allocation*
(after N. M. L. Barnes)

technical or financial failures, natural disasters, political influences, the supply of materials, or national disputes between trades unions and others). Purchasers in the developed countries rarely invite tenderers to be committed to complete their work regardless of these risks. The argument is that the tenderers would have to cover themselves by high prices, in excess of the most probable costs that might be incurred. In countries with less engineering expertise, governments and other customers at times ask tenderers to carry more risks. But the trend in industrial countries since early in this century has been to limit contractual risks to those that competent contractors should be able to foresee and control.

Avoiding disputes

There can be all sorts of minor and major disputes between the parties to a contract (for instance when trying to agree the payment due to a contractor for a purchaser's requests for changes). But at least only the parties are involved and, if necessary, such disputes can be settled eventually by reference to arbitration or to a court. More difficult for them to resolve are their responsibilities and liabilities that may result from delays, failures and any such events that seem to be beyond the control of either party. One hopes to anticipate such problems by selecting conditions of contract which include agreement on means of remedying disputes so far as is possible, but first the decision has to be made on the types of contract most appropriate to the work to be done and the uncertainties which may affect it.

Choice of contract type

Normally each purchaser has the option to choose the type and conditions of a contract. This is not true in the purchase of many mass produced goods, where standard conditions of sale are offered which purchasers normally accept in expecting these goods to be available on demand. But the larger plant, control systems, services and structures needed for a capital project are not available on demand and they usually need some or substantial design to suit a purchaser's requirements. For these contracts the following options are available:

(a) The contractor undertakes to provide a complete project or process capable of producing a specified product or service (such as a refinery or a steel mill) and is responsible for the design and supply of the systems, plant, structures and services for the project. This is known as the turnkey, all-in, design-and-build or package deal type of contract, but these are not precise descriptions: they depend, for instance, on the extent to which the purchaser or the engineer specify requirements and provide design information. Conditions of contract published by the Institution of Chemical Engineers are suitable for this type of contract; *or*

(b) The contractor undertakes to provide plant (such as a pump, ventilating plant, switchgear or other equipment) usually being committed to designing it to achieve a specified performance. Forms of contract agreed by the IMechE/IEE/ACE, the ECE conditions for plant and machinery for export, and the conditions for electrical and mechanical works published by FIDIC are designed for this type of contract; *or*

(c) The contractor undertakes to make or construct what is specified in drawings etc., issued to him by the purchaser or the engineer. The contractor is usually responsible for deciding his methods for how to do the work, but not what is to be the result. Conditions for civil engineering works issued by the ICE/ACE/FCEC and FIDIC are designed for this type of contract.

The consequential differences are in the role of the engineer and in the responsibilities of the contractor for performance and for mistakes.

The related differences are in the terms of payment.

Model conditions of contract

Published British models

The following models are published for use in engineering and construction contracts in the UK:

Conditions of Contract for Complete Process Plants, published by the Institution of Chemical Engineers (IChemE) in two model forms:

1 For lump sum contracts, the 'Red Book' (revised edition published in 1981)
2 For reimbursable contracts, the 'Green Book' (published in 1976 and, at the time of writing, under revision).

Conditions of Contract for the Purchase of Mechanical or Electrical Equipment, published by the Institution of Mechanical Engineers, the Institution of Electrical Engineers and the Association of Consulting Engineers (IMechE/IEE/ACE). A new version was published in 1988.

Conditions of Contract for Civil Engineering Works, published by the Institution of Civil Engineers (ICE), the Association of Consulting Engineers and the Federation of Civil Engineering Contractors (FCEC). The fifth edition (1979) was reprinted in 1986, incorporating some amendments. A new set of conditions is also being discussed at the time of writing.

GC/Works/1 Conditions of Contract for Construction, published by the Property Services Agency, a revised edition of which was published in 1971.

International models

Models for use in engineering and construction contracts outside the UK include:

General Conditions for the Supply of Plant and Machinery for Export, published by the United Nations Economic Commission for Europe (ECE). Alternative versions exist for supply with or without erection.

Conditions of Contract (International) for Electrical and Mechanical Works, published by the Fédération Internationale des Ingénieurs-Conseils (FIDIC). The third edition was published in 1988.

Conditions of Contract (International) for Works of Civil Engineering Construction, published by FIDIC. The fourth edition was published in 1987.

Various model forms for projects in developing countries, published by UNIDO.

Comparisons

These model conditions of contract vary in some important terms and in much of their detail. They also appear to have some similarities.

Figure 21.3 lists the principal clauses in which most of the above model conditions of contract differ. These sets of conditions also vary in the detailed wording of many clauses. The effects of their differences depend on the wording of each set of conditions as a whole, so that the list is only an indicator of how they vary and what they include.

The IChemE model forms include notes to explain the reasons for certain

	UNECE 188A 1957	*IChemE lump sum 1981*	*UNIDO turnkey 1982*	*IMechE/ IEE/ACE 1988*	*FIDIC mech/elec 1988*	*GC/ Works/1 1977*	*ICE/ACE/ FCEC 1973 rev 1979 & 1986*	*FIDIC civil eng 1987*	*ICE minor works 1988*
The Parties	the Vendor the Purchaser	the Contractor the Purchaser	the Contractor the Purchaser	the Contractor the Purchaser	the Contractor the Employer	the Contractor the Authority	the Contractor the Employer	the Contractor the Employer	the Contractor the Employer
The Engineer?	No	Yes	Yes	Yes, or the Purchaser	Yes	the Super-intending Officer	Yes	Yes	Yes
see clauses		1, 10	1, 6	1, 2	1, 2	7, 16, 49	1, 2, 7, 13	2	2, 3
Contractor's liabilities	9, 23	3, 5, 44	4	13, 44	8	2, 8	8,13	8,13	3
Suitability for purpose		3	28						
Excepted risks		43		45, 46	37	1	20	65	1
Design, drawings, patents	3, 12	4, 7, 19, 20, 21	7, 12	13, 15, 42	6, 16	4, 12, 15	1, 6, 7, 14, 28	6, 7, 28	3
Programme, progress	20	12, 13, 39	6, 11	14, 32–34	12, 25 to 27	6, 28	14, 41, 43, 46	14, 41, 43, 46	2, 4
Extension of time	20	14	19, 34	33	26	28	44	44	4
Payment	7, 10, 11	39	20, 23	39, 40	33 to 36	10, 38 to 41	55 to 5, 60	55 to 58, 60, 66	7
Bill of quantities						5	55 to 57	55 to 57	
Contract price adjustment			20		47	11G	separate optional clause	70	

Variations		9, 16, 17, 18	15	27	31	7	51	51	2
Extra costs, claims	13	5, 6, 14, 18	19, 32	27, 41	34	9, 53	12, 52	12, 52, 53	3, 6
Sureties, bond		37	21	8	10		10	10	
Sub-contracting		8	12	3, 33, 53	4	30	4	4	8
Nominated sub-contracts		9	12			31	59 A, B & C	59	
Safety	15	17, 27, 28	4	20	14	17	8, 19	8, 19	3
Vesting of plant, material		25, 41		37	32	3	53, 54		
Damage, insurance	10, 24	31, 32	24	43, 47, 48	40, 42, 43	25, 26, 47, 48	20 to 25	20 to 25	3, 10
Taking over	22	34	12, 18	29	29	28A	48	48	4
Defects maintenance	23	36	28	36	30	32	49, 50	49, 50	4, 5
Retention money			36			41	60	60	7
Bonus, liquidated damages	26	15	20, 27, 30	34	26, 27	29	47	47	4
Special determination		13, 41, 42		46, 49–51	24	44 to 46	40, 63	40, 63	2
Disputes		46, 47		2	50	61	66	67	11
		46 (Expert)							11 (Conciliation)
Arbitration	28	47	37	52	50	61	66	67	11
Applicable law	5, 13	2	36	54	51		67	5	12

Figure 21.3 *Differences in the scope and model definitions in model conditions of contract for engineering and construction projects*

clauses and to guide the preparation of the special conditions and schedules expected to be required for each contract.

Guidance notes on the use of the ICE conditions are published from time to time by the CCSJC (Conditions of Contract Joint Standing Committee representing the ICE, ACE and FCEC).

Similarities – the engineer

The model conditions of contract issued by the UK engineering institutions and by FIDIC have in common the employment of what all term 'the engineer'. Thus far, all these models are similar. Figure 21.4 illustrates the contractual relationships between the purchaser, the engineer and the contractor in these model conditions. This similarity is deceptive. The powers and duties of the engineer are different in each of these model conditions.

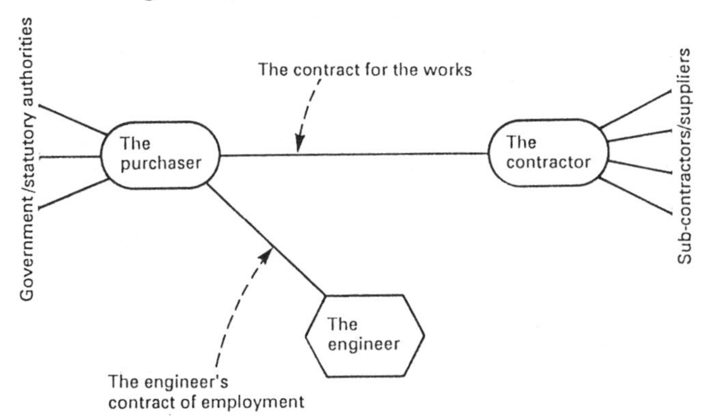

Figure 21.4 *Contractual relationships between purchaser, contractor and engineer*

International differences

Note that the law, legal systems and legal attitudes vary from country to country, even between England and Scotland. These differences, as well as different cultural and business practices, can complicate contracts with parties in other countries. In exporting or importing it is essential to define in a contract which country's law applies, which language will rule and where disputes are to be settled.

Terms of payment

Firm price contracts

In firm price contracts the prospective purchaser (or the engineer on his behalf) prepares a specification, drawings, conditions of contract etc., stating the requirements. One or more prospective contractors then tender for the works, accepting the risks inherent in their estimating of their direct and indirect costs. In these contracts the purchaser has usually agreed to pay most of the contract price in a lump sum on delivery of the plant, and to pay the remainder after a specified retention period (a period allowed for ensuring that the plant has performed to the contract requirements). Under most such contracts the engineer can order changes to the works (within limits), and the contract price may be varied accordingly.

Cost reimbursable contracts

In cost reimbursable contracts the purchaser pays the contractor the costs incurred for the works, usually plus a fixed amount or percentage for general management and profit. These contracts usually require the contractor to give the purchaser or the engineer access to all his accounts, so that the costs are known and can be verified. Payment is usually made at set intervals (monthly, for example).

Firm price contracts and cost reimbursable contracts are distinct in their possible effects on contractors' attempts to control costs. In theory, a firm price contract should induce care to control, whereas the agreement that all relevant expenditure will be reimbursed has no such incentive. Complete uncertainty in specifying his requirements at the start may force a purchaser to choose a cost reimbursable type of contract, but the following intermediate types are possible alternatives. The choice between them should be made depending on what can be specified initially.

Target cost contracts

Based on tender prices or on negotiations at the start to agree the probable cost of the works, target price contracts include an arrangement that the purchaser and contractor will partly or fully share any savings achieved or any excess costs in some specified way. This type of contract is intended to give the contractor some protection against uncertainties and some inducement to control costs. Its success in doing so depends on the validity of the initial agreement on the probable cost: yet such a contract is needed only if uncertainties at the start prevent agreement upon a firm price contract.

Schedules of rates contracts

Schedules of rates contracts are based upon specifications of each type of work to be done coupled with prices for payment to the contractor for each unit quantity of this work ordered by the purchaser (or the engineer). They are used for contracts for maintenance and other jobbing work, circumstances in which the purchaser (or the engineer) expects to be able to predict beforehand the types of manpower, repair work, plant dismantling and erection, minor construction, painting, etc. which may be needed but cannot estimate in advance which or how much of each will be needed. This type of contract is also used for purchasing design and development work.

For any of these uses of schedules of rates contracts the prospective contractors are usually invited to tender their rates (prices for each unit of specified work), to be paid according to the amounts ordered. Compared to the cost-plus type of contract, the prospective contractors can be chosen in competition for their rates and the contractor appointed has (at least in theory) the inducement to control his costs.

Bills of quantities contracts

These contracts are related to schedules of rates contracts, but in a bill of quantities (BoQ) the rates are expected to be based upon total quantities of each type of work. By this means tenderers are expected to try to optimize their rates by planning to achieve the economic use of manpower and equipment. This type of contract is common for civil engineering construction, and is also used in contracts for the installation of mechanical plant, piping and cabling. Flexibility is possible by agreeing in the contract that payment will be made for the actual quantities ordered during the contract, but at the tendered rates unless the purchaser (or the engineer) subsequently orders changes from the types or quantities of work specified at the time of tendering which would invalidate the contractor's planning and costing.

Compound contracts

Contracts are possible which combine two or more of the above choices. A compound contract might, for example, provide for payment on the firm price basis for definite work, but include also a schedule of rates (or a dayworks schedule) in which payment rates per hour (or per day) are stated for the use of additionally ordered manpower, machines, etc.

Other payment terms

The description *guaranteed maximum price* contract is used for one which combines the firm price and target price principles such that the purchaser and the contractor share savings, but the contractor meets all or most of costs incurred above a specified maximum.

When the requirements are initially uncertain, a *staged contract* can be used. In this, design or other preparatory work is undertaken competitively or by one contractor, but the resulting definition of the work to be done is used as the basis for inviting firm price tenders for all that remains to be done. By this arrangement the options affecting the prediction of final costs can be decided before commitment under a contract. Contracts of this sort, for design work only, can give a purchaser the option of going no further, or of changing to another contractor.

Contractors invited to supply and erect the major sections of plant for a project can also be asked, when tendering, to give rates for the work of incorporating other suppliers' items, rather than ask each of these other suppliers to provide resources for this. In addition, all can be asked to give rates for advising on the commissioning and operation of their plant when their contracts have not been inclusive package-deals for carrying out all work from design to handing over.

Convertible contracts usually start on a cost-plus or target basis of payment, but have an agreement to change to a fixed price when some definite state has been reached in the work (for example at the completion of design).

Figure 21.2 indicates important relationships between risk factors and terms of payment.

Contract policy

It is usual for the promoter of a project to purchase plant and equipment rather than make it himself when this is of advantage in securing the use of expertise, financial resources, equipment, management and other skills that are not required continuously by the purchaser. Contractors are therefore in business to provide these. In engineering, the purchaser usually chooses the type of contract. The choice should logically depend upon the nature of the work, its certainty, its urgency and other factors such as the relationship between an investment in new plant and the operation and maintenance of plant already in use. For instance, a contractor may have to be employed on a flexible arrangement so that his work can be rearranged from day to day according to priorities between installing new plant and the needs of the plant in use.

Conditions for the employment of labour and supervisors may also be important influences on contract policy, for instance to specify consistent payment systems between the purchaser's and the contractor's employees, or between those of one contractor and another working nearby. From the experience of projects in the UK it is quite widely agreed that standards of safety, working conditions, payments and managerial incentives to control costs should be consistent between adjacent or related temporary and permanent employers.

Strategy is also needed in timing when to invite contractors to tender. Rarely can all a purchaser's requirements be specified so that completely competitive firm price tendering is achieved without the risk of having to pay later for changes. Disruptions, delays and extra costs arising from some contracts have been so great that it seems wise to conclude that more time should have been used earlier to make final decisions before inviting tenders. But other experience suggests that decisions in some purchasers' organizations only become final when they are committed to contractors, so that the earlier the contracts are placed the better may be the results. More planning of the initial decisions coupled with greater control of changes may be most needed, not least because the initial decisions tend to be made casually if it is expected that changes will be allowed later.

Policy also needs to be deliberate in inviting tenders. Every tender adds to all parties' costs. It is commonly thought prudent to ask contractors to tender competitively in order to get minimum prices from them. Many purchasers believe that this is achieved but, increasingly, they also recognize that competitive tendering can go wrong by producing plant that is only initially the cheapest. To get value, a purchaser has first to specify quality, decide how performance will be measured, anticipate the exceptional conditions that may be needed, and apply the lessons of the real costs of long-term operation and maintenance of his and similar existing plant. Then tenders must be invited only from contractors who are known to have the financial resources, technical experience and managerial competence necessary for the contract.

The sad results of some projects indicate that purchasers need to establish self-control of their contracts through one manager who has experience of the potential conflicts of interest that can arise. One way of modelling a formal system for planning, awarding and supervising a construction contract is shown in Figure 21.5. Systems are needed by all parties for anticipating problems, motivating good performance, and for making and communicating decisions (these aspects of project management are dealt with in several other chapters).

Legal rights and commercial interests

Experience advises us to agree in a contract what shall happen, or what procedures will be followed, in the event of problems arising which will or could affect the parties' obligations. Although it is hoped that perhaps no problems (or only a few) will arise, the agreement in advance on what shall be done about them should not only help to get problems solved quickly and minimize disputes, but may also reduce the chance that they will be allowed to arise. Care and expertise is needed in drafting and managing contracts. Advice should be obtained on how to use legal means to achieve engineering ends.

The law and these agreements establish some legal rights (for instance for the payment of 'damages' to compensate one party for a default by the other). Typically this remedy is required if a contractor completes late or the plant does not achieve its specified performance and the fault is not at all due to other parties. In such situations, and on matters of detail which do not turn out as was expected at the start, the party who believes they have suffered a

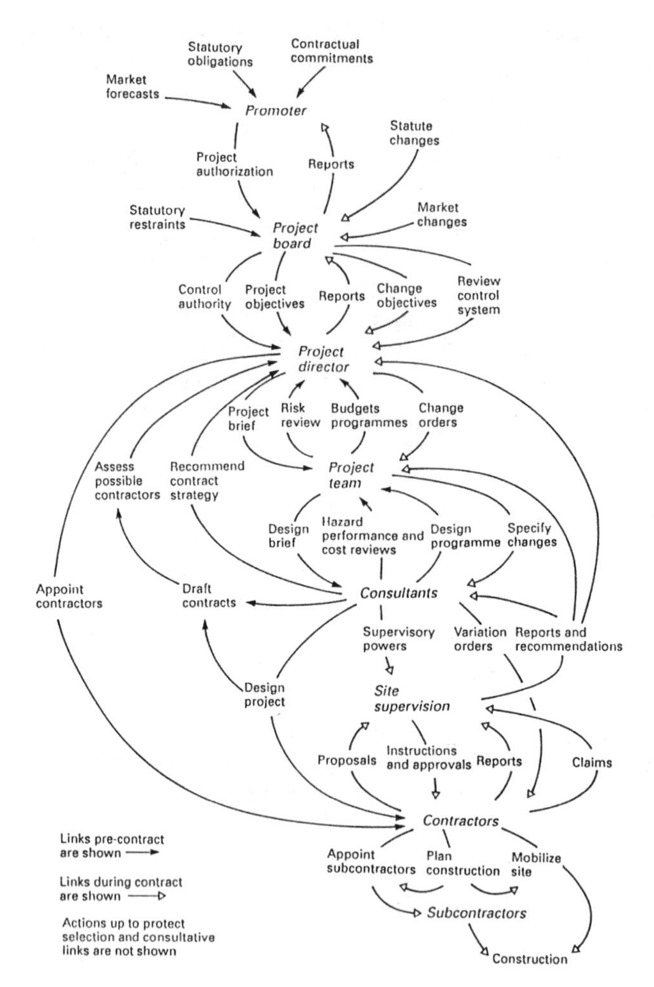

Figure 21.5 *An example of a system for planning, awarding and supervising a construction contract*
(Based upon studies by Ninos and Wearne)

loss should consider whether enforcing a right to damages or other consideration is in their best interests overall. One may wish to do so in order to show that one 'knows one's rights', and to set an example (for instance) by insisting on a standard of material specified in a contract even though another offered would be suitable. On the other hand, not all specifications can be exact, and no conditions of contract can anticipate all the combinations of problems that can occur beyond the control of the parties. Doubts may be referred to an independent opinion. Disputes may be taken to an arbitrator or to a court. But also of importance are the wider interests of each party, extending beyond the contract or project. For instance, a contractor may not wish to acquire a reputation for claiming every right, no matter how small in value. Or, a purchaser may not wish to be unreasonable in encouraging contractors to try to overcome unexpected problems.

Engineers and others who are representing any party to a contract should therefore consider both their rights and the potential effects of insisting on them. Sticking to one's rights is safer when there may be trouble and no support from the boss. Obtaining the boss's and other advice on alternative policies that should lead to successful projects is more professional.

Contractual negotiations to try to reconcile different views and interests can be lengthy and a great strain on everyone taking part. Cultivating goodwill when succeeding in an argument may prove valuable on later occasions.

Definitions

Precision and simplicity are rules to be followed in drafting contract documents.

The models published by the UK engineering institutions and others usually start with a list of definitions. Care is needed in using them because the definitions differ from one model to another. Also, various alternatives are used by engineers and others in publications and discussions. What matters in understanding a contract are the definitions given in that contract, plus the dictionary, custom and practice as the guides to all other words used, but here it is worth pointing out a few definitions that can cause confusion:

The purchaser means what is says, the party to a contract who is purchasing the goods or services. But note that alternatively the words buyer, owner, promoter, employer and client are used, this practice varying particularly in the model documents and commentaries on them published for the construction industry.

The contractor is the common term used in Britain, but the alternatives seller and vendor are also used. Note that 'the contractor' in a contract document means *the* firm which is a party to that contract: not any firm who are in business as contractors. The contractor can be a consortium of firms jointly acting as one.

The engineer in some engineering contracts has powers and duties defined in the contract. Traditionally an independent consultant or consulting engineer is appointed the engineer.

The resident engineer in many civil engineering contracts is the engineer's representative at the works. In mechanical and electrical engineering that title is sometimes used by contractors for their chief engineer on a site. In civil engineering in Britain the agent is the usual title of a contractor's manager on site.

Main contractor or *managing contractor* usually means a contractor who takes most or all of the commercial risks of undertaking the works, but employing subcontractors to do parts. *Management contractor* usually means a contractor employed to plan, help administer and supervise work to be carried out by other contractors.

The plant in some contracts is defined as the goods (machinery, etc.) being supplied, or erected (or both) under the contract. In other contracts the plant or equipment is defined as the machinery owned or hired by the contractor to do his work.

The works in most model documents means all the work to be done by the contractor under the contract. It can be the plant specified and all that has to be done to supply and erect it. It may just be design work, or a report, depending on what is stated in the contract.

Completion of the plant usually applies to its technical readiness for commissioning and operation. Contractual 'handing over' of its ownership may be separate.

A *firm* or *fixed* price may or may not be subject to change. For instance, there can be clauses in the conditions of a contract by which the scope of the works may be varied, or CPA clauses for adjusting prices partly or fully for escalation – changes in the costs of labour, material, etc. during a contract.

A *lump sum* is a price for the works complete or for major sections of it, not broken down into amounts for pieces or stages of work. Lump sum payments are common in contracts for mechanical plant. Payments for each identifiable piece of work are more common in the construction industry. (Note that the 'lump' in the building industry can mean a group of people who seek to be employed on site on the basis of being taxed, etc., as self employed individuals.)

The *construction* of a contract means the 'construing, explaining and interpretation of the text' (*Oxford English Dictionary*).

The essentials of a simple *contract* under English law are offer and acceptance, possibility, capacity, performance and consideration.

Abbreviations

These abbreviations are used in publications on British engineering contracts.

ACE	The Association of Consulting Engineers.
BEAMA	British Electrical and Allied Manufacturers Association.
BoQ	Bills of quantities
BAQ	Bills of approximate quantities.
BS	British Standard.
CASEC	Committee of Associations of Specialist Engineering Contractors.
CCSJC	Conditions of Contract Standing Joint Committee (representing

	the ICE, ACE and FCEC).
CIRLA	Construction Industry Research and Information Association.
CPA	Contract price adjustment of tendered rates or prices to allow or partly allow for escalation – changes in material, labour or other costs during a contract. (Confusingly, CPA is also used in network planning to mean critical path analysis.)
ECE	United Nations Economic Commission for Europe.
EEF	Engineering Employers Federation.
FCEC	Federation of Civil Engineering Contractors.
FIDIC	Fédération Internationale des Ingénieurs-Conseils (Lausanne).
ICE	Institution of Civil Engineers.
IChemE	Institution of Chemical Engineers.
IMechE	Institution of Mechanical Engineers.
IEE	Institution of Electrical Engineers.
JCT	Joint Contracts Tribunal for the Building Industry.
NECEA	National Engineering Contractors Employers Association.
NEDO	National Economic Development Office.
OCPCA	Oil and Chemical Plant Contractors Association.
OED	*Oxford English Dictionary*.
PCA	Price cost adjustment, as CPA.
PIT	Progressing, inspection and testing.
QS	Quantity surveyor.
RE	Resident engineer (on site).
RIBA	Royal Institute of British Architects.
RICS	Royal Institute of Chartered Surveyors.
SMM	Standard method of measurement, for instance as used in construction for measuring quantities of work for a BoQ.
SO	Supervising officer or Superintending Officer.
VO	Variation order.

Further reading

Booklets

Morrison, M. C., Book 3 in the series *Total Project Management*, Asset Management Group of the British Institute of Management (undated).

National Economic Development Office, *Guidelines for the Management of Major Projects in the Process Industries*, 1982.

Construction contracts

Abrahamson, M. W., *Engineering Law and the ICE Contract*, 5th edn, Elsevier Applied Science Publishers, 1982.

Hayes, R. W., *et al.*, *Risk Management in Engineering Construction*, Thomas Telford Ltd, 1986.
Institution of Civil Engineers, *Guidance Notes on Tendering Procedures and on the Role of the Engineer.*
Joint Contracts Tribunal for the Building Industry (JCT), *Guide to the Standard Form of Building Contract*, Royal Institute of British Architects Publications, 1980 (contains the JCT standard conditions of contract for local authority projects).
Ninos, G. E., and Wearne, S. H., 'Control of Projects During Construction', *Proceedings of the Institution of Civil Engineers*, Part 1, vol. 80, 1986, pp. 931–43.
Wearne, S. H., *Civil Engineering Contracts*, Thomas Telford Ltd, 1989.

International construction contracts

Sawyer, J. G. and Gillott, C. A., *The FIDIC Conditions: Digest of Contractual Relationships and Responsibilities*, 2nd edn, Thomas Telford Ltd, 1985.
Tendering Procedure for Evaluation Tenders for Civil Engineering Contracts, FIDIC, 1982.

Research

Hayes, R. W., *et al.*, *Management Contracts*, Report 100, Construction Industry Research and Information Association (CIRIA), 1983.
Perry, J. G. and Thompson, P. A., *Target and Cost-Reimbursable Construction Contracts*, Report 85, CIRIA, 1980.
Wearne, S. H., 'Contractual Responsibilities for the Design of Engineering Plant', *Proceedings of the Institution of Mechanical Engineers*, vol. 198B, no. 6, 1984, pp. 87–96.

Textbooks

Davis, F. R., *Contract*, 5th edn, Sweet & Maxwell, 1986.
Johnston, K. F. A., *Electrical and Mechanical Engineering Contracts*, Gower, 1971 (contains the ECE conditions of contract).
Marsh, P. D. V., *Contracting for Engineering and Construction Projects*, 2nd edn, Gower, 1981.
Marsh, P. D. V., *Contract Negotiation Handbook*, 2nd edn, Gower, 1984.
Scott, W., *The Skills of Negotiating*, Gower, 1983.

Acknowledgement

The author of this chapter gratefully acknowledges debts to the authors of publications and to former colleagues, bosses, lecturers, students and many others for their comments and advice on contract management.

PART SIX

Work Management

22

Planning and scheduling

Dennis Lock

It is generally true to say that any job which overruns its expected completion date will also overrun its cost budget. There are many strong arguments which support this claim, and there is every reason for ensuring that all engineering work is sensibly planned, scheduled and progressed. This chapter examines some of the available planning and scheduling techniques that can be used to provide a proper basis for effective progress management (progress management itself is the subject of Chapter 23).

Effective scheduling

How not to schedule

How familiar is this story? An engineering manager is asked to undertake a small design task, to result in a set of prototype drawings after about ten weeks. The production manager, who is also responsible for purchasing in this company, has been given a target date eighteen weeks away for completion of an assembled prototype, ready for testing. The plan looks like this:

	Start	*Finish*
Engineering	11 Jan. 1988	18 Mar. 1988
Purchase components	21 Mar.1988	8 Apr. 1988
Manufacture and assembly	21 Mar. 1988	6 May 1988

After about two weeks there is an informal meeting at which the engineering manager reports that progress is fine. After another four weeks there is a second progress meeting, at which the engineering manager thinks he might be running a little late. When pressed, he estimates a delay of one week. At the next progress meeting, when the design should have been finished, the forecast is that, owing to unforeseen difficulties, three weeks' design work remains. Eventually, at the end of fifteen weeks, the production manager gets his drawings, and is left with practically no time in which to manufacture the prototype. So, the prototype is late, and so also are other jobs on to which the engineers should have transferred earlier. Progress management was obviously less than perfect, but did the schedule give the project a fair chance from the start?

Characteristics of an effective schedule

By reviewing some of the management shortcomings in the above example a checklist can be assembled of factors that are either desirable or essential for the establishment of a schedule that is likely to prove an effective tool in managing progress.

Acceptance of commitments

All the dates in the example were imposed dates. Managers were told what their commitments were to be, without any discussion. In practice, estimates and target dates must be established by consultation with the managers who are to be responsible for keeping to them. If this practice is not followed, impossible schedules result, with failure, recriminations and frustration to follow.

Working methods

Another casuality arising from failure to discuss plans is the lost opportunity of finding more efficient working methods. In this example, it should have been apparent that some purchased components would have to be ordered as early as possible, so that the specification of these should have been a design priority. Another aspect of discussion at the early stage should be to look at the design proposals along with the production manager, in case a change of approach might be suggested that would save production time and money without detriment to product quality.

Estimating accuracy

Although the term accurate is used to describe good estimates, in reality there is no such thing as an accurate estimate. If project costs are estimated to be £500 000 and the accountants tell us at the end of the day that the final figure was £499 999, then the estimators and everyone else concerned can feel satisfied that they exercised good judgement, or good management, or even both. But estimates can only point to a likely outcome, and every new project is really a trip into the unknown. It is usually possible, however, to declare a degree of confidence (or lack of confidence) in an estimate by classifying it according to the possible degree of error (plus or minus so many per cent of the given figure). Obviously the confidence in any estimate will be higher if it is based on relevant past experience, and if the new work can be defined in careful and assured detail.

Estimates for planning and scheduling are mainly concerned with elapsed time, but with a need to estimate resource usage as well. The important thing in making estimates for schedules is to use the best possible judgement, backed up with statistics from past experience as far as these are available and relevant. The results can never be accepted as accurate estimates but, barring accidents, they will be good estimates. There is no evidence that sufficient judgement had been exercised in making the estimates used in the example.

Planning detail

It is often necessary to judge progress achieved on the basis of subjective judgement, but it is always better if actual events can be identified for this purpose, and ticked off as they are passed. A plan must be sufficiently detailed to highlight such events, which must not be separated too far in time. Even the separation of conceptual, layout and detailed engineering would have given more points of reference in the example.

A useful rule in scheduling, no matter how sophisticated the technique being used, is to consider highlighting as a separate planning event every occasion on which work must change hands. This means not only transfers from one department to another (such as the issue of drawings from engineering to production) but also the passing of work within departments. A good example here is the handover of drawings for checking. Engineers and draughtsmen have been known to claim (when pressed) that their design work is finished, while not being ready to release their work to the checker. A job is only finished when its following work can start, and the planning detail must identify such events.

Job sequence

In the example, the planning detail was so course that the sequence of the jobs listed was obvious. When planning is carried out in far greater detail, it is necessary to consider the sequence in which jobs should be performed, together with the constraints and interdependencies that exist between them. A simple listing of jobs with their start and finish dates is useless if these relationships have not first been taken into account. In order to do this, a suitable notation has to be found, so that such ideas can be expressed clearly on paper.

Planning or scheduling?

The simple plan given in the example was not a schedule. A plan only earns the title of schedule after it has been adjusted in line with the resources needed. In a busy company, there is certain to be competition between projects for common resources. Resources allocation takes place at several levels, which are:

1 The requirements of each project or subproject.
2 The integrated requirements of all projects taken together (known as multiproject scheduling).
3 Derived from the above, workloads for the company, and for departments or pools of particular resource skills.
4 The assignment of tasks to individuals or to machines, according to particular aptitude or capability.

This can give rise to many problems, with a seemingly impossible number of variables (Figure 22.1). But, in any problem with many variables the best way is to solve them one by one, and that is the approach recommended in this chapter.

Flexibility
No project of any size can be expected to run its course without at least one change of plan or design. When the inevitable change does happen, it must be possible to reschedule the project (and other projects too, if resource pools are affected) in order that working schedules remain relevant to the objectives and practicable to achieve.

Checklist
Here is a checklist of the factors identified. The list should be consulted whenever any new planning and scheduling technique is being considered for use on a project.

1 Have the schedule commitments been discussed and agreed with the managers who are to be bound by them?
2 Are the estimates of times and resources as reliable as possible?
3 Has planning been carried out in sufficient detail?
4 Is the level of detail too high? Is the plan too complicated?
5 Does the plan show the jobs in a logical sequence?
6 Are constraints or job interdependencies shown?
7 Are the relative priorities of jobs indicated by the plan?
8 Has consideration been given to the resources needed, the resources available, and the interacting resource requirements of other projects within the same organization?
9 Is the schedule flexible? Can it be revised or updated easily in response to any change in project requirements, resource levels or current progress?

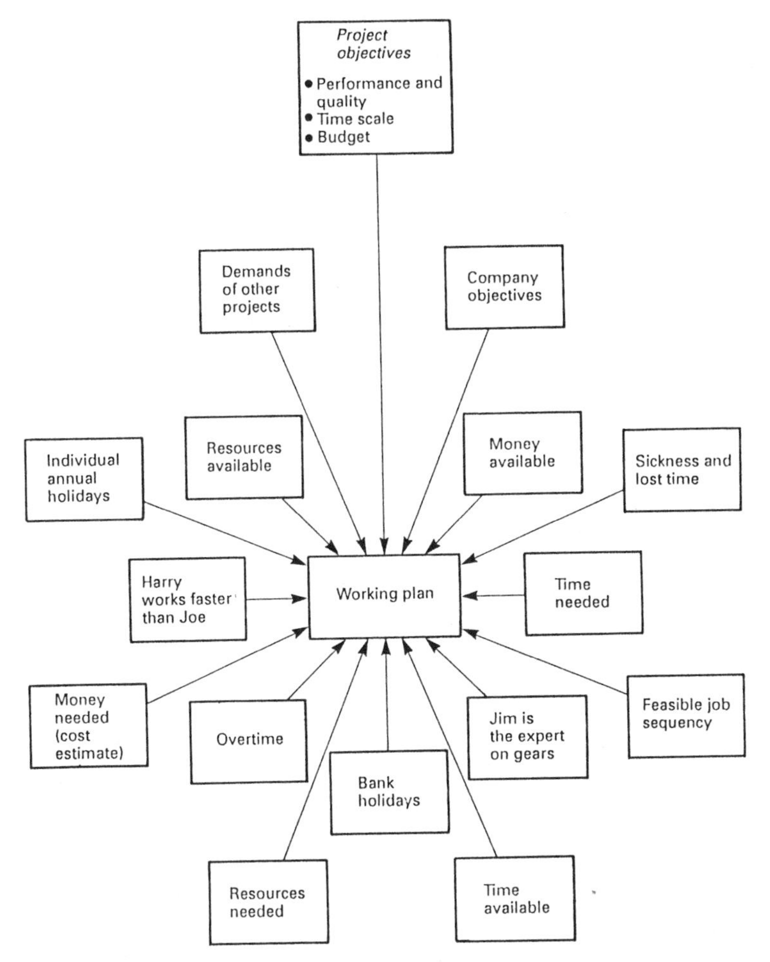

Figure 22.1 *Scheduling problem factors. At first, the number of variables involved in scheduling may seem daunting. But the problem can be overcome by a step-by-step approach in which the variables are eliminated one at a time*
(This illustration first appeared in *The Chartered Mechanical Engineer*)

Bar charts

Figure 22.2 shows a project plan expressed in bar chart form, which is probably very familiar to all readers. The example shown would fit the small engineering design case described at the start of this chapter.

Bar charts have a lot going for them. They show clearly the sequence of jobs on a readily interpretable timescale. No special training is needed in their use or application, and they are readily accepted on shop floors and construction

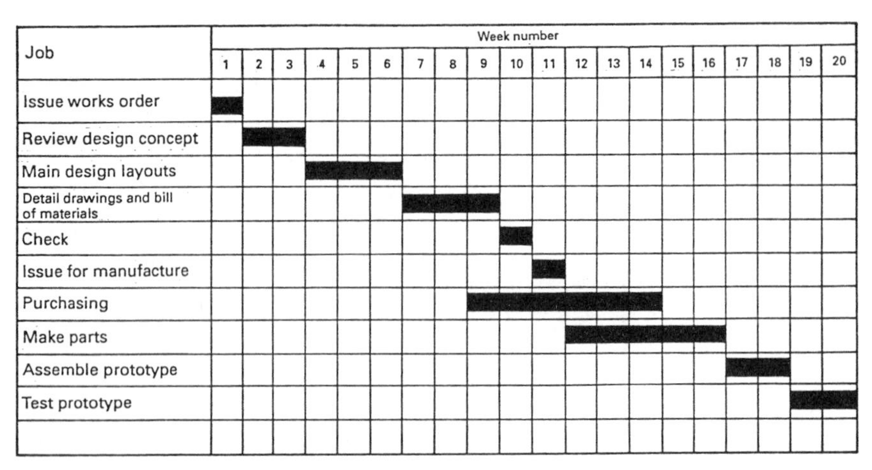

Figure 22.2 *A simple bar chart*

sites. In fact, they are so liked that it is often necessary to convert plans and schedules which have been produced by more advanced techniques into bar chart form, for the benefit of managers and supervisors at the workplace.

If each bar can be colour coded, or otherwise made to denote one unit of a particular resource type, then with ingenuity and patience it is possible to use the bar chart for resource scheduling. It is only necessary to count up the resources required down each vertical column of the chart, and move the bars to the right where overloads would otherwise occur. This can be done for several resource types on the same chart, and the daily usage requirement for each resource can be entered at the foot of each column.

The use of a bar chart for indicating daily resource requirements is illustrated by the simple example in Figure 22.3. This shows an attempt to calculate the workload requirements created by twenty-two different jobs, labelled *A* to *V* for simplicity. The situation represented is a small engineering department, in which the principal activities (and skills) can be broken down into layout design, detailing and checking. No attempt has been made in this case to adjust the plan to take account of the actual numbers of people available in each of the three resource categories. The resulting workloads are uneven, as highlighted by the histogram for detailing. This illustration will be referred to again later in this chapter, when the subject of resource scheduling is looked at in greater detail.

Variants of the bar chart are shown in Figure 22.4. These are designed specifically for allocating work, and the methods were in widespread use at one time for machine shop and other work centre loading. Large wall charts of this type can still be found in use for scheduling personnel duties in transport and

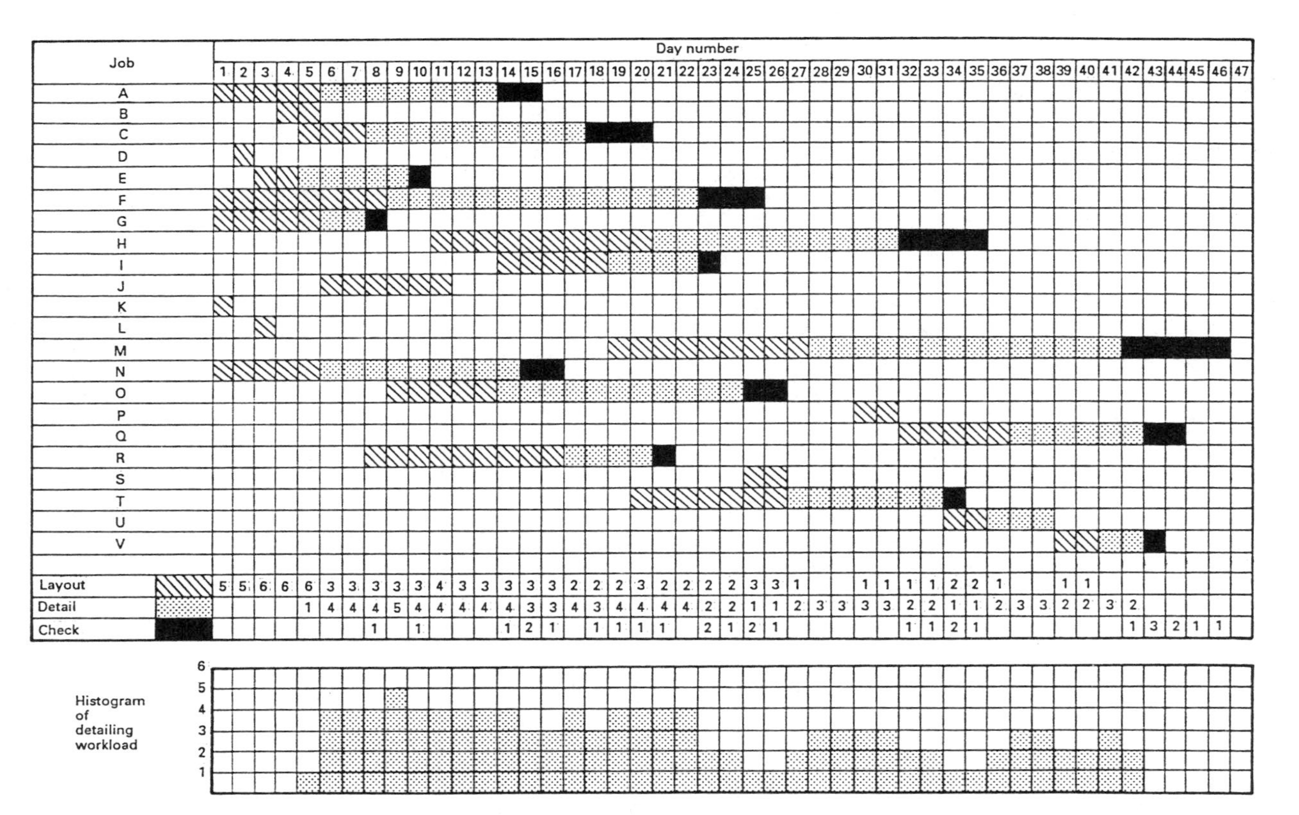

Figure 22.3 *Resource allocation by bar chart. By coding the bars to represent resource types it is possible to sum the resources needed in each planning period. In this example there are three resource types, layout engineers, detailers and checkers. The histogram at the foot has been included to show that the usage is uneven (using the detailers as illustration). If the chart is set up on an adjustable board, the jobs can be rescheduled until resource needs are levelled to a more practicable usage pattern*

service industries (such as security guards). They are useful for allocating jobs to individuals (or individuals to jobs) in an engineering or drawing office, although no reader should follow the practice once observed by this writer where a very senior engineer in a company (not in the UK) was seen to be writing individual names on a rolled paper chart that extended two years into the future – for practical purposes right out into the unknown. Two or three months is a sensible limit for an exercise in such detail. However, it is possible to use charts of this type for general longer-term manpower planning, provided that the level of detail is confined to expected workloads for complete jobs or projects on the one hand, and to staff numbers (not names) on the other.

A degree of flexibility can be achieved by setting a bar chart up on an adjustable board or wall chart. There are many types available commercially, using strips which plug into a peg board, Lego-like plastic pieces that fit on to the board, magnetic strips placed (and hopefully remaining) on a steel plate, or small magnetic rods that fit into parallel locating grooves on a steel board. For loading charts such as those depicted in Figure 22.4, slotted boards are available that can take small cards (in the manner of clock card racks).

Adjustable charts can absorb considerable planning detail, with depths extending to well over 100 rows possible. The number of possible column divisions may be limited only by the width of the supporting wall, since many charts can be extended by the purchase of add-on sections. Indeed, some wall charts even overflow on to two, or even three walls, wrapping themselves around the room (sometimes creating a claustrophobic effect). A large range of colours and shapes is possible for the adjustable pieces, making it possible to depict many different resource types or particular kinds of jobs or operations.

The limitation with adjustable charts is determined by the skill and patience of the individual planner. The more ambitious systems really hit problems when the time comes for a change of plan. The operators of very large and overcomplicated planning boards can easily be recognized by their red-rimmed eyes and planner's squint. Adjustable planning charts can be indispensible for short-term work allocation in an engineering department, but only if they are kept as simple as possible. Otherwise they lose their effectiveness.

Critical path analysis

Critical path networks approach the problem of planning from a different aspect to that of bar charts. They are not usually drawn to a timescale, and they cannot be used graphically to show or allocate resources. But they have two big advantages over other forms of planning notation:

Job number	Week 1	2	3	4	5	6
300	FRANKLIN				FRANKLIN	
321	CLARKE					
359	BRADFORD		CLARKE	BRADFORD		
368				CLARKE		
374	CRYSTAL					
375		CRYSTAL				
376			CRYSTAL			
377				CRYSTAL		
381					CRYSTAL +	BINGLEY
382						BINGLEY
390					BRADFORD	
396						CRYSTAL
401	RUSSELL					
401A					RUSSELL	
405	BINGLEY					
445	WRIGHT					
450					WRIGHT	
495	CLAR	KE				

Name	Week 1	2	3	4	5	6
BRADFORD	359		HOLIDAY	359	390	
BINGLEY	405			TRAINING COURSE	391	382
CLARKE	321 494		359	368		
CRYSTAL	374	375	376 377		381	396
FRANKLIN	300			TRAINING COURSE	300	
RUSSELL	401				401A	
WRIGHT	445			TRAINING COURSE	450	

Figure 22.4 *Job loading charts. The two charts here are really modified bar charts. Both have the objective of matching individual names to jobs waiting to be done. In (a), people are assigned to jobs. In (b), jobs are assigned to people. The second method is probably better, since it allows a reasonably constant framework to be maintained, with less need for frequent complete rearrangement. Charts at this fine level of detail should not be extended over a long timescale, which would be unrealistic considering the changes likely to occur and the relative inflexibility of these charting methods to cope with change*

1 They show job interdependencies very clearly.
2 They can be analysed to determine job priorities (which information can be used in subsequent resource allocation with a suitable computer program).

There are several networking systems, all of which are suitable for analysis by computer. For the purposes of this chapter, critical path activity-on-arrow networks will be described, since this is the system preferred by this writer. The principal alternative is the precedence system, and there are other variations that allow some degree of probabilistic estimation of timescale (see Lock, 1987 and 1988).

The arrow diagram

Many people are put off the idea of using networks because the diagrams seem strange, and are not as easy to understand at first sight as the more familiar bar chart. There is, however, no mystery. The notation, once explained, is simple and there are very few rules to learn. Figure 22.5 will serve as an illustration.

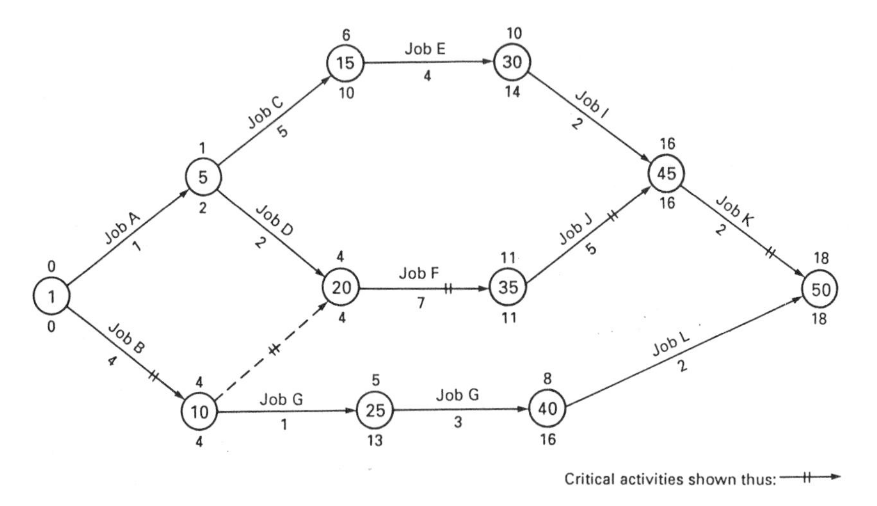

Figure 22.5 *Elements of a critical path network. This simple network illustrates the notation and provides an example of time analysis, all of which is explained in the next. Although this illustration only shows a small network, it contains all the basic elements necessary to construct a really large network*

Each circle represents an event in the life of the project, such as the start of one job or the completion of another (or both). The arrow joining any two events represents the activity necessary to progress the work from the preceding (left hand) event to the succeeding event. The convention says that no activity can start until its preceding event has taken place, which also means that all activities leading into the preceding event must have been finished. Thus, in the illustration, job *A* must be finished before event 5 can be considered complete, and only then can jobs *C* and *D* start.

Networks are not drawn to scale, and the length of activity arrows need bear no relation to their estimated durations. Another convention is that progress always takes place from left to right, and one should always try to draw networks this way (sometimes difficult when networks have to be modified).

The dotted arrow has been drawn to indicate that job *F* must wait until job

B has been finished (in addition to job *D*). The dotted arrow is a constraint only, and has no work content in itself and no duration. Such dotted arrows are known as dummy activities or, simply, as dummies.

The first aim in network planning is to think of all the activities necessary, and construct the arrow diagram by drawing them in a sensible sequence, with all the constraints shown in a logical way. In fact, network diagrams are sometimes called logic diagrams. Skill in drawing networks only comes with practice, and an organization using the tenchnique for the first time would usually benefit from having an experienced network planner on hand to guide the proceedings. The network does not have to be neat at first (it can be traced later or plotted by a computer). It will be found useful to have a stencil for drawing the event circles plus a straight edge for the arrows. A large sheet of paper, a soft pencil and an eraser (always needed as the logic is argued out) complete the tool kit.

An example will be given later in this chapter (Figure 22.6) to show the appearance of a network for a small project. When drawing any network, it is important to have on hand, in addition to the planner, the project manager or other senior representative of the work who can think of all the activities necessary and point to their interdependencies. Sometimes it is necessary to have several people present, representing more than one department. Those assembled should also be capable, subsequently, of estimating the duration of each activity.

The network notation allows those assembled at a planning session to express their intentions clearly and simply on paper. The process of drawing the network, even if no further use is made of it, can itself be of immense value. The outcome is a list of all the jobs to be done, set down in the sequence considered to be the most logical. As experience is gained from project to project it is quite likely that the discipline involved in the planning sessions will actually lead to more effective ways of working.

It will be seen that the network is liberally sprinkled with numbers. These can be a little daunting and confusing at first, but a step-by-step explanation should clear away any mystery.

The number written within the event circles serve as a convenient means for identifying the events. They could just as well have been numbered simply 1, 2, 3, 4, 5, etc., but it is usual practice to number events from left to right in jumps of five or ten. This leaves gaps for the insertion of new activities and events into the network should a subsequent change be necessary. It is not strictly necessary for events to be numbered in a series of ascending value, but this will be found convenient in practice, especially when referring to or searching for events in a large network.

Once the events have been numbered, then the activities connecting them have also effectively been identified. Job *A*, for example, can be described as the activity from event 1 to event 5 (shortened to activity 1,5 in practice).

A network can be drawn and then analysed mentally without any need to number the events, but numerical identification becomes essential when a computer is to be used for analysis. Large networks can be analysed by hand, but they become unwieldly as soon as any change is introduced (with many numbers to be erased and rewritten). Roughly speaking, a computer is needed for networks containing more than about 100 activities.

Activity durations

The numbers written below the arrows in Figure 22.5 are the estimated durations of the activities. Any units can be used (days, weeks, shifts, etc.) according to the nature of the project. For example each unit could be taken as one half day, so that a five day working week would be estimated as a duration of ten. Naturally the same units of duration must be used throughout the network. These estimates are only for duration (elapsed time) and are not related necessarily to the manhour content.

All activities occupying significant time must be included in the network. These include such things as the waiting periods for material deliveries, and the time taken administratively for the issue of drawings. Any delays expected while drawings await approval, or while authority must be obtained for the issue of purchase orders, must be recognized by the inclusion of appropriate activities on the network. These activities may not require any actual work to take place, and they may not consume any manhours to speak of, but they will take up valuable project time and must figure in planning and scheduling.

The estimates should be the most reasonable forecast that can be made of each activity's duration, given wherever possible by managers of other senior personnel from the departments who are going to be responsible for fulfilling the work.

When making these duration estimates there are rules for considering the relationship between available resources and the resources needed for any individual activity. Suppose that a job could either be done by two people in two days, or with four people in only one day. If it were known that only two suitable people could ever be made available, then the estimate must be entered as two days. If, however, four or more people were expected to be available with the relevant skills, then it is a matter for the estimator to say which method of working would be preferred as being the most efficient (four people on one job might turn out to be a bit crowded). The important thing is that only the activity being estimated is considered. No account need be taken at this stage of the possible demands of other activities on the same resources. Resource loading is a separate stage in planning, which comes later. This follows the rule illustrated in Figure 22.1, namely that it is necessary to take planning step by sensible step, eliminating the variables one at a time.

If a computer is to be used for subsequent resource allocation, it is necessary to note against each activity description the resources (number of people) considered necessary to finish the activity within the estimated duration. It is usual to identify each resource type by a simple code (so that 1L, for example, might mean one layout engineer, and 2 D + 1C could be used to indicate that an activity needed two detail draughtsmen and one checker).

Some authorities recommend that estimates should be added to arrow diagrams at random, rather than by looking at activities one after the other along any particular path. It is argued, possibly with some justification, that those making the estimates could have their judgement swayed if overall timescale trends start to become apparent when individual activities are being estimated.

Time analysis

Earliest possible times – the forward pass

Please refer again to Figure 22.5 and assume that the activity duration estimates on this network have been expressed in units of one day. Considering event 5, it is apparent that this event cannot be considered complete until job *A* has been finished. Since the estimated duration of *A* is one day, the earliest possible completion time for event 5 is end of day 1. This result is shown on the network by writing the figure 1 about event 5. It follows from this that the earliest possible start time for activities 5,15 and 5,20 (jobs *C* and *D*) is the beginning of day 2.

The process of calculating earliest possible times must be continued by adding the activity durations along each path in the network. Looking at event 15 (for example), adding the estimated durations of jobs *A* and *C* would give the earliest possible completion time for this event as 1 + 5 = 6 days.

The earliest possible completion for event 20 might seem to be on day 3 if jobs *A* and *D* are considered, but the dummy activity 10,20 provides another path leading into event 20, and this must also be checked. This second path is, in fact, longer. So the earliest possible completion time for event 20 is really given by adding the duration of activity 1,10 to that of the dummy activity 10,20: (4 + 0) = 4 days.

When all activity durations have been added through all possible paths, it is found in this example that the earliest possible completion time for the whole project (at event 50) is on day 18.

Earliest permissible completion times – the backward pass

In order to facilitate subsequent resource scheduling and progress control it is necessary to know not only the earliest possible start and finish times for any job, but also the latest start and finish times that can be tolerated without affecting the overall completion time.

If completion of the simple project depicted in Figure 22.5 is to be achieved in the shortest possible time, the lastest permissible completion time for event 50 must be taken as being equal to its earliest possible completion time, at day 18. This result is written below the event circle.

Staying with Figure 22.5, it is seen that the earliest possible completion time for event 40 is on day 8. But job *L* need not finish until the end of the project, on day 18. Since job *L* has only been estimated to take two days, its start could wait until the beginning of day 17 without affecting the overall programme. Day 16 is called the latest permissible completion time for event 40 (meaning the end of day 16, in order that the following activity can start at the beginning of day 17). This is depicted on the diagram by writing 16 under event 40.

The latest permissible completion time for all other events must be found. These latest times are written under the event circles. The process involves working through the network again, but this time travelling from right to left. The latest permissible time for an event is found by subtracting the estimated duration of its following activity from the latest permissible time of the next event.

Where there is more than one activity leading out of an event, all possible paths must be examined. In these cases the longest path will always determine the latest permissible completion time for an event (this will be the earliest of the various calculated times).

The critical path

When the forward and backward passes have been made through the network, it will be found that the earliest possible and latest permissible completion times for some events coincide. In Figure 22.5 this is seen to be true at events 1, 10, 20, 35, 45, and 50. The activities lying between these events are said to be critical, and the path which they trace is called the critical path. Although there is only one critical path in the example, it is possible to find more than one critical path through a complex network.

It is obvious that in any competition for resources or intensive management action the critical activities must have first claim, since none of these may be delayed at all if the project is to be finished at its earliest possible time.

Float

Activities which are not critical can be allowed to lag beyond their earliest possible dates, this being very useful when it comes to scheduling jobs in order to iron out unwanted peaks in resource loading. The amount by which any activity can be delayed without affecting the project end date is termed its float. Thus activities with zero float are critical.

Network analysts recognize several categories of float, some of which, for the purposes of this chapter, are a little academic. They are defined as follows:

Total float is the float possessed by an activity when all preceding activities take place at their earliest possible times (thus not eroding any available float along the relevant network path) and all following activities are delayed to their fullest permissible extent. It is calculated by the latest permissible succeeding event time minus the earliest possible preceding event time minus the activity duration. This is the most common type of float and, as an example, it can be seen that the total float of activity 15,30 in Figure 22.5 is (14 − 6 − 4) = 4 days.

Free float is also dependent upon all preceding activities taking place as early as possible, but with no delay permitted to following activities. It is calculated by taking the earliest possible succeeding event time minus the earliest possible preceding event time minus the activity duration. In Figure 22.5, activity 15,30 has a free float of (10 − 6 − 4), which is zero. If activity 5,20 is examined, it will be found to have a free float of one day (which in this case is equal to its total float).

Independent float is a relatively uncommon condition. It is found when, owing to intersecting network paths, activities are found which still have float, even when preceding activities take place at their latest permissible times and following activities start at their earliest possible times. For anyone interested, free float is found by taking the earliest possible succeeding event time, minus the latest possible preceding event time, minus the activity duration. There is no activity with independent float in Figure 22.5.

Remaining float is the amount of an activity's total float remaining after preceding activities have been delayed or scheduled later than their earliest possible times in order to level resource usage requirements.

In practice a planner can go happily and expertly through his whole career without worrying too much about these definitions. For the purposes of resource scheduling and progress management it is only really necessary to use total float and remaining float, both of which can be calculated by an appropriate computer program.

A practical network example

Figure 22.6, although simple, is a critical path network for fitting out a computer room. The room already exists, but is merely a shell, with no services or air conditioning. It is necessary first to discover the accommodation and environmental requirements specified by the computer company, and then translate these into plans for room layout and services.

This example, based on an actual project, does not include any resource considerations because all the physical work is to be contracted out. There are, however, several activities bearing the description 'delay', and these represent the minimum period of notice required by each contractor before he is able to commit his resources to starting on site after receiving a firm order. The activities concerned with ordering have duration estimates that allow for preparing the purchase orders and for mailing them to the contractors.

The time units used in this case are half days, so that ten units represent a five-day working week. The total estimated duration is seen to be 158 half days, which is 15.8 weeks. Critical activities are highlighted by bold arrows.

Integrated planning

Although this book is concerned with engineering management, it is not possible to consider the planning and scheduling of engineering activities in isolation from subsequent manufacturing or construction operations. It is well known that the interface between engineering and purchasing, for example, is often far from being straightforward, with continuous collaboration needed from initial enquiries through to final deliveries. For manufacturing operations, drawings are typically issued piecemeal, and it is important to attempt such issues in a sequence that suits production capabilities and priorities.

Critical path networks should, therefore, follow the work right through to completion, with the required completion date imposed on the final event in order to produce the appropriate time analysis and critical path. It is possible, however, to use a fairly broad brush approach to detail in non-engineering activities, limited to main assembles, subassemblies and the procurement of long lead items. This approach is acceptable and sensible, provided that the relevant engineering/purchasing and engineering/production interfaces are included, and that sensible duration estimates are given to the production activities. Detailed production scheduling is more of a production control function, to be carried out within the overall time framework indicated by the network.

In planning construction projects, there is often a tendency to draw large networks which do include detailed activities for engineering and site operations. It is also possible to draw networks that summarize the site activities, and to have those expanded in detail later. A common practice is to convert the detailed construction activities into a bar chart from the network, since bar charts are usually preferred by site managers and foremen. Bar chart conversions can be carried out with minimum effort using a suitable computer system and plotter, and some systems can plot lines linking the bars to indicate the network constraints.

Planning and resource scheduling by computer

Many programs exist for analysing critical path networks, and some of these can also schedule resources. Even microcomputers can handle networks containing thousands of activities. Although details must vary from one program to another, it is possible to summarize the basic steps.

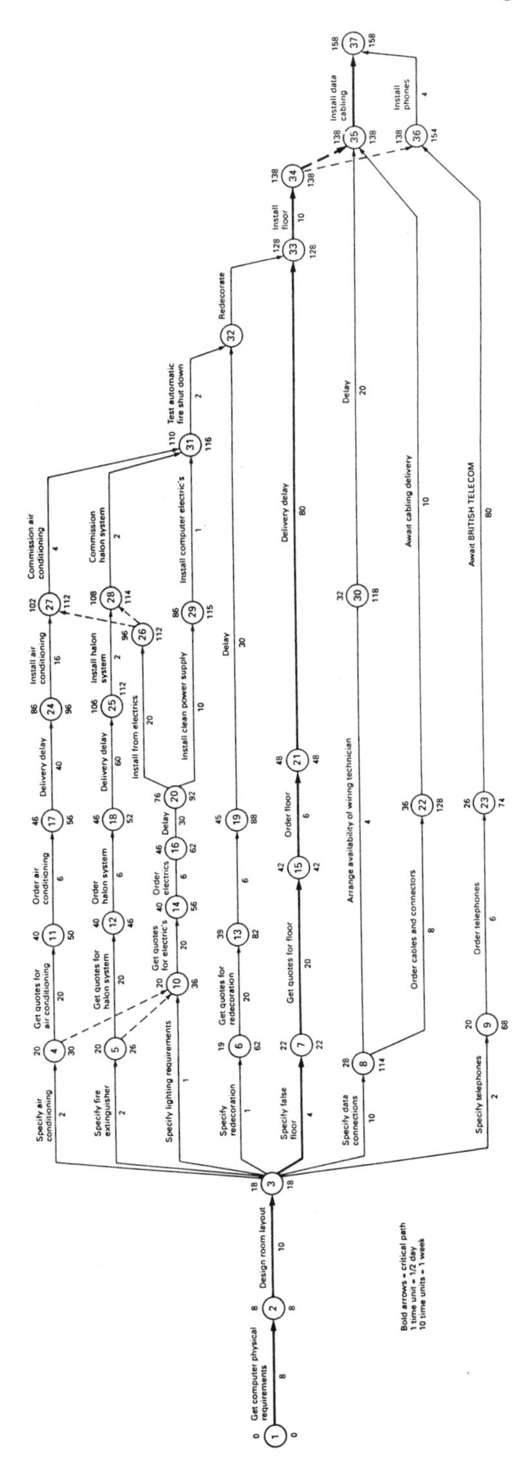

Figure 22.6 Critical path network for fitting out a new computer room

Input data

The input data really fall into two parts. On the one hand there is all the information needed to set the project scene. Complementing this are the data which define the network logic, consisting principally of one record for each network activity. The following list includes and explains most of the input categories needed to produce a resource schedule for a single project.

Network calendar
Unless the computer program is already in use for processing other networks, it will be necessary to set up a calendar in the computer. This will have to start from a date on or before the earliest start date for the whole project, and the total calendar span must obviously be sufficient to include the expected completion date. The calendar will specify which days are available for work, and those which are not (such as weekends and public holidays).

Some programs can accept more than one calendar at a time, which means that the planner is able to specify, for example, activities that can only be performed on weekdays, and others that can be scheduled over shift working or weekends.

Time units
The computer must be told the units of time to be used for estimated activity durations (hours, half days, days, weeks, etc.)

Start and finish events
Events which have no activities preceding them (there may be more than one on the network) and events which have no following activities have to be designated as start or finish events respectively. If this is not done the computer will detect these events during its error analysis, assume that the preceding or following activities have been omitted from the input by mistake, and print out unwanted error messages.

Schedule starts and finishes
It is possible to impose dates on events or activities. The most obvious cases are the project start date and the required project completion date, but it is also possible to place target dates on events or activities anywhere in the network. Suppose, for example, that it was known that the project site would not be available for work until 1 January 1990, although the project engineering and procurement can start much earlier. The network would have an event imposing this constraint in the logic, and 1 January 1990 would be fed to the computer as the earliest possible imposed date on that event.

Milestone events
Some programs allow events to be designated as milestones. These are simply

events that are considered to be of particular significance for progress reporting and project control. The computer is able, on demand, to print out reports on the planning and progress status of these milestones.

Resource codes

For each resource type to be scheduled it is necessary to use an identifying code. One or two letters are usually chosen. Thus, it might be decided to use *CE* for a civil engineer, *EE* for an electrical engineer, *DD* for a design draughtsman, *CC* for a checker, and so on.

It may be a mistake to attempt to schedule every type of resource. The usage rate of some resources can be expected to run at a percentage of 'mainstream' resources. This can be true of some engineering design functions (such as process control and instrumentation) and such factory operations as inspection. These approximations become more reliable when scheduling progresses to multiproject calculations (described later).

Normal resource availability levels

The normally available resource level must be stated for each resource type. Even if only one project is being handled in the company, the normally available level is not the same as the staff complement. Allowance must be made for absence, and for miscellaneous work on other tasks. As a rule of thumb, it is probable that a practicable schedule will result if only 80 per cent of total resources are made available for project work.

Threshold resource availability levels

During the process of resource allocation, the computer will attempt to produce the schedule that satisfies all imposed dates without exceeding the normal resource availability. Some programs allow the concept of threshold resources, which means that an extra layer of availability can be specified for some or all resource categories. The computer can then call up these additional resources if this is essential to meet the schedule requirements. In practice, this usually means the engagement of subcontract or agency staff.

Resource costs

The computer is able to calculate and report the cost implications of a resource schedule. This can be arranged by specifying a cost rate per unit per time for each resource category.

Scheduling priorities

Most resource scheduling programs will offer several options on priority rules for starting activities where resources are too scarce to allow two or more activities to start simultaneously at their earliest times. A useful option here is to give priority to activities with least remaining float.

The planner will also be asked to decide whether the schedule is to be time limited or resource limited. This rule is explained later, in the section headed resource allocation.

Activity records

Each activity record would always include:

1 Preceding event number.
2 Succeeding event number.
3 Description (or designation as a dummy).
4 Duration estimate (except for dummies).

Other information might include:

1 Number of resource units needed from each category.
2 Whether the activity is splittable of not. If it is, the computer is given permission to schedule the activity in two or more parts, interrupting it in order to divert resources temporarily to other activities with higher priority.
3 An estimated cost for the activity. This is not the same as specifying cost rates for resources. One application would be to put materials costs on all other activities, so that the resulting costed schedule would give the materials budget as the planned cost commitment against time. By placing the same cost estimates on delivery activities, which approximate to the invoice dates, the resulting cost schedule gives an idea of the procurement cash flow requirements.
4 Report or department code. A code can be added to each activity to designate the department or manager responsible for seeing that it is carried out. The computer will be able to sort on these codes for producing edited schedules that become very useful work-to lists.

Error analysis

Once the computer has digested, filed and merged all the input data, the first network analysis task is to test the logic. The program searches for and reports errors, or apparent errors. The most usual errors are:

Dangling arrows

Dangling arrows are seen when activities are left out of the input by mistake, or when start and end events have not been reported as such. The computer will report as a start dangle any activity not preceded by other activities or by an event which has been specified as a start event. Similarly, an activity with no following activities, and with a succeeding event that has not been specified as an end event, will be shown up as an end dangle.

Duplicated records
If two activities are input with the same preceding and succeeding event numbers, the computer will see them as the same activity and report them as duplicates. This situation can arise from a numerical mistake in the input data. It will also happen if two activities are drawn in parallel on the network between the same two events (in which case it is necessary to insert a dummy in one of the arrows).

Loops
A loop can be caused by transposing event numbers during data input. This causes the direction of an arrow to be reversed in the perception of the computer. The effect can be to produce a number of arrows forming an endless loop, which obviously wreaks havoc at any attempt to progress from left to right through the network. The more respectable computer programs are able to identify loops and print out all the activities contained in them, thus making error correction fairly straightforward.

Critical path analysis

Once errors have been eliminated, and the network logic is seen by the computer to be valid, forward and backward passes are made to calculate the float possessed by each activity and to find the critical path.

Earliest and latest start dates are calculated for all events and activities. The calculation takes place in terms of numbers, which the computer then converts to calendar dates.

Some planners never take their critical path planning beyond this stage, and are content to rely on reports which take no account of resources. These reports can list all critical activities, or list all activities for a department (for example). Activities would typically be printed in order of their earliest possible start dates, often with asterisks printed alongside critical activities. Typical column headings are:

Preceding event
Succeeding event
Description
Duration
Earliest start
Earliest finish
Latest start
Latest finish
Total float

Resource aggregation

Resource aggregation is the poor man's version of resource allocation. It consists of the computer scheduling every activity at its earliest possible time, and then adding up the resources that would be needed. The result is usually a workload with unacceptable peaks and troughs for most of the resource categories. The process can, however, throw out the total cost implications of the resources needed.

Resource allocation

During the process of resource allocation, the computer attempts to schedule each activity at its earliest possible start, but will only do so if this would not exceed the stated resource availability. In order to avoid overloading the resources, the computer will delay the scheduled starts of activities where necessary.

Often this process poses a dilemma. The computer must delay one or more activities beyond their remaining float, or it must schedule resources over the specified limit. The planner is able to decide which option must be followed by specifying the schedule as resource limited (in which case the timescale may have to be extended) or time limited (when the computer is given permission to exceed even the threshold resource levels if this is imperative to maintain the timescale).

It is recognized that computer resource levelling may not produce the optimum result, and there may still be peaks and idle periods that could be smoothed out. For very small networks (around 100 activities) it may be possible to produce a smoother schedule using an adjustable bar chart because the human brain has superior perception and is able to scan the chart and make decisions with more flexibility than a computer operating within a rigid set of rules. However, most projects are much larger than 100 activities and, fortunately, the computer scheduling is likely to improve as the network size increases. Experiments have been tried using different priority rules, and more powerful optimization programs have been developed that make multiple scheduling passes through the network. However, this author has always found that the priority rule of least remaining float has given very good results with networks ranging from a few hundred activities to over 10 000.

Multiproject resource allocation

To introduce the subject of multiproject resource allocation at this stage may seem like trying to run and jump before we have learned to crawl, but it follows logically from the previous section and is, in fact, not a difficult step to take once single project scheduling has been mastered.

The method involves the preparation of a network diagram, with resource estimates, for every project in the organization. The computer is set up with a long-term calendar, and each individual project is fitted into the time frame by means of scheduled start and finish dates. By regarding each network as a subproject, and the total organization's continuous workload as the project, the computer virtually contains a model of the company.

There are very many benefits obtainable from multiproject scheduling. The most important is that priorities are established for each subproject from the whole range of company resources, and a really practicable schedule should be produced.

Once the multiproject schedule is on the computer file, it is possible to use it for 'what if?' testing. By making very coarse summary networks of new projects under consideration or of possible customer orders it is possible to run test schedules to examine the feasibility of taking on the work, and to verify any delivery promises given.

Computer reports

Computer programs are available which allow the user very great flexibility in the scope and presentation of schedules and associated reports. Even if the computer is only used for time analysis, useful reports can be obtained. The real potential is only realized, however, when resources and costs are scheduled (preferably on a multiproject basis). The following list is a selection of the possibilities.

Schedules and work-to lists

Work-to lists are an extremely useful form of report. These print all the activities required from a particular department in order of their scheduled start dates. Column headings can include such items as:

1 Subproject code (in multiproject resource scheduling).
2 Preceding event number.
3 Succeeding event number.
4 Job number.
5 Activity description.
6 Estimated activity duration.
7 Scheduled start date (after resource allocation).
8 Scheduled finish date.
9 Earliest start date (from time analysis).
10 Latest finish date.
11 Remaining float (float left after resource scheduling).
12 Resources required (coded).
13 Activity cost.

The work-to list does not usually go down to the level of allocating jobs to individuals. That remains the responsibility of first line managers and supervisors (most possibly using simple short-term charts and broken down into finer detail). The detailed work allocation can, however, be carried out with the assurance that the work sequence, job priorities and the resource usage rate have all been taken into account to produce achievable scheduled dates.

In production departments, for example, the work-to list will probably schedule down to the level of subassemblies and major components, leaving the production planning and control system to load machines and work centres in far greater detail, but in the knowledge that the scheduled dates are practicable.

There are many other uses for work-to lists. Once example is their use for expediting by the purchasing department. If a report is prepared that lists only purchase delivery activities, and if these are printed out in order of their scheduled completion dates (i.e., in order of the required material delivery dates) then the expeditors have a very useful day to day control checklist.

Another application of work-to lists is in the control of work carried out by outside subcontract design offices, or indeed any other subcontract work. The relevant activities can be identified by their resource or departmental codes for this purpose, allowing work-to lists to be printed out that are particular relevance to the manager or liaison engineer with specific responsibility for controlling subcontracted work.

Resource usage tables

The resource usage table is an extremely useful planning and cost control tool. If the company is operating multiproject scheduling the computer will be able to produce a separate report for each subproject, plus a total resource and cost table for the whole company. Reports can be restricted to only one resource type or, if there is sufficient space on the paper, they can list all resources on one report.

A typical arrangement would print out one line for each day of the calendar, with data in columns under the following headings:

1 Date.
2 Resource type.
3 Resource availability.
4 Resources required (after levelling).
5 Resources unused (can go negative for a time limited schedule).
6 Repeat of columns 2 to 5 for other resource types.
7 Cost each day (reaching zero after scheduled completion date) computed from the resource usage multiplied by the appropriate cost rates per day.
8 Cumulative cost (reaching estimated total cost at scheduled completion date).

There are many variations possible, including the reporting of threshold resource levels.

Resource requirements can also be plotted as histograms against time, either using a plotter or, less satisfactorily, using character patterns formed by a printer.

Graphics

If a plotter is available, the computer can be programmed to plot an extensive variety of diagrams. These include resource histograms and curves of planned expenditure against time.

The network diagram itself can be translated from the digital logic and plotted. At one time, plotters were very expensive and the results clumsy and indistinct. With greatly improved software, very much cheaper plotters, and the use of coloured pens it is now possible to produce very clear prints. The computer is even able to plot the critical path in red, if required, and plot the network on a time scale.

Bar chart conversion from the network is a very useful facility, since such charts are usually appreciated on construction sites and by others who may not be familiar with the network notation. Bar charts are, of course, always drawn to a time scale. It is possible, given a suitable plotter, to arrange for bars to be colour coded, for example highlighting critical activities. The computer is also able to draw the network constraints on the bar chart, showing these as vertical lines in the relevant places (although the result may not always be easy to interpret).

Library networks

When experience has been gained over several projects it may soon be found that there is a good deal of similarity between the network logic and estimates. It should be possible to identify groups of activities that repeat throughout all the networks (such as specify), get quotes, order, await delivery). This effect can extend to such areas as the design, procurement and manufacturing of subassemblies. An example of such a standard group is given in Figure 22.7.

As more planning confidence is developed, a library of standard subnetworks can be built up, often complete with standard estimating tables. These library networks can be kept on file in the computer or stored as prints on self-adhesive transparent film. It is possible to develop this art to the point where quite large project networks can be assembled entirely from library network elements. The process can sometimes be carried out by relatively junior staff, starting from the project sales specification.

It is also possible, for some small projects, to draw a complete standard network, with the idea that it can be adapted by editing or by small additions to suit most of the company's projects.

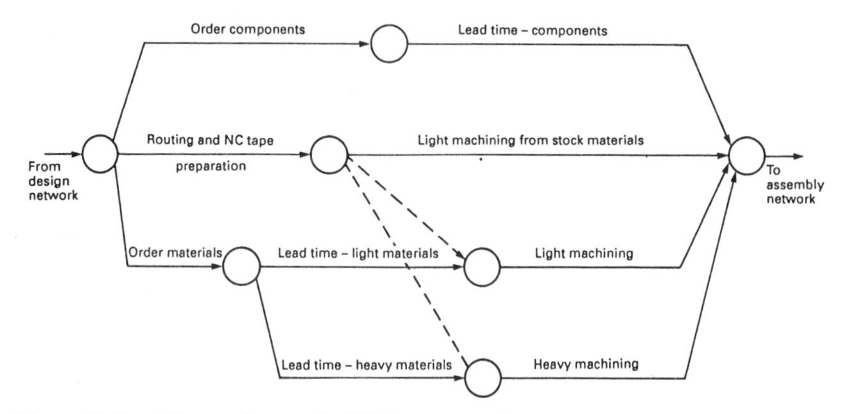

Figure 22.7 *Library subnetwork. This is an example of a network segment for provisioning and machining a subassembly as used by a heavy engineering company as part of a library of such segments. New project networks could be compiled by assembling the appropriate segments, adding standard estimates, and then having them reviewed and edited by the project manager*

It is obviously essential that the project manager should review any standard network or plan compiled from library segments, checking for any special logic constraints, deleting unwanted activities and verifying the estimates. There may, for example, be some design tasks that can be shortened or omitted owing to the existence of suitable designs from previous projects, or there may be commonality between different areas of the new project. Only the project manager can be expected to declare that the network adequately reflects the intended and most efficient working approach.

All this may seem far fetched and not all companies will be able to identify sufficient areas of standardization in their projects. But several companies have used these techniques effectively.

One heavy engineering company, for example, employed over 600 people on its permanent staff and had up to 150 engineers engaged at any one time on several projects. The entire multiproject resource scheduling was handled by a team comprising one full time planner and a clerk, assisted whenever rescheduling was required by another clerk and a planning expert. The company manufactured special purpose machine tools. Two standard networks were used for two ranges of machines, and networks for transfer line machinery were compiled from stick-on library segments. All activities were covered from design concept through to final assembly.

The resulting resource schedules were very effective for project control, and in the forward planning of all resources, including additional temporary engineering staff needs. Because the resources were all given cost rates, the computer reports were even used to provide total cost estimates for comparison with the estimating department's predictions, and differences averaged only 5 per cent.

Given the expertise that comes from experience, uncanny scheduling accuracy and control can be obtained. On one occasion for example the managing director asked, in May, 'If we were to get this new project order, when could we actually start work?' The answer, 'On 5 October', proved to be true (with the first design engineers actually becoming available on 5 October).

Summary of the planning and scheduling process

It is appropriate to review the techniques discussed in this chapter and put them in the context of the entire planning and scheduling process. This summary assumes the use of a computer and places the planning steps in sequence.

1. Plan the work sequence, preferably by drawing a project network diagram.
2. Estimate the duration and resource requirements of each network activity.
3. Feed the network into a suitable computer system. Supplement this with associated data, such as the required start and finish dates for the project.
4. Feed the computer with information on total resource capacities, resource cost rates, calendar information and similar relevant data. Do not declare total resources as available, but keep approximately 20 per cent back as a reserve. For the same reason, do not schedule overtime.
5. Attempt a time analysis run in order to detect the correct input errors for the project network.
6. Repeat steps 1, 2, 3 and 5 for every other project which is in progress or about to be started.
7. Carry out the computer multiproject resource scheduling run and produce the required output reports.
8. Use the resulting total resource usage schedules for forward company resource and subcontract capacity planning.
9. Company management or sales management can access the computer with coarse summary networks for possible new projects in order to test their effect on company workload and the fulfilment of existing projects.
10. Issue summary schedules and cost curves to the project managers for progress and control purposes.
11. Issue computer-plotted bar charts to construction managers or to other production workers if they need them.
12. Issue work-to lists to engineering and drawing office managers. Day to day allocation of tasks to named individuals is arranged by expanding the details of these work-to lists using simple planning charts, covering only a

few weeks ahead. Thus each manager retains responsibility and authority for day to day planning and work allocation in his or her own department. By committing only 80 per cent of each department's capacity to the multiproject schedule, there should be sufficient spare capacity to cover absences and to allow each manager to load miscellaneous tasks or carry out rework.

13 Arrange for new data to be fed to the computer in order to update, the schedules in line with progress, modifications or new projects.

Please turn back to Figure 22.1 and note how the logical planning sequence just described has taken care of the many variables. Steps 12 and 13 are discussed in greater detail in the following chapter.

Useful organizations

Association of Project Managers, Westbourne Street, High Wycombe, Bucks HP11 2PZ. This association organizes demonstrations and seminars on project management topics, especially concerning the application of computer programs for project network scheduling.

K & H Project Systems Ltd, Felco House, 72 Richmond Road, Kingston-upon-Thames, Surrey KT2 5EL. This company offers several powerful computer programs capable of resource scheduling and other project control functions using a range of commercially available computers.

Metier Management Systems, Metier House, 23 Clayton Road, Hayes, Middlesex UB3 1AN. With their Artemis program, this company specializes in very powerful computer programs that are sold principally with custom built hardware.

Further Reading

Lock, Dennis, (ed.), *Project Management Handbook*, Gower, 1987.
Lock, Dennis, *Project Management*, 4th edn, Gower, 1988.

British Standards

The following relevant British Standards are obtainable from the British Standards Institution, Linford Wood, Milton Keynes, Bucks MK14 6LE.

BS 4335: 1987: *Glossary of terms used in project network techniques.*
BS 6046: *Use of network techniques in project management.*
This standard is published in four parts as follows:

Part 1: 1984: *Guide to the use of management, planning, review and reporting procedures.*
Part 2: 1981: *Guide to the use of graphical and estimating techniques.*
Part 3: 1981: *Guide to the use of computers.*
Part 4: 1981: *Guide to resource analysis and cost control.*

23

Managing engineering progress

Dennis Lock

This chapter deals with the important relationship between engineering or drawing work and time. Effective control of progress is essential to any engineering business since habitual failure to meet schedules and delivery promises is a sure way of eroding profits, annoying customers and losing out to competitors in the quest for new orders.

Background requirements

Before proceeding to a detailed discussion of engineering progress management it has to be assumed that the work is taking place in a suitable organizational environment. The following summary lists the main conditions.

Supportive management

Management support for engineering or project management takes many forms. On a continuous basis, there must be a readiness to provide a reasonable level of staffing (but not overstaffing) for the scheduling and progress functions, together with suitable accommodation, equipment, facilities and training opportunities.

Occasionally, more acute action may be required, such as the resolution of interdepartmental disputes or high level approaches to external suppliers and contractors where routine expediting action or criticism from more junior managers has not produced a satisfactory response.

There has been more than one company where many of the rank and file engineering and drawing staff claimed, with reason, that they had never seen any of the company's senior managers and directors and would not be able to recognize one. In one busy engineering department there was a facetiously expressed opinion that the engineering director did not actually exist, but that he was represented only by the secretary, who guarded an empty office with his name on the door. These problems tend to be confined to the larger companies, but they are serious and feed discontent. Progress management in a badly motivated workforce is an uphill, and often futile struggle. An occasional informal visit to the engineering and drawing offices by a senior member of company management, with a genuine show of interest and the odd word of encouragement (or even praise, if deserved) costs nothing. It is, however, all part of management support and as such is usually very much appreciated.

Suitable organization and communications

Whether the organization is set up as a project team or is an arrangement of functional sections and departments, everyone must be clear about his/her organizational role in achieving the work objectives, and on how he/she fits into the work organization generally. Organizational ambiguities increase with the size of the company and with larger projects and it is usually necessary in those cases to define the organization in issued procedures.

Engineering is all about information, and the efficient transfer of information throughout the organization demands good communications within departments, between departments and with all external organizations participating in the work (including the customer). In order that information is properly directed and distributed, there should be one nominated addressee at each location, so that clear and controllable communication channels are established.

Sensible procedures

Engineering progress will be impaired if administrative procedures are either non-existent (leading to chaos) or too cumbersome (wasteful of time and money in themselves).

Competent participants

Obviously, it has to be assumed that progress management is being applied across a group of people whose qualifications, experience, training and aptitude fit them for carrying out the work.

Adequate work definition

Proper work definition is essential. There is nothing more embarrassing and frustrating for an engineering or project manager than to have to interrupt work owing to ambiguous instructions or, far worse, to discover on completion that the result is not what the customer intended.

Effective schedule

A practicable schedule provides the basis for the orderly issue of work and sets out the benchmarks or milestones against which progress must be measured. The essential characteristics of an effective schedule were listed in the previous chapter. The subject of scheduling is continued in this chapter in connection with the detailed attention necessary for progressing drawings and purchasing instructions.

Control of modifications

The management of modifications and engineering changes is sufficiently important to deserve a chapter of its own (Chapter 25). Uncontrolled modifications are one of the many factors that can set back or wreck progress and budgets.

A cooperative client or customer

Much engineering work depends on a cooperative partnership between the customer and the contractor. If planned progress is to be achieved, it has to be assumed that the customer will play his part. Among the customer's obligations are:

1 Early provision of information on such items as drawing numbering, drawing sheets, engineering standards and other general information relevant to the conduct of design and drawing.
2 The timely provision of technical information necessary for design.
3 Prompt attention to correspondence and to drawings submitted for approval.
4 Provision of facilities at the customer's premises, as necessary.
5 Avoidance of unnecessary changes.
6 Prompt payment of all properly certified invoices.

Work authorization

The first stage in progress management is the issue by company management of a document that authorizes work (and therefore expenditure) to start. This document might be called a project authorization or a works order, depending on the nature of the work and company practice.

It is customary to send a copy of the authorization to every department involved in the work, and to the accounts department. Data relevant to the authorization are:

1 The job or project title.
2 The job or project number.

3 The customer's name and address.
4 The customer's order number, or reference to other proof that an enforceable contract with the customer exists.
5 The work specification, or reference to a separate specification document, listing the specification serial number and its correct revision number.
6 The name of the manager or project manager with overall responsibility for the work.
7 The cost budgets, (specified for each department).
8 Outline schedule information (it is a fact of industrial life that delivery promises are made to the customer along with the quotation, so that subsequent detailed planning and scheduling is carried out not to decide how long the work should take, but to ensure that it takes no longer than the promises already given).
9 For manufacturing projects, a list of the items to be supplied, with the quantities stated.
10 Shipping or delivery instructions.
11 Pricing and billing instructions.

Starting work without a customer order

It is usually an inviolate rule that authorization to start work for a customer or client is never issued until an order or a signed contract has been received. There is an obvious risk of wasted effort and unrecovered costs if the expected customer order never materializes. There is a further risk that the contractor can prejudice his negotiating position if the customer believes him to be irrevocably committed to the project before the contract is signed.

In spite of the risks, there may occasionally be some justification for breaking the rules by pre-empting the order and starting work early. The initial few weeks of most large projects are usually taken up with gathering facts, planning and scheduling, and making procedural arrangements with the customer, such as ironing out the question of numbering systems, drawing standards, lines of communication, levels of authority, approval routines, and so on.

Such activities can be very time consuming, but they will only require the involvement of a few members of staff. The extent of any financial risk is likely to be considerably less than 1 per cent of the total estimated engineering costs, but the effort could save precious months on the overall programme.

An alternative way of viewing this somewhat controversial proposition is to consider the typical cost/time curve for a project (Figure 23.3). At the beginning of the project the curve slope is very shallow, with relatively light expenditure spread over several weeks or months while all the preliminary tasks and some of the conceptual engineering are taking place.

If a company does decide to authorize an early start, it is obviously essential that the authorization document defines the extent of work authorized and sets strict budget limits.

Issue of work to departments

Apart from technical instructions, the formal framework for departmental work must be the schedules prepared and issued by the planning department. The most useful form of such schedules is the work-to list, described in the previous chapter. If the work-to list has been properly prepared, every departmental manager should have the benefit of a schedule that is achieveable with the resources available (in-house or through the authorized use of subcontractors).

Work-to lists should be edited (preferably by computer) so that they are specific to individual departments. They must be updated and reissued whenever new orders or other changes would render the existing schedules unrealistic, but the frequent reissue of schedules for reissuing's sake must be avoided.

Issue of work within departments

It is unlikely that project schedules will contain sufficient detail to allow the allocation of work to individuals. This is seen in manufacturing, where production control is necessary for the loading of work stations. The principle is no less true in the engineering and drawing offices, where simple bar charts still have their place for the day to day issue of work to people by name according to the manager's detailed knowledge of individual capabilities.

The progressing and control of individual drawings is dependent upon drawing lists or drawing schedules, while the complementary (and equally important) control of purchasing instructions relies on some form of purchase schedules.

Drawing schedules

Clerical methods

A drawing schedule form is shown in Figure 23.1. This example was used for engineering mining projects, but all drawing lists or schedules share certain data, such as the drawing numbers, titles and latest revision numbers.

If the drawing numbering system is chosen so that each engineering discipline has its own identifying digit in the drawing numbers, then each chief engineer or departmental manager can be allowed to list the drawings required for the work, and number them within batches of serial numbers allocated from a central drawings register.

The drawing schedules for large projects can obviously consist of many

Smith Engineering Limited	Drawing schedule		Project number	Sheet number
	Plant	Plant section	Department	Revision number
				Issue date

Drawing no.	Dwg. rev. no	Drawing title	Progress data							Last issue				Client's approval reference	Remarks	Schedule revision
			Status						Target RFC date	To site		To client				
			S	H	D	C	R	B		Rev.	Letter ref.	Rev.	Letter ref.			

S Started H Drawing well in hand D Drawn C Checked R Released for construction B Drawing finalized (As built)

Figure 23.1 *Drawing schedule form*

separate sheets. These should be grouped into subprojects or work packages, further divided into sets for the various engineering disciplines.

Prints of photocopied sets of schedule sheets are issued to various members of the work organization according to requirements (certainly including the chief or other senior engineer of each discipline, and the central drawings registry clerks). It may be necessary for the customer to receive a set, and another set would normally be supplied when a construction site has been established.

Drawing schedules are reissued at intervals, updated to show drawings which have been added, revised or cancelled. Whenever an entry on a schedule sheet is revised, the revision number for the whole sheet must be updated. In order to avoid having to reprint and reissue possibly hundreds of pages at every schedule reissue, when perhaps only a few sheets have been revised, a company may decide only to reissue the changed sheets (leaving all the recipients to substitute the changed sheets in their sets).

There is controversy over this method, with some companies believing (sometimes with justification) that customers, site managers and other remote recipients cannot be trusted to make these substitutions correctly, so that the only safe method is to reissue the schedule completely each time, regardless of the printing and distribution costs.

In order to avoid the tedium of writing 'sheet *x* of *n* sheets' on every page, and then having to change all the sheets whenever sheets are added, it is necessary to provide a contents list at the front of the schedule. This need only list all the sheet numbers that make up the complete drawings schedule, with their current revision numbers. The contents list itself will carry a new revision number at each update, and this revision number is a convenient and foolproof way of defining any particular issue of the complete schedule.

For the purposes of progressing the preparation, checking and (where necessary) customer approval of drawings, it has been the practice in the past in some companies to write target dates against each drawing listed. A progress clerk or supervisor then uses a print of the schedule as a progress document. Considerable clerical effort was needed whenever there was a change to the project work, because even a change in the target completion date for a part of the work might mean that hundreds of individual drawing dates would have to be erased and re-entered.

For small jobs and projects clerically produced drawing schedules may still have their place, but computerizing this process brings many important advantages.

Computerized drawing schedules

At its simplest, the computerization of drawing schedules can be considered as a word processing exercise. Once on file, updating is simple.

If a program is available for sorting and editing the file, the schedules can be printed out according to specific requirements so that, for example, a list of all mechanical layout drawings could be run off.

The effectiveness of any drawing schedule for progressing work is greatly enhanced if an entry can be made against each drawing record that links it to its relevant activity on the network diagram. Now, suppose that the network schedule is updated and the scheduled dates for an activity are changed. It is possible to identify all drawings associated with that activity and change their scheduled dates using the computer. Now there is an effective drawings database file in the computer, which can be used with flexibility for work loading and for the administrative control of the drawings library or registry.

The use of a computer becomes even more effective with interactive databases.

One database (or file) lists all the drawing records, with the drawing number, title, progress status, revision number, originator's name or staff number and the network activity reference listed for every drawing. The drawing number itself, if the numbering system is sensible, will contain further encoded information, such as the relevant discipline, job number or project number, drawing type and size, and so on.

Another file is generated from the network scheduling program, using the techniques described in the previous chapter. This file will contain data on resource requirements, scheduled activity dates, float and departmental reporting codes.

Yet more files can be set up for cost information, and for purchasing requirements.

By combining information selectively from all these databases according to requirements, the computer can print lists and reports for a wide variety of progressing and control needs.

Now it is possible to print out complete drawing schedules to suit different uses: in numerical order for the drawings registry, and in order of scheduled dates for use as very detailed work-to and progressing lists.

Very powerful programs are necessary for interactive databases. These have been developed from network scheduling systems into fully interactive project management systems. Two of the leading companies in this field are K & H Project Systems and Metier Management Systems, whose UK addresses are given at the end of Chapter 22.

Purchase schedules

Bills of material and parts lists are produced as part of the drawing process (either clerically or using a computer aided design system). Their scheduling and progress is linked to the scheduling and progress of their associated drawings. But every sensible engineering and purchasing manager knows that a vital part of purchasing progress is to identify those bought-out materials and components which are going to have a long delivery time, get them specified as soon as possible, and order them without waiting for the final issue of a bill of materials (provisional bills of materials are sometimes issued for this purpose ahead of issuing the relevant drawings).

On major construction projects (especially including mining and petrochemical plants) the specification of purchased equipment is itself an important engineering function, to be planned and controlled rather in the same way as the preparation of drawings. An example of a purchase control schedule is given in Figure 23.2. The assembly, issue and reissue of such schedules follows exactly the same arrangements as those described in the previous section for drawing schedules. Indeed, it is usual practice in some companies for the drawing and purchase schedule sets to be issued and reissued together at the same regular intervals.

The major difference between drawing and purchase schedules is found in the numbering systems. Numbering tends to get a little more complicated with purchase schedules. At first, all known purchased equipment requirements are listed, and specification numbers are allocated (as for drawings, using a system with encoded information that identifies the project and the engineering discipline, plus possibly the subproject or work package involved).

The purchasing cycle goes through the stages of:

1 Write the purchase specification*.
2 Issue purchase enquiries* to prospective suppliers.
3 Summarize the enquiry results and choose the supplier.
4 Issue a purchase requisition* to the purchasing organization.
5 The purchasing agent issues a purchase order*.

Every stage marked with an asterisk involves the issue of a serially numbered document. If the same serial number can be used throughout for all documents associated with one piece of equipment, then correlation is easy and the purchase control documents are relatively simple to manage (e.g. specification 123XYZ results in an enquiry numbered 123XYZ, and the requisition and purchase order are also numbered 123XYZ).

It is rarely possible to realize this ideal, although some rationalization can usually be achieved. The difficulties arise as soon as attempts are made to get purchasing organizations (which may be external) to adopt a system other than their own. For example, a purchasing manager using his standard purchase order forms will usually want to issue them with sequential numbers, and the other form sets will almost certainly be prenumbered for this purpose. Such numbers can bear no relation to the specification numbers. Another problem is that one specification might result in more than one purchase order if it is decided to split the commitment, whereas an order for, say, a range of pumps

Smith Engineering Limited			Purchase schedule				Project number					Sheet number	
												Revision number	
			Plant:	Plant section:			Department					Issue date	
Spec. no.	Equip. no.	Qty	Description	Cost data – FOB		Reqn no.	Target dates/actual dates					Remarks	Schedule revision
				State currency:									
				Budget	Reqn		Issue enq	Issue reqn	Design data due	On board	On site		

Figure 23.2 *Purchase schedule form*

and valves might combine many specifications into one purchase order, when the benefits of large scale purchasing can be achieved.

It is important to cross reference all associated document numbers throughout each purchasing cycle, and this can be achieved more readily if the purchase schedules are set up in a database system. Similarly, the database approach is far more convenient for associating scheduled dates with individual specifications: these too can be linked to their relevant network activities, providing very flexible and detailed work-to lists for the preparation of specifications and the issue of requisitions.

This book contains a separate chapter on purchasing, but in the context of engineering progress it is important to mention the subject of drawings and other documents to be produced by suppliers. Some of these may require engineering approval, while other suppliers' documents become as important as the receipt of the goods themselves when it comes to installation and commissioning. Requirements for such documents must form part of the purchase order, and their expediting and progressing cannot always be entrusted to the purchasing organization but must be scheduled and followed up directly by the engineering progressing function.

Relationship between progress management and cost control

It is invariably true that any job or project which exceeds its allotted timescale will also cause expenditure to exceed budget. The converse is also true, namely that if work is effectively controlled so that it is finished according to plan, then the labour costs (and many other expenses) will tend to fall within the cost plan (budget or cash flow schedule).

The principal exception is the purchasing of materials, subcontracts and bought out services, where cost control depends largely on commercial expertise in getting the terms and prices right at the time of commitment, when the purchase orders and contracts are issued (it is too late afterwards).

When separate attention is given to cost control by writers, there is a tendency to expound methods for collecting and reporting costs, often compared against budgets. Some aspects of cost reporting are included in this chapter, since they go hand in hand with progress reporting. Cost reporting is useful if it can highlight adverse trends at an early stage, but it is really only the process of reporting past events: it is not directly a method of cost control.

The main elements of engineering cost control are effective progress management and sound commercial practice in the management of purchasing and contracts.

S curves and their use in progress management

The curve in Figure 23.3 shows how the costs of an engineering project can be expected to accumulate with the passage of time. Such curves, because of their characteristic shape, are usually referred to as S curves. The slope of the curve, apart from indicating the rate of expenditure, also approximates to the level of activity at any time. Thus it can be seen that right at the start of the project there is a very slow build up, with only a small amount of expenditure over the first few weeks or months.

As the project nears completion, the curve should be asymptotic to the budget, shown by the stepped horizontal line in Figure 23.3. The steps indicate additions to the authorized budget caused by customer requested modifications to the project specification.

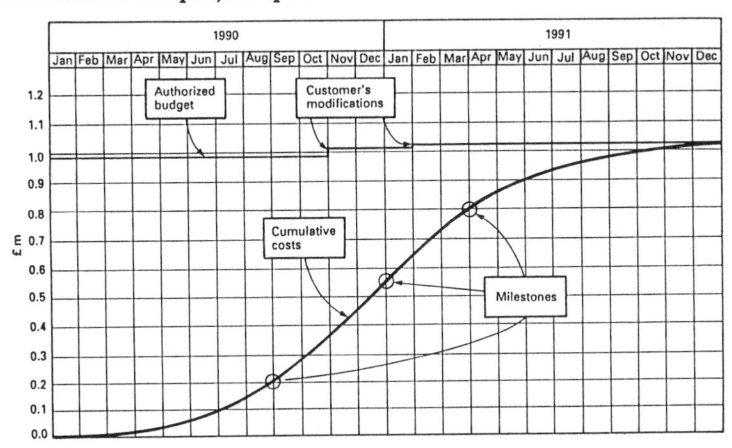

Figure 23.3 *Curve showing typical relationship between project costs and time*

If the curve is plotted to show total costs, including purchases, then there may be some distortion to the regular shape caused by the placement of one or two very large orders. It is important, for cost control purposes, always to think of purchases in terms of commitments; that is, to plot them on the curve at the time the orders are placed and not when the invoices fall due for payment. The only exception to this is when the project cash flow requirements are being calculated, in which case purchasing and subcontract costs will have to be plotted according the scheduled timing of payments.

The S curve can be plotted to show expenditure in terms of money (as in Figure 23.3) or in manhours, manweeks or any other units according to the particular application. Curves may show total project costs, materials only, or departmental costs. They can be used in a number of ways as an aid to cost and progress control and some of these will now be outlined.

Level of current activity

A very useful check on progress can be made early in a project by asking the question 'how much should we be spending at present' and then checking the actual rate of expenditure against prediction. The S curve can be used to gauge the rate or tables of day-by-day manpower requirements can be consulted.

In one case the question took the form 'how many subcontract draughtsmen are working on this project at the moment?' The plan predicted thirty draughtsmen, but less than half this number were actually employed. Yet engineering design was thought to be on plan, and all relevant activities were reported as complete. The reason was found to be that the appropriate layout drawings were still languishing in the engineers' drawers. Although 'finished', there was reluctance to release them owing to lingering doubts. Prompt management action got the show back on the road. Without the spot check on activity level, this problem might have been discovered too late.

Level of current expenditure

This aspect of progress and cost comparison deserves mention only as a warning of how not to use an S curve. Managers have been known to ask how much has been spent to date on their particular project, check this against the predicted expenditure on the curve, and declare themselves satisfied with progress when the two figures agree.

Although this error of interpretation may seem elementary, it is one that does happen. The truth, of course, is that only half the work may have been achieved, but at double the expected cost. Methods for comparing costs with achievement will be described later.

Milestone analysis

Suppose that the project schedule has been compiled in the manner described in the previous chapter, and that some network diagram events have been identified as milestone events. These are events singled out because they are considered to be easily recognizable as important stages in the general progress. Examples might be the start of work on site, the despatch from a factory of a planned consignment of equipment to the customer, or any event which has been agreed with the customer as the basis for a stage payment.

It is possible to use the schedule, linked to the estimated activity costs, to calculate not only when each milestone should be reached, but also how much should have been spent up to that time. An S curve for planned expenditure can be plotted, with the milestones highlighted (Figure 23.3). Then, a check can be made at the appropriate times to ensure that expenditure has not exceeded the plan when each milestone is achieved. This is a far more valuable check than simply comparing actual costs against planned expenditure without reference to achievement.

Predicting project costs to completion

Predicting the final costs of a project is an exercise that usually occupies a great deal of attention from those responsible for the control of progress and costs.

The first prediction is, of course, made when the project costs are estimated for the purposes of establishing the contract price and budgets. Such estimates may be subject to errors and omissions, but as work proceeds and the proportion of completed work to work remaining increases, so will the proportion of known actual costs increase in proportion to estimated costs remaining.

Thus, if the original cost estimates are revised continuously throughout the duration of a project, they should gradually become more and more accurate until eventually they merge with actual costs of completion.

It must be said that this account is somewhat simplified, since in practice there may be reasons why the actual costs of a project can never be accurately measured. These reasons can include complications through dealing in more than one national currency, contractual difficulties, modifications, or simply shortcomings in the cost accounting system.

There is also a danger, commonly experienced, that project costs will go on creeping up after the work has supposed to have been finished. The S curve (Figure 23.3) is drawn to show the final costs as levelling out to become truly asymptotic to the budget line, and this is the type of result that is usually expected whenever cost estimates are made or revised. But in practice the work can tend to drag on, with late commissioning problems, updating drawings to 'as built' standards, and a considerable amount of office work in putting all the project information safely on file.

One method of predicting costs to completion depends on evaluating the work actually achieved to date, and comparing the measured costs against the estimated costs for that work. If the two results agree, it could be said that there is a kind of efficiency factor at work equivalent to 100 per cent. Any discrepancy, resulting in a lower assessment of efficiency against the estimates, can be used to factor the remaining estimates to produce a revised cost to completion. The method is expounded more fully in the following section.

The results can be tabulated, or they can be plotted on the S curve, starting with measured costs to date, and then continuing with factored estimates, extrapolating the curve to completion. These methods can highlight the danger of overspending at a very early stage in a project. But if all seems well there is no room for complacency, for the reasons given in the earlier paragraph.

Evaluating engineering progress

One reason for wishing to know, at any given time, the value of work actually finished is in order to check the original estimates and carry out fresh predictions of final costs in the light of work experience. Another common reason is so to allow certified claims for payment to be made to the customer (when the contract terms provide for such stage payments).

In the physical work on a construction site the assessment of such values is usually based on the measurement of quantities by a quantity surveyor. For engineering work, although there may be drawings and purchase requisitions, the actual work output is less tangible, and its evaluation requires more imagination.

The method depends on the establishment of a project ledger, a simplified illustration of which is given in Figure 23.4 When the project starts, all the engineering activities are listed, with their cost estimates written on the debit side. As each job is finished, its estimate is entered on the credit side of the ledger. Thus, if a drawing or a group of drawings had been estimated to need 100 manhours, then the 100 manhours already written on the debit side would be added to a credit column, as the value of work done. It is easily seen that on project completion, when all the jobs have been done, the credit and debit columns should total to the same number of manhours.

The ledger must be updated by the addition or removal of jobs according to any change in the project scope.

In order to allow for work in progress, it is necessary to estimate the percentage completion of each job that has been started but which is not fully complete. An appropriate proportion of the estimate is written in the credit column. Although individuals' perception of work remaining tends to be optimistic, such errors quickly become diluted as more jobs are actually reported finished.

Dividing the value of work done by the total estimate, and multiplying the result by 100, gives an estimate of the percentage engineering achievement to date.

Suppose that the total engineering estimate for a project was 50 000 manhours, and that the total of jobs finished (including the assessed value of work in progress) amounted to 10 000 manhours. Engineering completion could reasonably be regarded as 20 per cent of the total project. It is now necessary to consider the measured time spent on the project up to the date of the progress assessment. Assume that this was 15 000 hours. This may have been recorded on timesheets in sufficient detail to relate actual costs to estimates for each job, but this is not strictly necessary and the method works as long as the total expenditure to date has been recorded.

In order to achieve work estimated (valued) at 10 000 hours, 15 000 hours have been spent. Thus, if this trend continues, all remaining estimates might be exceeded in similar fashion. The revised predicted total engineering expenditure at project completion can be worked out by simple arithmetic proportion, the result in this case being:

$$50\ 000 \times \frac{15\ 000}{10\ 000} = 75\ 000 \text{ manhours}$$

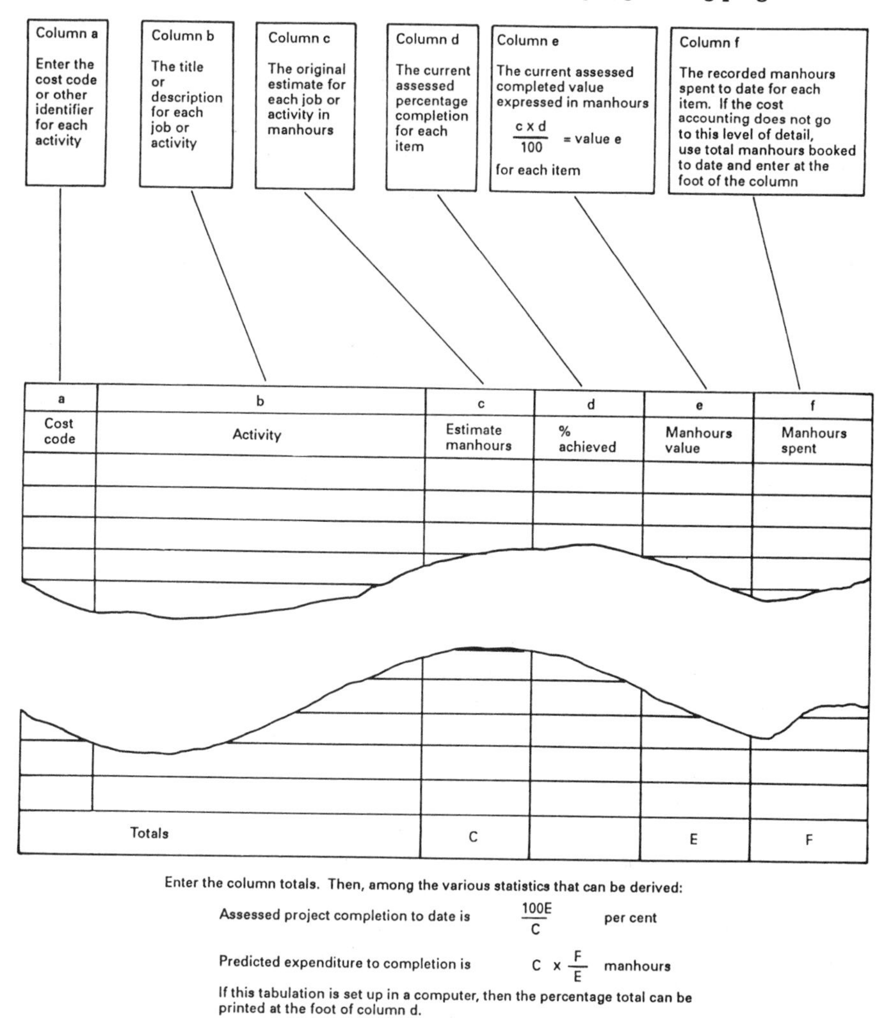

Figure 23.4 *Principle of the project ledger method for assessing progress and predicting costs to completion*

which should give cause for considerable consternation, questioning and action among the engineering managers responsible.

Although clerical methods have been used in this chapter to demonstrate the principles of progress evaluation, the project ledger is best managed using a computer. In addition to considering data for whole projects, the jobs and activities can be grouped into individual departments so that managers can be provided with trend calculations and predictions specific to the work that they control.

Progress meetings

Are progress meetings really necessary?

Consider an engineering project that has been specified with care and precision, is well planned, and lacks none of the other essential background characteristics listed at the start of this chapter. From day 1, every participant knows what is expected of him/her and gets on with the job. Competent managers issue work from sensible schedules, and progress is monitored on a day-by-day basis. When any job falls behind schedule, the urgency of any action taken is tempered according to the amount of float possessed by the activity. If a critical activity is in danger, then all stops are pulled out, with weekend working, extra staff, and any other measures needed to safeguard the programme.

It should almost be possible for a whole project to be managed in this fashion without the need ever to call a progress meeting. Certainly the experience of companies which develop and improve their planning and control techniques to a high level of competence is that progress meetings become rare events, rather than the time wasting affairs that tend to be inflicted on engineers and managers at frequent, regular intervals in many organizations.

Having questioned the need to hold progress meetings at all, it is relevant to list some of the circumstances in which they cannot be avoided. These are:

1 For one or more network planning sessions before the start of a project.
2 At the start of the project, when a 'kick-off' meeting has the essential purpose, through briefing and discussion, of ensuring that managers and senior engineering staff embark on the work with a clear idea of the project objectives and of their own responsibilities towards meeting them.
3 At the customer's request.
4 To control subcontractors, some of whom may have planning and control procedures that leave much to be desired.
5 When the project is large, spreading over a complex organization, so that meetings become necessary from time to time to coordinate the widespread activities and dispel ambiguities.
6 To resolve disputes, disagreements or other difficulties arising between two or more departments.

Conduct of progress meetings

Progress meetings tend to attract senior members of staff, and possibly senior outsiders, so that the time which they consume is both expensive and precious. Every meeting should therefore be organized and controlled as efficiently as possible. Some chairmen prefer to hold meetings towards the end of an afternoon, providing an incentive to keep the proceedings short.

The first requirement is that every department or organization involved in the matters to be discussed should be represented if possible.

Those invited to attend must have sufficient seniority to be able to speak for, and to agree commitments for, the departments which they represent.

An agenda should be issued well before every meeting, so that those attending are given forewarning of the matters to be discussed, allowing sufficient time for them to gather any pertinent facts and to have their answers ready.

Progress meetings and technical meetings are generally held on different occasions for different purposes. The chairman must not allow long technical discussions to develop among subgroups within a progress meeting, if these would waste the time of other senior members.

Interruptions should not be allowed. If there is a telephone in the conference room it should either be switched via a secretary or it should be programmed for outgoing call capability only.

When commitments are made, the chairman must ensure that they are clear and not ambiguous. For example, a promise to expedite a job so that it 'should be finished in a couple of weeks' is too vague. A date must be forthcoming that can be recorded in the minutes, allowing project management staff to follow up and check progress without having to wait for the next progress meeting.

Minutes must be concise. Preferably each item should only comprise one or two lines, and the agreed action must be shown together with the name of the person responsible for managing or taking that action.

Minutes should be issued promptly while the contents are still hot, and not one or two weeks later when some of the actions listed should already have taken place.

A copy of the minutes must be given to every person present at the meeting, and to any other person not present who may have had the misfortune to be named in the action list.

Progress reporting

Progress reporting takes place at several levels, inside and outside the engineering organization. At the lowest (but by no means the least important) level, progress data is collected against the work-to lists. This is often a word of mouth exercise, with activities being marked as just started, partly completed (possibly with a percentage achievement assessment) or finished.

Engineering departments of any significant size will usually employ one or more persons with responsibility for checking progress in this informal way, while in other organizations the planning and control functions are combined.

Very often, progressing is seen as part of the general supervision, adding to the burden of senior engineering staff. Whatever the arrangement, any discrepancy from plan must be brought swiftly to the attention of the relevant manager in order that prompt action can be taken to prevent a programme slip.

In the context of detailed work monitoring, it is important to note that the effective engineering or project manager will not be content to sit at a desk and expect all the information to appear automatically, or to accept that reports will necessarily be correct when they do arrive. People tend to be very optimistic when reporting progress, and it is necessary to check progress claims from time to time. One of the best ways for a project manager to do this is to take the occasional walk around the computer screens and drawing boards, getting a first hand feel for current progress and checking that the jobs in hand are those which are supposed to be in hand. When such walkabouts are conducted at regular intervals, the manager should be able to spot any job that does not seem to have advanced as it should. If the engineering or project manager's progressing responsibilities extend to production or construction, then the factory or construction site should also be visited (even if this means occasional international travel).

If a computer has been used for scheduling, it is important to realize that the computer can only provide the basis for work loading and progressing. The actual function of progress reporting and control is essentially a human one. If an activity on the critical path is found to be running a week late, it is apparent without any futher assistance from the computer that the whole timescale is at risk, and that immediate management action is called for.

Exception reporting

Progress reports may be produced for several reasons. Those intended merely to inform (or to reassure customers and shareholders) will necessarily contain an account of work done to date, supported by a wealth of statistics, charts and other illustrations.

Reports that are intended for purposes of progress control must, however, highlight any problems demanding action. This is the important principle of management by exception, in which managers are not swamped with routine facts demanding no action, but are instead told specifically of things which are going wrong, or are in danger of going wrong.

Updating and reissuing computer reports

Although the computer-produced network schedules are only a basis for control, and not a control method in themselves, they must be updated whenever changing circumstances intervene to render the existing schedules invalid. A change of project scope might be the cause. In very large projects it is likely that progress will not always follow the work-to lists exactly so that schedules may need regular updating.

The method of updating varies a little from one program to another, but the following kinds of data will probably be needed for a schedule revision:

1 The date from which the new schedule is to start (time-now) (which should be chosen to coincide with or follow closely the day on which the computer rescheduling run will take place.
2 A list of activities which have been finished, or will certainly have been finished by time-now.
3 A report of activities which have been started or will be started by time-now.
4 An estimate of the duration remaining after time-now for all activities reported as started (including any which were reported as started at a previous update, and which are still in progress).
5 Revisions to the network logic as changes become necessary in the light of experience. (One typical reason occurs when it is found that a new activity can actually be started even though the preceding activity has not been finished. The computer will see this as logically impossible unless the network diagram is changed to remove the constraint).

Most of this data can be gathered during routine progress monitoring. But when engineering work takes place in a widely dispersed organization, especially when overseas companies and agents are involved, the gathering of detailed progress information from such sources can be very difficult. This will lead to loss of control and inadequately updated schedules unless special arrangements are made to ensure that progress is reported regularly by telex or some other suitably reliable and rapid means.

In the past, early computer scheduling systems were always batch processed on large computers, often using bureaux services. Computers and suitable software were not plentiful, and the likelihood was that the engineering offices and the computer might be many miles apart. There were no data links, and the input data and the finished reports often had to rely on postal services or express road deliveries. The turn round time from input preparation to error free schedules frequently took a week or longer, so that the time-now date had to be chosen a week or even two weeks ahead of the time when actual progress data was collected.

The modern trend is towards systems that allow direct input from VDUs to local computers, or to central computers with datalinks. There is no need to batch the input data at all: the relevant records can be updated piecemeal, as soon as each item of information comes to hand. Rescheduling can therefore take place with far less delay, and the concept of a delay between collecting data and a future time-now date for the revised schedule is disappearing as turn round times shorten.

One final word on the strategy for updating schedules. Updating should not mean rescheduling for a later completion date. If one or two critical activities have slipped beyond all aid, the planners must get to work to find ways of pulling the programme back on course.

There may be a temptation under these circumstances to start reducing duration estimates arbitrarily, but this must only be allowed if it is really certain that the shorter times can be achieved. The more practicable course usually lies in re-examining the network logic, so that activities which were previously expected to follow each other in series (in single file, as it were) might be replanned to overlap or even run in parallel. Remember that one of the basic requirements for any effective schedule is that it must be achievable.

Formal progress reporting

A company may wish that formal reports are issued for circulation internally among its management in order that they are kept informed of the current progress status of all major projects. Customers for large projects will almost certainly expect to be told, on a regular and formal basis, exactly how the work is progressing (especially if they are receiving invoices for stage payments).

Reports to a customer must never mislead, and should be as frank as possible consistent with the security of company information.

If the work is running irretrievably late, the customer must be told, and the reasons should also be given (avoiding such phrases as 'unforeseen difficulties'). The customer may not be pleased, but at least he will be given a chance to replan his activities accordingly. This is far better than a process of issuing one optimistic report after another, only to have the customer discover that the programme is six months or more late when a scheduled delivery date arrives – and passes.

The need for honesty in reporting is even more critical when payments are claimed from the customer in stages linked to progress. Customers will usually require certification of progress against each claim, and independent consulting engineers or other professionals are sometimes employed for this purpose. Another option is to arrange for stage payments linked to milestones that are easily verifiable.

When a contract provides for progress payments, these become an essential element in the contractor's cash flow management. Valid claims must always be submitted promptly and followed up diligently. But the claims must be valid.

There are occasions when a customer will accept the contractor's own progress reports as the basis for stage payments. There may be temptation or pressure to submit such claims early. They then have to be supported by overoptimistic progress reports. This is a dangerous practice. Apart from any possible fraudulent aspect and the effect on customer relations, overbilling can lead to complacency in management accounting, overstated interim profits, and a very painful day of reckoning for all concerned when the truth emerges.

The contents of regular progress reports to a customer will obviously depend on the size, complexity and duration of the project, on what the customer wants, and on what the contractor thinks the customer should have. The result might be a sheet or two of paper, or a bulky package of documents.

The following account lists the contents of report packages that an engineering company compiled regularly for its clients in multimillion-pound international mining projects. These were often cost-plus projects, so that cost information was included in the reports. This list is supplied here not as a recommendation (it does tend to go over the top somewhat) but to provide a checklist which readers might find useful when designing reports for their customers.

1. Contents list.
2. Report text. A report combining progress statements collected from the managers of all departments working on the project, (eventually including the project site). After summarizing the principle jobs achieved during the reporting period, the report concluded with a brief statement of the work to be done in the following period.
3. Site photographs. Photographs would be included whenever prints were available that were relevant to progress achieved.
4. Revised drawing schedules.
5. Revised purchase schedules.
6. Table of engineering progress on drawings and purchase requisitions (see Figure 23.5).
7. List of drawings awaiting customer approval.
8. A tabulated list of purchase enquiries in progress for major equipment, annotated to show progress.
9. A tabulated list of purchase orders issued for major items, with progress information added.
10. A full list of letters and telexes sent to the client during the period, highlighting any for which answers were overdue. Correspondence in both directions was serially numbered owing to the possibility of loss in transit, and full lists were included in these reports at the clients' specific requests.

Project: *Report date:*

Drawings	*Discipline*					*Totals*	
	Civil and structural	*Mechanical*	*Piping*	*Electrical*	*Instru-mentation*	*This month*	*Last month*
Total required for project							
Number in progress							
Number released for construction							
Number released as percentage of total required for project							
Purchase requisitions Total required for project							
Enquiries in preparation							
Enquiries issued to purchasing department							
Requisitions issued to purchasing department							
Number of requisitions issued as percentage of total required for project							

Figure 23.5 *Tabulation of engineering progress. A simple method of indicating engineering progress in the issue of drawings and requisitions for a project progress report*

11 Complementing item 6, a list of letter and telex serial numbers received from the client, with notes on action taken in relevant cases.
12 List of outstanding project variations.
13 Tabulation of progress versus cost performance (see Figure 23.6).
14 Cash flow information, including updated forecasts of the timing of sums to be provided by the client for paying major suppliers, subcontractors and the engineering contractor's own invoices.
15 Network diagram revisions.
16 Computer printouts from the network schedules, edited to include activities involving the client (e.g. drawing approvals).
17 A covering letter from the project manager.

Summary closing report

Whenever a major engineering job or project is finished it is important that the experience gained during the work is recorded for future recall. The most obvious part of this exercise is to ensure that all the drawings have been filed or microfilmed in their final as-built state, and to safeguard calculations and other contractual information that might be needed in the event of operational problems with the finished project, or for use in extending or modifying the work at a later date. Retained engineering, whatever the type of project, is always a valuable part of a company's assets.

Where time and costs permit, some companies require that the project manager arranges the compilation of a summary project closure report. This is a package that includes one copy each of items such as:

1 Project authorization.
2 Project specification, as issued at the beginning of the project.
3 All project variations.
4 The project procedures or other definition of the organization, numbering systems used and the names and addresses of other participating organizations.
5 Final version of the drawing schedule.
6 Final version of the purchase schedule, with the purchase order number added against every item.
7 The final cost report.
8 Photographs.
9 A brief written summary of the work, highlighting any special events or difficulties that may have implications for the future.

Such reports are not very exciting to produce, at a time when work has been finished and all are anxious to get to grips with the next project. Some may

Project Report date:

a	b	c	d	e	f	g	h	i
Work package	Original budget	Approved variations	Current budget (b + c)	Costs to date	Percentage of budget spent (100 × f/d)	Percentage achievement	Predicted cost at completion (d × f/g)	Predicted variance at completion (d – h)
Project totals								

Figure 23.6 *Tabulation of project cost trends. Intended for inclusion in regular progress reports, this format is suitable for preparation by clerical methods or using a suitable computer program*

regard them as an expensive luxury. Quite apart from the benefit of retained experience however, such reports can be a vital first point of reference when, months or years into the future, it becomes necessary to reopen the files for any reason, legal, technical or commercial.

24

Managing bought-out engineering and design

Roland Metcalfe

There are occasions when it becomes necessary to use bought-out engineering and design in order to sustain a business and to maintain or enhance its competitiveness in the market place. The requirement to support a planned programme, meet target dates, or to respond to circumstances quickly may often demand some addition to existing resources on a temporary basis. If the necessary expertise does not exist within the company, it may be expedient to use the services of others who can provide the required knowledge and experience.

Planning and preparation

Planning to ensure that an engineering project will progress satisfactorily to its target will entail a number of 'make or buy' decisions. It is obviously necessary to formulate a policy for the engineering in terms of what is to be bought 'off the shelf' and what is to be designed and manufactured specially for a particular project. But it is also equally important to plan the company's resources so that these will be able to cope adequately with all the expected project work within the time available.

Planning and scheduling techniques were described in some detail in Chapter 22. The best chance of obtaining realistic resource schedules comes from the use of total (multiproject) scheduling, preferably computer assisted and based on critical path network techniques. But whatever the methods used, planning future requirements in the design and drawing office areas may reveal future deficiencies. These can conveniently be divided into two areas:

1 Problems relating to capacity: more work is forecast than that which can be completed within the timescale allowed using the resources expected to be available.
2 Problems relating to expertise: more technical or professional knowledge is

going to be needed than that which is currently available within the company.

Any deficiency in either of these areas will require that consideration is given to increasing manpower, either by recruiting additional people and increasing the permanent staff complement or, on a temporary basis, by the use of short-term contract staff, subcontractors or consultants.

When the decision has been made to use bought-out sources of help it must be borne in mind that without careful management there could be some loss of control in these areas. The overall management responsibility lies firmly with the company, since any future engineering difficulties or, worse, product failures, will be perceived as company problems, with all the consequences that these may entail.

Once a commitment has been made to the use of outside engineering or drawing involvement, management becomes largely a question of being able to exercise adequate control to achieve maximum benefits within reasonable cost and time constraints. The key to success is in accurately specifying what is required. It is in the best interests of all concerned to provide a clear, correct and comprehensive statement or specification for every bought-out task. The production of such a specification may be a fine balancing act between ensuring the required performance while yet remaining within reasonable costs.

It is seldom possible to foresee all the details and eventualities during the design and development of a project. However, time spent in thoroughly understanding the real requirements of what is to be achieved will generally help to facilitate later stages of the work.

Options for increasing design office capacity

The work flow in a design office seldom matches the steady capabilities of the staff. Normal planning will reveal peaks and troughs in workload that cannot always be smoothed to reasonable levels.

Overtime or weekend working is an option that might be considered for overcoming short-term overload problems. Although this is an often used solution to the problem, it is not one that should be sustained indefinitely. There is the risk with prolonged overtime working that the quality of work will deteriorate, creating subsequent problems at the later stages of production or customer use, or some other unintended lowering of product quality. Planning experts will always advise, with good reason, that overtime should never be regarded as a normally available resource: it should generally be held in reserve against the inevitable, unforeseen problems and crises common to all projects.

When major overloads are forecast through sheer volume of work or short deadlines, then the use of additional resources must be considered. These can take the form of employing:

1 Additional regular staff to work in-house.
2 Temporary staff, either through staff agencies or on direct short-term contracts. Such staff might work in-house or they might use their own external facilities.
3 Subcontractors, to work in third party premises.

The first of these options, which would increase the permanent staff complement, is obviously only a valid consideration if the projected work overload is to be sustained as part of a long-term expansion of the business.

Where the future is less certain, but with a possibility of the workload remaining high, the engagement of staff on short-term contracts may be an expedient solution. These limit the company's employment commitment to the specified period, but there is usually the possibility of converting the contracts later into permanent staff positions when the future workload position becomes clearer. Short-term contracts under these conditions can be regarded as prolonged interviews, reducing the unknown element of risk when the employees are eventually taken on to the regular staff.

Temporary increases in design staff will also depend upon the speed with which the additional people can be made available and, not least, upon the availability of suitable accommodation and equipment. The use of established subcontracting organizations working in their own premises is the usual method for dealing with this problem. In this way additional resources can usually be utilized very much more quickly. Occasionally it may be possible to use third party accommodation on a temporary basis: under-used facilities are at times available in industrial and academic places. Many higher educational establishments have professional facilities including computer aided design systems that are underused at certain times of the year. Although the use of third party premises can imply a loss of confidentiality it is generally possible to ensure reasonable levels of security.

Appointing subcontractors

Where to find adequate support at the right moment is a question best considered carefully, since the consequences of making a wrong appointment in haste will only add to the existing workload problems, if regular planning and scheduling are effectively carried out, last minute overload problems should be relatively uncommon, and the company will normally have sufficient time in which to investigate potential sources of outside help.

As experience is gained, a company can develop its own 'approved list' of subcontractors, letting work out to several agencies to spread the load rather than relying solely upon one organization. By this means it is usually possible to find reliable additional capacity at short notice when a workload crisis does arise unexpectedly (e.g. from a late design modification or from a welcome but unforeseen customer order). This approach has the advantage that the company's working practices and requirements in terms of administration, control and payments, will already be familiar to each listed subcontractor. It also allows the company to place work according to knowledge of special capabilities (e.g. instrumentation) demonstrated by individual subcontractors in the past, or to compare the costs, speed and quality of one external resource with another to obtain competitive advantage.

The starting point, if no suitable subcontractors are known or can be reliably recommended, has to be local trades organizations, local authority information facilities or trade directories (such as the *Yellow Pages* in the UK). It is obviously more convenient if the subcontractor's offices are within reasonable distance, but this is only one criterion. The likely quality of work is obviously the most important.

Following initial contact, which will probably be by telephone, a meeting must be arranged to discuss terms, the capacity needed, the capacity which can be offered, control procedures, and all the other administrative arrangements. It is preferable that this meeting should take place at the subcontractor's premises, which in any case should be inspected. This will enable a better assessment to be made of the likely capabilities, quality of work and capacity that the subcontractor can offer. Generally it is in the subcontractor's own best interests to be as open as possible, showing examples of previous work, other achievements and testimonials as far as confidentiality allows. The visit should also give some indication of the level of confidentiality that can be expected.

Some subcontractors may attempt to demand that a purchasing company agrees to use or reserve against guaranteed payments a minimum standing number of workers for the contractual period (usually for a year). The argument is that the subcontractor may not, unless this is done, be able to make people available when they are needed. If a minimum usage or reserve level contract is signed, the subcontractor will guarantee to give priority to servicing the company's work, if necessary diverting people from other subcontracts whose purchasers have not made any long-term commitment.

Such agreements are not generally recommended. A principal reason for opting to use subcontractors is to achieve a flexible manpower reserve. This benefit is forfeited by entering into any formal and rigid minimum workload commitment. If one subcontractor cannot provide the necessary capacity when the occasion demands, there will be others (on the approved list) who can. But this degree of experience and confidence in dealing with subcontractors may take a year or two to build up.

An engineering manager must never commit his company to reserving capacity contractually in the long-term. The long-term expected requirements should always be carefully planned (multiproject) and discussed informally with several subcontractors, so that a spread of future reserves can always be arranged without any contractual commitment.

If it is decided to go ahead with a particular contractor, then the working basis must be agreed. Suitable checks and controls must be agreed to ensure a satisfactory standard of work in a reasonable time span. A method of payment that rewards the subcontractor commensurate with effort will need to be worked out. These considerations are dealt with in a later section of this chapter.

All the working methods and commercial terms must be set out and agreed contractually in a clear specification of what is required. The responsibilities must be clearly understood.

Appointing consultants

A company in its natural progression and development may find that it is moving into technical areas that are outside its normal scope, or to levels of technology which are far higher than current experience within the organization. The new knowledge is perhaps less attainable in the short term through developing internal staff than by the involvement of specialist external consultants. Again, as with subcontracted design, it is essential that the requirements are clearly understood. There are, however, significant differences in the way in which specifications for consultants should be formulated.

In order for experts to give their best it is necessary that they are able to appraise a whole situation rather than a narrowly defined requirement. Only by having a comprehensive understanding of all the factors involved can good viable solutions be created. If a consultant is given a tight specification as perceived by a company at that particular time, the range of solutions that can be offered will be limited and the available expertise will not be used to the full.

A specification for consultants must therefore make the aims and objectives of what is required as clear as possible. Any constraints that are relatively inflexible should be clearly stated, but not in a way that becomes prescriptive to the solution.

It is not possible to be certain that consultants can or will always provide the answers that are sought. When considering the engagement of consultants there can be some difficulty in deciding who the ideal people are. Reputation, if it is generally known, is one important yardstick. In many instances the employment of consultants is more of an act of faith, for the very reason that discussions are based (to some extent) on the lack of relevant specialist knowledge on the part of those making the appointment. It is as well, therefore, to maintain effective control.

It is generally possible to divide the consultant's work into distinct phases.

An agreement that enables the work to be terminated if necessary at the end of any phase prevents a runaway situation and loss of control. Design work can typically be divided into the following phases:

1 Initial appraisal.
2 Feasibility study.
3 Product design.

Other types of work can be divided similarly. It is often difficult for either party to estimate with any accuracy how long a particular piece of work might take. The total cost of using consultants could therefore be open ended. The phased approach will limit the risk to some extent. It is as well to have a budget limit in mind which may, or may not, be revealed in the early negotiating stages.

An alternative that can be considered is to offer a royalty agreement. This is obviously dependent on the type of product and the envisaged level of sales. A general approach is to offer a fixed but small royalty on items expected to reach high sales volumes, or a sliding scale of royalties on shorter production runs. Large complex products should give a high royalty on the first few that are made in order to provide the consultant with a reasonable return in the short term. Cash flow problems are lessened by the use of royalties (as opposed to paying consultants' fees up front, long before revenues are seen from sales of the finished product). However, a good level of mutual confidence and openness is necessary for royalty methods of payment to succeed. A suitable agreement covering the royalty terms and allowing access to audited accounts and sales figures are normally all that is required.

Maintaining control

Control can conveniently be divided into three areas of responsibility:

1 Technical (quality).
2 Progress.
3 Financial (cost).

Safeguarding technical standards

The first essential step in achieving the expected technical results from any engineering designer or design team is to ensure that the requirements are made known in the form of a clear, unambiguous, properly quantified specification. The document may be formal or informal: a letter or memorandum will suffice in some cases. But the specification must be given, it must be dated, and it must relate to a package of work for which the cost estimates, cost budget and scheduled dates have been established. This specification is the basis for measuring all performance: without it adequate control will be difficult or impossible.

Whatever method of increasing design capacity is chosen, there will be some necessary learning while the new staff become familiar with the particular practices and methods of the company, in addition to absorbing the specific technical and commercial requirements for the allotted task. Learning time will be minimized by the use of recognized national and international standards and codes of practice wherever possible.

When a parcel of design work is to be put out to a subcontractor, to be executed in remote premises, the control of quality can pose a problem. Some form of control is essential in order to ensure that the quality of work is of a standard that is consistent with the company's own standards. The presentation of drawings should be such that it is not possible to distinguish the bought-out draughting from in-house quality. Obviously the styles of the individual draughtsmen will vary, but the general arrangement of layout, dimensioning, method of tolerancing and instruction notes should be normal to the company's methods of working.

An instruction to use specified national and international standards, codes of practice and other appropriate methods of presenting information will ensure the minimum level of confusion and ambiguity. If the company has developed its own standards, it will be necessary, and very sensible, to provide each subcontractor with at least one set of the relevant volumes or manuals, and ensure that they are followed. This will probably involve giving some instruction as to their interpretation and use.

With specific reference to the teaching and imposition of company practices on a new subcontractor, it has been found useful to follow the following procedure.

For each work package, the subcontractor is asked to nominate one of his team as a senior person to take day to day charge of the external group. The person chosen should be a suitably qualified designer or engineer, at least equivalent in experience and design capability to the employing company's own senior in-house design staff. The senior subcontract engineer is then asked to spend a period of several weeks in the main design offices, where he/she works under the direct supervision of the company's own engineering management. The outline design or layout drawings for the work package are developed at this stage.

The senior subcontract engineer is then able to take the work package back to the external office for the more labour intensive tasks of detailing and checking. The external engineer now has some familiarity with the required working practices and standards, and is in a better position to supervise locally the design, detailing and checking of the whole work package. The purchasing company would usually carry out its own quality check of the entire package before releasing the drawings for manufacture or construction.

The ideal form of regular technical control is to appoint a member of the engineering staff to be responsible for liaison with all subcontractors. One such appointment is generally sufficient for a medium-sized engineering operation, and the duties involved will probably not occupy the individual full time unless there is a large volume of subcontracted work and/or considerable travelling distances between the home base and the outworkers. Although there will be a loss of productive output from the person appointed, a much smoother flow of information to the subcontractors will be ensured, coupled generally with a more rapid response on their part. The identification of one person as a focal point of contact is of considerable benefit to subcontractors, who will be saved from the inconvenience of having to find ways through the company hierarchy in order to resolve day-to-day problems.

The person chosen as liaison engineer should be sufficiently experienced in company engineering and design practice, and of sufficient seniority, to be able to answer queries and resolve routine difficulties without reference to his superiors (acting on behalf of the company for most matters likely to affect the day-to-day work of subcontractors).

The liaison engineer should be mobile (a car driver is usually essential): the job cannot be performed well by a deskbound person and regular visits to all remote design offices are desirable. In this context, the duties of the liaison engineer might also include the delivery of new work or authorized modification instructions and the collection of completed drawings. Liaison engineers are often asked to combine progress control with their duties and to act as on the spot expeditors.

Commercial arrangements and controls

The first prerequisite in achieving good commercial control is the same as that for technical control, namely to agree jointly the arrangements that will operate and to record them in a clear statement or specification.

It is also very important to decide upon the size of each piece of work or task. By splitting the total task or project into manageable work packages, each with its own specification, the following are among the control advantages gained:

1 Control is simplified because it is easier to envisage the individual scope of relatively small work packages.
2 More points of reference or milestones are created, against which costs and progress can be measured and compared with budgets and schedules.
3 There will be better flexibility of job loading, it being easier to divide and spread the total workload over several subcontractors, or between subcontractors and internal resources as the occasion demands.

Pricing and methods of payment
The two most common pricing systems are:

1 An agreed contract price for a specified parcel of work. In this case, the company must obviously negotiate a price that appears reasonable according to the best possible cost estimates and associated budget.
2 A negotiated scale of rates (usually hourly) graded according to the seniority of subcontract staff required (e.g. design engineer, checker, draughtsman, tracer).

When negotiating hourly rates, it is important to agree what the rates do and do not include. It is also necessary to establish whether any overtime will be charged at the flat rate or at some premium. Some subcontractors will charge for time spent by their design staff in travelling. There are agencies which charge car mileage expenses for travel, and others that include these as overheads, covered within the flat rate. Hourly rates can therefore vary greatly from one subcontractor to another but, because of the differences in services included, the firm offering the lowest hourly rates might not result in the lowest final job cost. It is also not unknown for a design agency which quotes very cheap rates to compensate for this by claiming and charging for more manhours than a supposedly more expensive company would invoice for an equivalent job.

It is therefore necessary for any company engaging a subcontractor to weigh up the probable total job costs (invoiced time and expenses). Knowledge of the relevant cost effectiveness of different subcontractors really only comes from experience, which is a good reason for only letting small parcels of work out at first, and for giving more than one subcontractor the chance to show what they can do.

Both penalties and bonuses should be considered in the agreed payment arrangements. Any undue delay can have an undesirable knock-on effect on the business as a whole. Responsible subcontractors are well aware of this problem, and it is obviously in their best interests to keep any delay to a minimum. Although expense can be recovered to some degree, lost time unfortunately cannot. Suitable reward for completing the work before time could be considered as a means of ensuring that every effort will be made to keep to time (subject to proper quality control to avoid the danger of skimping).

A short-term job can be accounted for on completion. Work covering longer periods will require stage payments. These stage payments can be on a regular time basis (weekly or monthly) or they can be made on completion of progressive phases of the work.

The amounts actually invoiced are usually supported by timesheets which record the number of hours used in each accounting period for each grade or each named individual or both. A certain amount of trust is necessary in accepting the hours and expenses stated on invoices. These must be certified by a responsible member of the subcontractor's management. The purchasing company should make routine checks if possible, which are best performed by having the subcontract liaison engineer note the number of individuals actually working on the subcontracted work at each visit, and on gauging the progress made between visits.

Authorizing work

Jobs or work packages must be authorized individually, but it would be tedious to adopt a very formal procedure, with full blown contract documents and agreements, for every small task. One way of overcoming this problem is to enter into an agreement with each subcontractor as to the general working practices, terms and controls that shall apply between the parties. This agreement can be recorded in a formal contract or purchase order (which is itself of course a contract) that does no more than specify the terms and conditions. Each work item is then issued as required as a subpurchase order within the terms of the covering main contract or purchase order. It is sensible and desirable to place time and budget limits on the blanket contracts or orders, so that each of these has to be reviewed, reauthorized and reissued at suitable intervals (annually, for example).

The aim should be to introduce a system of subpurchase orders for subcontracted work packages which is simple to operate while still specifying what has been agreed and what is being ordered. The suborder forms should be as simple as possible, designed and printed specially to keep the clerical work involved to the minimum. Here is a checklist of items that might typically be included on individual orders for work packages that are to be charged at hourly rates:

1. Reference to the main purchase order or contract number that defines the blanket conditions and terms.
2. A job number.
3. The purchaser's cost code for the work package (which should ideally be identical with the work package job number, and should not need a separate entry).
4. The number of the main drawing to be produced (this might also be the same as the job number).
5. A brief but correct statement of the work required.
6. The date of issue.
7. The completion date required.
8. The start and finish event numbers of the activity, or group of activities

which embrace the job (obviously only relevant if the project has been planned by network).

9 The degree of timescale criticality, expressed in terms of the amount of network float available (optional, see comments below).

10 The budgeted manhours (optional, see comments below).

11 An authorizing signature.

Points 9 and 10 are optional. Some companies prefer to give this kind of information openly and trust their subcontractors to make sensible and honest use of it. Other companies feel that subcontractors are not to be trusted in this way, since they might be expected to take unwarranted advantage of the company's frankness.

If, for example, a subcontractor is told that there is a float of four weeks on the requested delivery date for a work package, it can be argued that this would be useful to him in indicating that overtime working would not be justified to pull back any short delay, but it might also easily lead to complacency. Perhaps the better way of dealing with this issue is not to release details of schedule float, but to have a general understanding that overtime premiums will not be paid unless the need for overtime working has first been approved by the company. Complementing this approach, the general importance of meeting scheduled dates should always be stressed, and any particularly critical job should be identified as such when the relevant suborder is placed.

If the job cost budget is revealed to the subcontractor this might act as an incentive to limiting the time spent on a job, but it might equally tempt the subcontractor to claim for all the hours budgeted when less had in fact been spent. Some idea of the expected hours involved must be known and discussed when jobs are placed, simply in order that the company and subcontractor can agree on completion date commitments. But cost control is usually best achieved through work supervision and progress control and the avoidance of unnecessary modifications. Provided that the associated terms are right the costs should then fall into line.

In the question of releasing precise details of schedule and budget data therefore the reader must weigh the pros and cons and choose his/her own path. The decision reached will be based on the working experience gained with each subcontractor. But caution is advised and, when in doubt, it is probably wiser to err on the side of caution and not volunteer this kind of information in the normal course of events.

Conclusion

The use of bought-out engineering and design is a question of expediency: of utilizing resources of the right type and of sufficient extent to meet the demands being made on an organization at a particular time. There should be no qualms about making the decision to use outside help, providing that the fundamental principles are adhered to. Always provide a clear statement of what is required. Always make explicit what is being agreed and, above all, organize to maintain control throughout the whole period of association.

25

Managing design modifications and concessions

Dennis Lock

No engineering project can be considered complete until every drawing has been revised to 'as built' status and put to bed in the archives. If, when that condition is reached, an analysis is made of the time spent in producing all the drawings used on the project, it might be found that the time could be apportioned as follows:

Initial design and drawing	60
Checking before issue	15
After-issue work	25
Total	100 per cent

These figures may vary considerably from one company to another, but the last category, after-issue work, will always account for a considerable part of the total effort.

The greater part of after-issue work comprises those activities in the design office which are concerned with design changes and drawing modifications, queries from the manufacturing, construction or commissioning team, and changes requested by the customer.

Although the example given above suggests that about a quarter of the total design effort could be spent on these tasks, in some companies the proportion is significantly higher. Add to that the effort spent in changes that take place *before* issue, and it is seen that perhaps a third of the engineering work on a typical project may be devoted to changes arising from errors, customer requests, or production queries.

Any group of activities which consumes up to a third of total design hours deserves special management attention. the subject can be approached from at least two directions. On the one hand, there is the possibility that the element of change could be reduced if the quality and accuracy of the initial design could be improved, which is all part of supervision and a company's attitude to quality management.

This chapter deals with the management of changes and after-issue work mainly from the procedural and administration viewpoint, the objectives being to ensure that:

1 Unnecessary or undesirable changes are not allowed.
2 Reliability, interchangeability and safety are not compromised by the introduction of changes.
3 The commercial effects are considered (timescale, costs, budgets, customer liability and profitability).
4 Approved changes are properly progressed.
5 Queries arising during manufacturing or construction are dealt with promptly, in order to prevent or minimize hold-ups through design errors or ambiguities.
6 Requests for manufacturing concessions are properly considered and progressed.
7 Emergency changes and marked-up prints are effectively controlled.
8 All changes are recorded in the master drawings.
9 Where changes are introduced during a batched manufacturing programme, that the build standard of any individual unit can be traced from its serial number.

Please note that the terms engineering change and modification are used synonymously throughout this chapter.

Cost aspects

There are several ways in which the need for engineering changes can arise. From the commercial point of view, the most important classification depends on whether or not a proposed change has been requested by the customer.

Customer changes

A change requested by a customer has the effect of varying the agreed project specification, which in turn varies the purchase order or contract. This means that any additional costs can legitimately be considered as the basis for a price increase.

It has to be mentioned that customers sometimes ask for changes that reduce the project scope. In these cases the customer may be able to claim a price reduction, but care must always be taken to ensure that the costs of any abortive work are recovered from the customer.

Figure 25.1 shows a form used by one mining engineering and construction company to record every customer request for a change in project scope (known as project variations in this case). Figure 25.2 illustrates a schedule used to summarize all variations, whether proposed or agreed, and a copy of this schedule would be submitted to the customer as part of regular cost and progress reports.

Unfunded changes

Strictly speaking the title 'unfunded changes' is a misnomer, because every change that costs money must be paid for somehow. If the costs of a change cannot be recovered from the customer or client, then the engineering company must fund the work out of its planned profits. However, the terms funded and unfunded are used in many companies to draw the distinction between those changes which are customer funded and those that are not.

Unfunded changes obviously require strict control, because they can threaten the agreed timescale and profitability. The risks are sufficiently high to lead some companies to place all proposals for unfunded changes under two headings. These are *essential* and *desirable* (assuming that *undesirable* changes are never proposed in a responsible engineering organization). A simple rule is then applied. Essential changes are accepted. Desirable changes are rejected. This approach may sound too severe for all occasions, but it emphasizes the need to keep unfunded changes to the absolute minimum.

Cost estimating and pricing changes

If a change in scope is proposed right at the start of a project, before any work has taken place, and if the change can be related to specific items in the original cost estimates, then it should be possible to re-estimate or delete the relevant items and declare the estimated cost difference attributable to the proposed change.

Cost estimates for any specific item of additional work can usually be estimated in straightforward fashion, and the timescale should also be plannable. Cancelled work demands greater caution, since there may have been some work already carried out, representing wasted expenditure in terms of design, manufacturing labour and materials, and administrative support. All of these considerations are important in reckoning the possible overall effect of any change on project costs, and very important for pricing modifications that are chargeable to the customer.

The most difficult circumstances for cost estimating come when a change is proposed for manufacturing work that is in progress. It may be extremely difficult to analyse a complex series of assembly tasks and identify separately any work that would be caused or cancelled by the change. A further complication arises from all kinds of possible secondary costs, such as lost or waiting time, scrap and rework, extended occupation time of production

LOXENGINEERING PLC Construction Division	PROJECT VARIATION RECORD

COST DATA UNITS:
Total increase / decrease in cost =

EFFECT ON PROGRAMME
Estimated extension to timescale (if any)
Expected revised completion date
This information is subject to verification by the Planning Department

LIST OF RELEVANT DOCUMENTS AND DRAWINGS

DESCRIPTION OF CHANGE (Use continuation sheet if necessary)

APPROVALS				
Originator Date	Project manager Date			Client Date

Client — Project —	Project No —
TITLE —	VARIATION No revision sheet

Figure 25.1 *Project variation record. Used to record and progress project variations requested by clients of a mining engineering and construction company*

LOXENGINEERING PLC
Construction Division

SCHEDULE OF PROJECT VARIATIONS

Variation number	Date	Description	Status A = Approved N = Not approved

Client

Project

Project No

Figure 25.2 *Schedule of project variations*

floorspace, and other overhead and administrative expenses. It may simply not be possible to estimate the true cost of a change to work in progress with any degree of accuracy.

Most changes have risk and nuisance value and, for the reasons given, it may be very difficult to gauge the extent of the cost and programme risks. If cancelled or additional work is not to be funded by the customer, then the only estimating question concerns the extent of the certain erosion of planned profits.

For customer-funded modifications the picture is quite different. Any change proposal still carries possible risk and nuisance value, but it also provides the contractor with the opportunity to obtain adequate compensation in the form of increased revenue from the customer. Once a main contract has been agreed, any proposed change by the customer presents the contractor with a big negotiating advantage. There is no third party competition and the contractor enjoys complete monopoly of supply. In these favourable circumstances, the contractor should ensure, at the very least, that the price for every agreed modification will be sufficient to cover his costs, any predictable risks, the nuisance value, and still yield a reasonable profit. Some contractors find that the revenue from customer-requested changes adds significantly to their contract revenue, providing a valuable boost to profits.

There are two sides to this coin. The contractor's own engineering managers will have to let or manage contracts themselves from time to time (including the placing of purchase orders). In this sense, the roles are reversed: the engineer becomes the customer, and the supplier or subcontractor has the negotiating advantage when a change is proposed. This destroys the protection that any fixed price arrangement may have provided. Any manager who has, for example, had to cancel a sizeable purchase order will probably have unpleasant memories of the costs involved. Engineers and their managers must strive for accuracy in specifying purchases and contracts, avoiding as far as possible the need for subsequent changes that could put their own company at commercial risk.

Scale-price modifications?

One electronics company was faced with the difficult problem of pricing a large number of minor modifications from a major customer. The equipment being developed was intended to carry out automatic testing of aircraft systems. The modifications were mostly changes in the programmed go, no-go limits, or other changes that required simple cabling alterations.

The contractor and customer agreed on a simple scale of charges, with a fixed fee for every software change, and a different fixed fee for every simple recabling request. Many hundreds of modifications were carried out and billed under these arrangements, to the mutual satisfaction of both parties.

This arrangement recognized the impracticability of trying to price each small modification separately.

This example has been included to illustrate how two companies adopted a commonsense approach to the difficult problem of pricing a large number of small modifications. It emphasizes that attempts to price small modifications accurately may simply be a waste of time.

Procedures for controlling engineering changes

Is a formal procedure always necessary?

When a designer, in a fit of despair or rage, tears up his drawing or clears his computer screen and starts again, there is obviously no need to invoke a formal engineering change procedure. Any design might undergo many changes before it is committed to a fully checked and issued drawing. This is all part of normal design development. Provided that the design remains within the requirements of the design specification, any changes before the formal issue of drawings are not generally considered to be engineering changes.

Some companies have a need to circulate early, pre-issue drawings for discussion or approval, and these are often distinguished from issued manufacturing drawings by giving them letter revision codes, so that they would be issued at rev. A, rev. B, and so on. The first issue of every manufacturing drawing would then appear at rev. 0, with subsequent revisions starting at rev. 1. A rule might, therefore, be suggested that formal engineering change procedures need only apply for drawing revisions after the first issue for manufacturing. But this rule might fall apart if pre-issue drawings were used to manufacture a prototype.

In general, as design work progresses towards the required issue date, the more inconvenient or risky any change becomes (the amount of wasted work is likely to be greater, and there becomes a higher possibility that the programme will be delayed). Some circumstances may even warrant a 'design freeze' date, after which no change can be allowed (except to correct errors). Now we are talking about engineering change formalities required before any drawing issue.

Another reason for invoking formal procedures is found when there is any intention to depart from the design specification (especially when the development work is taking place as part of a contract for an external customer). Again, here is a case for a formal change procedure before any drawing has been issued.

In all the cases mentioned, there is one characteristic common to those occasions when formal permission to make an engineering change is necessary. The question to be asked is 'Will the proposed change affect any instruction, specification, plan or budget that has already been agreed with other departments, a customer, or other external organization?' If the answer to that question is 'Yes', the probability is that formal permission will be needed. The engineering change request procedure provides a suitable and reliable system in these circumstances.

Originating an engineering change request

Figure 25.3 shows a form used in a manufacturing company to request a design change or a modification to actual production units. Since the form is used initially only as a request, it carries no executive instruction until it has been authorized. It follows then that any responsible engineer, or indeed any other responsible member of staff may be allowed to originate a request.

At the top right-hand corner of the form there is space for a serial number. This is required for progressing and recording purposes. The originator should leave this blank: the serial number will be added later by a coordinating clerk.

A change can easily affect several drawings, especially where a major assembly is involved. Every drawing affected must be listed, together with its pre-change revision number.

There are occasions when the originator realizes that an urgent change is necessary to manufacturing work that is already in progress, or to completed units that are in stock. Any such need must be highlighted on the form, identifying the relevant batches or units by job number, batch number or serial numbers. If the originator does not have this information, he must do his best to obtain it. If there is a prospect that work in progress is at risk, the originator will obviously warn the production management, although he/she may not have the necessary authority at that stage to halt the work.

Note that in the space marked 'reason for change' the originator is asked to specify any customer order number or authorization reference. Any such entry identifies the proposed change as customer funded, and this will greatly increase the chances of the change being approved.

If the change is being requested urgently in order to intercept manufacturing in progress, there may not be time for preparation of a formal cost estimate, but some attempt to indicate the cost region should always be made for the benefit of the authorizing committee.

Progress and coordination

It is strongly recommended that a clerk or secretary is given the responsibility for coordinating and progressing all engineering change requests. This same person will also feature in procedures (to be described later) for dealing with

LOXENGINEERING PLC Manufacturing Division	**ENGINEERING CHANGE REQUEST**	Serial No.

Project or works order:

Drawings affected

Number	Rev.	Number	Rev.	Number	Rev.	Number	Rev.

Emergency action — Units in production or stores which require immediate action (if any)

Part number	Batch or serial numbers affected

Details of this change:

Reason for change: If this is requested by customer, enter the order number or other authorization reference in this box →

____________ (originator) ______ (date) Estimated cost on this project/works order £

Change committee decisions: Point of embodiment, action on stocks, limitations:

This change is rejected / approved ____________ (Committee chairman) ______ (Date)

Figure 25.3 *Engineering change request form*

various engineering requests and queries originating in the manufacturing and inspection departments. For the purposes of this chapter, the person responsible will be referred to as the coordinator.

When the originator has completed and signed an engineering change request form, it is handed to the coordinator. The coordinator enters the change in a register, which might be similar to the format shown in Figure 25.4.

A serial number is allocated to the form, from the left-hand column of the register. The coordinator keeps a copy of the change request and passes the original to the chairman of a committee, known as a change committee or modification committee, which has the authority to approve or reject every change request.

The change committee meets on a regular, frequent basis to consider all pending change requests. If a request is particularly urgent, a special meeting might be called. Alternatively, the request forms might be circulated among the committee members, each committee member being asked to mark the change as recommended or not recommended. The committee chairman makes the final decision, and completes the final section of the request form accordingly. The committee passes the form back to the coordinator for further action.

If the change has been rejected, the coordinator completes the relevant line in the register and rules it through so that it is still legible. A diagonal line is ruled across the original change request form to signify that it has been cancelled. The coordinator sends a copy of the cancelled original back to the originator for information, and files the original. All other copies are destroyed.

If the change has been approved, the coordinator distributes copies of the request according to a standard list. This will include the chief engineer (to initiate drawing changes and revised manufacturing instructions) and one or more members of production management and quality control staff (for advance warning).

The register is used as a progress aid, since the coordinator will follow up any change request where the committee has delayed its answer. Blank spaces in the final column highlight those changes that are still pending.

In some systems, the coordinator's responsibility extends to progressing each change right through to the issue of revised drawings, in which case the register sheets would have the appropriate extra columns.

The file of original change request forms and the register together form a valuable record of all changes introduced on a project. They must be retained in central records after the engineering project has been finished. The descriptions on the register are entered with this long-term aim in mind, so that although they can be brief, they should summarize the key part of each change.

ENGINEERING CHANGE REGISTER

Project:

Serial number	Originator	Part or drawing number of item affected	Summary description of change	PROGRESS		
				Date sent to committee	Approved or rejected?	Date distributed

Figure 25.4 *Register for recording and progressing engineering changes*

Updating drawings

The rules for updating drawings to incorporate engineering changes are generally well known, but the principal points are given here for completeness.

Once a change has been made to an issued drawing, it has to be reissued at the next higher revision number. It is customary to write the revision number in an inverted triangle alongside any changed dimensions or redrawn areas.

The drawing title block should have a table for listing revisions. If the revision has been the result of an engineering change, the change serial number should be entered in the table alongside the new revision number.

Additionally, drawing sheets may have a column at one side for summarizing the details of each change.

Some companies which have to issue drawings externally, to a remote construction site for example, list the issue dates for each revision on the drawing title block, possibly showing the date issued to site and the date issued to the client.

The typical procedure for modifying a drawing is to withdraw it on loan from the central drawing files (where original drawings should always be stored after their first issue) carry out the change, enter the revision details, and return it to the files via the print room. The print room will arrange the reissue before passing the drawing on for filing. That system, to some extent, imposes a self regulating discipline on all concerned to ensure that the revision number is raised.

The procedure for computer produced drawings is very different, since the plotter produces a completely new drawing after every change. Thus a number of originals can be created and, if care is not taken, several can exist simultaneously. Worse, it is easily possible to make a small change to a drawing, replot it, forget that it has been issued, and fail to update the revision number. Special care and supervision is necessary under these circumstances.

Some engineering companies like to keep a copy of every revision of a drawing. This can be useful for repair and maintenance of manufactured products in the field that were made to earlier drawings. Such earlier revisions are best kept on microfilm aperture cards.

If a company microfilms all drawings as soon as they are issued, or even issues drawings for manufacturing from microfilm, the original will still be kept on file for possible revision. When such revisions take place, the microfilm aperture card can be withdrawn from its file and used as the loan card for the original drawing. The microfilm aperture card is placed in a 'drawings on loan' tray, and this prevents new prints of the drawing being made from the film and issued while the modification is in progress.

Finally, it is necessary to state a rule that should always be applied to any manufacturing modification which takes place after manufacture has started, and which will not affect earlier production output. If the proposed change will render the modified units or components non-interchangeable with those produced earlier, the drawing number must be changed. In other words, non-interchangeable parts must not bear the same part number.

Change committee composition

However a change committee is appointed, and whether or not it meets formally or operates on the basis of circulating documents, there should be a clear understanding of its functions. Since a proposed change might affect reliability, interchangeability, safety, manufacturing, costs, and delivery schedules, it is essential that each interest is represented on the committee at a level of authority able to give or withhold approval.

A typical committee might comprise:

1 A chief engineer or other engineering manager acting as the *design authority*.
2 A senior representative of manufacturing management.
3 In appropriate cases, a commercial manager or contract manager, to consider the question of funding and the customer's liability to pay for the change.
4 It may be necessary to have a senior representative of the quality control function (such as the chief inspector) present to consider some changes.
5 Specialists, coopted as required to advise the committee.

Change commiteee decisions

The two most obvious decisions that a committee can give are a straight yes or no. If a change is rejected, the reasons should be given.

When a change is introduced after the relevant drawings have already been used for manufacture, the committee must define the point of embodiment of the change. The decision must depend on engineering advice: are products in stock or already in use at risk if they are not modified? A good example of this problem is seen when manufacturers have to recall motor cars for modification for reasons of safety.

Depending on the circumstances, the committee will choose from the following options:

1 Leaving existing stocks, work in progress and products in the field unchanged. Introduce the change at a point convenient to the manufacturing department, when a specified production batch is loaded, or to products from serial number *nnnnn* upwards.
2 Modify work in progress and future production only.
3 Modify work in progress and future production, and scrap or modify stocks (whichever is most economic).

4 Modify all work in progress and stocks, and future production, and introduce the modification to all units returned from the field for servicing or repair.
5 Modify all units retrospectively, whether in store, in manufacture or in use. Recall units from customers or modify them on the customer's premises.

Modification labels on hardware

A practice is sometimes adopted of attaching modification record labels to equipment. The labels list the serial numbers of all modifications incorporated in each unit, both during manufacture and added since. Such labels can be invaluable to service engineers. If additional modifications have to be made, the serial numbers are also punched on the existing label.

Build schedules

A build schedule is really nothing more than a list of all drawings and specifications that were used to manufacture or build a product or project. The important feature is that the revision number of every document is given to show its status after all changes arising during the works and final testing or commissioning have been incorporated.

For a major civil engineering or construction project, the build schedule is represented by the final issue of the drawing and purchase schedules, which should be updated to list all the drawings and specifications at their 'as built' revision status.

For manufacturing jobs and projects, build schedules (alternatively known as master record indexes) may have to serve a slightly more complicated need. This occurs when more than one item is produced, and when there can be a difference in the design content between two or more of the products. This difference need not be so great as to affect performance or interchangeability, but it is often important that the exact build content should be known. Possible reasons are:

1 For servicing or repair instructions.
2 For the supply of replacement parts.
3 To facilitate investigation in the event of any failure in use.
4 So that the starting point for any future in-service modification is known.

The first prerequisite is that each product must be provided with a unique identifier. This might be a serial number, production batch number or a mark number.

Build schedules need not list anything more than the relevant document numbers and revision status. It may be necessary to list every sheet of every document. There may be hundreds or even thousands of documents listed, so brevity is important.

If a logical drawing numbering system is in use, the simple process of listing all drawings in drawing number sequence should automatically group them into a family tree structure. The schedule can, if required, be split into separate parts of the family tree in order to reduce the bulk (each separate assembly and subassembly having its own subbuild schedule).

One build schedule must be compiled for each variant, and each must obviously denote the units to which it applies, giving the relevant serial, batch or mark numbers. It is also usual to list the modification numbers that the build schedule incorporates.

A clerical version of a build schedule sheet is shown in Figure 25.5 to illustrate the typical content, but in practice this has become a very suitable application for a word processor or computer. Advantages from the use of a computer system in this context include:

1 Greatly reduced clerical effort.
2 Ease of sorting into drawing number sequence.
3 Preservation of numerical sequence when new drawings are added.
4 Availability of edited reports, so that, for example, lists could be produced giving only main assembly numbers, bought out parts, or any other selection provided that suitable data have been encoded in the input file.

Manufacturing queries and concessions

The relationship between a manufacturing organization and the engineering team responsible for designing its products depends on the size of the company, the physical distance between the departments, the nature of the product and management attitudes. At one extreme communications might be very relaxed and completely informal. At the opposite extreme a very formal relationship might exist, perhaps exaggerated by a special quality assurance requirement in the contract. These considerations can be illustrated by describing what happens in two different companies, each representing one of the two extremes.

The informal extreme

Suppose that a company, employing a total of only fifty staff, manufactures a small range of garden tools and equipment. Now imagine that the company

LOXENGINEERING PLC Manufacturing Division	BUILD SCHEDULE	Sheet No of sheets

Product: Catalogue/assembly number:

This build schedule applies only to serial/batch numbers:

Drawing/spec No	Sheet	Rev	Drawing/spec No	Sheet	Rev	Drawing/spec No	Sheet	Rev

Modification numbers incorporated in this build (enter only on sheet 1)

Figure 25.5 *A manufacturing build schedule. Whether it is produced by hand or using a computer, a build schedule defines the modification and hardware content of a complex manufactured product, listing all drawings and their revision numbers. When a range of variants is manufactured, a separate build schedule is needed for each case, with the relevant units given unique identification by means of serial numbers, production batch numbers, or design mark numbers*

has embarked on a project to produce a new wheelbarrow, and that the first production batch is just going through the workshop.

If any production difficulty were to crop up as a result of a drawing ambiguity or error, there should be no communication problem. The workman with the query would either go himself, or ask his foreman to go along to the tiny design office. The grubby shop print would be produced, designer and workman might pore over it for a few minutes, the problem would be discussed and a remedy produced. Alternatively, the designer might be called into the workshop, and the discussion would take place there.

If any change to the drawing proved necessary, this would take place without a formal change procedure. If one or more wheelbarrows had to be manufactured in a slightly different fashion, or with alternative materials or parts, to get them through production, this would simply be done on the designer's instructions (again without formality).

When the batch of wheelbarrows finally emerged ready for despatch or placing in a warehouse, it would not matter too much if the product differed slightly from the manufacturing drawing. The inspector would not write out a reject report if the red plastic handle grips had been substituted for black rubber in order to overcome a supply shortage. There would, in fact, be no inspector other than a semiskilled checker/packer, whose check would be simply to ensure that one item resembling one complete wheelbarrow finished up inside each pack or parcel.

There is no implied criticism here of this very informal arrangement, which would probably work very efficiently in a small company manufacturing such a simple product. There is no need for many of the paper procedures used in larger organizations, and the workforce can be regarded as craftsmen who identify with the company's success, take a pride in their finished work and can be trusted to spot and rectify any substandard work before the product reaches the end of the line.

The formal extreme

For the example of a company in which extreme formality exists, the product is again a vehicle. But this time instead of a wheelbarrow it is an aircraft. The company employs thousands of people, the engineering offices are remote from some of the manufacturing and assembly areas, and an independent inspecting authority exercises constant vigilance to detect any nonconformance with the manufacturing specifications and drawings or the purchasing specifications.

Reliability and inspection discipline demands that build schedules are combined with strict quality control of all components, whether bought-out or manufactured, so that in the event of any failure in service even the tiniest screw used in assembly can be traced back to its purchase order number, manufacturer, inspection release note, and the production batch.

Just as the lack of formality could be justified in the simple case of the wheelbarrow, so the need for very strict procedures can be explained in an aircraft project where production methods are complex and involve high technology, tolerances are tight, adherence to materials specifications is vital, and the safety of passengers is at stake.

Procedures

It is unlikely that any of the procedures described in the remainder of this chapter would be appropriate to the very small, informally managed company. They become increasingly necessary as the complexity of the organization and product increases, being essential in the case of companies working on weaponry, naval, aviation and other government and major projects in high technology manufacturing.

The procedures described are all related, and they are not mutually exclusive. All three have been known to exist in one company, but in other cases two or more of them can be combined in order to reduce the number of different forms in circulation. The most likely approach is that the first two procedures listed, the engineering query and concession request procedures, should be combined.

All three procedures described here share the following characteristics:

1. They represent a formal request to the design authority from a manufacturing department for some kind of decision arising from a difficulty in complying with issued manufacturing drawings or specifications.
2. Each of the procedures can result in the design authority granting permission for a manufactured item to be accepted by the inspecting authority, even though it does not comply in some specified respect with the issued manufacturing instructions. In such a case the design authority is said to have granted a *manufacturing concession*.
3. The procedures are formal and fully documented, so that every concession is recorded and kept with the relevant archives (important in the case of possible investigations into the cause of a subsequent malfunction or failure in the field).
4. Analysis of the reason for any concession could lead to the introduction of a

future engineering change, if it proves that the original manufacturing instructions were unnecessarily restrictive or impracticable to achieve.

Request for concession

The most straightforward cause for a manufacturing concession occurs when a manufacturing department finds that it cannot easily comply with the drawing or test specification instructions. Perhaps it is not possible to obtain the correct materials in the time allowed, in which case a superior but more expensive alternative is proposed. A bearing may not be available from a particular manufacturer, but another company may be able to supply an equivalent item.

An example of a concession request form is given in Figure 25.6. The method of use is self evident from the form itself. In practice, the form should be treated with urgency by the design authority, since it is likely that production is being held up.

The requests can be handed to the coordinator responsible for engineering changes (described earlier in this chapter) and he/she can register, progress and record each concession using exactly the same procedure as for engineering changes. Thus each concession will be given a serial number, and such serial numbers will be shown on any relevant inspection release certificates.

Engineering query procedure

Figure 25.7 shows a form used to present the engineering department with queries that arise during manufacturing. This procedure implies a very formal relationship between the engineering and manufacturing departments, where a telephone call or visit would seem to provide a more efficient and cost effective approach to this problem.

One explanation for adopting such a formal approach relates to the manager of an engineering design department, who despised blue collar workers and refused to allow any machine operative or foreman to tread the new carpets in the engineering offices. If that sounds far fetched, please believe that the example is taken from real life.

More constructive reasons for using a formal engineering query procedure are that:

1 It provides for queries to be registered and progressed by the coordinator so that, although some delay is introduced into simple queries, no query is forgotten and all are followed up and progressed.
2 Some queries may result in an instruction by the engineers to deviate from issued manufacturing instructions, in which case the query is converted into a manufacturing concession.

Inspection reports

If an inspection operation identifies a manufacturing or test non-compliance

<table>
<tr><td>LOXENGINEERING PLC
Manufacturing Division</td><td colspan="3">CONCESSION REQUEST</td><td>Serial No</td></tr>
<tr><td colspan="2">Part or assembly</td><td>Drawing number</td><td>Rev</td><td>Works order No</td></tr>
<tr><td colspan="2">This requests a deviation from the above named drawing in respect of the production batch or serial number specified only.

Details of concession requested:

Requested by</td><td colspan="3">Batch or serial number

(Date)</td></tr>
<tr><td colspan="5">Reason for request</td></tr>
<tr><td colspan="2">DECISION
The concession described above will not affect reliability, interchangeability or safety</td><td colspan="3">Concession approved/not approved
__________ (Chief engineer)
__________ (Chief inspector)</td></tr>
</table>

Figure 25.6 *Manufacturing concession request and record form*

<table>
<tr><td>LOXENGINEERING PLC
Manufacturing Division</td><td colspan="3">ENGINEERING QUERY NOTE</td><td>Serial No</td></tr>
<tr><td colspan="2">Part or assembly</td><td>Drawing number</td><td>Rev</td><td>Works order No</td></tr>
<tr><td colspan="4">To the Engineering Department. We have the following query regarding the manufacture/testing of the above named item:

Department: Raised by (foreman) (date)</td><td>Is manufacture or testing held up?
YES NO</td></tr>
<tr><td colspan="5">Answer:

(Senior engineer) (Date)</td></tr>
<tr><td colspan="5">CONCESSION APPROVAL (if relevant)
The above instructions to deviate from issued manufacturing instructions will not affect reliability, interchangeability or safety
__________ (Chief engineer)
__________ (Chief inspector)</td></tr>
</table>

Figure 25.7 *Engineering query note. This version is designed to double as a manufacturing concession request and record*

LOXENGINEERING PLC Manufacturing Division	INSPECTION REPORT	Serial No

Part number: Rev: Batch/serial numbers:

Test spec : Rev; Works order number :

The above unit(s)/assembly/part(s) have failed to satisfy the requirements of the relevant drawings and specifications in the following respects

(Inspector) (date)

REQUEST FOR CONCESSION (to be completed if relevant)
The discrepancies described in this inspection report do not affect reliability, safety, or interchangeability. Other remarks:

(Chief engineer)

DISPOSAL INSTRUCTIONS	Batch/serial numbers	AUTHORIZATION
Scrap and remake		
Rectify and reinspect		(Chief inspector)
Concession granted		(date)

Figure 25.8 *Inspection report form. An inspection report form which provides for the consideration of a possible concession to accept a product that has failed the inspection standards set*

that is considered by manufacturing management to be borderline or unavoidable, the inspection report can be submitted to the design authority for consideration as the basis for a manufacturing concession. Once again, the coordinator is responsible for progressing the application through to approval or rejection. A suitable form is shown in Figure 25.8.

PART SEVEN

Quality and Value

26

Quality assurance

S. J. Morrison

There can be no doubt about the critical importance of quality and reliability in manufactured products. During the last decade these determinants of international competitiveness have been the subject of a number of a studies (see list at the end of this chapter) looking at the relevant shortcomings of industry and at the opportunities that exist for improvement. In the UK, the government added its weight by launching a National Quality Campaign. But it is one thing to be critical of industry and quite another to say what should be done to set matters right. What is needed is a wider practice of the structured improvement process advocated by Juran (1984), and a greater acceptance by industry of the dictum that quality, in its essence, has become a way of managing a modern organization (Feigenbaum, 1984).

Some definitions

It has been said that the failure to differentiate between 'quality assurance' and 'quality management' is a major problem that has held back British industry for the last thirty years. The usage of these terms in this chapter is consistent with the definitions in current BSI and ISO specifications. *Quality assurance* is an all-embracing term that covers every activity and function concerned with the attainment of quality. *Quality management* identifies the managerial aspects of such activities and functions. Within the general framework of quality assurance the term *quality control* is used to describe the operational techniques and activities that are used to sustain the product or service quality. Some of these are based on statistical methods – hence the term *statistical quality control*.

Historical and cultural background

Looking to Japan as one of our principal foreign competitors we can see a nation that has advanced from feudalism to become one of the greatest economic powers in the world in the remarkably short space of just over one hundred years. The economic, political and social aspects of this dramatic change are summarized by Richie, 1978. In the present context it is interesting to note that the Japanese are nearly 100 per cent literate, that the level of universal education is very high, that in earlier schooling the curriculum is broad (with more teaching hours devoted to the arts and physical education than in any other country), and that lifelong education is practised on a scale found in no other country. Richie concludes, 'In Japan, more than in most countries, education counts. Good jobs depend upon good schools and knowledgeability remains a mark of culture.'

It was against this national cultural background that Japanese industry conducted what must have been the world's largest training programme in quality assurance. The programme began about forty years ago with a top-down approach at the level of chief executives who were given training initially in statistical quality control and latterly in quality management. The spectacular and much publicized quality circle phenomenon was a later development, when the need to train supervisors and foremen came to be appreciated. It is a fact that Japanese company executives are in general thoroughly trained in quality assurance, and that quality training permeates across all branches and downwards to all levels of company structure. This is something that we in the West have to come to terms with.

Economists will argue that Japan's economic success is due to the use of nearly 30 per cent of national income for capital investment, the adoption of new technology, and higher labour productivity. These factors alone could not have succeeded without the attention that has been paid to quality. Japan is highly dependent on international trade, and it is the combination of high levels of quality and reliability with reasonable price that has given modern Japanese products their competitive cutting edge in world markets.

It must be recognized that the totally integrated Japanese approach to quality reflects a unified professional discipline. The body responsible for promoting quality assurance is the Japanese Union of Scientists and Engineers. This one body embraces all of the scientific, technological, and managerial issues of quality. As it happens, in Japan a high proportion of company executives are engineers.

In contrast, there is a more fragmented approach in the UK. Science and technology tend to be separate, and subdivided into distinct disciplines – mathematics, physics, chemistry, etc., on the one hand, and civil, electrical, mechanical etc., on the other. Many of the practitioners of these disciplines have little or no training or expertise in industrial quality assurance. This is even more true of business executives and managers, many of whom are not even trained in the basic technologies of the industries they are managing. It reflects a weakness in an industrial situation when quality assurance experts have become yet another specialist group, with their own professional body.

There is a need for bridges to be built across the interdisciplinary boundaries of quality assurance. The Institution of Mechanical Engineers (IMechE) has been attentive to this need. Support was given to the National Council for Quality and Reliability in the 1960s. Various papers on quality assurance have been published from time to time, including one by Plunkett and Dale (1985) on an important management aspect of quality and another by this author on quality management (Morrison, 1985).

The award of a James Clayton Grant by the Institution of Mechanical Engineers made it possible for the author of this chapter to complete a programme of research into quality assurance in the machine tool industry (Morrison, 1984). The evidence gathered during a postal survey and the subsequent follow-up showed that quality management in the machine tool industry as a whole tended to be weak (or even non-existent in some firms). To counter this, the outline of a quality management model for the industry was drafted and a paper was circulated to firms participating in the survey, for comment. The comments received were favourable and encouraging, and further work was done to develop certain aspects of the model more thoroughly. The model is given here in more general form, in the belief that it will be widely applicable in the engineering industry, and possibly elsewhere.

Quality management model

There appears to be a quality management gap in the standard literature. Management texts usually make only passing reference to product quality, by comparison with the attention given to financial management, personnel management, production management, marketing management, etc. The new generation of texts on operations management pays more attention to quality but the treatment tends to be technique-orientated towards sampling plans and control charts rather than to management principles. The same is true of the quality assurance literature. Even the excellent *BSI Handbook 22* (1981) falls short of what is needed.

The quality management model that has been developed is based on the argument that quality management is not fundamentally different from other operational areas of management, but that it differs only in the details of practical application. It has been developed by considering the broad principles of general management as they are usually presented in the textbooks used in management education, selecting those that are particularly relevant to quality management, and developing their application in that context.

In the management literature the principles are often presented within a conceptual framework of planning, organizing, directing and controlling. The relationship between these four elements is that planning provides a basis for organizing which in turn sets the stage for directing and controlling. It is convenient to deal with quality management in the same way.

Quality planning

There are two reasons for underlining the importance of management planning: its primacy from the standpoint of position in the sequence of management functions, and its pervasiveness as an activity that affects the entire organization. Both these considerations apply with full weight in quality planning. Any organizational structure that does not relate to clearly defined quality objectives and policies is unlikely to be effective, and the direction and control of such an organization will be a fruitless task. The factors affecting product quality are so widespread that it is difficult to identify an area of management that is not involved in some way. They are to be found not only in the technical areas of design and production, but also in marketing, purchasing, personnel, finance, legal and secretarial, and indeed in every sector of company activity.

It follows that quality planning must not be considered in isolation, but has to be approached in the context of overall management planning. For the purpose of this model the writer has assumed that a hypothetical industrial company will already have created a corporate plan for survival and long-term profitable growth. At the strategic boardroom level of planning, the importance of quality and reliability will have emerged from a comparative assessment of the corporate strengths and weaknesses of the company and its principal competitors, and the pursuit of quality will have been recognized as one of the most important company objectives. If that point has not already been reached, an established company will probably have no future in competitive markets. Given that it has been reached, the company will be ready to develop a suite of detailed tactical plans, one of which should deal specifically with quality assurance.

It might be argued that quality can safely be subsumed under something else – production most likely – but that would be unrealistic in view of the recognition that must be given to quality as a prime determinant of competitiveness. Equally, the quality assurance plan must not be regarded as a substitute for other important elements in the suite of tactical plans. The quality function must act as a catalyst between other operational functions, as well as exercising authority in its own right.

The details of a tentative quality assurance plan for a typical industrial firm will now be set out in terms of objectives, policy and procedures.

Quality objectives

The prime objective should be to achieve a high degree of customer satisfaction, with due regard to quality costs. In this connection 'quality' must include any aspect of the product or service of which the customer may, within reason, take a critical view. The customer is entitled to be critical of design quality if the product specification falls short of his requirements. Quality of conformance will be an issue if the product, as supplied, does not meet the specification. Quality of performance may leave something to be desired. Higher precision and accuracy are increasingly being expected of many products. Reliability is important at all times if the product is functional, but especially if it represents large capital investment. Quality of service can be a sensitive area, in human as well as technical terms. The impression given by individuals can do much good, or much harm, to the company image.

Stating the quality objective in terms of customer satisfaction brings all these issues into a single focus and serves as a constant reminder that the customer who has grounds for dissatisfaction on any of them is at liberty to seek better satisfaction with a competitor.

In any particular company the quality objective will be subsidiary to the corporate objective, which (in general terms) should be a statement of intent to survive and flourish and the means thereto. In most companies the quality of product and/or service will play a vital part in securing that objective. It is no bad thing for the statement of the quality objective to embody the words 'customer satisfaction' (or similar) to concentrate people's minds.

The precise formulation of the quality objective will vary from firm to firm depending on particular circumstances, as seen in the following examples. One European manufacturer makes a point of reminding its customers and its employees that one of the fundamental ideas of the company's founder, more than half a century ago, was that, 'we must never leave the customer in the lurch'. A big multinational company declares that their operating philosophy is, 'to meet customers' needs and expectations'. In the quality handbook of a major UK electronics manufacturer the increasing interest of customers in quality is recognized, and the various authorities whose specifications must be satisfied are identified. A manufacturer of high quality entertainment electronics equipment reminds itself that its products are designed to be used in the home and that they must be problem-free and easy to operate. The quality objective of a famous manufacturer of precision electronic measuring equipment begins with the statement, 'the quality of a project is the degree to which the project meets the requirement of the user'. There is no end to the variety of statements of quality objectives – the one thing that is common to them all is the expression of concern for the satisfaction of the customer.

What is surprising is that in some companies, even among those with recognizable quality systems, there does not appear to be any written-down statement of the quality objective (or if there is, it seems to be a state secret). The absence of such a statement must be taken as evidence that the top management have not really thought their way through the formulation of quality policy, but have left it to others further down the line to do the best they can.

It would be naive to assume that each manufacturer should strive to become the quality leader in his own industry, in every sense of the term. In particular cases there may be sound commercial reasons for settling for something less. A distinction must be drawn between the *quality* and the *grade* of a product. There is a market for the Mini as well as for the Rolls-Royce. What is important is that the quality objective for the specific grade of product should be well chosen in relation to the company's present capabilities and future prospects, and that it should be clearly defined and widely promulgated with the authority of the chief executive, so that everyone within the company can understand it, grasp it, and work towards a common goal.

Quality policy

In terms of good management, quality policy must be more than a statement of intent – it must be manifest in a specific course of action designed to achieve the quality objective. This will involve considerations of finance, of personnel, of hardware, of procedures and systems, of education and training.

Of these, finance is probably the most critical, for financial limits will automatically impose constraints on the others. It is a salutory fact that the writer found that many firms were more vague and less specific on the financial side of their quality operations than on the technical side. For example, in the machine tool industry very few would admit to operating a modern quality cost system along the lines of BS 6143. There is some evidence of the quality function being regarded as a non-productive area suitable for cuts and economies, rather than an area in which productive investment can be made. It appears that some firms have not progressed very far beyond the traditional engineering inspection concept of quality towards the modern concept of quality management, with all that this entails in terms of financial investment in quality.

It is difficult to see how such an important function as quality assurance can be properly managed without adequate budgetary provision and financial control. The fact that no formal provision may be made in a company does not mean that there are no quality costs – it simply means that they are hidden and are not susceptible to management. One of the first decisions to be taken in quality policy-making must be how much to spend on running quality assurance operations. It may be necessary, at the outset, to make some rather crude estimates of the magnitude of the existing costs of failure, and of appraisal and prevention, to reach a global figure. Whatever figure is arrived at, some appropriate amount of working capital should be set aside immediately for the development and operation of a quality cost system. If money is to be found to develop quality management as a separate management function, it should be expected to pay its way alongside other management functions, and it should be accorded the financial services necessary for good management.

Once the initial financial decision is taken, other policy-making decisions will follow. It will be necessary to decide what sort of quality assurance system to adopt, how it is to be organized, and how it will be staffed and equipped. There may well be a conflict between what is desirable and what can be afforded. The resolution of such conflict may rest in phasing the development of the system over a period of time.

Finally, it must be said that the policy-making decisions are of such fundamental importance that some of the more important ones can only be taken at director level. It is axiomatic that product quality begins in the boardroom even before the product reaches the drawing board.

Quality procedures

In modern quality assurance practice it is necessary to create a system of standard procedures to service the quality function, and to determine the way in which it interacts with other functions. These procedures are managerial as well as technological and they extend far beyond the scope of the traditional statistical aids of inspection, such as control charts and sampling plans. They tend to form a complex highly interconnected system, but the complexity can be clarified by relating each procedure to one or more of three characteristic dimensions of the system. These are the production line dimension, the product life-cycle dimension and the management dimension. Each of these will now be expanded.

The production line dimension is self-evident, but it must be extrapolated beyond the physical boundaries of the plant into the market in which the firm's products are being sold and into the suppliers' market from which the firm is drawing its raw materials, components, or supplies.

The following issues need to be dealt with:

1 Current market requirements.
2 Current design/market requirements gap.
3 Product liability.
4 Field failures.
5 Quality of customer service.
6 Commissioning tests.
7 In-plant finished product tests.
8 Component and subassembly tests.
9 Acceptance tests of raw materials and supplies.
10 Approval of suppliers.
11 Purchasing standards and specifications.

The product life cycle dimension relates to the modern 'total quality' concept of designing quality and reliability into the product, rather than attempting to 'bolt them on' afterwards. For this to be done the quality function must involve itself with the design and development of new or improved products, and not simply confine its attention to the current product range. The issues involved include the following:

1 Future market requirements.
2 Market opportunities.
3 New product target specifications.
4 New product design specifications.
5 New product production specifications.
6 Process capability studies
7 Product reliability analyses.
8 Product safety and liability studies.
9 Product quality forecasts.
10 New quality operations and techniques.
11 Quality cost budgets for new products.

In *the management dimension* the procedures are those which are necessary for the effective management of the quality system and the coordination of quality activities within all the various operational areas of the company. These deal with the following issues:

1 Field failures.
2 In-plant failures.
3 Supply problems.
4 Service problems.
5 Quality costs.
6 Quality budgets.
7 Quality reporting.
8 Quality audits.
9 Education and training.

The design of a suitable quality system for an individual firm must relate to the quality problems and requirements of that particular firm, but the issues involved are commonly those outlined above. It is important that each procedure constituting an element in the total quality system should be properly written up in a standard format and made available for reference on a company-wide basis. No essential element of the system should be allowed to exist as an unofficial document that is the private property of any one department, or (worse still) as an abstract concept of 'normal practice' in the minds of individual employees. Knowledge of each procedure has to be freely available not only to the individual(s) responsible for execution, but also to others with whose duties the procedure will interact. The complete collection of procedural documents will constitute the quality manual or handbook and will have the status of in-house company law. As well as the basic elements of the quality system itself, the manual should carry suitable background information about the company, so that the role of the quality system in relation to overall company operations is clearly identified.

There are many different ways of writing quality procedures, but a good format to adopt is to allocate a single A4 sheet to each procedure using the name of the procedure as a heading along with a numerical reference code and the date of issue. The text can be presented under suitable side headings such as description, purpose, system connections and references. The text must be grammatical, succinct, and unambiguous. If a procedure appears to require more than one page of description, then almost certainly it will be capable of being broken down into separate elements, or else insufficient use is being made of reference sources to fill out detailed information. A hypothetical example is shown in Figure 26.1 to illustrate the layout.

Not all the procedures will be technical like the example given. Many, or even most, may be predominantly managerial because of the need to identify the relationships between individual departments right across the whole span of company organization and beyond the bounds of the company to external organizations.

Provision must be made for rapidly updating and reissuing individual procedure documents as the need arises. This will certainly be the case when a quality system is being developed from scratch. It will go on being the case if the industry is one in which technological change is rapid, with radically new products and processes appearing as part of the normal way of life. For this reason the practice in some companies of issuing the quality manual as a bound book printed in several colours on glossy paper is questionable. It may impress some customers, but it may not be the best way of providing back-up for the day to day running of the quality system.

All that is necessary for in-house use is to prepare the master documents on a word processor and to issue photocopies to be held in looseleaf binders. In some companies it may even be more convenient to read the quality manual on a visual display unit rather than lift a file down from the office shelf.

Quality organization

A structure of organization is created by identifying *roles*, by assigning *responsibility*, by delegating *authority*, and by creating *accountability*. It will be convenient to consider first the role of the quality assurance function within the company, before deciding the details of the final organization structure.

The network of procedures that has already been described and which is necessary for the full development of a comprehensive quality plan extends far beyond the confines of production management. Apart from production, the branches of management in which essential quality-related activities have been identified include marketing, sales, service, purchasing, design, finance,

Procedure No 07.22 June 1988

INSPECTION OF BOUGHT-IN COMPONENTS

Description

All bought-in components must be approved by Materials Inspection before issuing to Production. Batches of 50 or fewer components will be subject to 100 per cent inspection. Others will be subjected to sampling inspection. Inspection criteria are laid down in the component specification issued by Design Engineering Department. Classification of defects into critical/major/minor categories will be decided by the Production Department. The responsibility for negotiating with suppliers through Purchasing Department to reach agreement between Design, Production and the supplier on AQL values will rest with the quality engineer. Continuous inspection records will be maintained on form XYZ 1234. Inspection will be by single sample at inspection level 2 unless otherwise authorized by Quality Engineering. Switching between normal, tightened and reduced inspection will be at the discretion of Materials Inspection, using the rules in MIL-STD 105D.

Purpose

To protect the quality of the finished product in the market place and to minimize the cost of dismantling and rectifying finished units rejected at final test.

System connections

1.07 3.24 5.12

References

Component specifications
MIL-STD 105D

Figure 26.1 *Example format for a quality procedure*

accounting, personnel, as well as legal and secretarial. Many of these activities properly belong in the departments in which they are located, and it would be foolish to transfer them or to duplicate them in a quality assurance department. The dominant role of the quality assurance function is therefore that of coordination, to ensure that all these activities are brought within the framework of an integrated system dedicated to achieving that most important objective – customer satisfaction. Responsibility and accountability for the individual tasks must still remain with the departments in which they are performed, but their responsibility will include that of collaborating with the quality department. The quality department will, of course, have unique duties of its own to perform.

The quality assurance role is managerial in its essentials, which are to plan, organize, direct, and control all quality-related activities in the company. The basic task of the quality manager is that which is common to all of management. It is to create and maintain an environment in which individuals can work together to accomplish common goals.

Given that quality assurance must be found a place in the management organization chart, where and how should it be accommodated? The resolution of conflicting departmental quality interests will be an important part of the quality manager's duties. It cannot be assumed that conflict of interest will be less likely to occur in quality matters than in any other aspect of company activity. It is in the best interest of the company that quality management should be impartial, and independent of parochial departmental interests, and its importance to the survival of the company is beyond doubt. It follows that quality assurance should be a top-level management function, reporting directly to the board and enjoying equal status with other top-level management functions (production, design, sales, marketing, etc.) as in Figure 26.2. The traditional practice of not giving the quality function any specific identity on the organization chart and leaving it to be subsumed under the supervisory role of the chief inspector, as in Figure 26.3, is unsatisfactory and outmoded.

The question of which director the quality manager should report to is one that must be resolved within the particular company in the light of their own circumstances. The directors of production, design, marketing are possible contenders, but it is unlikely that the directors of personnel or of finance would be appropriate. In large companies it can even be appropriate to appoint a director of quality to the board.

The argument that a company may be too small to afford a quality manager is not sound. No company of any size can afford to be without a quality manager. In small companies it may be necessary for individuals in the management hierarchy to 'wear more than one hat'. It is possible that the individual wearing the quality hat may have to carry other responsibilities as well. In such a case the identity of the quality function must be preserved by making a clear definition of the various distinctive functional roles.

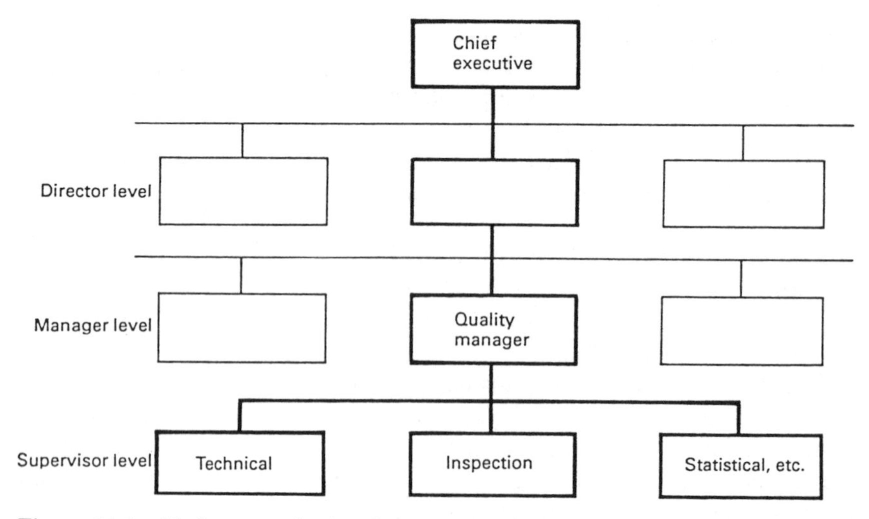

Figure 26.2 *Modern organization for quality assurance. Quality assurance should be identified as a management function above supervisory level, reporting direct to the board and chief executive*

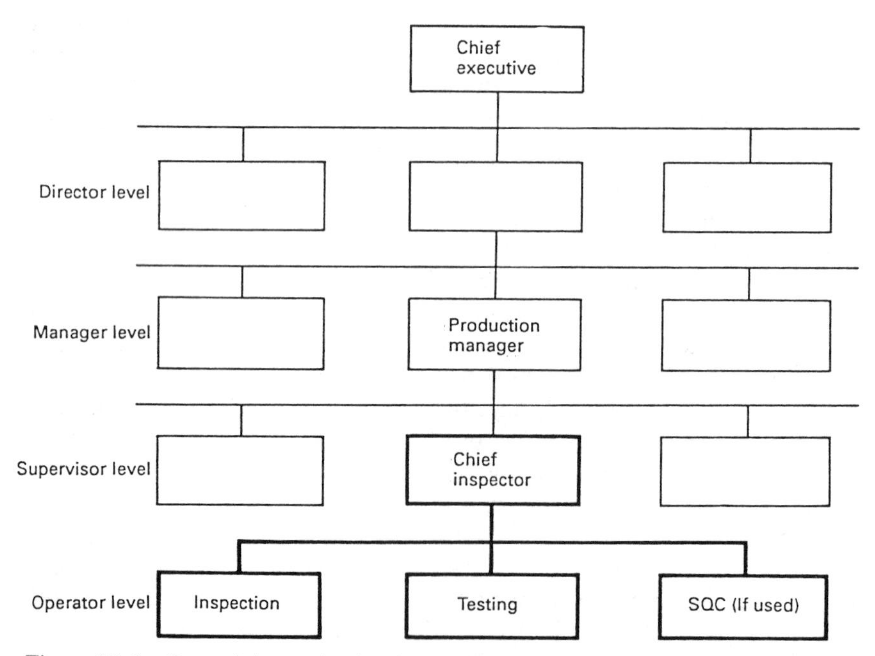

Figure 26.3 *Outmoded organization for quality control. It is unsatisfactory for quality management not to be identified on the management organization chart*

In large companies a matrix organization structure is sometimes adopted as a means of overcoming the inherent difficulties of hierarchic organization by combining the advantage of functional specialization with conventional departmental organization. Quality assurance is well suited to matrix organization because it does span the whole of management organization. A possible matrix organization embodying quality assurance structure is shown in Figure 26.4.

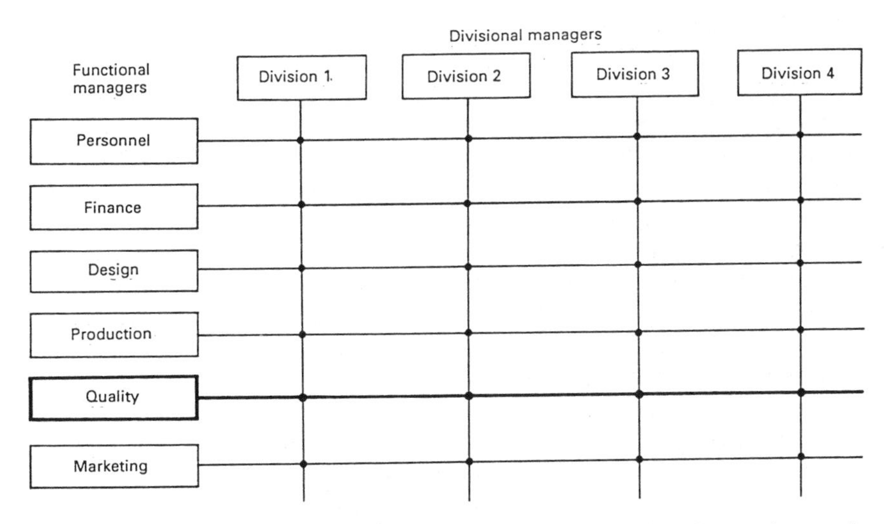

Figure 26.4 *Quality assurance in management matrix organization. Intersections on the grid imply that individuals or groups have a dual responsibility – one to the appropriate divisional manager, and one to the appropriate functional manager*

Turning now to the responsibility and the accountability of the quality assurance function, the duties of a quality manager can be enumerated as follows:

1 To manage the quality function in such a way that the company's quality objective of customer satisfaction is achieved.
2 To accomplish 1 with due regard to the control of quality costs.
3 To develop and maintain the company's quality system.
4 To encourage employee motivation by whatever means may be appropriate (e.g. quality circles).
5 To be responsible for issuing quality specifications.
6 To take charge of metrology.
7 To take charge of inspection and testing.
8 To take charge of quality engineering and to ensure the effective solution of outstanding quality problems.

9 To plan and implement the quality elements in the new-product development programme.
10 To act in an advisory capacity in the fields of standards, statistical methods, quality control and any other quality-related specialized field.
11 To organize quality training for personnel in any part of the company.
12 To monitor the company's quality performance and quality costs.
13 To report on the discharge of the above responsibilities.

There remains the question of the internal organization of the quality assurance function. Given that the quality manager is a key executive, what sort and size of department will he require to support him? Will he be expected to discharge all the duties assigned to him personally or will he be given a staff of subordinates and personnel to whom he can delegate?

The quality assurance personnel can be divided into two groups. There must be an operating group engaged on measuring, testing and inspecting, and their numbers will be dictated by the scale of production and the requirements of the product specification. There may also be a managerial/technical group which will include people engaged on solving quality problems, the clerical staff necessary for the maintenance of records and the administrative staff of the department.

The key to the size of the quality department should be that small is beautiful. The quality manager will operate under the same regime as other operational managers, and must be prepared at all times to justify the existence of his group. He will have an operating budget at his disposal, but he will have to give a strict account of expenditure. The quality department must be seen to pay its way in terms of productivity and customer satisfaction.

Finally, it must be recognized that the formal organization chart is only part of the organization story. Individuals in company management structures usually have dual roles – executive roles and representative roles, and these are often confused. The executive roles identify with the formal organization chart. The representative roles are sometimes catered for by creating a separate representative system, but when that is not done the representative roles still exist even if they are not formally recognized. Working across departmental boundaries so much of the time, the quality manager has to be very careful to recognize when people are acting in their executive role, and when in their representative role.

Directing the quality function

At a time when many industries stand desperately in need of a turn-around of their fortunes, the appointment of a quality manager to lead and direct quality operations is probably one of the most critical steps that can be taken by any individual manufacturer. To face international competition in world markets that are becoming ever more quality conscious, manufacturers have to match their foreign competitors with the best in quality management.

Management of the national economy is, of course, a matter for government. But Chatterton and Leonard (1979) are undoubtedly correct in saying that the cure for poor economic performance requires positive action at individual factory level. It is interesting to note that among the six prime ingredients of their prescription, four identify with good quality management. These are:

1 Leadership.
2 Lateral communication.
3 Representative project groups.
4 Job satisfaction.

Leadership

What sort of an individual is required to administer the medicine? What are the criteria for selecting a quality manager?

A certain minimum level of technical competence is required, but the personal qualities are of far greater importance. The point has been neatly put by Duerr (1971) in his pen-portraits of two types of professional managers. 'Mr Beta is the man with a business school background who, at the drop of a hat, will quote Herzberg, Maslow, McGregor, Likert, and tell you where Taylor and Urwick went wrong. Mr Able can read a balance sheet, assess a marketing campaign, has a shrewd idea of what he can get out of a computer, but above all he knows people and can handle them.'

Quality manager Able has to be, above all, a leader with special qualities of leadership. The task of running his own small department will be quite a minor one. The real leadership challenge will come in projecting the quality ethos into all the other departments where the principal quality tasks are carried out. That is where his interpersonal skills will be tested to the limit.

His style of leadership must be both authoritative and participative – the dichotomy between these two that has crept into social science thinking is a nonsense. His authority will stem from his professional commitment to quality and from the strength of will that he displays in not allowing the quality operation to be turned away from its objective. At the same time he must inspire participation in all quarters and at all levels, for he can accomplish nothing by himself – he is totally dependent on others for the success of the quality programme.

Finally, he must have the ability to do all of these things without getting up other people's noses. As Duerr says, he must know people and how to handle them.

Lateral communication

Quality assurance depends more than most other industrial operations on lateral communication. It is a simplistic view, but none the less true, that product quality begins 'at the top' in the boardroom. But if every quality problem had to be referred upwards to the chief executive most of them would never get solved and a lot of other company business would suffer. The prime quality responsibility of the board is to create a management structure within which the quality department can work with other departments across departmental boundaries over the whole company organization.

The collateral relationships between the quality department and other departments involve actions which operate in both directions. The quality department provides services to the other departments, but at the same time it has responsibilities for taking initiatives and promoting activities in these departments and for monitoring their activities. It follows that a two-way system of lateral communication is essential. The effectiveness of the communication system can be judged by the extent to which it promotes the acceptance of responsibility and commitment to quality at all levels in other departments.

The need to recognize the importance of informal as well as formal communication is of paramount importance. At any point in the system, be it in the design office, the production shop, or the market place, there will be some individual who knows more than anyone else about an issue that is of vital importance. Such information is not always captured effectively in a formal system of communication, yet it may be essential to the solution of a problem. To use a mining analogy, not all the coal-face problems can be solved on the surface back at headquarters.

Project groups

As an extension of the general principle of not involving the managing director in every minute quality detail, the quality manager and his opposite number in other departments should encourage their subordinates to use their initiative and to work together across departmental boundaries without referring everything to their departmental heads. To promote this, it may be useful to set up project groups to deal with important issues, with group members drawn from several departments and charged with the responsibility of working together to solve specific problems and then reporting back to management.

It is important that project groups should not become self-perpetuating oligarchies soaking away precious manpower for all time. Each project group should have temporary terms of reference and must be disbanded when its task is complete.

It should be noted that a project group, as defined above, is a formal (though temporary) extension of the management structure. As such, it must not be confused with a quality circle. The project group leader will usually be a manager, membership will be mandatory, the task will be delegated by management, and the project group will belong to the company. The essential characteristics of quality circles, as they are now developing in the UK, have been described by Robson (1982). Circle leaders are first-line supervisors, membership is voluntary, they address themselves to things that go wrong at the members' own workplace, and the members 'own' their circle.

Quality circles should never be seen as an alternative to a full-blooded quality management programme. To encourage quality circles to develop without a quality management framework is likely to promote a 'them and us' atmosphere – the very thing that quality circles are intended to avoid. To encourage quality circles to develop alongside good quality management can be beneficial to both.

Job satisfaction

Improved job satisfaction can come both from job enlargement and job enrichment, and good quality management can contribute to each of these. One of the basic steps in modern quality assurance is to enlarge the production operators' jobs by placing on them the responsibility for checking their own work. Job enrichment comes from every member of the workforce being given a more active part to play in a company-wide quality assurance programme. Quality is everybody's business. The routine production of a high-quality reliable product has to be seen in terms of human achievement, and there is no denying the satisfaction which that entails.

Certain industries such as the machine tool industry have always been repositories of craft skills of the highest order. Modern quality assurance practice does not render traditional skills obsolete, but turns them to better use. With good quality management it should not be difficult for individuals to perceive that their personal goals are in harmony with company quality objectives.

Controlling the quality function

The control of quality assurance, as distinct from the control of product quality, completes the circle of quality management. In simple terms, management control involves comparing the actual outcome of events with the planned outcome, in order that intervention can be made where necessary. Sometimes corrective action will be needed to drive the actual outcome closer to what was planned, sometimes it will be recognized that the plan was defective and should be modified. Management control is an essential continuing part of quality management, to ensure that the quality assurance function is maintained on course and does not get deflected away from its objective.

It will be the responsibility of the quality manager to design and implement a quality management control system as an extension of the quality planning function. The planning system will set the targets, and the control system will monitor progress towards the targets.

It is desirable that the quality information system needed for control should form part of an overall management information system. For this to be so, an input of quality records and production records will be necessary, for which the quality manager and the production manager must be responsible. Because quality is everybody's business, this data should be accessible to all departments through a centralized management information system. Moreover, quality costs are a prime necessity for the quality manager and he should be drawing his quality cost data from the same source as all other departments so that there will be no inconsistencies. It will, of course, be a responsibility of cost accounting to provide the input of cost data, but there must be close liaison between the accountant and the quality manager to ensure that the cost data will be in a format suitable for quality cost management.

It must be recognized that, of all the quality data available to managers for control purposes, cost data is inherently the most imprecise. Unlike the physical quality characteristics of the product which can be measured in absolute terms, quality costs (like all other costs) are abstract data based on certain conventions and assumptions. Whether these are suitable for the task in hand must be a matter for debate between the quality manager and the cost account. Some of the pitfalls of quality costing are dealt with in a paper by Daisley *et al.*, (1984).

Increasingly, computerized management database facilities are becoming available, even for quite small firms. This trend should be encouraged since the more sophisticated statistical analysis and reliability analyses that are needed by quality assurance are expensive to perform manually, but can be catered for readily in computer software.

Conclusion

The quality management model set out above is essentially a statement of the principles of good management applied to the quality function in manufacturing industry. It could equally well apply in a service industry.

The model is offered as a definitive model, but only for the purpose of discussion. It is not argued that this is the only acceptable model in all circumstances. It may well have to be adapted in particular cases.

If the model assists industrial management to think its way through its quality assurance problems, and to reach a satisfactory conclusion, it will have served its purpose.

What can be asserted without fear of refutation is that the lack of managerial identity accorded to the quality function in many organizations constitutes a serious hazard to the competitiveness of the product or service. **No industry can afford to take quality for granted.**

Further reading and references

Items marked with an asterisk (⋆) refer to the studies mentioned in the opening paragraph of this chapter.

⋆Advisory Council for Applied Research and Development (ACARD), *Facing International Competition*, HMSO, 1982.

British Standards Institution, *BSI Handbook 22: Quality Assurance*, 1981.

Burgess, J. A., *Design Assurance for Engineers and Managers*, Marcel Dekker, 1984.

Chatterton, A. and Leonard, R., *How to Avoid the British Disease*, Northgate, 1979.

Daisley, P. A., Plunkett, J. J. and Dale, B. G., 'Quality Costing in the UK', *Proc. World Quality Congress 1984*, vol. 1, Institute of Quality Assurance, 1985, pp. 557–567.

Deming, W. E., *Quality, Productivity and Competitive Position*, MIT-CAES, 1982.

Duerr, C., *Management Kinetics: Carl Duerr on Communication*, McGraw-Hill, 1971, p. 41.

⋆Egan, J., 'The Jaguar Obsession', *Proc. World Quality Congress 1984*, vol. 3, Institute of Quality Assurance, 1985, pp. 14–18.

Feigenbaum, A. V., 'A Way of Managing the Modern Company', *Proc. World Quality Congress 1984*, vol. 3, Institute of Quality Assurance, 1985, pp. 5–13.

Feigenbaum, A. V., *Total Quality Control*, 3rd edn, McGraw-Hill, 1985.

Hutchins, D., *Quality Circles*, Pitman, 1985.

Juran, J. M. Gryna, F. M. and Bingham, R. S., *Quality Control Handbook*, 3rd edn, McGraw-Hill, 1979.

Juran, J. M. and Gryna, F. M., *Quality Planning and Analysis*, McGraw-Hill, 1980.

Juran, J. M., 'Adventures in Improvement', *Proc. World Quality Congress 1984*, vol. 3, Institute of Quality Assurance, 1985, pp. 28–29.

⋆Lockyer, K. G., Oakland, J. S. and Duprey, C. H., 'Quality Control in British Manufacturing Industry', *Quality Assurance*, vol. 8, no. 2, 1982, pp. 39–44.

⋆McKeown, P. A., 'Quality Through Technology', *Quality Assurance*, vol. 5, no. 2, 1979, pp. 37–47.

⋆Morrison, S. J., 'Importance of Quality and Reliability', *IEE Proc. 128*, A, 7, 1981, pp. 520–4.

*Morrison, S. J., *Quality Assurance in the Machine Tool Industry*, University of Hull, Department of Engineering Design and Manufacture, 1984.

*Morrison, S. J., 'Quality Management', *Proc. Institution of Mechanical Engineers*, vol. 199, no. B3, 1985, pp. 153–9.

Oakland, J. S., *Statistical Process Control*, Heinemann, 1986.

Price F., *Right First Time*, Gower, 1984.

Plunkett, J. J. and Dale, B. G., 'Some Practicalities and Pitfalls of Quality-related Cost Collection', *Proc. Institution of Mechanical Engineers*, vol. 199, no. B1, 1985, pp. 29–33.

Richie, D., *Introducing Japan*, Kodansha International, 1978.

Robson, M., *Quality Circles, A Practical Guide*, Gower 1982.

In the UK, the Institute of Quality Assurance publishes the following periodicals: *QA News* (monthly) and *Quality Assurance* (quarterly).

In the USA, the American Society of Quality Control publishes the following periodicals: *Quality Progress* (monthly) and *Journal of Quality Technology* (quarterly).

Acknowledgements

This chapter is based upon (with substantial additions) the author's paper *Quality Management*, which was published in the *Proceedings of the Institution of Mechanical Engineers*, Part B, 1985. The relevant material is reproduced by permission of the Council of the Institution. Acknowledgement is also due to the Department of Engineering Design and Manufacture at the University of Hull for assistance in the preparation of the text.

27

Drawing office standards

Trevor C. Ashton

The situation of the drawing office is essentially between design and production although usually it belongs to the engineering function. There is a clear distinction between the objectives of the two departments but the difference in their working methods often results in the interfaces being complex, and difficult to understand by the uninitiated. Uniformity in the product of the drawing office is therefore important as it can be most effective in overcoming the language difficulties which sometimes occur between research/development and the shop floor.

Apart from the drawing office association with production and development engineers, there are continuous inputs and outputs to the purchasing function, stores, inspection and manufacturing departments. There is also a need for consultation with commercial departments such as sales, marketing, contracts and publicity. With such a wide variety of contacts communication must be direct and unambiguous. Drawings and specifications must be clear and in terms that all will understand no matter how remote from the source. In this respect the need for standardization is obvious. From international down to in-house levels standard terminology, abbreviations and symbols are essential. Chapter 28 refers to the organization and means for achieving this standardization. This chapter will describe how the standards can be used particularly by the drawing office. There was a time, which does not seem long ago, when one might also have included the need to observe a variety of customer standards. Thankfully this is now less frequent as most of the large companies and institutions have realized that contribution to the national standards effort is to the advantage of both manufacturer and user alike.

Drawing offices vary widely in their organization and structure but while in many ways dissimilar in one respect they have a common purpose, that is to convert design ideas into manufacturing instructions. They are professional communicators. The opportunity is also presented to be creative but discipline and organization must be exercised to control the extent of this creativity and avoid unnecessary complexity. There is also a danger that in the search for speed insufficient thought is given to planning and simplifying of designs or instructions.

The draughtsman may take much of the responsibility for the mechanical and aesthetic design of a product. He has the major influence on the choice of materials, components and finishes as well as the functional design which will have to be economically manufactured. In most instances the product design will be followed by the design of tooling and product planning for the use of machine tools or processes. If the wheel is not to be continually reinvented then the adoption of design and manufacturing standards is implicit. This requires the creation and availability of in-house standards, in the drawing office known universally as 'The DO Handbook'.

Variety reduction and standardization procedures

An awareness of the need for standardization is constantly impressed on daughtsmen. Minimum variety is sought for many reasons but particularly to maximize the use of stocked components and materials and permit the use of established processes to optimize manufacturing efficiency and improve quality control. Standardization guidelines must be spelt out for the following purposes.

Raw materials

There should be a catalogue of the materials available within the organization and additions to the range should be controlled.

Bought out components

As with raw materials, all bought out components should be catalogued and additions vetted before extending the range. Even with the introduction of computer systems into areas such as purchasing, stock control and accounts, each component carries an appreciable overhead.

Codes of practice

Often repeated procedures should be written down as codes of practice. Attention to creating and maintaining these should not be neglected. Drawing formats, limits and fits and standard plating or painting finishes usually come into this category.

Manufacturing standards

Components and assemblies must be designed within the limitations of the manufacturing facilities and skills available or to be made available. For example, the capability and capacity of manufacturing facilities should be recorded and the designer informed.

Coding

With increasing use of computer databases, coding systems should be designed with the collaboration and cooperation of potential users. Those who might be involved are:

- Marketing, for advertising if product type numbers are involved.
- Engineering development, if extensions to the range are necessary.
- Purchasing and stores, if the item is to be purchased or stocked.
- The accounts department, to assist in information retrieval and costing throughout the system.

Developing a drawing office handbook

The drawing office handbook should include guidance or instructions on all of the topics listed above.

Handbook format

There may be sufficient justification for company standards to be available in the drawing office in several forms. Figure 27.1 shows a typical set of codes of practice which have been created in an electromechanical and electronics manufacturing company. These comprise:

1 A drawing office handbook which contains information required by the draughtsman on a daily basis, such as number systems, standard hole sizes, tolerances and finishes.
2 Separate codes of practice covering various aspects of the company's operations for production, machine capacities, administration procedures etc. These are more likely to be required on an occasional basis.

The drawing office handbook should be issued in a loose leaf binder so that additions and amendments can be made on a regular basis. Documents should be printed on international A4 size paper as this is usually the most commonly available size for both binders and copiers.

Separate codes of practice in regular use may be made up by printing four pages per A4 sheet, folding in half and stapling to make a small A5 size booklet which can be issued separately. These are useful for codes of practice such as

Figure 27.1 *A collection of drawing office codes of practice*
(Courtesy of GEC Measurements Limited)

development procedure, capital procedure etc., where the distribution may be wider than just the drawing office.

Specifications in the drawing office handbook should be uniform, formal and official. They must not be subject to written amendments or scribbled additions. It is most important that when changes are made they are incorporated quickly by all users. They should be numbered so that issues can be recorded and the document indexed. They should carry a date of issue (which should be updated as amendments are made) and they should carry an authorizing signature. A preface sheet is placed in the front of the volume to record the amendments made. The drawing office handbook is usually issued one per person to ensure that draughtsmen are notified immediately of amendments and have constant access to reference data. The contents will vary depending on the nature of the industry. For example a company producing computers would have different routines and standards from one which designs and manufactures turbine generators.

Handbook content

Regardless of the details, specification of the following aspects is a common requirement of most engineering concerns:

- Administrative procedures.
- Materials and components.
- Processes.
- Tools and equipment.
- Drawing office practice.
- Revision control and records.
- Ad hoc procedures.

Administrative procedures

Administrative procedures include, typically, codes of practice for:

- Development procedures, which describe the stages for embarking on a new development.
- Accounting procedures, which set out methods for charging out and recovering costs, as well as for standardizing the raising of requisitions for capital expenditure.

Development procedures

A typical standardized procedure for planning development projects could be arranged on the following lines:

Definition of project type
Development projects can be classified as:

- Minor development.
- Fundamental investigations.
- Main development.

Minor developments are limited cost projects such as the appraisal of a proposed new project or tests on competitive products if this is feasible. Short-term small developments may come under this heading or evaluation of new ideas, components or techniques.

Fundamental investigations may be carried out to pursue studies in technological advances which could be of importance to the company's products. The significance of this category is that no detailed commercial justification is available at the time of initiation.

Main developments are significant and may be long-term projects which are carried through to create a new product or to improve an existing one. Once embarked upon these projects may have four stages:

Stage A Preliminary investigation to decide whether the project is worthy to proceed to the next stage.
Stage 1 Technical development to the point of producing bench tested models.
Stage 2 Design; the preparation of detailed drawings for manufacture, production and evaluation of prototypes and completion of detailed drawings.
Stage 3 Production development, which is the provision of all the necessary production facilities and the installation of a pilot production line (if appropriate).

Instructions should be available on the procedures to be followed for initiating a new project. These can cover the responsibilities of the commercial department and the production and development departments.

Specification of the documentation required and the authentications required at various stages should be included and the whole system may be charted on a critical path network. From the drawing office management point of view, as well as others, knowledge of the status and stages of the development work is essential in planning workload and knowing the deadlines for the various outputs of the office (which may include specifications, test instructions and maintenance manuals as well as manufacturing drawings).

Capital procedure

Another commonly used procedure, particularly in a large company, is the system for applying for funding of capital equipment. A standardized capital procedure is a valuable aid to efficient budgeting. It should define the limits (whether they are financial or length of life of the item) and the start and finish of the financial year. The contents may include events culminating in provision of a capital item, compiling the capital budget and raising the application for capital expenditure. There is ample scope for standardization in these areas. The benefits are to be seen in the ability to control the budget by strict definition of categories and in the uniform presentation of the applications. Both these are useful for comparison when trying to establish priorities.

Materials and components

Guidance is provided in an efficient drawing office on the use of preferred materials and components. Separate catalogues of company standards are produced for materials, bought out components, fasteners and common hardware such as pillars, spacers, labels and nameplates. The materials catalogue will also include selection guidelines.

Materials

A catalogue of the materials that are already being used in existing designs should be available to the designer. Many of the common engineering materials such as steel, aluminium, copper, strip, rod, bar, etc., can be specified by reference to British Standards, which in themselves are increasingly complying with international standards.

Use of these specifications directly is the most likely way of ensuring availability to a known standard of performance and quality. However there may be other factors which will have a strong influence on the material to be specified (for example, processing requirements such as special punching and bending properties). In these cases the national standard will require further qualification. It is also unfortunate that the policy of stockholders does not always keep pace with international standardization. It is therefore frequently necessary to modify requests for a particular specification in order to be able to use a superseded grade that happens to be held in stock. This type of activity is regular and requires a flexible specification system to accommodate the changes and retain full control in the design office.

The drawing office handbook must also provide guidance on the use of preferred materials, grades, sizes and tolerances and should standardize the method of specifying both on the drawing and on the purchase order. This will require collaboration between the designer who requires the mechanical performance properties and the production engineer who also has to consider processing properties such as punching, bending or machining.

The production engineer will also be involved in deciding the stock material size as this will depend on the production method, the orientation of parts made from strip, the standard collet sizes available or other practical requirements.

When these technical specifications have been decided the buyer can more confidently embark on his task of finding a suitable supplier.

Components

Bought out components

The need for a catalogue of bought-out components is similar to that for raw materials in that the use of an existing item can save the introduction of unnecessary variety and its resultant costs. Specification in the component area is frequently further complicated by the lack of national or international standardization to the same extent as it exists on raw materials. Recent advances in international and national standards for electronic components however is an indication of the work being done to achieve better standardization in this area. The more common components (including resistors and capacitors) are well catered for in the IEC, CECC and BS 9000 series of standards. Fasteners, particularly screws, nuts and washers, are likewise covered very thoroughly by national standards. The necessity in these areas is therefore for in-house variety control.

The draughtsman requires a list of preferred components that he can select from with experience of their performance and characteristics in the knowledge that there will be no problems of availability. Even with a British Standard specification available it is not sufficient to allow full choice from the complete range. ISO metric screw threads in BS 3643; Part 1 for instance lists thirty-four first choice nominal diameters from 1 mm to 300 mm. It is unlikely that one manufacturing company would need the whole range.

In the areas where national standards do not exist it is necessary to organize one's own in-house standards procedures. The means of doing this is detailed more thoroughly in the chapter on engineering standards (Chapter 28). It should be noted that the personnel involved with proprietary bought-out components are likely to be different from those influencing the selection of raw materials. The development engineer is more likely to be responsible for the initial selection and evaluation and will need to remain involved together with the draughtsman and the buyer.

Standard hardware

There are bound to be a number of components that are common to a range of products, items which are custom designed rather than bought out. Pillars, posts, spacers, escutcheons, mounting brackets and nameplates are just a few examples. If uncontrolled one can eventually discover that similar objects have been drawn and manufactured under different identities over and over again. This kind of repetition is extremely wasteful on resources right down the line. The extra cost is not confined to the production of a drawing – if it were then the precautions required to avoid it might prove uneconomical. The costs however usually extend to creating new tools, new production planning paperwork, costing, production scheduling and time studies, culminating in higher inventories, excess stock or unnecessary shortages.

Instructions or guidance must be available to prevent such repetition. These may be in the form of a classification and coding system. An alternative is to rely on a catalogue of parts, compiled on a continuous basis, and arranged so that parts with similar features are grouped together. Classification and coding systems are sometimes incorporated into drawing numbers for this purpose, but this has been proved to be time consuming and not sufficiently flexible in most production drawing offices.

The principal disadvantage of having common drawing/classification system based on physical characteristics is that a change to a feature of the part would change the code. In an organization making maximum use of standard parts this change would be reflected in all the assemblies using the part. Changes could therefore become very costly in drawing office time. This disadvantage can be overcome by giving component drawings an additional significant code describing its features. This code can be independent of the drawing number. A computer can then be used for compiling a register of standard parts, cross referenced between the drawing number and the code. With an appropriate program the file can then be interrogated to identify any items already drawn which incorporate the features required before embarking on creating a new drawing.

Processes

It is not usual for new product designs to have to start entirely from scratch, using the very newest materials and employing novel manufacturing techniques. Design is more commonly the application of available knowledge, skills and materials. The knowledge and skills need not be the designer's own. He relies on using the skills of the production engineers and the operatives and the capabilities of the processes already available in the factory before resorting to additional or special methods. Development in manufacturing techniques has been considerable over the last few years and looks likely to continue as the use of robotics increases. The designer must therefore keep note of the limits and the requirements of the processes that he quotes on drawings or specifications. Likewise the production engineer must take the responsibility for providing this information and data on the developments in manufacturing technology to the drawing office.

Surface coating

One of the most common industrial processes is surface coating. The standards of finish, whether painted or plated, can depend both on the nature of the product and on the conditions under which the product is to be stored or operated. A company standard can be beneficial in making sure that minimum standards are quoted and that the capability of these standards can be measured. The company standard may quote:

- The type of component.
- The material of the component.
- The climatic category.
- The recommended surface coating.
- The production process necessary to achieve the recommended coating.
- Related standards (national, international and company).

The standard may also specify cautions or precautions to be taken when using finishes in applications which are not the norm. Such instances may arise as a result of extraordinary environmental conditions, customer specification or even government regulations. So an awareness has to be maintained. If colour needs to be specified then reference should be made to BS 4800 to quote approved reference shades.

Other processes

Codes of practice may be compiled for all the common processes in the organization. In an electronics company this could mean wiring, assembly, printed circuits, and soldering. In other spheres the requirements will be different. For example, structural steelwork, welding, pressure vessels, spring design, pipework, metal castings and plastic mouldings may have varying codes of practice based on their own technology. Codes of practice are not only of use to the designer or draughtsman but are, as well as being production instructions, essential to the quality control process. They are the substance of quality audits, providing a record of the procedures in use. Changes to written procedures have the signal advantage of being able to be controlled in a formal manner, keeping everyone informed of technological or administrative developments.

Tools and equipment

Effective employment of machine tools and measurement equipment requires standardization at the design stage.

Drill sizes

The range of hole sizes may be restricted so that maximum utilization can be made of the drills or punches available. Standard drill sizes are specified in BS 328. It is an advantage if hole sizes are based around these preferred sizes even if a punch is to be used rather than a drill. This then allows flexibility of choice in the manufacturing method as well as minimizing the number of gauges required for checking purposes. The list of preferred sizes should iterate the various applications: for example, tapping drill sizes, clearance holes, rivet holes, counterbores and countersinks. Tolerances may also be standardized for the various materials and processes available.

Punch sizes

A company engaged on sheet metal work will be concerned in keeping and maintaining a minimum number of standard punches. These punch sizes, which may be round, oval, square or rectangular, should be listed and used to rationalize the range and avoid special tooling.

Equipment

Metrology equipment deserves as much consideration as manufacturing plant when trying to control costs. Close tolerances should be avoided and not specified unnecessarily. The management of design for economic production was the subject of a special study under the jurisdiction of the British Standards Institution and its findings have been published as PD 6470. It includes the nature of design costs as well as the costs incurred in accuracy and complexity.

Drawing practice

BS 308 *Engineering drawing practice*

Drawing is a universal language. It has been said '. . . a picture is worth a thousand words'. This is not to say however that the nature of pictures cannot be improved as a means of communication. Engineers have worked for many years in developing a universal language in the form of an engineering drawing. With the adoption of the metric system a gigantic step forward was taken in Europe. Since its introduction in the UK in the 1970s standardization has kept apace in the form of BS 308, which reflects the international work done in this field by the International Standards Organization (ISO). Any drawings made based on these standards are understood and can be used around the world. The first edition of BS 308 was produced by the British Engineering Standards Association (BESA) on an initiative from the War Office in 1924. It claims to be the earliest surviving British Standard specification not concerned with a product or material. BS 308 is the standard for engineering drawings in the manufacturing industries associated with mechanical, electrical, electronic, hydraulic, pneumatics and nuclear engineering disciplines. The requirements of the building, architectural, civil and structural engineering and construction sources industries are covered by BS 1192. BS 308 does not cover special disciplines and there are some standards published using BS 308 as a basis with additional recommendations and particular conventions: for example BS 4301 for optical elements and BS 2774 for laboratory glassware. Engineering diagrams for the same industries using BS 308 are catered for in BS 5070.

BS 308 is published in three parts:

Part 1 *Recommendations for general principles.*
Part 2 *Recommendations for dimensioning and tolerances of size.*
Part 3 *Geometrical tolerancing.*

The latest edition of the above standard is recommended as the basis for drawings in any of the appropriate industries. In it will be found recommendations for all the basic drawing requirements. The contents are:

Part 1 *General principles*

- Drawing sheets: ISO A series drawing sheets, standards for borders and frames, grid systems, centering marks, print folding marks, scale bar and title block.
- Scales: this part specifies how a scale be quoted and gives recommended enlargement and reduction scales.
- Lines: line thickness, density and types of line.
- Lettering: style and character height.
- Projection: orthographic, first and third angle.
- Views: number and choice of views, partial views, auxiliary and enlarged views, representation of symmetrical parts, repetitive features.
- Sections: hatching, thin material, longitudinal, half, local, removed and successive.
- Item references.
- Symbols and abbreviations.
- Conventional representations: adjacent parts, developed views, plane faces on cylindrical parts, knurling, splines and serrations, screw threads, bearings, gears and springs.

Part 2 *Dimensioning and tolerancing*
This part deals with the application of dimensions and tolerances and other related information in size and finish. It covers interpretations and tolerancing principles and the indication of surface texture on drawings.

Part 3 *Geometrical tolerancing*
Geometrical tolerancing includes an explanation of the symbols for geometrical form and positional error. The system is international in that the symbology avoids the use of written explanations.

PP 7308
BS 308 is accompanied by PP 7308, which is a publication specially conceived for educational use.

Drawing sheets

Sizes
Two sheet formats are recommended. These are landscape, intended to be viewed with the longest side horizontal, or portrait, to be viewed with the longest side vertical. Other than for specialized applications it should only be

necessary to specify one of these formats for each size. Preferred sheet sizes are those in the international A series, as specified in BS 3429 *Sizes of drawing sheets*. A table of these sizes is given in Chapter 13 (Figure 13.1).

BS 3429 also includes special or exceptionally elongated sizes based on multiples of the standard A series. It is usual to have drawing sheets preprinted including a frame to enclose the drawing as shown in Figure 27.2. The border may be used to include grid references, centring marks, print folding marks, scale bar, symbols indicating method of projection and units of measurement.

Title block

The title block should preferably be in the lower right-hand corner of the drawing frame. With a progressive tendency towards single part drawing and the economics of plain paper copying in the drawing office the A4 size (210 mm × 297 mm) is being increasingly used. A4 also has the advantage of being able to be incorporated directly into publications. For these reasons it may be considered more functional to rearrange the title block on this size so that it is on the top of the sheet (see Figure 27.3). Note also that the grid references may be considered unnecessary on the A3 and A4 sizes. Because many drawing offices have the facility to reduce A3 (297 mm × 420 mm) originals to A4 prints, one may use two formats for A3: landscape (with the title block in the bottom right-hand corner) for actual size printing and portrait (with the title block at the top) for reducing from A3 to A4.

The title block should be designed and preprinted to include space for the following:

- Name of the company.
- Copyright clause of symbol. This must be accompanied by the company name and the date.
- Drawing number.
- Date of drawing.
- Revision issue information block.
- Signature block.
- Original scale.
- Title of drawing.
- Sheet number.
- The number of sheets or the next sheet number. (Using the next sheet number avoids having to reissue all the sheets to update the number of sheets shown when a sheet is added or deleted from the set.)

Lettering sizes used in the border and title block must not be less than the sizes required by BS 5536 *Preparation of technical drawings for microfilming*. The drawing number may be repeated in the border on an opposite edge of

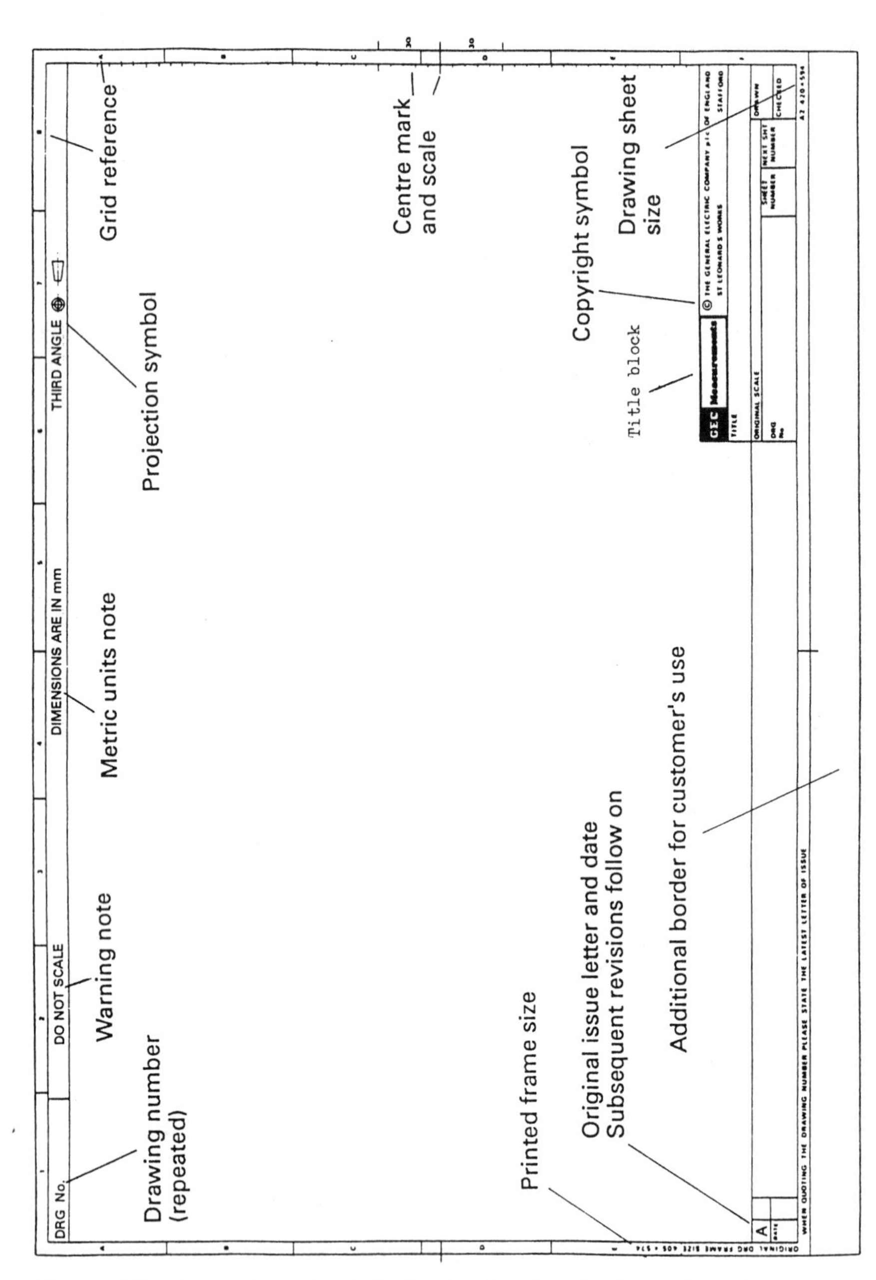

Figure 27.2 *A typical preprinted A2 size drawing sheet (landscape format)* (Courtesy of GEC Measurements Limited)

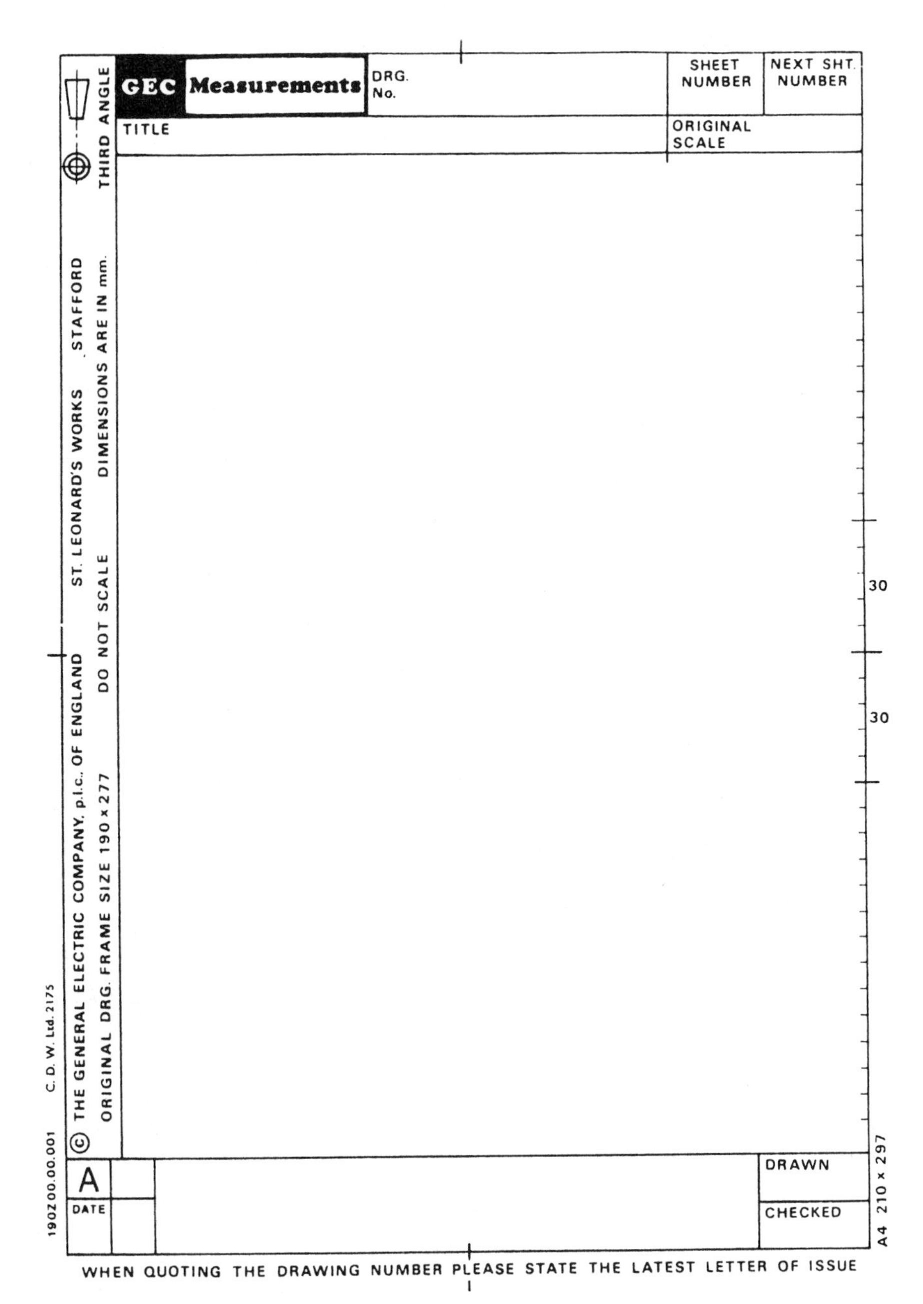

Figure 27.3 *A typical A4 sized drawing sheet in portrait format*
(Courtesy of GEC Measurements Limited)

larger drawings to help in identification of folded prints and in the retrieval or drawings from planfiles.

Sheet materials

When choosing materials to be used for drawing sheets, it is necessary to consider the medium, the methods available for reproduction and storage needs. These three aspects have all been extended in recent years by the development and use of computer systems for drawing, the introduction of plain paper copying as well as dyeline into the print room, and the increased use of microfilming for storage and backup.

There are basically two types of material: paper and plastic. Modern plastic drawing materials with etched and treated surfaces can be used to draw with ink and pencil manually or by machine plotter. Their translucency also makes them preferred for speedy dyeline printing and they are much more durable both for amending and handling. Plastic is however up to ten times more expensive than paper, which therefore remains a common material in drawing offices. The lack of durability of paper is offset by the ability in the modern office to produce a replacement 'original', either by replotting from the computer (if produced by computer aided design) or by printing onto plastic film from the microfilm (if this method has been used to back up drawings).

Drafting paper

Paper base materials include natural tracing paper, detail paper and cartridge paper. Natural tracing paper is translucent and as well as being useful for tracing through is capable of being printed at faster speeds by the dyeline process. Its disadvantages are its hardness (which causes pencil drawings to smudge) and brittleness (with a tendency to crack when folded, particularly when aged). It is a practical drawing material if used with ink, stored flat and handled infrequently. Tracing paper (Gateway) is available in weights of 60 to 112 gsm (grams per square metre). Detail paper is the most commonly used drawing medium either as preprinted drawing sheets or in plain rolls. It is a light material in the 50 to 65 gsm range and suitable for pencil and ink. Detail paper can also be used when dyeline printing is required. Cartridge paper is very heavy (up to 160 gsm) and is used for artwork which does not need to be reproduced by dyeline printing (as it is too opaque). It is not usually used for technical drawings which have to be reproduced and updated but may be required if the drawing has to be coloured by hand.

Plastic film

Plastic drafting films are polyester based, available in thicknesses of 0.002, 0.003, 0.005 or 0.007 in. They are naturally translucent and can be bought with single or double matt surfaces. Ink or pencil may be used for drafting but the pencil should preferably be one of the special plastic lead types for non smudging. Problems likely to be encountered with inferior grades are 'ghosting' when erasing, lack of ink adherance and cracking if folded. Polyester drafting film is a very durable material and can be used both for tracing and drawing. Its translucency is obviously a benefit to high speed diazo printing. Its cost may prohibit its use exclusively in the drawing office but its durability is essential to drawings which are subject to constant revision or frequent handling.

Computer aided design plotter materials
Both paper and polyester films are now available for use on computer aided design (CAD) plotters. The same arguments are applicable for using either paper or polyester, cost is again the critical factor against using film. The choice of writing medium includes felt-tip pens and roller ball as well as ink pens, and they are available as refillable or disposable. If choosing refillable pens then extra care must be taken with maintenance and cleaning. Roller ball pens may be used for check plots but are not recommended for finished drawings because of the lack of line definition. Felt tips also have a tendency to lose line thickness with use and may have inadequate density for diazo printing.

Revision procedures

Drawing records and issue control

Specifications and drawings are a vital part of quality control. There is usually however a lack of attention to their maintenance in storage areas outside of the originating office. It is almost certain that initial quality audits in a department will reveal redundant or superseded paperwork in these areas. Such a situation detracts from the efficiency of the department and the ability of the company to control quality. Information going into a working section requires very diligent handling by that department to remove superseded documents when adding new. This effort can be reduced by better control of the input, that is by making sure that vital information is coming from an official source. The practical way of doing this is to have all changes to manufacturing drawings or specifications originate from the drawing office who have the resources and the traditional skills to do this effectively. It involves proper identification of instructions and the setting up of detailed recording systems. These consist of an index of documents and a means of recording the date of issue and revision.

A drawing system should record the following:

- Drawing number.
- Title.
- Name of originator.

- Date of original issue.
- Superseding (if replacing another drawing).
- Superseded by (if replaced by another drawing).
- Total number of sheets.

The above information is general and relates to all sheets on a drawing. To supplement this an additional record should be created for each drawing sheet. A common record for a multisheet drawing is only permissible if all sheets are to be re-issued together when a change is made.

Drawing sheet records should list the following:

- Drawing number.
- Sheet number.
- Number of associated sheets on same drawing number.
- Size of original (and size of prints if different).
- Title.
- Original issue date.
- Revision issue date.
- Change notice number.
- Distribution.
- Archived date (when withdrawn).

Such records are necessary for traceability in the event of changes being required and are retained indefinitely.

Change procedures

Two change procedures are required. The first is for recording permanent changes to drawings or specifications. The second is for temporary changes or departures from instructions which result from day to day production problems. The following account is directed specifically to a drawing office handbook: for a fuller and more general treatment of this subject please refer to Chapter 25.

Permanent changes

All changes must be notified in writing and properly authorized. The drawing office handbook should include a list of persons authorized to approve changes. Change notices are usually issued in advance of the revised drawing being available so should be written in the appropriate tense with this in mind (e.g. 'Dimension 3.5 mm should be 3.4 mm' rather than 'Dimension 3.4 mm was 3.5 mm'). Different categories may be required, for example, warning, hold or change. The notice should indicate the nature of the change and the extent of the action to be taken. It should indicate whether it affects stock, work in progress or future work only. To make such notices official requires documents to be numbered and dated including the revision date. Handwritten corrections on drawings and specifications are not acceptable, so any change notice should be accompanied by a reprint and redistribution of affected documents.

A standard form is usually produced for change notices to ensure that correct information is included and to aid recognition on the shop floor. It is an official document in the same way as a drawing, so is subject to the same record requirements. It should be identified by a reference number (e.g. CDN 296) and carry a date and authorizing signature.

Temporary departures from drawings and specifications

From time to time provisional changes may be necessary. These could be as a result of temporarily not being able to obtain the specified material, or a component accidentally manufactured outside of tolerance which can be compensated for by changing the mating part. In such instances a temporary authorization to change is required. They do not justify a permanent change to drawings but nevertheless must be subject to similar authorization procedures by the design and production departments. A standard proforma is also required to keep on record the departure from specification and the extent of its approval. Temporary departures must be confined to the work in hand and should not be allowed to apply to all subsequent work without resorting to a permanent change. A date of expiry included on the form is useful to avoid this happening.

Ad hoc procedures

A section should be reserved at the rear of the handbook to include additions which may be required immediately. These may take the form of locally issued memoranda on drawing office practice issued on the sole authority of the chief draughtsman. They may comprise completely new procedures or temporary amendments to existing ones. It is necessary to have this arrangement in order to maintain the integrity of the handbook, which should only be altered by the democratic process of consultation and authorization, while still allowing a facility for communicating temporary departures.

Drawing for special requirements

Microfilming

The small space occupied by microfilm has the advantage of saving filing space and reducing costs of postage particularly overseas. Printing costs also can be reduced by using the microfilm copy to make prints which are less than the original size. This can frequently be an advantage to the user who may find the smaller print easier to handle in a restricted workspace. Approximately 40,000 microfilms individually mounted in aperture cards can be stored in a multidrawer filing cabinet occupying less than a square metre of floor space.

Microfilming is dealt with at some length in Chapter 13, but it is relevant here to discuss drawing office standards relating to the preparation of drawings that are to be filmed. Standardization of original drawings is a prerequisite to obtaining good results from microfilming. This starts with the sizes of drawing sheets and extends into the details of the drawing such as minimum line widths and text sizes. Current drafting practice to BS 308 in general caters for the needs of microfilming. Additional requirements are covered in BS 5536 *Preparation of technical drawings and diagrams for microfilming*, which recommends minimum line thickness of 0.3 mm and minimum character heights of 3.5 mm for A0 and 2.5 mm for A1 down to A4.

Materials

Normal drafting materials are suitable if they have a matt surface and adequate contrast between drawing matter and background. If translucent materials have been used then it is possible to improve the quality of the microfilm by backlighting the original during filming. Backlighting is also beneficial where printing has been applied to the back of the original as well as the front or where the drawing has been creased. Polymer type pencil leads are superior to graphite in providing a non-reflective surface.

Drawing sizes

Drawing sizes must be to the national standards mentioned earlier in this chapter to avoid complications at a later date when trying to reproduce from the microfilm on to standard print paper. Drawing sheets larger than A0 should be avoided because of the limitations of microfilming equipment. A centring mark in the form of a line on each of the four sides of the drawing is essential to position the original drawing correctly in the frame of the film.

Scale of orientation

The camera reduction ratios at which originals are photographed are important to the subsequent user of the microfilm. They are restricted to the ratios 1:30, 1:21.2, and 1:15. The tolerances in BS 4210: Part 1 are arranged to permit a smaller than standard image and prohibit a larger than standard image. if the choice of reduction ratios for filming from A sizes of drawings is restricted to those shown in Figure 27.4, then the microfilm will project to the maximum viewing size using a 15× lens as standard. Thus A4, A3 and A2 sized drawings will project back to their actual sizes. A1 and A0 drawings will be projected at A2 size. A2 is the usual screen size of a standard viewer which has a limited number of lenses.

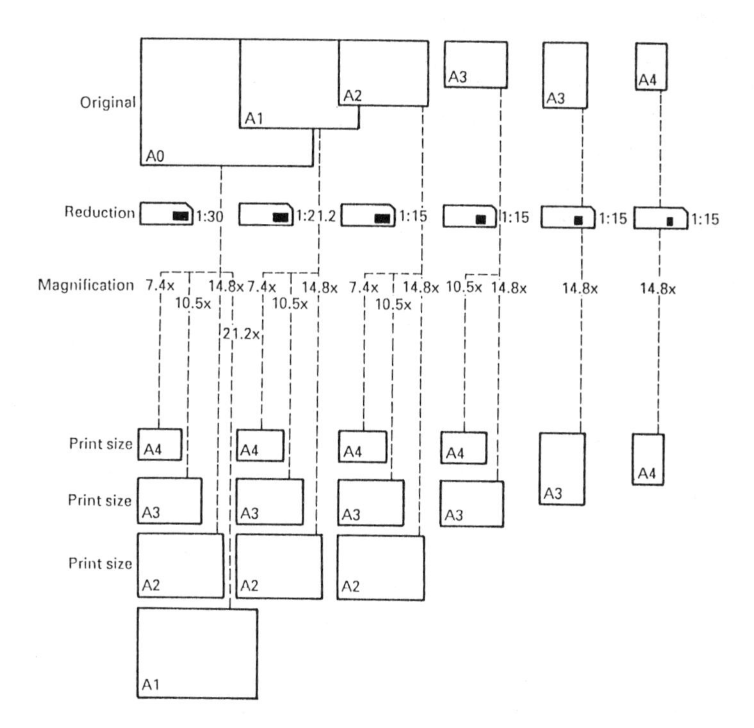

Figure 27.4 *Standard reduction and enlargement ratios for microfilming drawings*

For similar reasons the orientation should also be standardized so that drawings are right reading when projected on a viewer. It is usual to film originals singly so that they can be updated individually. However where independent alteration is unnecessary more than one image can be microfilmed on each frame. The position of originals in this instance should still meet the requirements for right reading and orientation. Standard positions for multiple imaging can be found in BS 4210: Part 1.

Since the size of the printed copy from the microfilm will almost certainly be different from the size of the original reference to the scale used must be on the drawing. A numbered scale may also be included in the border to indicate the original drawing size.

Computer aided design

Computer aided design (CAD) is part of the wider concept of computer aided manufacture (CAM) or as it is now more widely referred to, CIM (computer integrated manufacture). CAD systems can be categorized into those designed for specific design areas (such as printed circuit design) and more general packages for mechanical drafting. The mechanical drafting systems can also be classified as powerful three dimensional drawing systems capable of stress calculations etc., or more economic two dimensional drafting systems operating on PC compatibles. As the cost to power ratio decreases there is a large growth in their use in drawing offices.

Apart from the automation of the drafting process, benefits can be obtained by creating a library of standard symbols or commonly used parts. This speeds up the dafting process considerably. The quality of the drawing original can also be improved by choosing the correct plotter. This should result in controlled line thickness and letter heights and the use of standard founts.

There is initially a considerable amount of work to be done in creating a standards library for CAD applications. Once established, however, standardization is implicit in CAD. This applies to the drafting standards which control the quality of the final drawings and to the design checks that can be incorporated if the parts put into the standards library are controlled.

Simplified drafting

While an engineering drawing must accurately specify the manufacturing and design requirements, the presentation of data to the operator making or assembling the parts should be considered carefully. Complex assemblies particularly can present a frightening and unhelpful picture. Various means have been used to simplify the drafting process and clarify the drawing. Among the methods used are photography or three dimensional drawing. Freehand drawing has also been attempted to reduce the drawing time but without much success. Simplified drafting is an alternative method of drawing which should benefit both the drafter and the user of the drawing. The aim is to avoid unnecessary or misleading detail and therefore make the drawing easier to read as well as to produce.

In its basic form, simplified drawing involves using single lines and simplification rather than accurately depicting complex detail. An example is shown in Figure 27.5. Note that, in this example, all profiles have been drawn with single lines, and the positions of screws are shown by centre lines only. Radii have been replaced by sharp corners. The purpose of this drawing example is for assembly, and the simplified drawing shows unambiguously the position and orientation of the parts. Item references have been simplified by removing the balloon except for subassembly references. Compared with the cost of producing a photograph or the time to draw full detail, this is a cost effective drafting method. Associated with the trend separating the bill of materials from the pictorial aspects of the drawing, this type of presentation lends itself to computerized drafting or desk top publishing at minimum cost. Other standard simplified presentation methods are given in BS 308: Part 1.

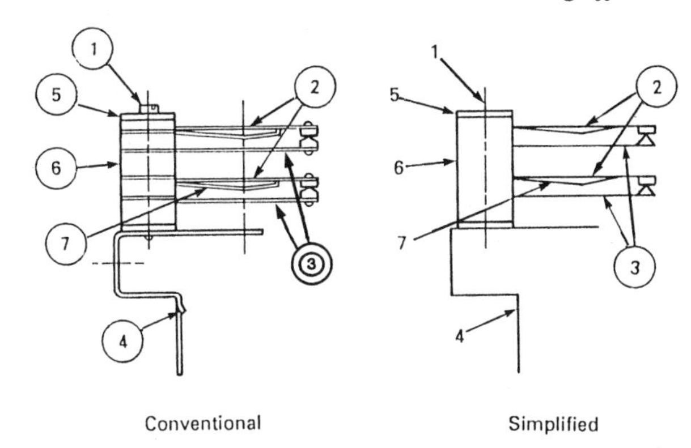

Figure 27.5 *Contact set assembly: comparison of conventional and simplified drafting methods*
(Courtesy of GEC Measurements Limited)

Standards referred to in this chapter

BS 308: Part 1 *Engineering drawing practice, general principles.*
Part 2 *Engineering drawing practice, dimensioning and tolerancing.*
Part 3 *Engineering drawing practice, geometrical tolerancing.*
BS 328: Part 1 *Drills and reamers, specification for twist drills.*
BS 1192 *Construction drawing practice.*
BS 2774 *Specification for drawing conventions for laboratory glass apparatus.*
BS 3429 *Specification for sizes of drawing sheets.*
BS 3643: Part 1 *ISO Metric screw threads, principles and basic data.*
BS 4210: Part 1 *Specification for 35 mm microcopying of engineering drawings, operating procedures.*
BS 4301 *Recommendations for preparation of drawings for optical elements.*
BS 4800 *Specification for paint colours for building purposes.*
BS 5070 *Drawing practice for engineering diagrams.*
BS 5536 *Specification for preparation of technical drawings and diagrams for microfilming.*
PD 6470 *The management of design for economic production.*
PP 7308 *Engineering drawing practice for schools and colleges.*

All the above publications are available from the British Standards Institution.

28

Engineering standards

Trevor C. Ashton

Experience has accumulated within many organizations, gained over many years through investment in research or simply by involvement. The extent of such common knowledge goes virtually unrecognized, so that it is frequently undervalued and underutilized. A wealth of material can be made available by tapping these resources. Many wide-ranging definitions have been coined for the term 'standardization' but the concept of standardization being used to make more use of existing knowledge and experience is one of its more important aspects. It is also one that is, in principle, easy to apply. The manager who is able to recognize and use most effectively all the available assets of his company avoids the wasteful process of repeatedly attempting to solve the same problems.

History and origins of the British Standards Institution

The need for engineering standards on a national scale was recognized in the UK with the publication of the first British Standards in 1901 by the Engineering Standards Committee. Understandably it was the development of transport and the industrial revolution that prompted the formation of the British Engineering Standards Association (BESA) which subsequently became the British Standards Institution (BSI). It is a tribute to the work done at that time, and to the subsequent efficiency of the Institution, that standards of that era are still traceable and have formed the basis of current standards. For example, BS 1: 1920, although obsolete, is now listed in the catalogue as having been replaced by BS 4, last revised in 1980. Also, interestingly enough, the major motivation behind modern standardization is still communications and the developments of technology. A strong answer to those critics who accuse standardization of being inhibitive to development.

Up to 1930 BESA work was almost entirely confined to engineering but it was the decision to integrate chemical standards into the programme that gave rise to changing the Association's name to the British Standards Institution.

The Institution has since grown to become supported by over 20 000 subscribers, a grant-in-aid from the Government and sales which produce a total income of over £30 million. This revenue goes towards maintaining over 10 000 current standards and supporting over 9000 standards projects covering all spheres of engineering, chemicals, building and multitechnics.

The British Standards Institution is now divided into four groups: Standards, Quality Assurance, Testing and Technical Help to Exporters.

Standards

The British Standards Institution is recognized as the UK authority for the preparation and promulgation of national standards and this was further strengthened in 1982 by the signing of the memorandum of understanding between the UK Government and the British Standards Institution which recorded 'their joint commitment to enhance, strengthen and maintain the national standards system . . .', '. . . and confirm the status of British Standards as agreed national technical criteria . . .'.

The importance of this memorandum was in the extra support and commitment offered to the national cause by the public sector, which had hitherto gone its own way and sometimes enjoyed a credibility often in direct competition with British Standards. This applied particularly to the armed services and Ministry of Defence specifications. Since the agreement British Standards are receiving more support and they are progressively being developed and adopted to supersede existing public sector standards. This collaboration has strengthened the national cause as well as having contributed to a reduction in variety which will reduce costs.

The Standards division has the responsibility for laying down requirements in the form of standard specifications. Under a committee structure they rely on the cooperation of all those with substantial interest in particular topics, wherever possible through the views of an industry, sector, trade or other interest. The normal means of reaching decisions is by concensus. This implies that objections have either been met or withdrawn before a standard is issued. In BS 0: Part 2 can be found a more comprehensive explanation of the organization and functions of the British Standards Institution. It outlines the procedures governing the preparation of British Standards and UK involvement in international standards.

Quality assurance

The importance of quality and its kinship with standards gained further recognition in the mid-1970s. The newly appointed Director, Technical's previous association with the Ministry of Defence Procurement Executive had considerable impact in bringing the experience and philosophy of the Ministry of Defence standards into national standards (particularly electronic components). During the 1970s there was a proliferation of quality assurance standards by various purchasing and third party organizations. A manufacturer could find himself spending a disproportionate time in demonstrating his ability to meet these varying specifications. Past president of the British Standards Institution, Sir Frederick Warner, pointed out the need for a British Standard that would rationalize these requirements in his report *Standards and Specifications in the Engineering Industries*, published by the National Economic Development Office (NEDO) in 1977. Since that time a series of British Standards has been reproduced in a set as *BSI Handbook 22*. The handbook contains all the published standards for quality assurance, including BS 5750 *Quality systems*, and is recommended to directors and senior managers, of purchasing and manufacturing industries as well as being a useful reference for educational establishments.

The British Standards Institution operates two certification marking systems. The kitemark used for certification since 1903, and the safety mark introduced in 1974 to demonstrate a compliance with a British Standard specifically related to safety (see Figure 28.1). The Certification and Assessment Service designs and operates schemes which incorporate the registration of firms of assessed capability, including the registration of individual firms (QUASAR). This department also operates the BS 9000 system for electronic components and undertakes work for other certification bodies including the Canadian Standards Association.

Figure 28.1 *BSI certification marks*

Testing

The test house provides a wide range of testing facilities embracing electrotechnical, mechanical, chemical, building and telecommunications equipment. The work is concerned with factory inspection and assessments; these services being generally available to industry.

Technical Help to Exporters (THE)

This service has been set up to provide technical information and assistance to all sectors of industry engaged in exporting. THE can help with anything from an answer to a question such as 'What is the domestic wiring colour code in Chile?' to more technical problems, such as obtaining test certificates for exports. The service also covers the identification and sale of foreign standards.

National standards in countries other than the UK

Standardization is prevalent in the majority of countries, with some developed countries such as Germany having very refined and influential standards which are not only widely adopted in their own country but also throughout the world. The USA has a plethora of standards, standards organizations and regulations. There is evidence that such standards and regulations can result in barriers to trade and the unwary can run into considerable obstructions and delays through insufficient attention to this area. It is obviously important that a designer or manufacturer be aware of the relevant standards when producing for export or competing with imports. He would be well advised to make use of the British Standards Institution and its THE division (see above) when considering a venture into export markets or competing with foreign imports.

International standards

There are two bodies responsible for international standards. Electrical standards can be traced back to the formation of the International Electrotechnical Commission (IEC) in 1906. The non-electrical standards are the concern of the International Standards Organization (ISO) which was formed much later from the International Standards Association (ISA) in 1947. Both have their central office in Geneva. For any country to participate it is necessary for that country to form a 'National Committee' which is representative of the various interests such as manufacturers, users, supply authorities, research and development authorities and government. The task of drafting new standards is undertaken by technical committees, subcommittees or working groups. Once the draft is prepared, an established and formal procedure is laid down for its approval and adoption as an international standard. In spite of steps taken to reduce the time, it takes some years to produce an international standard from concept to publication.

Progressively there is a far greater commitment to international standards by the national standards bodies. It is now usual for a standard to be created internationally by concensus before submitting it as a national standard. If accepted at international level, members are obliged to adopt the standard nationally. This has generally resulted in better and more uniform standards. Many British Standards are now issued with a dual number indicating that an international standard has been adopted: for example BS 2782: Part 3:

Method 335A: 1978 ≡ ISO 178. (≡ means an identical standard). Other such internationally equivalent standards can be identified from the BSI catalogue.

Outside the fields covered by ISO/IEC there are other international bodies concerned with standards. Most of these are intergovernmental organizations and many are specialized agencies of the United Nations (e.g. the International Organization for Legal Metrology OIML). Between the international scene and the national bodies there are also regional standards bodies such as the European Committee for Standardization (CEN) and the European Committee for Electrotechnical Standardization (CENELEC) which are West European. Separate from CEN and CENELEC, and bringing together the electrical approvals authorities of a number of West European countries including the UK, is the International Commission for the Approval of Electrical Equipment (CEE).

Company use of national and international standards

Structure

Standardization can be envisaged as a three-tier structure, with international standards being the top tier (see Figure 28.2). The company is the base of the pyramid representing the grass roots 'user'. The lowest level (company standards) is where the higher standards should be put into action. It is here that the guidance and rules are needed for design and manufacture. A company standard should not be created without reference to existing national standards. This is a policy which will also be adopted at national level. Each level should look above itself for guidance. The higher the level of standardization, the more likely it is to produce rationalized procedures and avoid barriers to trade.

Application

The International Federation for the Application of Standards (IFAN) is an association of representatives of the national organizations concerned with the application of standards. Associations such as the British Standards Society (BSS) and the American Standards Engineering Society (SES) have sister organizations all over the world. They have a strong interest in helping companies make the best use of international standards. Recognizing the importance, a working group was set up and worked for ten years studying the implementation of international standards at company level. Their findings are published in PD 6515: 1986; *IFAN-Guide 2*. Their recommendations were six fold:

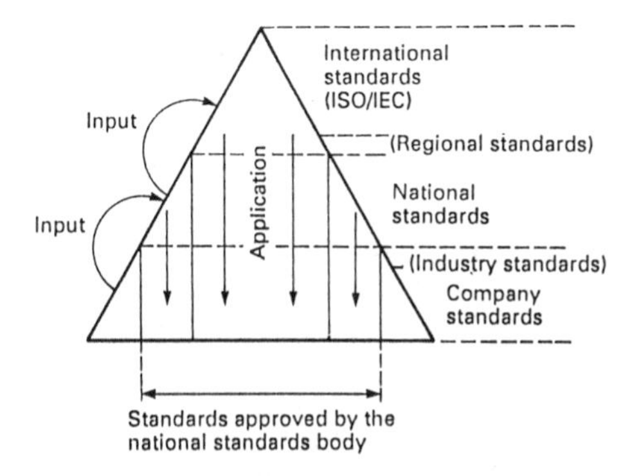

Note: The diagram can be expanded or simplified to give varying national or industrial conditions. In some areas regional standards are important. In some industries standards are issued by trade associations('industry standards').

Figure 28.2 *The use of international standards*
(Courtesy of British Standards Institution)

1 It should be noted by the international and national standards bodies that companies prefer to make indirect use of ISO and IEC standards through the medium of national standards.
2 There is a great need for more international standards to be published by industrially influential countries as identical or at least technically equivalent national standards.
3 Countries voting for international standards should adopt them nationally.
4 The exact relationship between a national standard and its international counterpart should be clearly stated and a system of symbols of equivalence should be agreed internationally. Where deviations from the international standard are made they should be clearly indicated.
5 Greater alignment, preferably complete, is needed between IEC and ISO in the production and presentation of standards.
6 In international standards the stating of alternatives or the provision of different grades should be adopted rather than a compromise which satisfies no one.

Considerable progress has been made by international and national standards bodies in complying with the above recommendations. This is a direct result of the work done at the grass roots by members of the Standards Engineers Sections all over the world who continually contribute to such work. It is encouraging to them that cooperative effort through standardization is demonstrably effective.

Standards are unlikely to satisfy all the needs of technological development. They may be hampered by competition and, by their nature, have to wait for stabilization before being written. However, at the earliest possible stage standards should lead in essential governing aspects such as interchangeability, reliability and safety.

Export and import

International agreement on standards has widened the markets of the world and hence has created opportunities for the manufacturer, but in so doing has of course widened the competition. To take advantage needs not only efficiency in production, but also an awareness of the purchasing opportunities for importing parts or raw materials that conform to international standards.

Contractual standards and regulations

There are several government regulations which are worthy of note. First, the Health and Safety at Work Act 1974 and subsequent statutory instruments such as the Safety Signs Regulations 1980. Also, the Consumer Protection Act 1987 covers virtually all consumer goods and services rather than the more restricted scope of the safety regulations. It permits the courts to take cognizance of 'any standard produced by any person'. There is therefore an onus on the British Standards Institution, trading standards officers and manufacturers alike to cooperate in producing common standards.

Depending on the industry there will be a number of other regulations which must be observed. They may be DEF specifications if dealing with the Ministry of Defence, MIL standards for American customers, European directives if you export to the Continent, health regulations if dealing with food, etc. A good example in the electrical contracting field is the IEE Wiring Regulations for domestic electrical installations. The list is extensive. Fundamentally many can be classified as customer standards. Obviously the first requirement is to be aware of the appropriate regulations governing a sphere of operation. This may be done via the relevant trade association or help can be sought from one of the various sections of the British Standards Institution. Having identified the requirements it is then necessary to decide how to store them. In a one man business this may be possible by opening a file and popping everything in. The larger the business, however, the more difficult the problem becomes, not only in storing the documents but also in two other respects. These are communicating the requirements to all in the organization who are concerned with their implementation and keeping the standards and regulations up to date.

Information sources

There are various organizations which can provide assistance. The following list is not exhaustive.

Technical Help to Exporters – BSI
If it is a technical export/import problem or one of testing to another country's regulation Technical Help to Exporters (THE) will advise. In some cases (such as testing for the Canadian Standards Association) they can even provide a testing service. There is a nominal fee for joining this particular branch, but having their service available and the promptness and willingness of the experienced staff is well worth it, not only to the small company without the resources to justify its own team of experts but also to anyone moving into an area outside his own experience.

Professional institutions and associations
Institutions such as the Institution of Electrical Engineers provide advice and assistance, sometimes with publications such as their booklet on *Health and Safety Regulations* which gives advice on engineering liabilities for contract and negligence, health and safety at work, consumer safety and product liability. Other associations such as the Production Engineering Research Association (PERA) and the Electrical Research Association (ERA) can also be helpful with similar services.

Her Majesty's Stationery Office
Her Majesty's Stationery Office (HMSO) is a useful source of reference material. Worthy of mention is the *Health and Safety* series of publications covering a variety of processes and materials. The list is very extensive and a visit to the HMSO bookshop, if near, can often turn up a wealth of reliable information.

Local sources
Often overlooked are the resources closest at hand. Many local authorities run information cooperation schemes, usually in conjunction with the borough or county library. It is worth checking whether the information you require is available either from the local library on loan or from a fellow sufferer in the locality.

Storage and maintenance

Having located the information you need in the way of standards and regulations precautions must be taken to ensure their proper use. This means:

1 Keeping them in a secure place.
2 Making sure that they are kept up to date.
3 Communicating their contents to those concerned in the organization.

This can be done more efficiently if carried out in a routine manner. This is why it is essential to have the responsibility for standards recognized in any organization. The appropriate communication should be by means of a company standard.

Developing in-house standards

Producer or user

Standardization effort at company level is essentially divided into two distinct sections. Product standards, including all the standards that govern the final product which is being marketed and user standards for all the components and materials which are to be purchased and built into the product (see Figure 28.3).

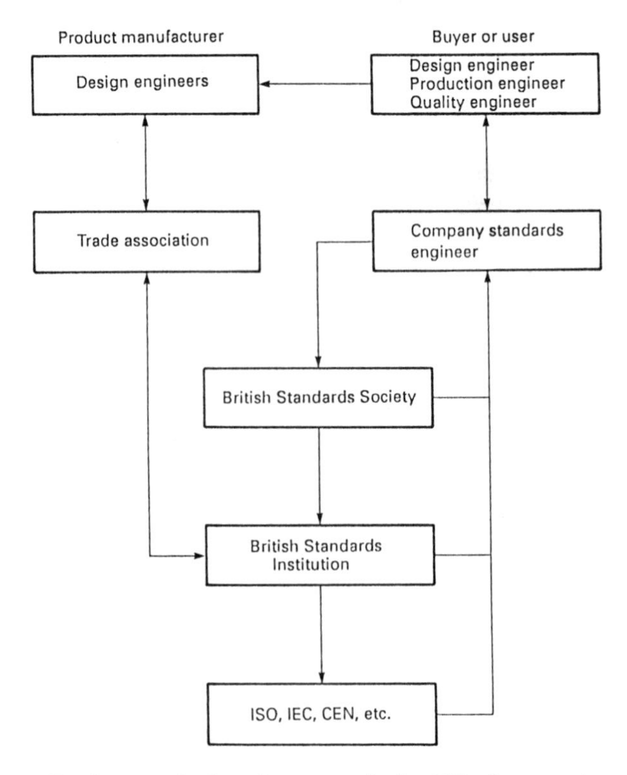

Figure 28.3 *Product standards and user standards. This diagram shows the flow of information to and from the standards makers, the direct involvement of the manufacturer and the input to the designer who uses purchased parts*

Product standards will involve contractual, international and national standards. As they are a direct influence on the profitability or even existence of the company one must invest proportionally in making sure that research and development is in line with them. This should certainly mean being an active member of an appropriate trade association and perhaps representing that association on national standardization committees. If the product is exported, then this may also require representing the UK on international committees. A limited number of committees will be involved so it is feasible that the contribution in this direction be made from the research and development team rather than by a standardization specialist. The involvement of senior development engineers attending international and national standards meetings is a costly business.

The best way of influencing this kind of standard is to be involved with the drafting. Initial drafts are rarely prepared by a committee. They are best written by one person and then submitted to criticism. This applies to international, national or company standards. To be considered a world leader, or not to be too much under the control or influence of others, one must support the national standards body and in one's own interests take responsibility for the initial draft whenever the opportunity occurs.

The company standards specialist may, on the other hand, be required to voice the company's interests as a user (for parts and materials or for codes of practice for such things as screw threads, graphical symbols or drawing office practice). It is not reasonable to expect a manufacturer to support the national standardization activity on the nuts and bolts of his business to the same extent as he would for his products. This perhaps explains why many standards committees are top heavy with manufacturing rather than with user interests. If one wishes to be kept informed on the development of standards in all areas then it has to be done on a more global basis. Standardizers facing this problem have organized themselves with assistance from the British Standards Institution into an association, now known as the British Standards Society. Through them it is possible to keep in touch with all aspects of standardization, to share problems with others working in the same field and to influence standards committees by nominating their own representative who will be guided to put the user view. Such a specialist with this support is in a better position to develop in-house standards for his company.

Documentation

Availability of information makes an important contribution to a company. In a one man business this could mean the proprietor always being personally available to give instructions or advice. Larger concerns require a more permanent arrangement. Documentation can take many forms, from the scribbled note to a directive from the management board. Standards documentation is intended to be formal and provide a permanent record which is capable of being updated. More than this it is designed so that the status of each document can be identified. The following details should be included on each.

1 A number to facilitate storage and retrieval.
2 A date so that the latest issue can be recorded.
3 A title for indexing.
4 A signature of authorization.
5 Contents in a uniform format, which gives the document credibility and makes it easier to read.

They should preferably be issued as part of a coherent system of company standards. A comprehensive volume of specifications cannot be achieved in the short term. The best organizations have patiently built up their standards over many years. Time would well be spent, therefore, initially in designing a credible system of presentation as this will be essential to encouraging the use of standards. They must be easily recognizable as having a status and authority in the organization.

The component part of a standards system, the vehicle for transmitting all decisions and directives, is the specification. By necessity it is inclined to be in slightly 'legalized' language. The important thing is to avoid ambiguity. Terms such as 'generally to', 'may be' and even 'approximately' are used sparingly, if at all. A good guide is available in BS 0 for drafting an in-house standard.

The following kinds of specification are popular generally in a manufacturing environment.

Purchase specifications

Purchase specifications are used for identifying and describing bought out equipment, parts or materials. There is considerable merit in standardizing purchase descriptions whether they are short enough to be included totally on the purchase order or extending into several pages. Establishing the correct purchase description at the development stage can avoid duplication and transcription error as well as allowing the opportunity for comparison with existing standards. In a modern organization with electronic data processing a database can be created which allows the standard description to be used directly on the purchase order.

The trend is to adopt available national standards and avoid lengthy in-house purchase specifications. The latter are counterproductive in terms of requiring the supplier to continually check for special requirements. This increases the cost and there is a real danger that an in-house specification is being quoted due to some oversight, when a proprietary product or one to an even higher national standard would be cheaper by virtue of its common use.

Material specifications (sometimes called data sheets) may be used as an alternative to the purchase specification. They are compiled in a similar fashion but are internal control documents and not sent to the supplier. They are generally used to specify a preferred range of components or materials, giving the user or buyer a list of types and approved suppliers. The buyer then purchases the recommended item to a British Standard or, if approved, the supplier's type number.

Process specifications

The purpose of process specifications may be production control or education. They are written in a manner which gives precise instructions to the operator in the method and materials to be used. Their existence is a precaution against deterioration in quality when the regular operator is not available and also ensures that the operator is advised of appropriate hazards and safety precautions. In addition they can be used as the basis of payment in a piecework system.

Performance specifications

A product may be specified in terms of its physical characteristics, size, shape, etc., or its ability to meet certain functional tests. For example, in preference to dimensional accuracy the speed of operation, ability to carry load or accuracy of function may be specified. The performance specification usually provides greater freedom to the designer or purchaser but generally has to be accompanied by more elaborate methods of test.

Test specifications

It may be necessary to have an extensive range of test specifications. They may be needed to cover everything from materials and component testing at the goods inward stage to the testing of the product before despatch. Methods of test are extremely important as without a standard comparisons at a later date will be impossible. A great deal of work has been completed both nationally and internationally on methods of test and these should be adopted in preference to creating in-house standards. They will help to avoid the problems that can be encountered between suppliers and customers who have no previous agreement on test methods.

Codes of practice

Codes of practice are created to record sound methods and procedures which have been developed on the basis of practical experience and scientific investigation. The codes of practice available from the British Standards Institution consist of recommendations rather than mandatory instructions.

However, in the manufacturing industry the term is generally applied to procedures which have been documented in a single standard to avoid repetition on drawings. In many instances they are produced because of the impracticality of including them on the drawing. For example a code of practice could be produced for an electronics workshop on soldering and wire termination methods. In the machine shop a code of practice may be produced for the use of lubricants. There are also many instances of administrative procedures being controlled by codes of practice. Prime examples of these are drawing office practice and quality assurance procedures.

The nature of the manuals will depend on the organization but the following are typical.

Handbooks

A document numbering system should have been developed and adopted which puts specifications into groups or classes. These may be design, manufacture, assembly, commercial or any other convenient grouping. It is usual however for users in a particular geographical locality to require a number of specifications which are randomly distributed across those classes. In these circumstances it is necessary to compose a handbook of related standards. For example a printed circuit assembly handbook will contain documents covering design, drawing office procedure and production processes and codes of practice. Other handbooks which may be required are drawing office procedure, packaging, wiring and assembly, health and safety etc. This list is not exhaustive and can be extended according to local needs.

Bought out parts catalogue

The parts catalogue is essential to the control of bought out parts in a company. It has several functions. First the items in the catalogue are classified into similar groups and given an in-house code. This has the advantage of not restricting the source of the item to a single supplier. The code can also be used as a drawing number so that it becomes compatible with the items on a drawing which are manufactured in-house. Being a similar shape and size the code is familiar to the other members of the organization who use it. It can be used for parts explosion, purchasing records or accounting. It will fit into any electronic filing systems used in the organization and, being familiar, will be less prone to error and misinterpretation than the supplier's code. In designing an in-house code for these purposes care must be taken not to build in too much significance. The more descriptive a code becomes the more it is prone to alteration as characteristics are altered, either by the supplier or by the purchaser.

As the parts are grouped together in a catalogue there is less likelihood of duplication. Similar items can be recognized immediately. The methodology adopted in producing the catalogue sheets is an aid to consistent specification. It is usual to record the following details as a minimum against each item:

1 The in-house code.
2 Purchasing description.
3 Outline dimensions.
4 Basic parameters or characteristics.
5 Related standards.
6 Supplier's name.
7 Supplier's type number.

The use of international or national standards in the purchase description allows the possibility of maximum sourcing and interchangeability. When multisourcing, however, evaluation should be supported by testing in-house or by using one of the approved certification schemes run under the auspices of the standards institutions (e.g. BS 9000, CECC or IECQ for electronic components).

Communication

The development of a standards system in the above manner will gradually build up into a respectable collection. As the collection grows the task of maintenance also increases, so there is good reason not to over-elaborate the content of specifications. However, as it will become a permanent task the aim should be to maintain standards in as natural a way as possible (perhaps similar to the methods by which drawings are kept up to date by direct feedback from production). The effort is justified by the fact that in the long term standardization will result in less documentation than the unique (and therefore more diverse) specification of individual items.

Those organizations employing a standards and quality system also encourage an effective communications network. Recognition of the authority of written instructions and procedures is established and this forms the basis of quality assurance checks. There is a constant feedback from this source to the standards section, which keeps the specifications up to scratch. As well as receiving information and issuing specifications and standards there is also a need to keep everyone with even a marginal interest informed. For these reasons responsibility for standardization should be allocated to a particular person. It may be a full or part time task, but a recognized control and distribution point must be created. The individual responsible will obtain standards, distribute them and keep them up to date in the organization. An awareness bulletin giving a brief description of standards progress and listing relevant new, revised and withdrawn specifications should be produced and circulated regularly.

It is noticeable that this style of management tends to provide a spin off in terms of access to information generally. Contacts are made within and outside an organization which can be used more extensively than pure standards. The various institutions, trade associations, standards bodies, libraries, etc., all have considerable information resources which can be invaluable.

Rationalization of parts, materials, equipment and processes

The standardization process consists essentially of identification, classification, specification and variety reduction. The first part of the exercise, identifying the subject, means collecting all the pertinent data such as the range of items involved, their individual descriptions, stock value, re-ordering level, and forecast usage. A description of their purpose or application may also be included. It should be noted that not all parts having the same title have the same use and vice versa. Classification is the next step and takes cognizance of this fact. For example certain bushes, spacers, sleeves and washers, although having different titles, may be under one classification.

Electronic components are a category which have particular problems in this respect, their standard of reliability and performance often needing a great deal more explanation than is at first obvious from the data sheets. This was recognized over twenty years ago primarily by the joint services (Ministry of Defence) who compiled a list of preferred components for their subcontractors' use. It was accompanied by a series of specifications which related not only to size and performance but just as importantly to specified quality levels. This system was eventually incorporated into the BS 9000 system which is a series of British Standards that covers electronic components of assessed quality. It is also supplemented by a certification and approval system with a list of approved components and suppliers.

The word rationalization is often used synonymously with standardization but must not be confused with the more restrictive function of variety reduction. Standardization has more extensive value in design, production and quality assurance (see Figure 28.4). Variety reduction is a follow up exercise which hangs on the results of the standardization effort. Its implementation is eased considerably if the functions described above are accompanied by a standard specification.

Specification of bought out parts and materials is a most lucrative and beneficial area in which to apply standards. It must begin at the earliest stage of development to provide the development engineer with a guide to preferred components. Surprisingly the benefits of this frequently go unrecognized. The development engineer is often left with the whole market place to choose his components from, with subsequent cost in delays to production owing to unavailability or inadequate quality. It is always worth first considering the

Function	*Method*
Structured and authoritative documentation	Standards manual
Collaboration to solve interface problems	In-house standards committee
Influence on national/international standards	Representing trade association
Awareness and dissemination within the company	Standards engineer and awareness bulletin
Identification of parts and equipment	Classification and coding
Variety control	Vetting additions

Figure 28.4 *The scope of standards*

components or materials which have been used previously. They are not only purchased economically and available in stock but they have a known reliability and quality factor. Why create problems? It will be said that having a mandate to consider existing stock first is inhibitive to designers. This is a good argument for their more direct involvement with the choice of standard components.

The standards committee

One frequently hears the committee being criticized as not being able to achieve anything. In spite of this it is recommended as the most effective way of achieving democratic standards. A standards committee should be constituted with very clear responsibilities and objectives. Most importantly discussion should not be allowed to ramble on without direction and should be concluded with a formal agreement, whether this is positive or negative. Members of the committee should be chosen with care by senior management. They should be fully responsible for the decisions taken and be able to compromise when necessary. The committee should be totally representative of the interests involved.

Consider, for example, a committee for standardizing electronic components. First, its objectives should be clear. In this case the committee may be required to compile and maintain a list of preferred components. The objectives will have been set by a senior manager, probably prompted by the production director (who has a vested interest in the availability and cost of parts).

The committee chairman will be appointed at the highest level possible. It is the chairman who will plan how the objectives are to be achieved, and over what timescale. He should be able to pick his team (the members of the standards committee). For this electronics components example this would probably comprise a development engineer, a buyer, a production planning engineer or a draughtsman and a quality engineer.

Responsibilities should be defined (such as who is to take notes) and a decision made on how the committee's decisions are to be presented. This may best be accomplished by compiling a specification for each of the components or ranges of components approved. Initial discussions will be devoted to the extent of each specification but, once established, meetings should have more purpose and direction, working towards the drafting of a specification which can be used as a control document. This document records the decisions made and allows further development. A list of these specifications eventually develops into a standards manual to meet the original objectives by providing a catalogue of preferred components which can be maintained as the business continues.

The value of standards

The value of standardization and its formal implementation is in the provision of control over the design and manufacturing processes. The results are a direct contribution to the quality and reliability of the product and the efficiency of the company.

Many attempts have been made to quantify the benefits of a standardization activity. Of particular note is the *IFAN Guide 1* published by the British Standards Institution as PD 6495. The standardization process is however a collaborative effort and has many qualitative advantages which are difficult, if not impossible, to quantify in financial terms.

Modern influences on standards have come from consumer needs, health and safety requirements and, more momentously, from the efforts to improve the level of quality on a national basis. The results extend into our everyday living. How frustrating if you travel abroad and your shaver or hair dryer will not plug into the local system? But how much more damaging it would be to the potential exporter who is not aware of these things, or even to the established trader, when he finds that somebody else has changed the standard. The moral is to keep involved. It is dangerous to the future prospects of a business to leave standardization to anyone else.

Further reading

Published by the British Standards Institution (Milton Keynes):

BSI Catalogue (annual)
BSI Buyer's Guide (annual)
BS 0 *A Standard for Standards*
  Part 1 *General Principles of Standardization*
  Part 2 *BSI and its Committee Procedures*
  Part 2 *Drafting and Presentation of British Standards*
PD 3542 *The Operation of a Company Standards Department*
PD 6470 *The Management of Design for Economic Production*
PD 6489 *Guide to the Preparation of a Company Standards Manual*
PD 6495 *IFAN – Guide 1, Methods for Determining the Advantages of (Company) Standardization Projects*
PD 6515 *IFAN – Guide 2, Company Use of International Standards*
BSI Handbook 22, *Quality Assurance*
THE, *International Certification and Approval Schemes*

Available from Her Majesty's Stationery Office or HMSO bookshop

The Health and Safety at Work Act 1974
HS (R) 6 A Guide to the Health and Safety at Work Act
HS (R) 7 A Guide to the Safety Signs Regulations
The Consumer Protection Act 1987

General publications:

Douglas-Woodward, C., *The Story of Standards*, The British Standards Institution, London, 1972.
Preston, R. P., *Standardization is Good Business*, Standards Council of Canada, 1977.
Sanders, T. R. B., *The Aims and Principles of Standardization*, The International Organization for Standardization, Switzerland, 1972, Marcel Dekker, New York, 1983.
Verman, L. C., *Standardization, a New Discipline*, Archer, 1973
Warner, F., *Standards and Specifications in the Engineering Industries*, National Economic Development Office, 1977.

Acknowledgement

Extracts from PD 6515: 1986 are reproduced by permission of BSI. Complete copes can be obtained from them at Linford Wood, Milton Keynes, Bucks, MK14 6LE.

29

Management of reliability

Patrick D. T. O'Connor

Reliability is the ability of a product to function correctly, not just when new, but throughout the expected time when it will be in use. Reliability is affected by many activities and people involved in specifying, designing, testing, production, use and maintenance. A high standard of reliability can be achieved, therefore, only by an integrated effort that ensures that there are no weak links in the total process of development, production and support. Such a comprehensive approach can be applied only from the top of the organization concerned. Quality and reliability must therefore be key elements of the corporate strategy. This in turn implies that the chief executive understands the causes and effects of unreliability, and the methods to prevent them.

This chapter presents the main methods used to prevent failures in service. The references describe the methods in more detail. It is important to note that there is a large overlap between the methods used to ensure reliability in service and those used to prevent failures during production. Production quality is an essential element of any reliability programme, and a good design can be unreliable in service due to poor quality control of production. Production quality control is explained, within the scope of the overall quality assurance function, in Chapter 26.

Costs and productivity

The effects of quality and reliability, like those of any other product features, must be measured in financial terms, since this is the ultimate measure of business success. However, financial success is not measured only in short-term performance, but in long-term considerations such as survival of the enterprise, market share, future product and marketing strategies, and security and quality of life for the employees. Quality and reliability have a major effect on these long-term business objectives, and only long-term strategic action can ensure that quality and reliability match the requirements of modern markets.

A product that is designed by an integrated engineering team motivated and equipped to prevent design weaknesses and to detect and eliminate them as early as practicable in the development programme, will be less costly to bring to the production stage. In other words, design and development for reliability improves engineering productivity.

Finally, failures in service lead to costs of warranty repair, loss of reputation, and the maintenance of expensive service and spares organizations.

Therefore the old concept of an 'optimum' level of quality and reliability, less than zero failures, as shown in Figure 29.1, is misleading and defeatist. The modern approach recognizes that the only optimum is continued improvement towards perfection, as shown in Figure 29.2.

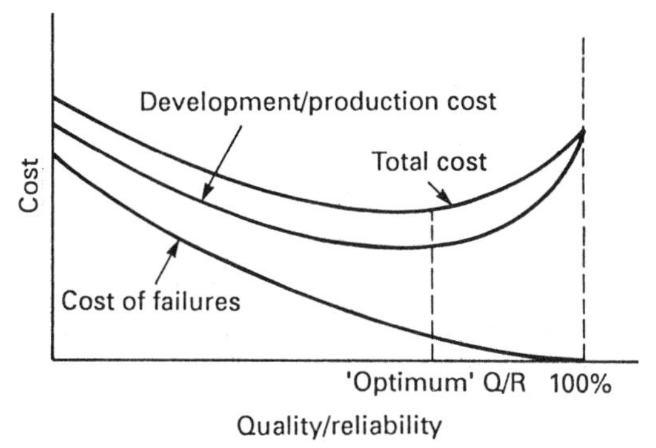

Figure 29.1 *Quality, reliability and cost (old view)*

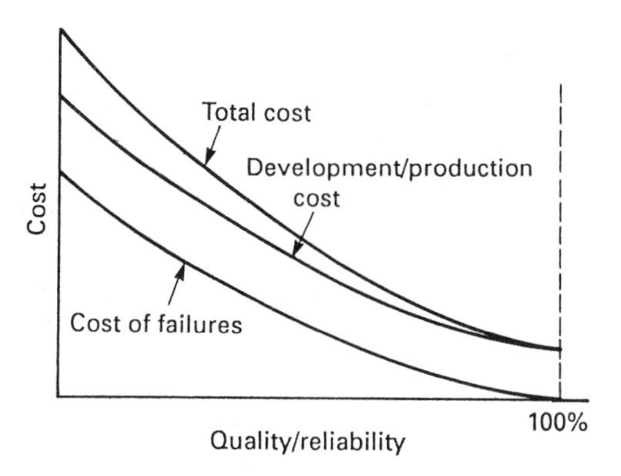

Figure 29.2 *Quality, reliability and cost (modern view)*

Risk

The level of risk involved in the failure of a product can differ enormously, depending on the type of failure and its consequences. At one extreme is a failure of a nuclear power plant leading to major radioactive contamination, or a major structural failure of an aircraft or bridge. At the other is a failure of a fuse on a domestic electrical appliance. Perceptions of risk relate to the consequences in terms of safety, cost, legal and statutory aspects, market forces such as product reputation and competition, and other factors particular to certain products. The perception of risk also varies between different people and at different times. The rate of occurrence, or the probability, of failures also affects the perceived risk.

Recent developments in product liability legislation, increased consumer and industrial user expectations for quality and reliability, and international competition have all strongly influenced the perceptions of risk inherent in the development of new products.

Historical development of reliability engineering

The discipline of reliability engineering as a formally recognized response to the growing complexity of new systems, particularly in the military area, was developed initially in the USA in the 1950s. The US Department of Defense and the electronics industry set up a joint task force, the Advisory Group on Reliability of Electronic Equipment (AGREE), which recommended that special disciplines be imposed during the development of new sytems, to ensure that they would be reliable when introduced to service. These methods included design analysis techniques and formal reliability demonstration testing. These techniques became standardized in documents such as US Military Standard 785.

Similar standards have been introduced in the UK and elsewhere. For example, there is now a NATO reliability standard, and reliability standards have been produced for non-defence work, notably by the British Standards Institution (BS 5760) and by organizations such as NASA and large utilities and companies.

Recent developments in reliability engineering and management have tended to reduce the emphasis on quantitative techniques for predicting and demonstrating reliability, with more emphasis being placed on design analysis and on testing to discover potential weaknesses so that improvements could be made early in the development cycle. This approach has always been the one used by the best commercial industries, and the purchasers and developers of large systems (particularly the military and defence equipment manufacturers) are now realizing that they are more effective than the methods brought into use in the 1950s after the AGREE report. This chapter stresses the modern approach to achieving high reliability.

Reliability concepts and patterns of failure

The reliability of an item which can only fail once, such as a transistor or light bulb, is usually defined as the probability of no failure over a specified time period, which might be the expected life of the item (see BS 5760 *Reliability of Constructed or Manufactured Products, Systems, Equipments and Components*). During this period the instantaneous probability of failure is called the *hazard rate*. Other reliability characteristics, such as the *mean time to failure* (MTTF) or the expected life by which a specified proportion, say, 10 per cent, might have failed, are also used.

For items which are repaired when they fail, the reliability is the probability that there will be no failure in the period considered, when more than one failure can occur. For repairable items and systems, the reliability can be characterized by the *mean time between failures* (MTBF), but only when the failure rate is constant in time.

Hazard rates can be constant, or can vary with time. Knowledge of the hazard rate trend can be very revealing in helping to identify and understand underlying causes of failure.

A constant hazard rate is usually characteristic of failures caused by random applications of loads that exceed the design strength, occurring at a constant average rate.

Decreasing hazard rates are seen in items that become less likely to fail with increasing time. This pattern is characteristic of failures due to production quality problems, such as the production of a proportion of weak components which fail early, leaving a population of good components with longer life to failure. The 'burn-in' of electronic components is a good example of how the knowledge of a decreasing hazard rate is used to improve the reliability of a product, by deliberately causing weak items to fail without damaging good ones.

Increasing hazard rates are characteristic of items which suffer strength deterioration with age, due for example to wear, fatigue, corrosion, etc.

The combined patterns of failure can be shown on the 'bathtub curve' (Figure 29.3).

The failure rates of repairable items can also vary with time, and these patterns can also be illustrated by use of the bathtub curve. Similar deductions can be drawn on likely causes of failures as for non-repairable items, but additional aspects must be considered, such as the effects of repair and the likelihood that the distributions of times to successive failures are not identical.

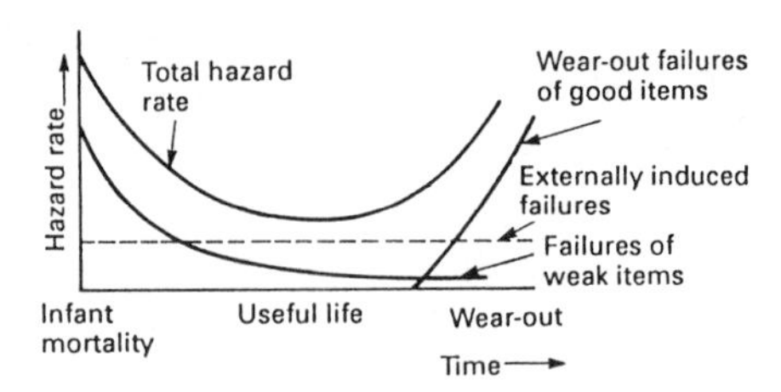

Figure 29.3 *The 'bathtub' curve*

The mathematical methods used to analyse failure data in order to identify the underlying failure patterns are described in BS 5760 and O'Connor (1985).

Variation in engineering

All processes and materials used in the manufacture of products are subject to variation. For example, the diameters of machined parts, the purity of silicon crystal, and the forms of solder joints all vary. If these variations are sufficiently large, the items concerned might fail in use. In all of these cases close control of the variation could prevent failure.

The environment in which products must operate (or be stored, maintained or transported) also varies. Typical environment variations which can affect reliability are temperature, humidity, shock and vibration. If these variations exceed the ability of the item to withstand them, failure will occur, either immediately or as the result of cumulative damage.

Finally, people working as producers, maintainers, handlers or operators of products, introduce further sources of variation which can affect reliability. Examples are incorrect assembly or inspection on a production line, incorrect maintenance such as neglecting lubrication, and electrostatic damage to electronic equipment due to handling by personnel carrying an electrostatic charge.

Variations can cause failure when, for example, the combined effects of a weak component and a high stress level are such that the component cannot withstand the stress. Such interactions of variations are more difficult to evaluate than are the effects of single variations, but they are important sources of failure in products.

In reliability, we are concerned with sources of variation which are time dependent. For example, chemical processes such as corrosion and physical processes such as wear and fatigue can result in an item which is initially very reliable becoming less so with age, due to gradual weakening, until the strength is no longer sufficient to withstand the applied stress. Time-dependent variation leads to increasing hazard rates, as described earlier.

Control of variation

To ensure that a product will be reliable, we need to control the sources and effects of variation. In principle, we need to:

1. Design the product so that it will withstand the expected environmental variation.
2. Design the product so that the expected variation due to production processes will not create items that will be too weak to withstand the extremes of environmental variation.
3. Control the production processes so that the design requirements are met.
4. Provide sufficient endurance against progressive weakening, by design to withstand fatigue, corrosion, wear, etc., for the expected life in the expected environments. This includes maintenance.
5. Provide for the expected variations in human behaviour, as producers, maintainers and users.

Obviously it is not always practicable to design and make products which will always be inherently failure-free in all environments. There will be mistakes, and practical considerations of cost, weight, etc. might limit the extent to which total reliability can be assured. However, systematic attention to these points can lead to extremely reliable design and production, at realistic costs. The methods by which these objectives can be realistically attained will be described later.

Design for reliability

A first principle of design must be that the design is inherently reliable, in the expected environments and for its expected life, and taking account of the expected production processes. This requires that the designer, in addition to understanding and taking account of the usual design features such as stress, weight, appearance, etc., must also fully appreciate the range of environmental conditions throughout the life of the product, and the production methods which will be used in its manufacture.

With the very short development timescales now typical of many modern products, and the importance of the design being correct the first time, the traditional approach which led to reliable design through experience and gradual refinement as a result of problems found and corrected is no longer adequate for most modern products. Therefore it is necessary to adopt more disciplined approaches to design, to ensure that the high demands of the modern market for reliability are met. This section describes the main methods used to ensure that designs are inherently reliable.

Stress analysis

In order to ensure that all environmental and operating stresses and stress combinations have been properly considered, a formal stress analysis should be performed. The review should be performed by a team, including the designers and by specialists who can advise on aspects such as environmental resistance of materials used, test results, etc. The review should be documented, with a record to show the stress conditions reviewed and the results.

The variations of environmental conditions and production variability inevitably combine to affect reliability. Product designs therefore must take account of this.

Tolerance analysis

Tolerance analysis is a statistical method of optimizing the tolerances of parts which must fit together or operate together, when the output value or specification is to be met as cost-effectively as possible. Tolerance analysis can also be very useful in optimizing yield from production, for example of electronic assemblies.

In many cases, particularly in electronic circuits, the component values and tolerances are set by the component manufacturer. Therefore, the circuit designer cannot apply the statistical tolerancing techniques to the same level of refinement for most component combinations. In such applications it is necessary to assess which are the critical parameter combinations from the tolerance and parameter variation point of view, and to apply statistical tolerancing where such combinations can affect yield or performance. Modern electronic circuit analysis software, such as SPICE, enable this type of assessment to be performed economically and effectively.

Where multiple tolerance possibilities exist, methods such as statistically-designed experiments and subsequent analysis (analysis of variance, ANOVA) can be used to optimize designs, by showing which are the most significant sources of variation and by enabling these to be controlled at their optimum values.

Robust design

Robust design is a term used to define a design approach pioneered in Japan by Genichi Taguchi. The Taguchi method combines elements of control theory and statistical design of experiments to optimize product designs in relation to their ability to be 'robust' against variations in the environment and in component and other parameter values.

The Taguchi philosophy also includes the concept of the 'loss to society' of output values being displaced from the target. The traditional approach to output tolerances is that so long as the product is within specification, it is 'good', and there is no loss. Conversely, any product out of specification is 'bad', and results in a loss of a fixed amount. However, the Taguchi approach includes the derivation of a 'loss function', which is a quadratic expression centred on the target value, and which increases rapidly for any deviation. It is not considered necessary to derive precise values, provided that the concept is accepted that all deviations incur losses, either directly costable (such as rework or scrap) or indirect (such as loss of reputation, higher risk of processes going out of control, etc.). The concept of the loss function is fully compatible with the modern trend towards continuous improvement in quality, reliability and productivity.

The Taguchi approach has been used widely in Japan for several years, and is now generating interest in the West. It seems to be at least part of the reason why the Japanese have been able to create intricate designs and to put them into mass production at low cost and with high reliability. The method provides a very systematic, disciplined and practical way of ensuring that a design is as robust and reliable as possible, and at the same time economic to produce. The Taguchi design approach, and statistical process control methods to keep production variation within the tolerances identified as necessary by the analysis, are very powerful techniques when used together in a totally integrated approach to design and production.

Failure modes, effects and criticality analysis

Failure modes, effects and criticality analysis (FMECA) is a design analysis method in which each mode of failure of every component or function within a system is assessed for probability of occurrence, effect of failure, and criticality in terms of successful operation, safety, maintenance, etc. The traditional approach to FMECA involves tabulating this information on standard forms, and the work can be tedious and expensive. However, computerized FMECA methods are now available, which enable the analysis to be performed much more effectively and economically.

FMECA is a very useful technique for evaluating the safety and reliability of a design, to highlight potential problems early in the development cycle so they can be corrected with the minimum of expense and delay to the project. It requires a disciplined approach, using competent, independent engineers working in close collaboration with the designers.

FMECA is also useful for evaluating other design features such as testability, and as an aid to preparing diagnostic information.

FMECA can also be performed in relation to production processes, to assess the effects of failures of these rather than of components or functions in the product. Such *production FMECAs* can be a valuable way of reviewing the production processes in the same disciplined, documented way in which the design itself is evaluated, and there can be useful cross-checking between the two approaches.

Fault tree analysis (FTA)

FTA is a reliability/safety design analysis technique which starts from consideration of system failure effects (referred to as 'top events') and proceeds by determining how these can be caused by single or combined lower level failures or events. It is a top-down analysis, and takes multiple failure modes into account. Standard graphical symbols are used to construct the fault tree picture, by describing events and logical connections.

FTA is a more powerful method than FMECA for safety analysis since it can take account of multiple failures (including dormant failures, human failures or incorrect actions) and complex system logic such as redundancy. If probabilities are assigned to the events, the FTA can be used to evaluate the probability of the defined top event(s) occurring, and the combinations of events which must occur for this to happen.

FTA evaluation becomes complex for even quite simple systems, and therefore computer programs have been developed. These evaluate the top event probabilities and identify the single and combination events which would cause them. Some of these programs can also construct the FTA diagrams, using suitable graphics plotters.

FTA is the standard safety analysis method used in industrial contexts such as nuclear power and safety-critical process plant.

Reliability prediction

Methods have been developed for predicting the likely reliability of new products. These are based upon databases of past failure information, mathematical models of hazard or failure rates as affected by applications and environments, and the use of mathematical models of the system reliability logic. For example, FTA can be used to evaluate the overall reliability of a system, and other techniques are available. However, quantitative methods for reliability prediction are very controversial, and can give very misleading results. They all ignore the fact that a product or system can in principle, and often in practice, be designed and manufactured to be inherently reliable. Therefore the methods imply that significant improvements in reliability are not feasible, and result in pessimistic conclusions being made.

It is significant that the techniques are not used in many industries which nevertheless achieve extremely high standards of reliability.

Reliability testing

Testing for reliability should be planned as part of an integrated test programme, which should include basic functional testing to ensure that the specified performance requirements are met, formal environmental tests, and safety tests, as well as specific reliability tests. The objective of reliability testing is to generate failures, so that design or process weaknesses can be identified and corrected.

All testing provides reliability information, and therefore it is essential that an integrated approach is used, including a common system for failure reporting and investigation.

Since reliability is a function of time and variability, reliability tests must be performed over suitably long periods of time to ensure that important failure modes, particularly wear-out failures, are generated. Also, sufficient numbers of items must be tested in order for the effects of product and process variability to be evaluated. Obviously it is necessary to balance the cost of test time and test articles against the benefits to be derived, but experience shows generally that comprehensive reliability testing leads to lower total development costs, reduced costs of warranty and support, and reductions in other direct and indirect costs of failure.

In theory, if high reliability design principles as described earlier are used, and if high production quality is assumed, a reliability test programme should not be necessary. However, if there are any elements of risk in the design, such as new technology, techniques, materials or processes new to the design and production team, or any other perceived risks, a reliability test programme should be set up to verify that the product is capable of operating reliably under the likely conditions of environmental and other stresses.

Reliability demonstration

One of the main features of the reliability methods introduced by the AGREE report was the concept of reliability demonstration by testing a product over a period of time and measuring the achieved reliability (MTBF), taking account of the statistical uncertainty involved. However, these methods have remained controversial and are not much used in most commercial industry, since they are not compatible with the modern philosophy of excellence in design and production, and continuous improvement. Therefore the methods described earlier (in which the objective is to generate improvement by forcing failures, rather than testing to generate statistics) are much more widely used.

Specifications and contracts

Reliability can be specified in quantitative terms, such as a minimum MTBF. However, quantitative specifications should only be applied when they will be supported by appropriate contractual conditions. Quantitative reliability specifications must also be meaningful and fully understandable by all parties. For example, there is little point in specifying an MTBF if there is not a clear definition of what constitutes a failure, and the system to be used to collect and analyse the data.

Quantitative reliability specifications and contracts are appropriate in cases when good data collection systems exist, and when the product is sufficiently complex to justify the effort. For example, an airline, a television rental company or a company buying a fleet of cars might consider the approach to be an effective way of providing motivation and of ensuring rapid corrective action should this prove necessary. Quantitative reliability specifications, often supported by contractual conditions, are common in the procurement of military equipment.

There are several types of reliability contract. The simplest is the straightforward warranty, extending over a defined period. Warranties can be extremely effective, due to their simplicity. No reliability measurement is involved, so there is motivation to continuous improvement. More complex approaches include support contracts and reliability improvement warranty contracts, both of which involve supplier support for the product in use over an extended period, typically four or five years, for an agreed payment. Such contracts also sometimes include MTBF clauses, requiring the supplier to take corrective action if the measured reliability is lower than that specified. More complex reliability contracts such as these are commonly applied by customers of large systems such as power generation systems, military equipment, etc.

Reliability audit

Some customers carry out audits of the organization and methods for reliability applied by suppliers. These audits are typically based on the approaches described in documents such as national or military quality standards, or international standards such as ISO 9000. Internal standards are also used; for example car manufacturers audit the reliability systems of their suppliers using their own internally generated procedures.

It is essential that a company which is subject to external audit performs its own internal audits to ensure that all procedures and methods are in order. Internal audit is good practice even if no external audit is to take place.

Internal audits should be performed by senior staff, who can correct deficiencies. The common practice of delegating audits to low-level staff does not encourage effective control, and can lead to a bureaucratic approach being adopted, in which compliance with procedures is given more attention than the development of better methods.

Customer involvement

Some customers, notably for defence equipment and other large systems, become closely involved with the management of the reliability programme.

For example, they might require that the contractor complies with standard procedures as mentioned earlier, and they might also set specification and contractual requirements for reliability-related activities and achievement.

Organization for reliability

Since product reliability is affected by design, test, production, control of suppliers and subcontractors, and by usage factors such as application and maintenance, it is essential to apply a coordinated approach to all of the activities that can influence reliability. This in turn requires that top management creates the right environment and motivation, as well as an effective organization, to ensure that very tight control is maintained during the design, development, production and in-use phases.

The exact form of organization for reliability can vary, depending upon the type of product. The reliability engineering team can be part of the engineering organization, or part of the production quality assurance department. The reporting line is of less importance than the degree of integration with the other functions involved, and the extent of top management involvement.

Ideally, reliability should be seen as an aspect of the total (or 'company wide') quality effort. Organization, motivation and training directed at ensuring reliable design and production should all reflect this top-down approach.

Further reading

BS 5760, *Reliability of Constructed or Manufactured Products, Systems, Equipment and Components.*

Gunter, B., 'A Perspective on the Taguchi Methods', *Quality Progress*, June 1987.

O'Connor, Patrick, D. T., *Practical Reliability Engineering*, 2nd edn, John Wiley, 1985.

30

Simultaneous engineering

Thomas J. Lindem
(Introduction by *Dennis Lock*)

The best management decisions are based on all relevant facts, sound estimates and professional advice. This is seen, for example, in investment decisions for new manufacturing plant. No responsible manager would limit his decision parameters to the purchase price and rated performance of the proposed new equipment. Instead, all the costs of installation, maintenance, training, operating and so on would be taken into account and compared for each possible course of action. This total cost approach is aimed at avoiding the expensive mistake of concentrating only on the immediate, narrow aspects of an investment or organizational change without considering all the wider issues and consequences. Viewed from the perspective of the maintenance manager, the name terotechnology was coined to describe the total cost approach to the management of plant and buildings. Economic life cycle costs is another term used in this context.

In engineering design, an engineer or a design team may be tempted to concentrate their creative verve on apparent design excellence, without due regard to the wider issues such as manufacturing cost. Obvious aspects of this approach are seen when high cost materials are specified in preference to lower cost and equally suitable alternatives, in calling for unnecessarily close manufacturing tolerances, and in the provision of design features that are difficult or expensive to manufacture. Some of these design/production cost considerations will be obvious to designers: others are not and depend upon collaboration with the manufacturing department during design if they are to be identified and considered. Value engineering and value analysis are examples of techniques which recognize these cost aspects, providing a formal procedural framework within which design and production management can work together.

The engineer has to realize that design and manufacture are a partnership and that it is the total cost of these activities which is important. There may be occasions when it is economic to spend more time and money on the design of a product in order to reduce the subsequent manufacturing costs, again aiming at the lowest total cost solution. Just as the engineer sets out to design a product that should satisfy the customer, so he should regard his own company's manufacturing organization as another customer to be satisfied. The development of computer aided design and manufacturing systems is, of course, a means of producing greater collaboration between the engineering and manufacturing departments.

Ingersoll Milling Machine Company, in Rockford, Illinois, create high grade heavy and special purpose machinery. Their product range includes one-off transfer line machining systems for components such as automotive cylinder blocks, heads and gear cases. For many years their engineering designs have been pursued on a total cost basis, using retained engineering wherever possible, and involving Ingersoll's manufacturing management in design engineering discussions from the earliest stages of each project.

Ingersoll have always worked in very close collaboration with their customers before and throughout the project design stages in order to arrive at solutions which will ensure that every machining system will deliver the required performance. In more recent years, Ingersoll have taken this close customer collaboration a stage further, involving the customer's component design engineers in a simultaneous engineering programme in order to achieve the optimum combination of component configuration and machining system.

This chapter describes simultaneous engineering at Ingersoll. Although this case is provided by the machine tool industry, it offers all engineers designing products for manufacture an example of what can be achieved through a logical and analytical total cost approach to component design and manufacturing method.

Simultaneous engineering at Ingersoll

This account is designed to explain clearly and concisely what simultaneous engineering is at Ingersoll Milling Machine and why it is so exciting. Simultaneous engineering is also known by some as 'concurrent engineering', 'life-cycle engineering', or other terms that connote a team or partnership approach to product and process design. The concept and methodology of simultaneous engineering is still evolving, however. Although there has been development of specific analyses and techniques, full-scale efforts between manufacturers and suppliers are only just beginning. The advantages of simultaneous engineering are so great, however, that it is crucial for the reader to be kept informed as the process is created and defined.

Simultaneous engineering – why do it?

The answer is simple. Simultaneous engineering will help machine tool customers (manufacturers) to survive and prosper in competitive world markets. To compete, manufacturers must make quantum leaps in performance – improvements of 30–40 per cent over present practice. They must find ways to introduce new products faster, at lower cost and higher quality, and at the same time be more adaptable to changing market conditions and environmental factors. Simultaneous engineering addresses this task.

Doing it will not be simple. It will be hard work. Incremental improvements, as proud as manufacturers or machine tool suppliers might be of them, are not enough to compete in today's world market. Major changes are required, and an entirely new approach to product and process design must be developed. There is no need to remind the reader of the recent and dramatic decline in the size of US manufacturing and the US machine tool industry. It is therefore incumbent on both manufacturers and machine tool suppliers to protect and enhance their competitive positions. One way to do so is through simultaneous engineering.

Simultaneous engineering – what is it?

Simultaneous engineering at Ingersoll is the concurrent designing of a product and its manufacturing process by teams of engineers from Ingersoll and a customer in a partnership arrangement.

The way things used to be – and still are in most cases – product design, process design, and implementation were done sequentially by different people in separate organizations. Manufacturers would design a product and a process, and come to machine tool suppliers with specifications defining how to make it. Machine tool suppliers would struggle with this assignment and submit competitive bids, and the low bidder would win. In extreme cases, and after a great deal of effort, manufacturers might change their design in some limited way, but significant change was rare. More often, suppliers simply overpowered any process problems with dollars and technology, provided customers with a manufacturing system, and left. Then it was the manufacturer's turn to struggle – to make it work, to get the bugs out, to solve reliability problems and finally get product through it year after year. It might be thought of as a single strand of rope or wire, with the supplier's effort spliced in between the product design and implementation efforts of the manufacturer. The results of this sequential process are well known:

- Long lead times to get new product to market.
- High manufacturing costs.
- Low manufacturing quality.
- Products that are not competitive.

In simultaneous engineering, process design is done in parallel with product design, starting at the front end of the product design activity where design changes are still the norm. But simultaneous engineering is much more than parallel design activities. The processes of product and process design themselves become integrated in real time as product and process requirements iteratively feed back to each other, and the compromises and trade-offs necessary to result in an optimized product-process system are made. The result: competitive products that are designed in less time with lower manufacturing costs and higher manufacturing quality. This might be thought of as many parallel strands of rope or wire winding together and becoming one – far stronger than either strand separately or simply added together.

Simultaneous engineering at Ingersoll is a comprehensive, structured approach to making integrated product-process design and implementation with manufacturing customers happen. The approach used is unique in its scope, analytical rigour, and attention to detail. Although this approach is unpatentable, it is regarded as some of the company's most valuable intellectual property and as a significant competitive asset.

A common database and understanding

Sequential product design-process design-implementation has resulted in a lack of common understanding and goals within manufacturers regarding their new product introductions, and also between manufacturers and their machine tool suppliers. The major problems with the traditional way things have been done are:

1. There has not been a common, detailed data base that all can use as a reference point.
2. Communications among all the relevant parties have been limited.

This is inherent in the sequential mode of doing things in which product design occurs in relative isolation, is handed off to process design which accepts what is given and again occurs in relative isolation, and is again handed off to the operations people who are left to implement the process and manufacture the product.

Part of this problem is technical. Manufacturers may be quite familiar with the detailed operations that are performed to make their products, but they do not always have the data to understand the implications of those operations for product cost, quality, timing, and adaptability, or how one operation affects another. They know that product design can affect manufacturing cost, but the specific cause-and-effect linkages between design feature operation and product cost, quality, timing, and adaptability are not apparent. The data has not been available or, if available, it has not been organized or processed to use in this way. Simultaneous engineering first of all means developing a common data base at a level of detail which enables quantitative analysis – not just opinions and beliefs – to determine an optimal product-process system. Simultaneous engineering second of all means using the common data base with manufacturers to build and communicate a shared understanding of the product-process design in pursuit of a common goal.

Through simultaneous engineering, the linkages between design feature operation and cost, quality, timing, and adaptability are made explicit so they can be easily understood and appreciated at all levels and in all functions of management. This data is aggregated and used to communicate alternative options in making trade-offs among product design features and process cost, quality, timing, and adaptability. This allows conscious decisions regarding product-process design to be made, rather than just having many things kind of inevitably go forward as often happens now. Through simultaneous engineering one can begin to optimize at the product-process systems level, balancing among design features, operating costs, quality, development time, flexibility, capital and start-up costs, etc., rather than sub-optimizing for any single aspect of the system.

Measuring the results

The operational objectives of simultaneous engineering are major improvements in quality, cost, timing, and adaptability. To know how much simultaneous engineering has improved the product-process system, there must be measurement; in order for there to be measurement, there must be a bench mark or base line from which to measure. Defining the appropriate bench mark is crucial. The most recent product design and the most modern or highest efficiency plant of a customer can be a bench mark; a preliminary design and off-the-shelf process to make it might be a bench mark.

Measurement of product cost and its components is particularly important, since many of the variables that are being optimized are captured in this measure. Engine cylinder block costs, for example, should be broken up into components of casting block material, capital equipment (depreciation), direct labour, indirect labour, overhead (heat, light, rent, scrap and rework, etc.) and supplies. As a result of simultaneous engineering one can expect a 30–40 per cent reduction in total product cost and corresponding improvements in quality, timing, and adaptability. Simultaneous engineering may also result in changes in the percentage composition of the cost components. For example, the proportion of material costs might increase because of near net shape casting, while the proportion of equipment costs might correspondingly decrease.

Costs should similarly be measured at the level of the design feature – the cost to machine each specific feature to perform its function. Feature costs would similarly be made up of material, equipment, labour, overheads, etc.

components. The sum of the feature costs plus the cost of the raw casting block being machined should be the total product cost.

The approach to simultaneous engineering

Simultaneous engineering is done in four integrated ways (see Figure 30.1):

1 Optimizing by feature analysis.
2 Optimizing by analysis of advancing technology.
3 Optimizing by analysis of equipment reliability and cost.
4 Optimizing by analysis of operating environment requirements.

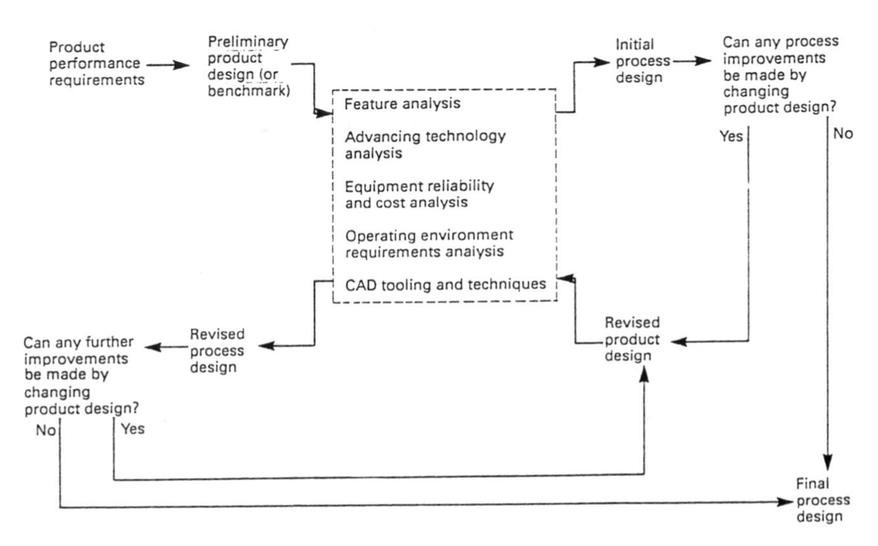

Figure 30.1 *Flow chart for the simultaneous engineering process*

The procedure begins with performance requirements for a new product being established by a manufacturer. These performance requirements are translated into a preliminary product design (in many instances building upon previous designs) with design feature specifications. Based on the preliminary product design, the four analytical techniques listed above are used to arrive at an initial process design. These are complex processes that optimize design features and product cost, quality, timing, adaptability and related variables. Once the initial product-process design has been established, it is continuously improved and refined through the further application of the analytical techniques. A final product-process design, optimized at the systems level, results from this series of optimizations.

In addition to the analytical techniques of simultaneous engineering, the entire range of available computer aided design tools are, of course, brought to bear on the new product design. These include finite element analysis, cutting parameter models, solids modelling and other computer graphics tools that analyse and evaluate shapes, weights, parts clearance, mechanical stress, thermal stress and other aspects of product design.

Feature analysis – definition of a feature

The fundamental unit of analysis of simultaneous engineering at Ingersoll is the design feature *operation*. The design feature is the focus of data collection that forms the data base from which the initial product-process design optimization is derived. *A design feature is a single specific and purposeful man-made geometric shape created on a solid object.* A design feature is defined by its surface dimensions and locational position on the solid object, and by *the machining or casting/forming operations which can produce it.*

A hole in an engine block, for example, is a feature, defined by its mid-point position, diameter, and depth. It may be the product of a single operation – drilling – or the product of multiple, sequential operations (e.g., first drilling, then boring or reaming). If the hole is deep, and for purposes of product flow the hole is drilled at more than one work station to greater and greater depths, it is still a single design feature and the product of a single operation – drilling. If the hole were formed in the casting process rather than by machining, it would still be the same single design feature.

If the hole is composed of more than one geometric shape, however, it would be more than one feature. If, at the bottom of the initial hole a smaller diameter hole was subsequently drilled, that would be a second design feature, also produced by a single operation-drilling. If a tapping operation were performed on the bottom third of the hole, this would be a second design feature, produced by two operations – drilling and tapping.

It is important to reiterate that the unit of analysis in feature analysis is the *design feature operation*. For features produced with only one operation – e.g., drilling – only a single cutting operation is analysed. For features produced with multiple operations (drilling and tapping, for example) each operation is analysed separately and in sequence.

If there is more than one hole with exactly the same dimensions, but occurring in different positions on the workpiece, these would be distinct design features because some of the variables on which data are to be collected (such as quality, flexibility and feature surface) may depend on position. If variance due to position is not a problem, data may be aggregated in feature analysis to include all holes with the same dimensions.

Feature analysis – database development

Feature analysis consists of data collection and analysis at the level of the design feature operation, and optimization at the level of the product-process system. Database development activities in feature analysis are shown in Figure 30.2

For each design feature, certain initial design data are taken from the preliminary design – e.g., feature dimensions, position, tolerances, name, etc. For example, in an engine block, cap bolt holes are a design feature with initial diameter, depth, and centre line position dimensions and tolerances given. The first key task to be addressed by the simultaneous engineering team is to determine the current state-of-the-art machining technology for producing the feature. This might involve single or multiple machining operations. An additional degree of complexity is added if there are alternatives for the solid object material that must be considered (e.g., if the engine block can be cast iron or aluminum).

Once the state-of-the-art machining operation or operations are determined for the given design feature, a number of data points are calculated or estimated for each operation. The most important data point for each operation is cutting time – the time it would take to perform the operation in producing the design feature. Cutting time is the basis of comparison in most of the analyses which follow.

Capacity related data is also calculated/estimated for each machining operation, including such factors as surface speed, revolutions per minute, feed rates, number of blades, feed stroke, depth of cut, horsepower per spindle, thrust per spindle, etc. Although the state-of-the-art cutting time may represent the lowest operating cost for an operation, in subsequent steps product flow rates, capacity requirements, and equipment costs need to be taken into account in order to optimize at the systems level, not suboptimize at the design feature operation level.

For each design feature, data on a number of other independent variables are also indentified, calculated, or estimated. These data include:

1 The *feature surface*. The surface of the solid object on which the feature is to be located must be identified. For example, on an engine cylinder block these surfaces would include top, bottom, left side, right side, front, rear, left bank, right bank, left under bank and right under bank.
2 The *feature type*. This is identifying the function of the feature. On an engine block these would include oil related, coolant related, attachment device, precision locating device, fuel system related and precision boring related.
3 A *flexibility index*. This indicates the probability that the manufacturer (Ingersoll customer) will want to be able to change the dimensions or

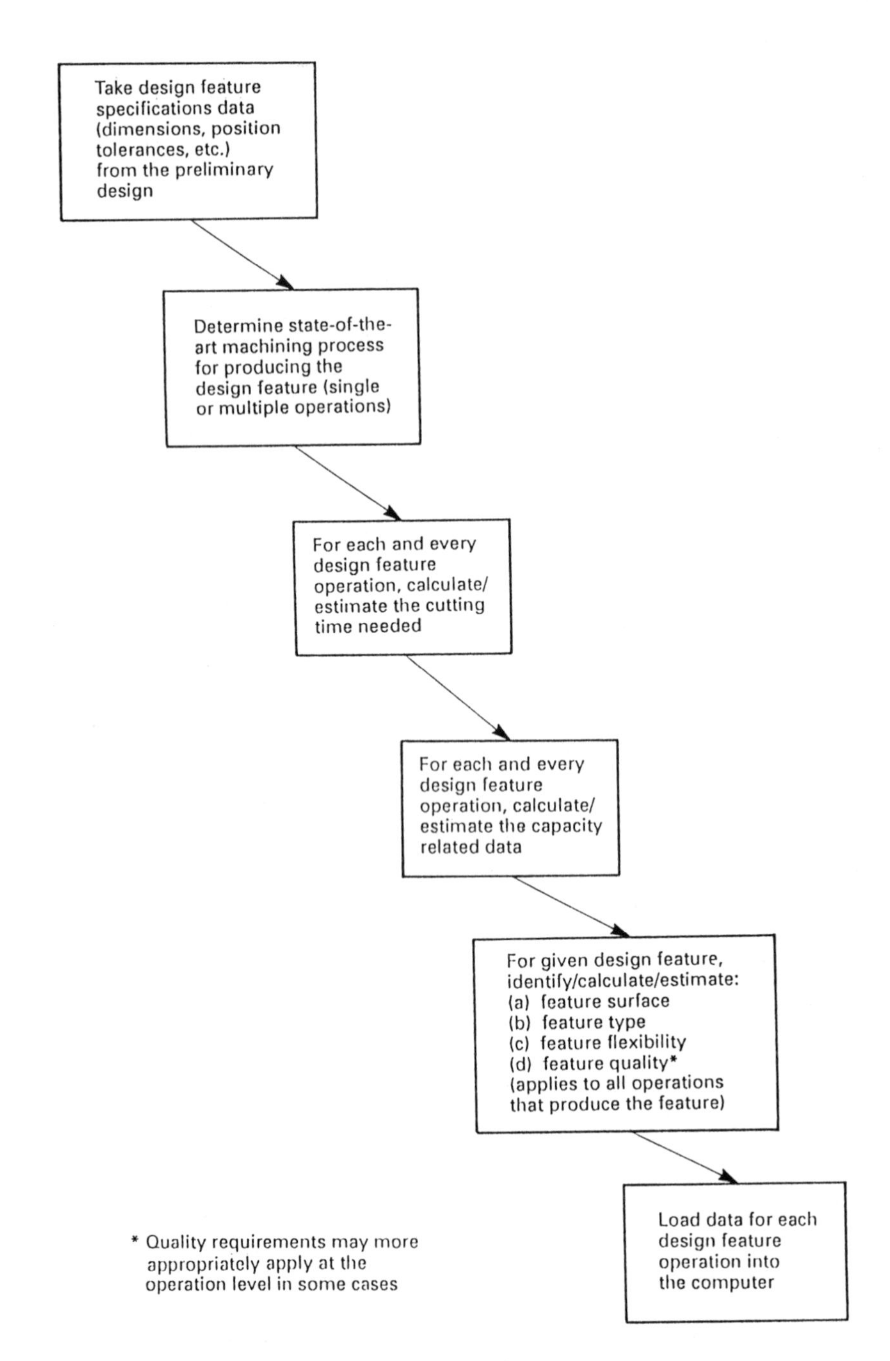

Figure 30.2 *Feature analysis – database development*

position of the feature over the expected life of the product. For the engine block example, the cylinder head mounts feature will have a low probability of change, while the dipstick hole might have a high probability of change.

4 A *quality index*. This indicates the degree to which the machined feature must meet the design specification in order to be cost effective. For an engine block the dimensional and positional tolerances of the cylinder holes and crankshaft borehole would have high quality requirements, while the dipstick hole would have low quality requirements.

These design feature data apply to all machining operations which are utilized in producing the feature. There may be some instances in multiple operation features, however, in which the quality index might apply only to the final machining operation rather than to all operations – i.e., the final machining operation to produce a high quality design feature would have to also have high quality requirements, but preceding operations might not.

The above data are contained on a 'feature analysis sheet' – one for every design feature or set of design features that are exactly the same except for position, and where no change in any of the other variables is caused by the different positionings.

For the typical engine block, there may be over 500 design feature operations, and the only way to handle this amount of data is by computer. Data for each design feature operation are taken from the feature analysis sheet and loaded into the computer, thus creating the simultaneous engineering database.

Feature analysis – data analysis and interpretation

Once the database described above is constructed, a number of different processing manipulations become possible. Those currently being performed at Ingersoll are described below, but more may be identified in the future:

1 Array all design feature operations against cutting time. Cutting time gives a prioritization of those operations requiring the most attention or offering the most opportunity for improvement.
2 Sort design feature operations according to any of the other variables and array the sorted groups against cutting time. This gives specific looks at the cutting time distributions of, for example, engine block top feature operations, engine block high flexibility feature operations, engine block oil related feature operations, or engine block high quality requirements feature operations.
3 Sort design feature operations by two or more variables and array the sorted groups against cutting time. This gives specific looks at the cutting time distributions of, for example, high flexibility – oil related feature operations and low flexibility – oil related feature operations.

4 Sort design feature operations by variables and specific machining operations or other preliminary design data and array against cutting time. This gives specific looks at the cutting time distributions of, for example, all drilled feature operations on the right side of the engine block or all oil related feature operations having a surface speed greater than 65 feet per minute.

Based on these types of data processing manipulations, feature analysis includes a number of specific analyses. These analyses include, but are not limited to, the following:

Operations analysis
The operations analysis provides a general survey of the state-of-the-art machining times of all design feature operations, which impacts the 'natural' production rate of a process designed to make the product. It identifies the most time consuming design feature operations for further analysis and study. All features requiring machining operations to produce them are not equal – some take more time, and therefore are more costly to produce, than others. Some design feature operations, because of the length of cutting time required, can control the throughput rate of the 'natural' process (i.e., before capacity requirements, cycle time, and work station operations are taken into account in process design) and therefore require special analysis and study. The feature operations data can be sorted in any number of ways and arrayed against cutting time, as discussed above, to examine the priority feature operations of any sorted group.

Flexibility analysis
Over the expected lifetime of the product-process system which is being designed, certain design features may be subject to changes in dimensions, position, quality requirements, etc. due to unanticipated events, model changes, market demand changes, or other reasons. It is important to know which features are likely to be subject to no change as opposed to those which may be subject to great change, so that flexibility can be built into the operations which are necessary to produce the latter features. These high flexibility requirement operations, once identified, sorted if desired, and arrayed against cutting time, will be the subject of special analysis and study. If major discontinuities (e.g., new applications of the product, product obsolescence, technological breakthroughs in performance, raw material changes, etc.) are anticipated during the expected lifetime of the product-process system, these could (of course) affect the flexibility requirements of the entire product-process system.

Quality analysis
Quality in this case refers to the tolerance requirements of a design feature to

perform its function, *not* to the ability of the process to meet consistently the design specifications. The issue addressed in this analysis is the appropriateness of the given tolerance requirements of the design feature for the function it is to perform. Two possibilities of inappropriateness exist: tolerance requirements could be either too high or too low for the intended purpose the design feature is to perform. The first case can result in unnecessary processing costs; the second in product problems for manufacturers in their markets. The high quality requirements operations, once identified, sorted if desired, and arrayed against cutting time will again be the subject of special analysis.

Tooling analysis

Tooling analysis addresses the tooling requirements for the cutting operations needed to produce the design features. Total cutting time required for each tool (e.g., a $\frac{1}{2}$ in drill bit) can be examined, as well as subtotal cutting time by feature surface, feature type, etc. for each tool. As a result, the number of tools required for the process, the variety of tools required, the number of tool changes required, etc. can be calculated or estimated. These results will be used to identify and evaluate the potential benefits to be gained from standardized tooling, combining feature operations and tooling, or modifying product design to simplify tooling requirements.

Dedicated tooling analysis

There are three generic types of process system than can be considered in design efforts:

1 The conventional transfer line, with high volume throughput and low flexibility, dedicated operations.
2 A flexible machining system (FMS) or machining cell/centre, with high flexibility operations and low volume throughput.
3 A line with a combination of dedicated and flexible operations.

In pursuing option 3, the design feature operations database will, for example, be used to identify patterns of feature operations that can be:

(a) conveniently combined into dedicated heads
versus
(b) machined on flexible machining centres with pot heads, head exchangers, head indexes, etc.

The constraints that limit either of these alternatives (such as feature spacing, different machining speed and feed requirements, etc.) and the benefits from each will be analysed and evaluated.

Other analyses

Three additional analyses in the product-process design optimization flow were shown in Figure 30.1. The first of these is an *advancing technology analysis*. In this analysis advances in process technology are anticipated – particularly in areas such as cutting tool performance, casting techniques, machine spindle speeds and positioning elements, etc. – that might reasonably be expected to occur by the time the process is to be implemented as a result of continuous technology development or of a new, concerted development effort. The impacts of such technological advances, if incorporated into the process design, on product cost, quality, timing, and adaptability are then calculated or estimated.

The equipment reliability and cost analysis takes into account the fact that the cutting machines and tools selected may be more or less reliable with corresponding varying impacts on product cost, quality, timing, and adaptability. As future process technologies identified in the advancing technology analysis are considered, equipment reliability and cost impacts become an even more complex issue. As part of the equipment reliability and cost analysis, up-to-date reliability standards, reliability requirements, and reliability tests for processing equipment are defined, as appropriate for the process being designed.

The *operations environment requirements analysis* includes all of the concurrent support functions required to make the process operate smoothly and efficiently. State-of-the-art equipment today (let alone advancing process technology of the future) requires magnitudes of performance improvements over the past in such areas as:

- Plant maintenance (preventative and emergency, spare parts inventory, etc.).
- Tooling management (ordering, tracking, maintaining, etc.).
- Quality management (inspections, methods, records, etc.).
- Job classifications (breadth and variety of responsibilities, etc.).
- Plant layout (efficiency of work flow, etc.).
- Physical plant (foundation stability, temperature range, etc.).

This analysis includes an examination of the manufacturer's current operations environment and the identification of improvements that would be needed to operate the process being designed smoothly and efficiently.

System optimization

Based on the above analyses, the volume/capacity requirements demanded by the manufacturer, and operations sequencing constraints imposed by the timing relationships of one operation to another in producing a design feature, an optimized process system is determined. The process is optimized only for the given set of design conditions however, and must undergo iterative reoptimizations as the analysis of the process and product designs continue.

In optimizing the process, costs per design feature (including cutting time and tooling changeover, part positioning, machine manoeuvring, and other inefficiencies for multi-operation features) can be calculated for the different equipment alternatives identified in the dedicated tooling analysis (see above). The operating costs and throughputs of equipment alternatives can be calculated and compared, and equipment purchase costs can also be factored into these calculations. Costs can then be related to different aspects of the product – the individual design feature, the feature surface, feature type, etc. – for purposes of further analysis and study related to reducing product costs.

It is clear that this structured approach to feature analysis is a very powerful tool for integrated product-process design and for communicating design alternatives and trade-offs to manufacturers. This allows an understanding of where 'waste' – unnecessary costs – is in the system and results in a prioritized, focused attack on waste, wherever it may be found and whatever the reason causing it. Simultaneous engineering teams from Ingersoll and its customers, including functionally based people involved in product design and manufacturing, can simultaneously attack all high priority sources of waste, rather than attacking waste first in designing features, then in machining time, then in tooling purchasing, and so on.

The feature analysis database points to the right questions to ask ('why . . . ?' questions) for improving the initial product-process design through a series of iterations. One can ask 'what if . . . ?' questions, and get the cost, quality, timing, and adaptability answers through the computer. For example, why are the quality tolerances on this feature so high? Do they need to be that high to perform the function intended? What if they were lowered – could a different machining operation then be used? Could we combine operations? What would be the impacts on cost, quality, timing, and adaptability? Through simultaneous engineering there is the capability truly to begin to optimize product-process systems for manufacturing.

31

Applying computers in engineering

Andrew Kell

The rapid development of computer technology in recent years has brought with it an increasing number of possible applications in industry. These advances have been particularly dramatic in engineering. Early computer systems found limited use in engineering because of their relatively slow operation and high cost, but the great improvements in price and performance, coupled with the development of the computer's ability to handle geometrical information and display this graphically, have provided the engineering function with powerful tools which can give significant benefits to a company. This chapter outlines the scope of using computers in engineering and discusses the requirements for successful introduction.

The task

In an increasingly competitive commercial world, a business must be continuously seeking to improve its products and operations. The successful application of relevant new techniques and tools is one of the approaches with which engineering management can gain business advantage.

While the technicalities are the province of skilled employees or external specialists, the engineering manager must be responsible for ensuring that the investments made by the company contribute effectively to achieving the aims of the business. He will, therefore, be increasingly involved in the decision-making processes surrounding the selection, implementation and management of computer-based systems.

The task is not simple. In recent years companies introducing computer-based engineering systems have had mixed degrees of success. Some have undoubtedly been successful, but others have not realized anywhere near the full potential of the systems they have installed. Experience shows that there are a number of common characteristics in a successful installation. These are:

1 The systems are selected and developed to enable key business objectives to be achieved.
2 A systems strategy is prepared at an early stage.
3 Procedures are simplified and good disciplines are in place throughout the organization before introducing computers.
4 Systems are planned in detail at each stage of development before resources are committed.
5 Progress is monitored and reviewed against the requirements of the business and corrective action is taken if circumstances change or targets are not being achieved.

From an engineering management viewpoint, the most significant task in applying computers in engineering is the planning and control of projects introducing the technology. Where computers are already being used, the task may be to improve their contribution to the company – but again the planning is crucial.

It is important to remember that every project must aim to improve the wealth of the business. Also no project, however innovative and original, can be judged a success until implementation is completed: that is, completed on time, within budget and with all the technical and commercial targets achieved or bettered. The introduction of computers into engineering is no exception.

There are six basic steps (see Figure 31.1) in adopting computers in engineering:

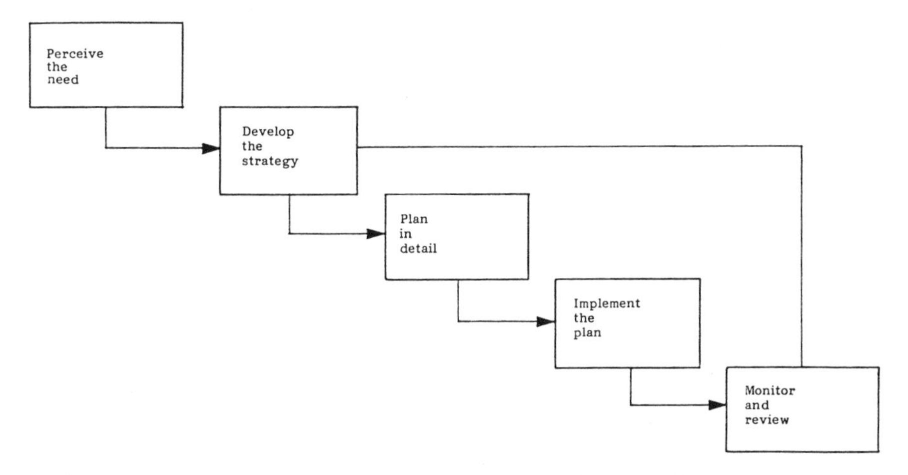

Figure 31.1 *Six steps in the introduction of computers in engineering*

1 Perceive the need.
2 Develop the strategy.
3 Simplify and integrate.
4 Plan the detail.
5 Implement the plan.
6 Monitor and review.

Perceiving the need

The first step towards the successful use of computers is to identify the business drivers for implementation. These drivers will vary from industry to industry, and from product to product, and will also depend on the current and future position of that company in its marketplace.

In order to put these 'drivers' into perspective it is necessary to have a good understanding of the marketplace, products and manufacturing position of the company. The business strategy is the starting point which gives a direction, not only for the introduction of computer aided design, but also for other initiatives which are aimed at improving company performance. It is only within a perspective of the direction of the business that the needs for improving engineering systems can be truly assessed. With an understanding of the current business situation, and of how future events might effect the position, the drivers for change can be identified and quantified.

Generally the drivers for applying computers in engineering fall into two categories:

1 To gain competitive advantage.
2 To reduce costs.

Gaining competitive advantage may necessitate:

- Increasing technical performance.
- Improving product quality.
- Maximizing reliability.
- Reducing lead time.

Reducing costs may be aimed at:

- Product costs.
- Project costs.

The driver for using computers in engineering may not arise from the engineering department itself, but may be a key part of achieving improvements in manufacturing or sales and marketing. A company which has to prepare tenders may employ computer aided design for generating preliminary design information and estimates of product costs. Reducing manufacturing lead time may depend upon the production of computer numerical control (CNC) machine instructions, and the definition of part geometry in the design office using computer aided design would lead to significant savings in the production engineering and manufacturing areas.

The reason for installing computer systems in engineering may not therefore be to improve the engineering function but to provide the base information which can then be used to greater benefit in other areas of the company's operations (see Figure 31.2).

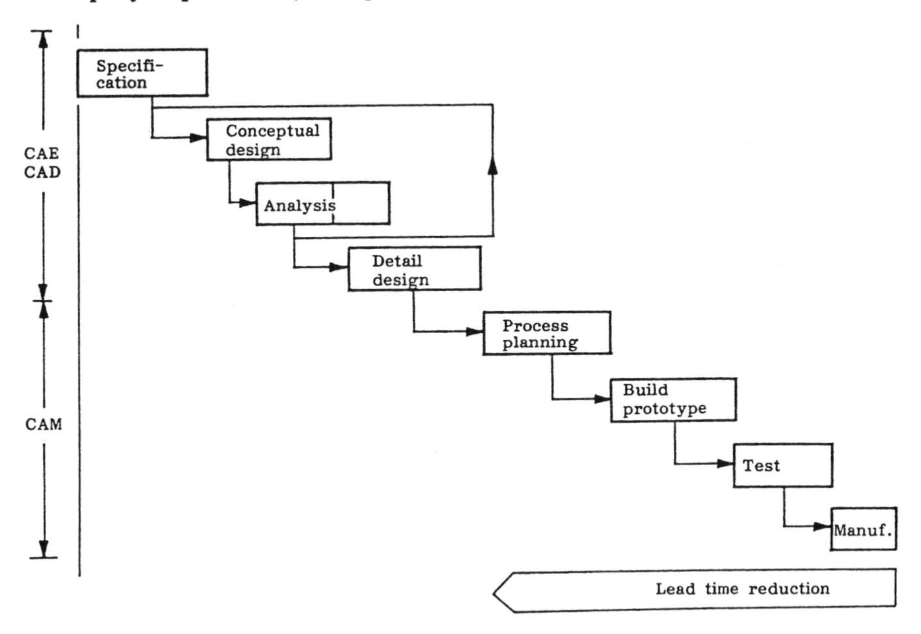

Figure 31.24 *The design-manufacturing process. In order to reduce the time taken from specification to manufacture, it is necessary that the whole process of sequential activities is examined*

To view computers only as a tool for the engineering function may be overlooking the real needs of the company and underestimating the areas where the main benefits are to be gained. Any initiative in engineering must therefore be considered as part of the overall business operation.

Ultimately the driver is improved financial performance, whether through improved sales of more competitive products or through cost reduction. The benefits are unlikely to be directly measurable either in the engineering departmental cost budget or by narrow investment appraisal techniques related to labour cost reduction.

The first step in a successful project will be identifying and quantifying the business drivers which will lead to improved business performance.

Planning the strategy

The introduction and development of engineering systems should be covered by a strategic plan. As the systems are likely to be introduced over a period of time, and will also need to be complementary to other developments in the company, it is important that the overall objectives and requirements are well understood and documented. Some key guidelines in developing the strategy are:

1. Identify the core activities critical in the company's operations, and those that will have an effect on achieving the improvements that are necessary.
2. Consider processes and information flows as a single system throughout the company and not as individual departmental activities or needs.
3. Establish a carefully considered conceptual approach for the business.
4. Establish a suitable systems outline plan.
5. Determine the steps necessary to reach the desired objectives and decide how many steps can be taken at one time. Put the steps into a priority order which meet the short- and long-term needs of the business.
6. Acknowledge that the plan will change.
7. Find ways of overcoming constraints.
8. Settle for nothing less, in the long term, than an integrated approach. Aim to develop in a well structured modular way.
9. Avoid becoming boxed in. If the plan is so inflexible that it cannot be changed future responsiveness and flexibility may be reduced or destroyed.
10. Seek objective advice, although the responsibility for installing the system and making it run cannot be abdicated.
11. Be prepared to invest substantial time, effort and resources in getting the systems to work.
12. Devote time to planning each step in detail.

Part of the strategy development will be to determine the feasibility of the changes which are being considered, both technically and commercially. This will require an assessment of the available technology, and identification of gaps which need to be overcome. An assessment of the technical risks involved may be required so that decisions can be made with a full understanding of the possible implications of success or failure.

Budget costs and the expected benefits should also be assessed so that the commercial viability of the project can be tested. It is unlikely that a justification based on individual 'stand-alone' investments will give a true reflection of the benefits to the company as a whole. If the whole package is not justifiable, then a critical appraisal of the approach may highlight simpler ways of achieving most of the benefits at a lower overall cost.

Simplifying and integrating

Many company practices and procedures have evolved over a number of years and some may be totally inappropriate to the current needs. In addition, the increased flexibility offered by the use of computer-based systems may offer opportunities for enhancing the methods of work and the way activities are carried out.

In many cases a review of current practices and a critical appraisal of their relevance to the company's requirements will highlight a number of opportunities for improvement. The purpose is not just to find out what goes on so that an equipment specification can be written, but to identify and change traditional practices which may in themselves be inappropriate to current and future needs. Wherever possible, procedures should be simplified and made more effective before the introduction of a computer-based system is considered.

At this stage, the value of information can also be questioned. All information costs money to produce. Sometimes the cost of preparing information is disproportionate to its value and, in many cases, the volume of information makes practical use impossible. Focusing on essential information that contributes to the aims of the organization can considerably simplify the flow of information and the supporting computer systems.

In the engineering function, as in manufacturing, non-value-adding activities should be eliminated (for example the checking of drawings and the re-interpretation of information).

In many instances the flow of information can also be improved by bringing together, or integrating, the activities in the engineering process. Design and production are two activities where a closer organizational link may be appropriate. Bringing design and production engineering together and understanding the interrelationship between the two functions can give significant benefits in ensuring the smooth passage of information from design into manufacture. Additional benefits arising from the interaction of design with production at an early stage in the overall process will help ensure that product designs are better from a manufacturing viewpoint.

Applying computers alone will not always provide the improvements desired. In many cases applying computers to traditional manual based practices is difficult and inappropriate. Simplifying existing practices will greatly enhance the chances of implementing a system successfully. It may even remove the need for a sophisticated computer system. A computer is not essential for better performance: in some cases manual methods can be just as effective.

Planning the detail

There is no short cut to implementation. Each step in the strategy will need to be planned in detail. It is at this stage that the system is turned from a concept into realistic statements of requirements. Specifics replace generalities and facts replace opinions. Detailed planning involves three stages:

1 User specifications.
2 Functional specifications.
3 Evaluation and selection.

The success of the project will be directly related to how thorough, practical and accurate the detail planning has been. Ambiguities and omissions can be the main source of problems in the later stages of implementation and development.

The definition process should normally begin with the production of a user requirements specification. This is a description, in plain language, of the functions the system is to perform. The detail planning phase may be split into a number of smaller projects, so consideration should be given to the other elements of the systems laid out in the strategy. It is particularly important that the interfaces with other areas are well developed so that future requirements are taken into account.

The user requirements will be converted into a specification defining the various elements of system software. This should enable different packages to be identified. Wherever possible, proven software packages should be used but, should bespoke software be needed, the requirements should be written so that this can form the basis of a software specification.

The user requirements definition enables a functional specification of the system to be prepared. This will form the basis of enquiries to potential suppliers. The functional specification is a translation of the user requirements into a complete technical description of the systems.

Consideration should also be given to non-technical selection criteria. The ease of use of the system, the level of support available and training, are all features which are perhaps more important criteria for selection, because it is these elements which will be critical to the success of the system during implementation and subsequent operation.

The functional specification step should be completed before any formal contract is made with suppliers. It is easy to be diverted from the original objectives when faced with the offerings of the supplier. The original objectives alone should dictate the selection criteria. The preparation of a comprehensive specification of requirements built up from a sound definition of the company's objectives and user requirements will ensure that the evaluation of potential systems and suppliers is as objective as possible, and it provides a basis for comparison.

The process of evaluating and selecting a suitable system and supplier will probably comprise several stages. With a large number of systems to choose from, a full evaluation of all the alternatives would be lengthy and costly. During the first stage, the objective will be to generate a short list. This can be achieved by extracting from the functional requirements a list of essential key features which form an enquiry document against which an initial evaluation can be made. To ensure that account is taken of the relative importance of the selection criteria, a system of ranking can be used to weight those features which are essential against those which are 'nice to have'. Final selection will be a more lengthy stage where the conformance of the system and supplier to the full functional specification will be tested. Of particular importance during the final selection are the areas identified as being critical to the operation of the system, or where there is a significant technical risk. These areas should be tested by the use of practical examples to ensure either that the system can perform the task, or that the implications of non-conformance are clearly recognized and understood.

Implementing the plan

It is during implementation that the investment in good planning will begin to pay off. However, the success of the implementation will depend largely on the users. The most cleverly designed system with admirable technical qualities will not achieve the desired objective if the people will not work it or they are inadequately prepared.

The emphasis in the implementation stage is therefore on preparing the people and creating an environment which is ready to accept changes. Keeping people clearly informed and canvassing their views throughout the planning phases will help to develop a common understanding of what is to be achieved and why certain actions and steps need to be taken. This will also help in the early identification of possible problems which will have to be overcome before the systems are installed. Some key points to consider for implementation (see Figure 31.3) are:

1 Make people aware of the strategy and the planned longer-term developments. This helps put some of the preliminary steps into perspective.
2 Ensure that objectives and expectations are clearly understood at each stage in the development. Relate these back to the basic needs of the business and how this benefits the individual users. Allow adequate time

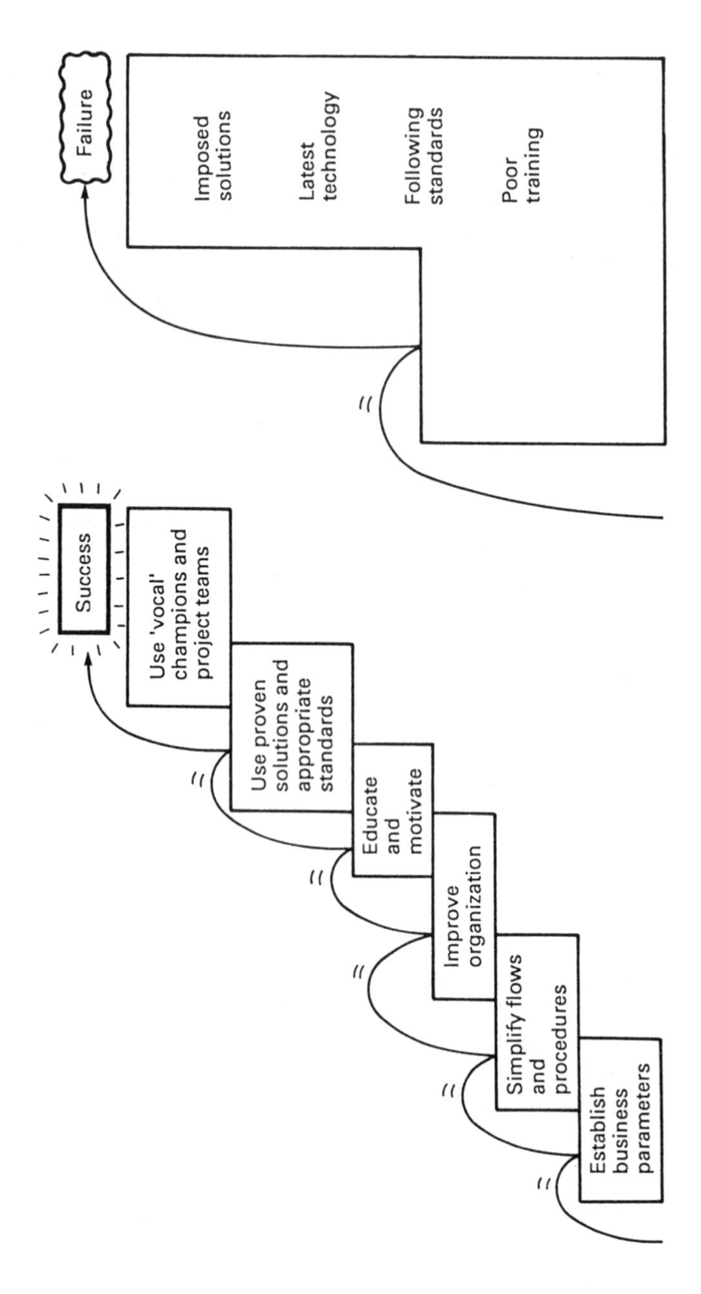

Figure 31.3 ***Simple, small steps towards successful implementation***

and resources for training and increasing awareness of the technology being employed.
3 Implement the changes in small steps with recognizable achievements at the end of each step.
4 Set project milestones and performance indicators which measure the success of the project against the basic business objectives. Do not lose sight of the original intention of the project.
5 Hold regular reviews with the project team. Monitor the project against 'milestones', and identify and solve any problems before moving on to the next stage.
6 Integrate the systems into the normal working environment. Make those who will be the day-to-day users responsible for as much of the implementation as is practical.

Monitoring and reviewing

During implementation, and when the systems are operational, there should be a regular review of the systems performance, against the requirements of the business.

It is very easy for the original objectives to be lost as more experience is gained and other capabilities of the system are discovered. This does not mean that the system should not be fully exploited and experience employed to enhance the system's use. However, all potential developments should be assessed against the strategy to ensure that resources are not being wasted on technical developments which do not contribute to attaining the fundamental business objectives. If a change in emphasis is required, review the strategy and update (see Figure 31.4). The full implications of changing direction should be understood before action is taken.

The scope of using computers in engineering

Systems

In a manufacturing company systems can be considered in four major areas:

1 Business systems. These include, for example, accounting, sales order processing, decision support and management reporting.
2 Material control systems, such as production scheduling and stock control.
3 Production systems, including (for example) process control, shop floor data collection and CNC machine control.
4 Engineering systems, including computer-aided design (CAD), process planning and project management.

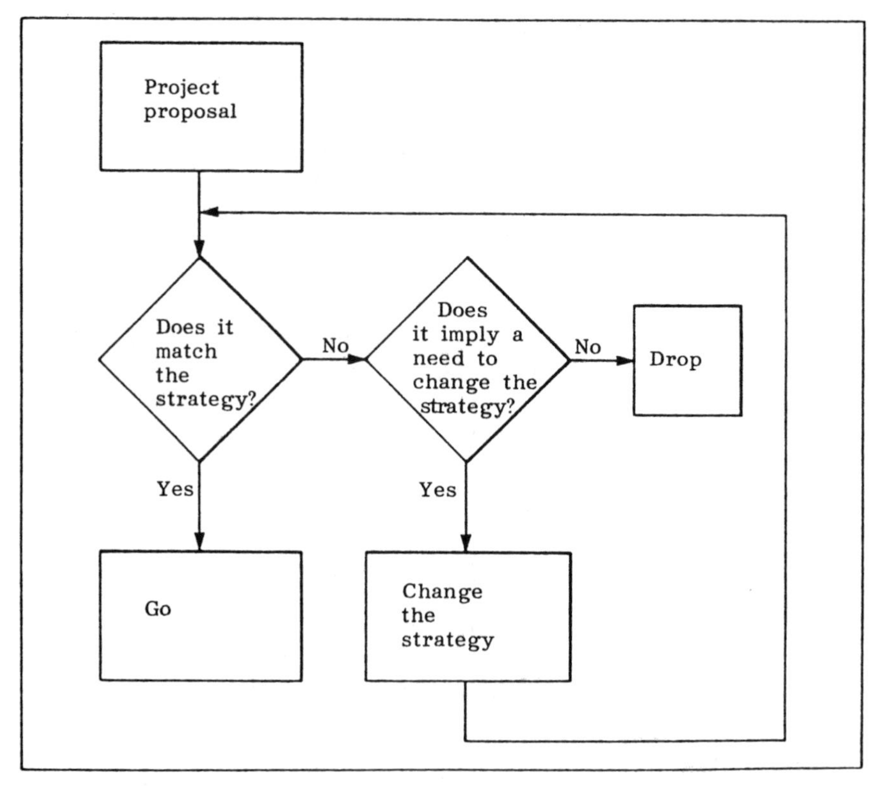

Figure 31.4 *Providing for change*

These areas are related to each other in that information is required to flow between them if the total system is to function. The principal information flows are shown in Figure 31.5

Engineering systems, as shown in the diagram, are concerned with providing information to the material control and production functions by showing how a product is to be made. This involves the provision of bills of material, process plans, and information on components and assembly which specifies the make-up and configuration of the product and how it is to be made. The major function of systems employed in engineering is ultimately to generate and supply this information.

Within engineering there are a number of activities which are carried out to translate orders from a customer requirement into the information required for manufacture. The nature of these activities will depend upon the products, and on the market place being served (Figure 31.6). For example, the engineering activities of a make-to-order jobbing shop will differ significantly

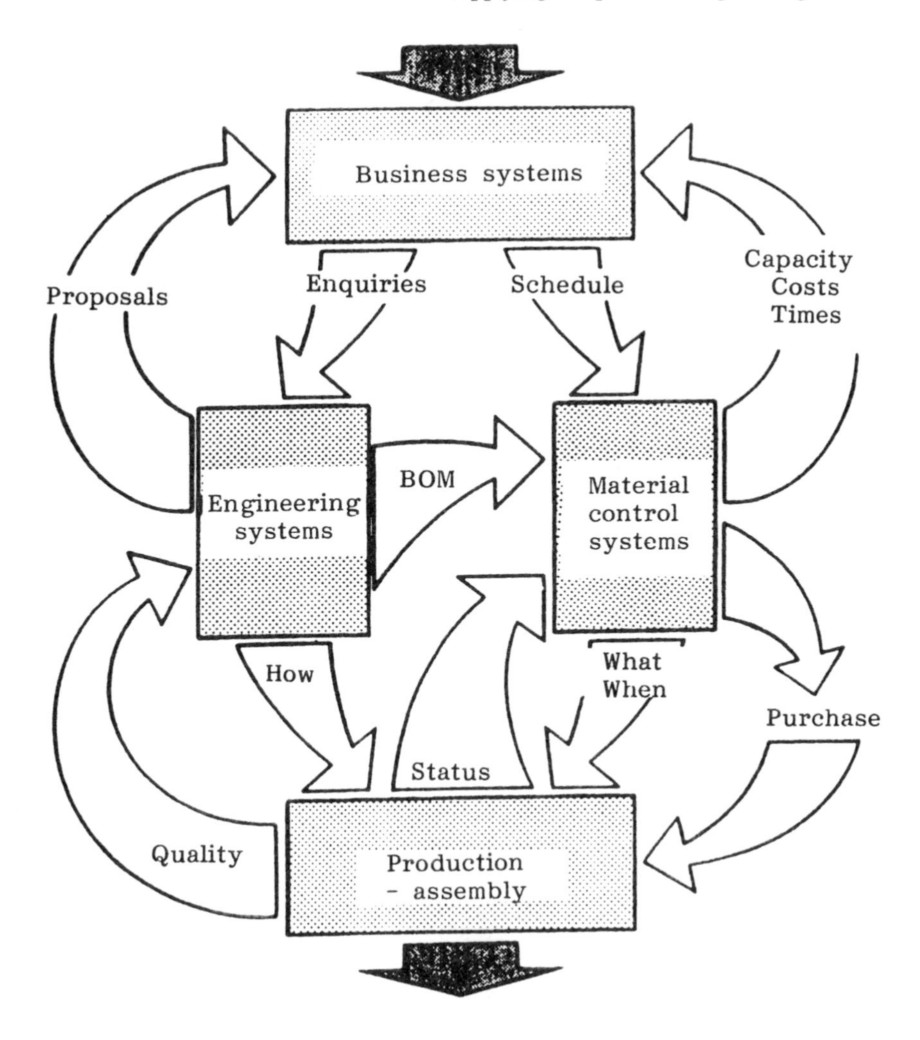

Figure 31.5 *Principal flows of information in a company*

from those of a volume manufacturer in, say, the electronics industry. These differences will alter significantly the kind of systems which would be relevant to improving the engineering function in a particular organization.

Applications for computers in engineering

Obviously computers can be used to assist in different engineering activities in different companies. The following is a summary of the principal areas which may be relevant in an engineering environment.

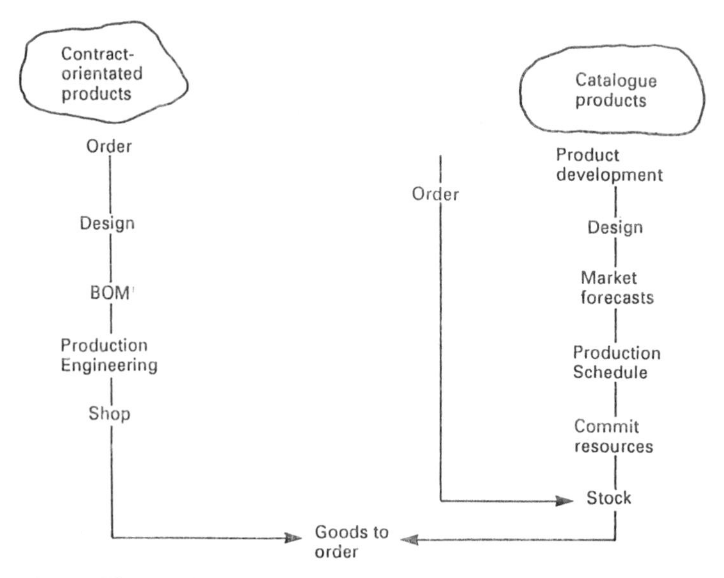

Figure 31.6 *The route from order to delivery. This chart illustrates how the route varies for different companies and products*

1. Draughting – the production of two-dimensional drawings for piece parts, schematics (electrical, hydraulic, etc.), general arrangements, and so on.
2. Design analysis. The calculation of areas, volumes, masses, surface areas, etc., stress analysis, fluid flow and heat loss calculations. Analysis of mechanisms, kinematics and dynamics. Complex spatial arrangements. Simulation of electronic circuits.
3. Industrial design. Modelling of product aesthetics, packaging, ergonomics and the effects of different colours and finishes.
4. Bills of materials. Generation and maintenance of material lists.
5. Engineering records and change control. The generation and maintenance of records relating to engineering, part numbering and modifications.
6. Standards. Introduction and maintenance of standards for components, materials, processes, features and codes of practice and, also, generic designs which can be adapted to specific applications.
7. Information retrieval. The storage and retrieval of information on previous designs, suppliers' literature, standards, etc.
8. Numerical control (NC) programming. The generation, post-processing and proving of CNC machine programs, inspection machine programs and process control data. Robot programming.
9. Jig and tool design. Design and drafting of jigs and fixtures, maintenance of standard components, parts listing and costing.
10. Process planning. The generation of process plans and synthetic times.

11 Technical publications. Drawing and editing diagrams and text for use in operating manuals, spares catalogues, etc.
12 Tendering and estimating. Preparation of tender documents, drawings and designs, and the calculation of product cost estimates.
13 Project management and control. The planning of projects using critical path techniques, time and resource planning and cost estimating.

It is likely that in any one engineering organization a number of the above will be applicable. However, the relative importance and sophistication required for each application will vary, depending upon the nature of the engineering function. For example, a pure research and development environment will possibly place a greater emphasis on the use of design analysis tools utilizing complex three-dimensional models. Project management may also be relevant, particularly on larger projects. A company primarily involved in sheet metal fabrication offering a quick response to customer orders will find a two-dimensional design system with NC programming and direct numerical control links more appropriate to the main activities of the business. In most companies there will be a core of applications making up the mainstream of the engineering activities with a number of supporting or complementary engineering tasks. The solutions for the core application areas may be very different from those required for some of the supporting activities. This may mean that more than one solution is appropriate in a single organization.

It is also likely that where a number of the above activities are carried out there will be some commonality of information between them. For example, a circuit simulation system may require access to circuit diagrams and component information in order to function. It is this sharing of information which dictates whether the two activities need to be part of an integrated system, either utilizing the same software package with a common database, or linking together two packages through a mutually understandable interface.

The need to use integrated or linked systems will depend upon the importance of transferring information from one activity to another. In some cases, where the volume of data is small and transfer is infrequent, a fully automated interface may not be absolutely essential.

The technology

Software

There are many software packages available for applications in the engineering field. The choice available is sometimes bewildering to the potential user. This makes it essential to have a clearly defined specification of what is required, if the relevant features and facilities offered by different vendors are to be distinguished from those which are unimportant to your application.

This section outlines some of the key features which differentiate the types of system available.

Computer aided design and drafting

At the core of a computer aided design (CAD) system is the software which enables the object or drawing to be described. This allows geometrical items to be understood by the computer. There are different types of 'core' software which are usually described as two-, two-and-a-half, and three-dimensional.

Two-dimensional systems most closely resemble the method of working on the drawing board, with the pen and paper replaced by a screen and electronic input device, and graphic elements represented in the computer only as two-dimensional items. They are not simply electronic drawing boards, as it is possible to develop more sophisticated applications on a two-dimensional system. Printed circuit board design, NC drilling and plate profiling are three examples of how a two-dimensional system can be used as more than just a drafting machine.

Two-and-a-half dimensional systems enable the computer to recognize that an object is multidimensional, but these systems are limited in that the third dimension is either constant in section (as in an extrusion) or a constant section revolved around an axis (such as in a turned component). These types of system were developed primarily for NC programming applications for turned components.

Three-dimensional systems can generate a model which defines the total shape of the component. This type of system falls into three categories, which can be distinguished by their increasing level of intelligence about the form of an object (see Figure 31.7).

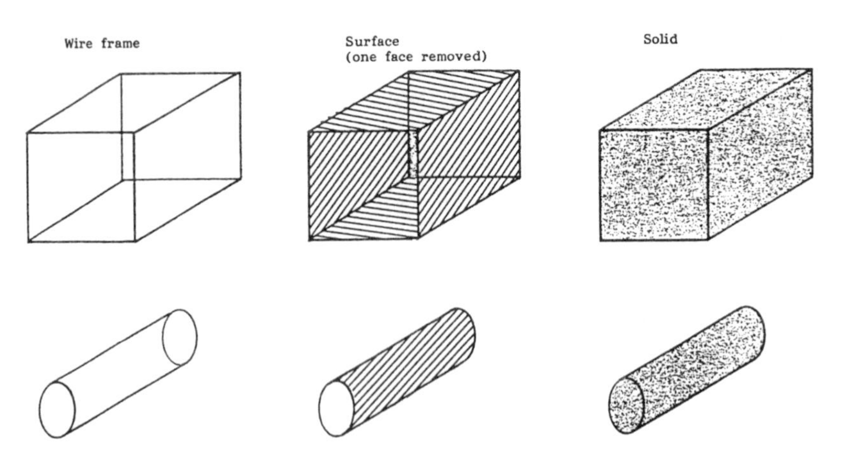

Figure 31.7 *Three techniques for three dimensional modelling*

The first category, wire frame models, can describe the edges of an object but are unable to identify the boundaries, that is, the surfaces which contain the object. A model of a cube, for example, will be made up of twelve lines.

The second group, surface modellers, represent objects by describing the enclosing boundaries. A cube would be described by six surfaces. In this case the computer is not aware of which parts are solid and which are free space. For example, a cylindrical surface could be a hole, or it could be a rod.

The third type are solid modellers which describe the objects in a way which allows the computer to identify 'substance'. It recognizes which parts of the model have mass and is able to distinguish between holes and rods. The solid model is the most complete way of describing a real object on a computer.

Most three-dimensional systems also allow for the creation of drawings directly from the three-dimensional model, although this process is not automatic. The user can take two-dimensional views of the object and add or remove lines to create a conventional dimensioned perspective.

The choice of system – two-dimensional or three-dimensional, wire, surface or solid model – will depend on the application, as there are advantages and disadvantages associated with each. Generating three-dimensional models can be very time-consuming and this must be considered against the benefits which are to be gained.

Other features which can be found in CAD systems are:

1 Parametric programming – the ability to program the computer to generate a component drawing or model automatically by specifying key dimensions. This is particularly useful in manufacturing a range of components which are similar in shape and where the variables can be specified as a set of dimensions.
2 Symbols. Common shapes can be drawn once and stored complete – for example, electrical and hydraulic circuit diagrams.
3 Attributes. Attributes allow items of non-graphical information to be stored with the drawing or model. This could be text information, such as part material type, component name; or numbers such as part number, revision number, etc.
4 Programming languages. Many systems are supplied with a programming language which allows the user to tailor the system to his own requirements by writing programs to create graphics, or handle non-graphic information.
5 Macros. These enable the user to define a sequence of commands to be executed in one instruction – a useful aid to speeding up repetitive routines.

Computer aided engineering

The term computer aided engineering (CAE) generally refers to systems which enable some form of design analysis to be undertaken. In many instances these systems use the geometry of the object generated in a computer aided design system as the basic information on which the analysis is carried out. An example is finite element analysis, which uses a geometrical model of the component as the basis for the determination of stresses. Many modelling packages have direct interfaces with design analysis software which enables the part geometry to be utilized directly.

There are also software packages available which are not based on CAD. These are programs written to carry out engineering calculations and they rely on the parameters being specified as numeric or text variables.

Computer aided manufacture

Computer aided manufacture (CAM) systems cover activities which convert the design of the product into instructions that define how the component will be made. This includes systems for generating machine tool programs and process plans.

The systems for machine tool programming vary in sophistication, with those available ranging from relatively simple two-dimensional applications, such as turning or drilling, to more complex software for five-axis machining. These systems are largely graphics-based and can be used to generate the toolpaths directly from a design which was defined in a CAD system. Many systems are capable of creating both the product geometry and the machine tool instructions, but some rely on transferring geometric information between two different systems by means of a direct translator.

Process planning packages also vary in the way in which they handle the information required. They range from simple record and replay systems to more intelligent packages which can generate process plans based on descriptive attributes of the component.

Interfacing between different CAD packages

Transferring information from one CAD system to another is a feature which many applications may require, particularly where information is to be passed to customers or suppliers, or between different systems within the same group of companies. Hardware links are generally well proven, as is the transfer of text; and standards in these areas, whilst not totally reliable, are well advanced.

Transferring graphics information between two different systems can be achieved in two ways: through direct translators which convert the data of one system to that which can be understood by the other; or through a standard interface definition. One such standard is Initial Graphics Exchange Specification (IGES). However, because of the complexities of transferring graphics and the different interpretations of the standard by different vendors, the IGES interface can be unreliable. In addition, not all types of geometrical information have as yet been incorporated into the standard and so some types of model cannot be transferred via IGES.

Hardware

In general, engineering systems comprise a number of basic elements, namely:

1 Processors.
2 Screens.
3 Input devices.
4 Plotters and printers.
5 Storage devices.

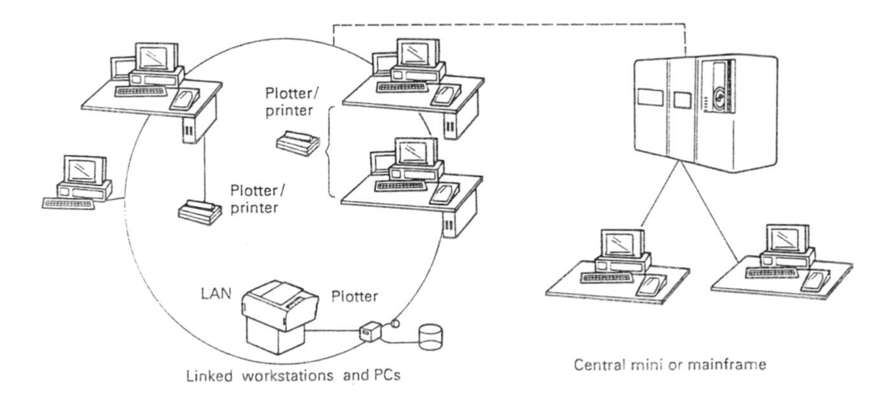

Figure 31.8 *System configurations*

There are two main approaches to configuring a system (see Figure 31.8). The first utilizes a central processor supporting a number of workstations. Each workstation shares the facilities provided by the central processor. The second type is where each workstation has its own processor and can run independently. These workstations fall into two groups, namely systems based on personal computers (PCs) and those which are commonly called engineering workstations. The difference between these two groups has become blurred as advances in computer technology have removed some of the distinguishing features. However, engineering workstations are generally high-performance computers which are capable of handling the more demanding, processor-intensive tasks which frequently occur in engineering applications. Most types of processor can be linked together by a network to allow the sharing of data or the sharing of peripherals such as plotters, printers and storage devices (Figure 31.9).

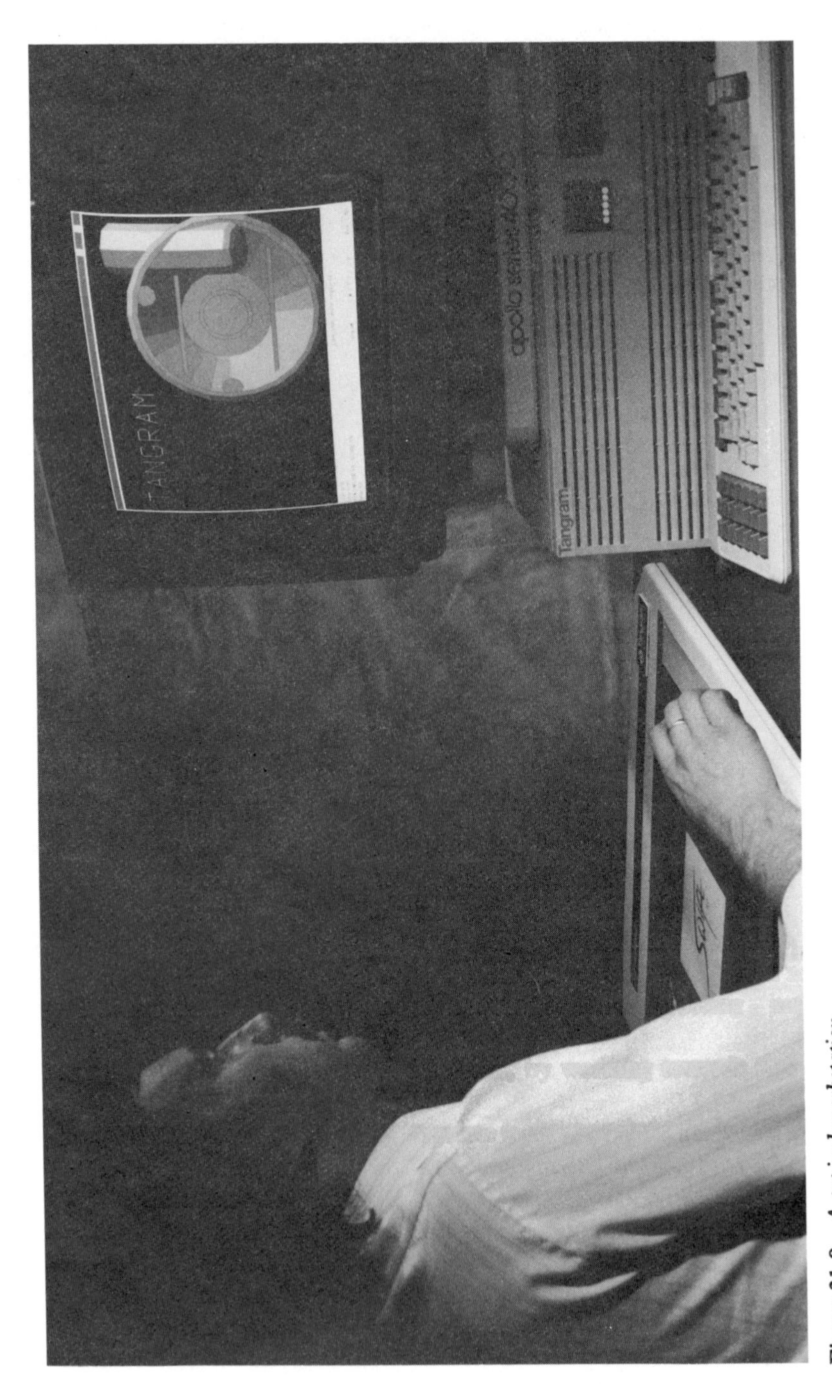

Figure 31.9 *A typical workstation.*
(Courtesy of Tangram Computer Aided Design Limited)

The screen is used to display the information and, where only alpha-numeric data is displayed, low-resolution devices are adequate. However, where graphics are used, high-resolution screens are employed as they give a clearer picture, particularly when the graphics are complex. The screens are generally larger, permitting a greater area to be viewed without loss of definition.

There are different types of input devices available which enable the user to interact with the system. The keyboard is still used for text input, but most systems can be driven by a mouse or digitizing tablet. These allow commands to be selected and a cursor to be moved across the screen to construct geometrical features.

The most common form of output from a CAD system is drawings produced by plotters. These can be either pen plotters or electrostatic devices. Many systems can now output to other devices such as laser printers, ink jet plotters, dot matrix printers or directly on to photographic film. The type of output device chosen will depend upon the type of hard copy required, the quality of print and the number of prints to be produced.

Storage devices are required for the files generated by the system. For personal computers and workstations, there is typically storage local to the processor, or a common storage device provided through a network. On central systems, the storage devices are normally attached to the central processor. The use of networked or central storage provides greater flexibility and the rapid transfer of information between users. The use of storage devices requires careful consideration, particularly for data which is common to all users, which is a 'master copy' or which requires confidentiality. Procedures need to be designed so that the integrity and security of the information is maintained at all times. The use of networked or central storage devices, with an adequate operating system, will allow these procedures to be controlled automatically.

Conclusion

The key task for engineering management is to contribute to the wealth of the business. There are a number of computer-assisted techniques available which can be used to help in this process. It is the responsibility of the engineering manager to ensure that advances in computer-based systems are effectively used within a company and that they provide a worthwhile contribution. The success of any project involving the introduction of computers will depend upon the ability of the manager to:

1 Identify the key business drivers for improvement.
2 Develop a coherent strategy which achieves these improvements, taking into account the total activities of the organization.

3 Seek out new ways of doing things and challenge traditional methods.
4 Plan in detail the requirements and implementation of the systems.

The use of computers in engineering can give significant benefits, but it is the selection of appropriate, well-proven systems and the subsequent successful implementation which will yield the required results. This cannot be achieved without a sound strategy aimed at meeting key business objectives.

32

Designing for excellence and visual quality

Dan Johnston

Works of genius are, of course, rare. They stand out in history and quite often they break the rules – at any rate the supposed rules when the work was created. This chapter is about something less than that. Nevertheless, excellence is a word of high commendation, a word that indicates great achievement. Whenever it is used in relation to creative work – literature, music, art, architecture, capital goods engineering, or even the design of products for the consumer market – a number of things will be implied.

Factors affecting good design

The concept

The starting point will have been an idea, a worthwhile concept, something capable of realistic development and involving a degree of innovation. If the concept proves good enough, it will still call for the highest quality in the writing, painting, building, engineering or design that result in the final product. The variety of characteristics to be looked for in the working up of a concept will vary according to its nature. Some of them will seem obvious. Added on to the merit of the original concept they may result in excellence overall.

Good function

In engineering, and generally in the design of products large or small, the first of these considerations must be efficient function, efficient achievement of the purpose for which the product is intended. Without that, other desirable qualities of design, however meritorious in themselves, will be irrelevant.

This overriding requirement for good function implies, if possible, that the new product will be better than the competition (if there is any). It involves, if a machine, its rate of production, the precision with which products are turned out, its reliability, its flexibility in relation to the processes possible, ease of setting up and maintenance, cost and cost of operation, etc. At the other extreme the main purpose of a designed product may be decorative – for example – a carpet. But, as well as pleasing people with the nature of their colour and pattern, carpets for modern homes, hotels and offices have a functional purpose. They must have highly desirable qualities in relation to use – good pile resilience, cleanability, warmth on the floor and reasonable fastness of their colours to light. Even at the decorative end of the spectrum of designed products, good function comes high up the list of required criteria for good design.

Safety

Safety is clearly a vital consideration with many products, important with most machines and the same thinking is equally applicable to children's toys.

Ergonomics

Ergonomics is the study of human factors, the man–machine relationship or (alternatively stated) the consumer–product relationship. Obviously, good solutions to ergonomic problems are important with the design of an aircraft pilot's instrument panels. But equally important, in a different way, is the ergonomic approach to the design of a typist's chair.

Materials used

The most appropriate choice of materials for product manufacture is another important design factor (the best plastic to use for a car component, or perhaps the packaging for cosmetics, are examples).

Method of manufacture

The choice of the most appropriate (or the best available) means of manufacture.

Cost

The last two considerations, materials and manufacturing method, will have a major impact on function. They will, however, also influence cost – value for money – which is itself a vital factor in evaluating the merit of most products and their design.

Appearance

Last in this listing, but certainly not least in importance, consideration must be given to appearance design – what the 1949 Registered Designs Act calls 'features of shape, configuration, pattern or ornament applied to an article by an industrial process or means, being features which in the finished article appeal to and are judged solely by the eye . . .'.

This legal definition of so-called 'aesthetic' design covers a wide span. 'Features of shape' are at the opposite end of the list to applied features of 'pattern or ornament'. Indeed it can be argued that products involving an elegant engineering solution to a problem create an aesthetic appeal of their own. The neatness of the way in which good performance has been achieved may be quite sufficient in itself. There will be no need for applied ornament; indeed it would be totally out of place – a tasteless gilding of the lily.

Nevertheless, when it comes to the design, for example, of domestic appliances, ingenuity in respect of function, neatness of detailing and convenience in use may still not be enough. Sales appeal, and the marketing men's assessment of it, will usually call for carefully considered use of colour, inbuilt lighting, nicely chosen (but different) knobs, catches, switches and indicators. There may be something new by way of surface treatments, textured finishes. Still further, graphic treatment will be deemed important – clear but distinctive typography, lines (in the right places) and even simple patterns may be thought necessary to break up plain surfaces and give the product a less than clinical appearance.

Varying taste

Of course, taste comes into all this. The dictum 'form follows function' (with form meaning design) is clearly true, so far as it goes. But the first impression made by a designed product can be crucial. Its looks may be the deciding factor in sales, however illogical this may seem in relation to function. Also people vary greatly when their judgement is applied to intangible matters of taste. Many people, if not a majority certainly a large minority, feel happy (reassured) if they are surrounded at home and at work by things they know, things they understand, products whose appearance does not seem strange. They like products to have nostalgic appeal. Where technical change or even complete technical innovation must be accepted they prefer that it should be disguised by an outward appearance inherited from quite different products of years ago.

There is another minority whose enthusiasm for change is extreme. It sometimes seems that appearance design is required at all costs to be different; an impression must be made rather than the contribution of a new but practical design solution.

A middle view takes the more sober line that modern technology should be clearly linked with industrial design solutions of the same era; that each age should express its confidence in the products it produces by giving them an appearance that relates to the contemporary environment. In short, without striving too hard, modern products should be given a modern look.

Good and bad in modern design

It is, of course, true that giving a product a modern look will not ensure that it is of good appearance design. It may be a pale imitation of what someone else did a few years before. The detailing may seem not to have been well managed. It may seem to be clumsily proportioned. If decoration is involved it may have been overdone. Perhaps there is too much trim. Ostentation is usually the antithesis of good design.

The challenge to all designers is that the products resulting from their work should show that a genuine contribution to new thought has been made – and in a controlled, well thought out way.

As a footnote to all of this it must be said that both popular and educated taste in design are constantly changing. Popular taste in particular is notoriously fickle. It is no surprise, therefore, that by no means all products that are acclaimed by experts as being of good design prove themselves as best sellers.

Visual appeal and engineering products

What has been said in the last few paragraphs about appearance design relates mainly to consumer products. Many engineering products do not need to have visual appeal. Such products have been referred to in government papers concerning possible changes in copyright law as being 'purely functional' or in a more extended form as 'three dimensional articles where the appearance of the article does not influence the purchaser, who buys the article only in the expectation that it will do the job for which it is intended'. Some spare parts for motor vehicles (also spare parts for domestic equipment and industrial machinery) are indeed made to designs that are 'purely functional'. A good example is the often replaced exhaust system for popular cars, the subject of much copyright litigation in the 1980s. As it was brought out strongly in the courts, such components have to meet design criteria in the same way as more glamorous products. But the criteria do not include appearance qualities.

However, it needs little argument to show that most engineering equipment benefits from having a good, purposeful appearance. It is easy to say (and obviously true) that the sale of production machinery, heavy outdoor equipment, or delicate scientific instruments should be determined almost wholly by practical considerations. But logical thinking in its design should extend to good and appropriate choice of colour, the avoidance of cluttered surfaces, the choice of easy to read dials and well executed lettering. Appearance values, well considered, result in a businesslike look, with a better impact at exhibitions, in publicity brochures, and at the point of sale. Finally, there is a psychological benefit for all concerned when the product is in use.

Industrial designers

Where the industrial designer should fit in

The designer responsible for the product's ultimate appearance will usually be an industrial designer. And the fact that 'aesthetics' has been considered last in the catalogue of factors affecting good design does not mean that he should be brought in when all the other problems have been solved. That this often happens is a bitter complaint made by many industrial design consultants. They feel that they have been seen as *only* concerned with appearance; that they are being asked to do 'a tidying up job'. In reality they are anxious that good appearance should be married in with good function and all the other characteristics of good design. They will feel that they could have contributed much more if they had been seen as a full member of the design team from the beginning.

Sometimes the industrial designer may be the person who inspired the original concept. In any case, he should certainly be given the chance to influence the way it is developed, particularly by thinking from the user/consumer's viewpoint (ease of operation, ease of maintenance, ease of replacement of parts). Along with other members of the design team he will look for simpler, more elegant design solutions. This may lead to cost saving through rationalization of components. He may also have a major influence on cost through his interest in the materials to be used and their decorative possibilities.

Engineering and the industrial designer

Other chapters in this book have dealt with the nature of engineering design and the qualities and skills expected from the engineering designer. Before explaining what is expected from the somewhat misleadingly entitled industrial designer, it must first be acknowledged that for many engineering projects he is either not needed at all or his contribution can quite reasonably be absorbed into the work done by a good engineering designer.

However, due to the vast expansion of scientific and technical knowledge, specialization is the order of the day. Indeed, the industrial design profession is itself an amalgam of specializations. Probably more than half of the profession are graphics designers – designers for print in all its manifestations. Then there is the world of fashion and textiles; designers who specialize in interiors, the theatre, films and TV; product designers who work in pottery,

Figure 32.1 *Inca Showcases by Click Systems Limited. The aim of the designer (Sandy Webster) was to give 'the aesthetic excellence of an absolutely clean glass box'. The cases also have high security rating, but can easily be opened (by the right person), readily maintained and are flexible in use. Chosen in the durable consumer and contract goods section of the 1987 British Design Awards programme*
(Courtesy of Click Systems Limited)

glass, toys, jewellery and many other industries. And then those industrial designers whose work relates especially to engineering.

Consumer products such as domestic appliances and hand and garden tools have long been seen as requiring the skills of the industrial designer. So also have those products of engineering industry whose use is close to the lives of all ordinary people in a modern society (for example, cars, trains and ships). An increasing number of industrial designers are also employed in areas of capital goods engineering where the general public is less directly involved.

With processing machinery and outdoor equipment the designer's concern will be with working convenience and appearance values relative to the indoor or outdoor environment.

The training of industrial designers

The great majority of industrial designers in the UK are trained in art and design colleges or at polytechnics. Their basic educational requirement before entering college to do a degree course will have included a range of GCSE and A level passes (or their equivalents) in subjects appropriate to their proposed specialization – for product design in all probability, mathematics, physics and engineering drawing. Their college courses will include training in the characteristics of materials and in industrial workshop and drawing office techniques. Students will be given some insight into marketing and managerial methods. Ergonomics will obviously be an important subject but the main emphasis in the training of an industrial designer is on building up his creative skills. Design is seen as a problem solving activity so actual design projects are given high priority.

Students who have done the full course will hope to emerge with an honours degree – BA. Some will go further. They may do master's degrees or attend postgraduate courses at the Royal College of Art.

At the conclusion of their design training most young designers seek work as staff designers either directly in industry or with older designers who have established themselves as design consultants. Others work abroad as a way of broadening their experience.

However, it is important to note that a significant minority of industrial designers come from quite different backgrounds. They may have trained as engineers, or architects or, less frequently, in mathematics, electronics or computer sciences.

The Chartered Society of Designers

The professional body to membership of which most British industrial designers aspire is the Chartered Society of Designers (CSD), renamed in 1986 having previously been known as the SIAD (Society of Industrial Artists and Designers). The Society was founded in 1930 but was virtually reconstituted after the war. Its original membership was heavily biased towards the visual arts and architecture but over the years its involvement in engineering and related fields has grown steadily. Engineering designers were given a category of membership in 1980 alongside the full range of industrial design specializations from fashion and graphics to interiors and three dimensional objects of all kinds.

A fundamental principle on which membership selection is based is that the would-be member must be able to show evidence of actual work done. This is,

Figure 32.2 *Vitalograph-COMPACT spirometer. Spirometers have long been used to test lung function. Designed by a Vitalograph team, with consultant Julian Abrams of David Stubbs Design Associates, this unit has had great success because of its portability, and the fact that it is user friendly, allowing quick tests without distress to the patient. Chosen in the medical equipment section of the 1987 British Design Awards programme*
(Courtesy of The Design Council)

of course, important in the context of design. People may have been trained well. They may advocate all the right things. But if they have not put their training and their thinking into practice they may not have measured up to the problem of reconciling idealism with the demands of the marketplace.

The Society has a code of conduct which emphasizes the professional nature of the designer's work in relation to employers, clients, fellow designers and society in general. For example, work done for an employer must be treated as being in proper confidence. Also, a consultant designer must not accept similar commissions from competing firms without their full knowledge and agreement. It is a condition of membership that members should abide by the code and members failing to do so may in the last resort be expelled from the Society.

The CSD includes as an affiliate body the Design Business Group. The Group's membership consists of design practices ranging in size from the largest groups employing some hundreds of staff down to quite small partnerships.

An important service offered to industry and commerce is recommendation by the Society of a short list of designers whose capabilities will have been matched against any enquiry made.

The Society has some 8000 members; Fellows are entitled to use the affix FCSD, Members, MCSD. There are also Honorary Fellows, Associates and Diploma Members. The Society has links with other similar bodies worldwide and it provided until recently the secretariat for BEDA (The Bureau of European Designers Associations).

The CSD's address is the Chartered Society of Designers, 29 Bedford Square, London, WC1B 3EG.

Briefing the industrial designer

Carefully prepared briefing is of prime importance for all industrial design work, whatever the designer's relationship with the company (member of staff, freelance or consultant). If the designer's briefing is inadequate, time will certainly be wasted, an opportunity may be lost or, worst of all, a poor design may result.

Staff designers

If an industrial designer is employed in a staff capacity, he should be briefed for his particular task, and he should contribute to it from his special knowledge and skill, in exactly the same way as his engineering colleagues. Like the rest of the team he will do a better job (and he will have more job satisfaction) if he is given a good general picture of the project on which he is to work. He should also be given copies of any preliminary marketing or other studies that are relevant.

A point which not infrequently arises with staff designers is whether or not it is acceptable that they be free to undertake additional work for other people in a freelance capacity. If agreed, such work should not normally be allowed to involve any conflict of interest with the business of the principal employer. There is a possibility, of course, that the designer's value to his main employer will be enhanced by reason of the greater experience and reputation he will gain by undertaking outside work.

Consultant designers

If the industrial designer is employed as a consultant he will be unlikely to know as much as a staff designer would about the company's production methods or its marketing potential. On the other hand, apart from his design ability, the outside consultant will be expected to bring to his task a wider experience and an open mind. To get the best possible result it will be necessary for the management of the company not only to brief their consultant very thoroughly, but then, in every possible way, to integrate him as fully as possible into their design and development team.

The consultant designer's briefing should start with the product concept, then the market aimed at, the manufacturing method proposed, the price range intended, the target date for launch. The briefing should be backed up by any relevant marketing reports and by an opportunity to study the firm's production facilities. It may be that the consultant should meet likely purchasers of the product proposed. The main points of the briefing should be confirmed in writing. A senior member of the company management should be nominated as the consultant's day to day contact.

Planning the design work

For his part, the consultant (or the staff designer) will usually plan his work in phases:

1 Consultations to establish complete understanding of the project proposals.
2 Submission of preliminary sketch ideas to clarify thinking on the part of both the designer and the company.
3 Further development of preferred ideas, supported by visual presentations or first models (to be narrowed down to one as soon as possible).
4 Production of more detailed drawings, leading to a working model or first prototype (according to the nature of the product).

Contact will be maintained with company management throughout and it is usually advantageous (with consultants) that visits be exchanged. The method of making models or prototypes will be a matter for agreement according to the availability of facilities. (Consultant designers who do not have model making capabilities of their own usually have close contact with specialist model makers.)

At this stage in the design/development process it will be necessary to have a formal meeting at which the designer presents his design proposals – hopefully, for authoritative agreement as to further progress. It may be that modifications will be required. And, if a range of related products is intended, further drawing office work (and model making) will be agreed.

Final engineering drawings for production of the product will be undertaken by the consultant designer (if employed) or it may be that the company's staff engineers will take over. It is highly desirable that the designer (staff or consultant) be given the opportunity to cover the final development stages right up to full production. When it comes to planning for the launch of a new product it may be that the designer will be able to contribute usefully to the preparation of sales and user instruction literature.

Consultant designer contracts

A consultant designer may be retained by a company in a more or less permanent role, in which case he will be paid a monthly or an annual fee to

Figure 32.3 *Super Snorkel by Simon Engineering (Dudley) Limited. Designed by Denis Ashworth and David Johns, these high-rise work platforms are used for a variety of industrial purposes, and for fire fighting and rescue. The Super Snorkel and Super Sixty platforms were chosen in the engineering products section of the 1987 British Design Awards programme* (Courtesy of Ivanhoe Service Industries Ltd)

cover general advisory work. Extra fees will be paid if he is called on to undertake design work on specific projects. He will not be able to accept design commissions from other companies if they conflict in any way with the interests of the firm retaining him.

Alternatively a consultant designer may be employed in a one-off capacity – possibly having had no previous connection with the firm. He will be asked to do a specific job and he will be asked to quote for doing it. This he will do by quoting a lump sum, supported by his estimate of the time he will spend on each phase of the job and details of the work he will put into it. He should clearly identify the stage to which he will progress the design, for example whether or not he will produce a model or perhaps go on to final production drawings. He should give starting and completion dates.

A second method may be for the designer to estimate the time likely to be taken by the job and to quote hourly or daily rates of fee, perhaps at different levels of skill. He will not be bound by his estimate of total cost and will submit monthly accounts as the job goes on.

Sometimes royalty arrangements are made – whatever the method of employment of the designer. Such arrangements will be seen as being in lieu of, or by way of reducing, fees.

It is essential that consultant designers be engaged on a clear basis of understanding between them and their clients. The design task involves imponderables. If success is to be achieved good personal relationships are essential. For this reason it is important that contracts of engagement should be clear; not likely to lead to misunderstanding. They should be in writing and have a legal basis. It is desirable that such contracts should cover intellectual property rights – patents, registered designs copyright, or design right. If such rights are to be vested in the employer this should be stated in the contract. As regards copyright or design right, signed and dated drawings should be handed to the client and the right assigned to him when the final fees for a project are paid.

Industrial design partnerships

Some consultant industrial designers work as individuals – from home or from a small studio/workshop. At the other extreme are quite large businesses; there are some two dozen or so large partnerships, some employing perhaps two or three hundred staff. Most of these large groups cover all aspects of industrial design but the emphasis seems to be on graphic design and planning for the retail trade. Design partnerships specializing in work for engineering-based products usually have two or three principals and a total staff of twenty or thirty. Some such groups have small but impressive workshops so that they can back up their design work with good quality model making. Several have associates who are not directly designers but have special, design related skills

– such people as computer specialists, electronic engineers, etc. In the graphics area such back-up people may include copy writers and multilanguage specialists.

As compared with the individual consultant larger groups have an advantage ensured by their greater resources. They will also have a greater spread of experience – perhaps abroad as well as in the UK. The actual design work may be done by assistants but they will be supervised by people of established ability.

However, many very competent people prefer to work virtually on their own. Such people will usually be very keen – they supply their own motivation! And there may be some satisfaction for their client in feeling that he is dealing with the boss throughout.

The Design Council

The Design Council was set up in the UK by the war-time coalition Government in 1944. Its purpose was (and still is) 'to promote by all practicable means the improvement of design in the products of British industry'. More than forty years ago it was realized that Britain was bound to face an increasingly competitive future in the industrial world. It was rightly felt that good design would be one of the keys to successful international competition. That is still being said just as loudly as ever.

The original name chosen was the Council of Industrial Design (the CoID). It was funded by the then Board of Trade but encouraged to earn other revenue by charging for its services wherever possible.

The name change to The Design Council came in 1972, marking an important change of policy. The range of the Council's activity was to be greatly enlarged so that engineering design would be covered equally with industrial design.

The Design Advisory Service and 'Support for Design'

The principal activities of today's Design Council include its Design Advisory Service, staffed by engineers and designers chosen to cover most relevant aspects of industry. They are geographically based so that they identify as nearly as possible with the whole of the UK. For some years the Design Advisory Service embraced 'Support for Design', a scheme separately funded by the Department of Trade and Industry. Specific design projects were agreed following application by small and medium-sized companies. Design consultants, appropriate to the task, are then chosen, with their fees paid for a limited period. The firms receiving the help had to pay a proportion of the design cost. Apart from giving an immediate boost to the development

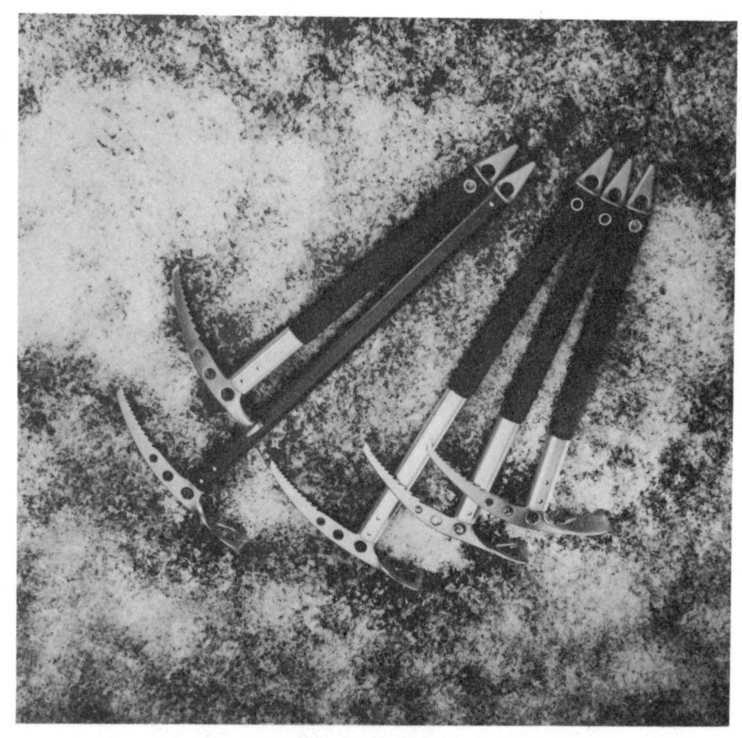

Figure 32.4 *Ice axes designed by Hugh McNicholl. These ice axes by Mountain Technology (Glencoe) Limited are designed for various uses from winter hill walking to climbing in severe ice and snow conditions. The heads are drop forged in nickel chrome molybdenum steel, the shafts are drawn high tensile aluminium alloy, the end spike is an aluminium alloy drop forging with a hardened steel insert, and the assembly is bonded with toughened epoxy resin adhesive, with stainless steel rivets for maximum security. This range of axes was chosen in the durable and contract goods section of the 1987 British Design Awards programme*
(Courtesy of The Design Council)

of new products, another benefit foreseen was that firms would be encouraged to use design consultants in the future. This scheme has now been absorbed into the Enterprise Initiative Scheme of the Department of Trade and Industry (DTI).

The Designer Directory and the Engineering Directory

The Council's Design Advisory Service is supported by two offices responsible for recording the work done by design consultants. The Designer *Directory* covers the full range of industrial design consultants working individually or in partnerships and groups. There is an annual publication, which lists them together with their specializations. The second record, the

Engineering Directory, covers engineering design consultants, again with details of their special skills. It also includes research organizations and university departments able to offer help on technical problems related to design.

The two records make recommendations of designers or supporting specialists against enquiries received from industry. They also select consultants as part of the DTI's Enterprise Initiative Scheme.

Publicity for design

Publicity relating to good design is a very important part of the Design Council's total activity. Accordingly there are two monthly magazines, *Design* and *Engineering*. Publicity is also achieved through contact with the press, radio and television. The focus for much of this publicity is the well known Design Centre in London's Haymarket. The Design Centre has been open six days a week since 1956. As well as showing a constantly changing exhibition of well designed products of British manufacture many thematic displays are mounted covering particular aspects of design. The Design Council has smaller versions of the London Centre in Glasgow and in Cardiff.

The success of the London Design Centre has inspired the opening of similar centres in several European cities and indeed throughout the world.

Other activities

Other areas of Design Council activity include the publication of books on design subjects, and a broadsheet for schools, *Designing*, which is produced once a term. This is intended for teachers and students of crafts, design and technology. The Council organizes courses and conferences on design subjects relevant to industry and commerce. Finally, there are the Design Awards, which are sufficiently important to warrant their own section in this chapter.

British Design Awards

The choice of these awards is one of The Design Council's best known activities. The Council's official brochure says that:

> The very best of British design is highlighted throughout the world by the British Design Awards. Given annually to a select few products, these awards are recognized as the premiere mark of design achievement for British manufacturers, designers and their products. The awards cover consumer, contract, and engineering goods as well as medical and motor products and computer software.
>
> The award winners are presented by HRH The Duke of Edinburgh with certificates to mark their achievement at a special presentation ceremony. The winners also gain from the widespread publicity that the British Design Awards obtain in the press, special exhibitions, and in other media.

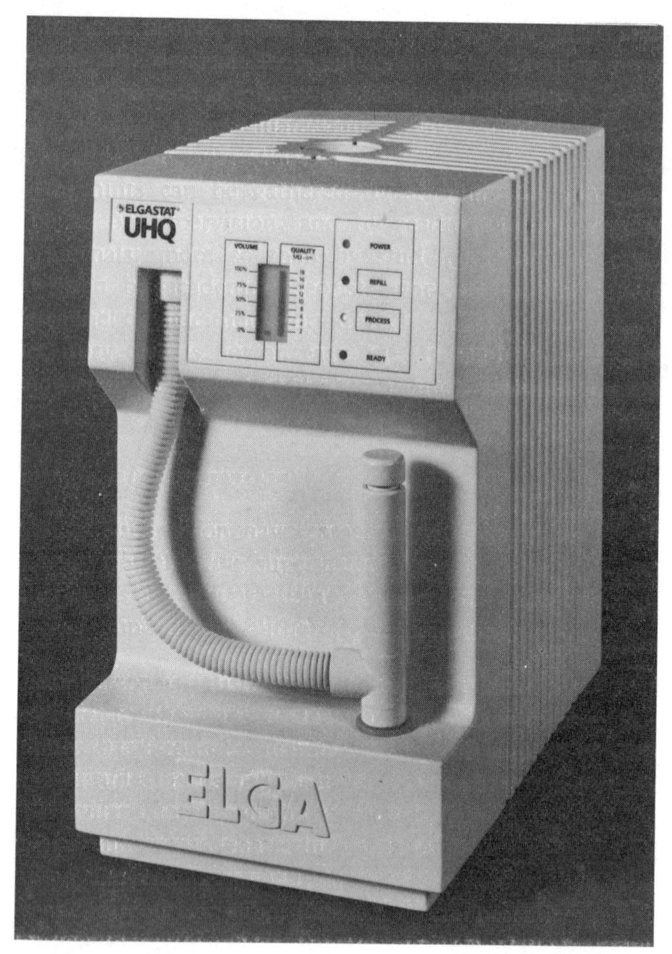

Figure 32.5 *Elgastat UHQ water purification system. This bench top unit produces ultra pure water efficiently and economically. Ultra high purity water is essential for critical laboratory analysis. This product was designed by Elga Ltd, with consultants Anthony van Tulleken and Sean Greer Perry. It was chosen in the engineering products section of the 1987 British Design Awards programme*
(Courtesy of Elga Ltd)

In addition, firms winning awards can publicize the fact in their own promotional programmes.

The Duke of Edinburgh's Designer's Prize

The awards scheme is topped up by The Duke of Edinburgh's personal Designer Prize which is chosen every year from the designers associated with that year's British Design Awards. The choice is made personally by the Duke of Edinburgh in consultation with the chairmen of the several judging panels. The prize goes to the designer or design team leader who has made the greatest contribution to design. The prize is itself a design challenge because the winner is asked to use the money to commission or design and make an object to mark his/her achievement. It is presented by the Duke at the awards ceremony in the year following.

The selection system

The various judging panels are chaired by notable people from business, the academic world and from engineering or design. Their panel members likewise give a balanced representation of these interests and (where appropriate) members are also drawn from consumer interests, from retailing and from journalism. The Design Council suggests that the judges should look for the following in products submitted to them:

1 Design innovation in the concept of the product.
2 Benefits in manufacturing, particularly in the effective use of materials and resources.
3 Outstanding appearance, ergonomic features, accompanying literature and user manuals.
4 Ease of use and maintenance.
5 Good functional performance and reliability during an acceptable period of service use, substantiated by user reports.
6 Good commercial performance and value for money.

The rules applied are very much as would be expected in a judging process of this importance, including the requirement that judges should declare an interest in the design of any product or its manufacture. It may be that a company will be asked to defer its entry for one year. The panel members are given help as necessary from technical advisers and may seek user opinion. The Design Council itself must ratify each panel's selection of products.

All products considered must be predominantly of UK design and manufacture.

How to enter

A company may submit any number of products and an entry form must be submitted for each of them. The closing date for entry is the end of August but there may be some variation from year to year. Judging begins in September,

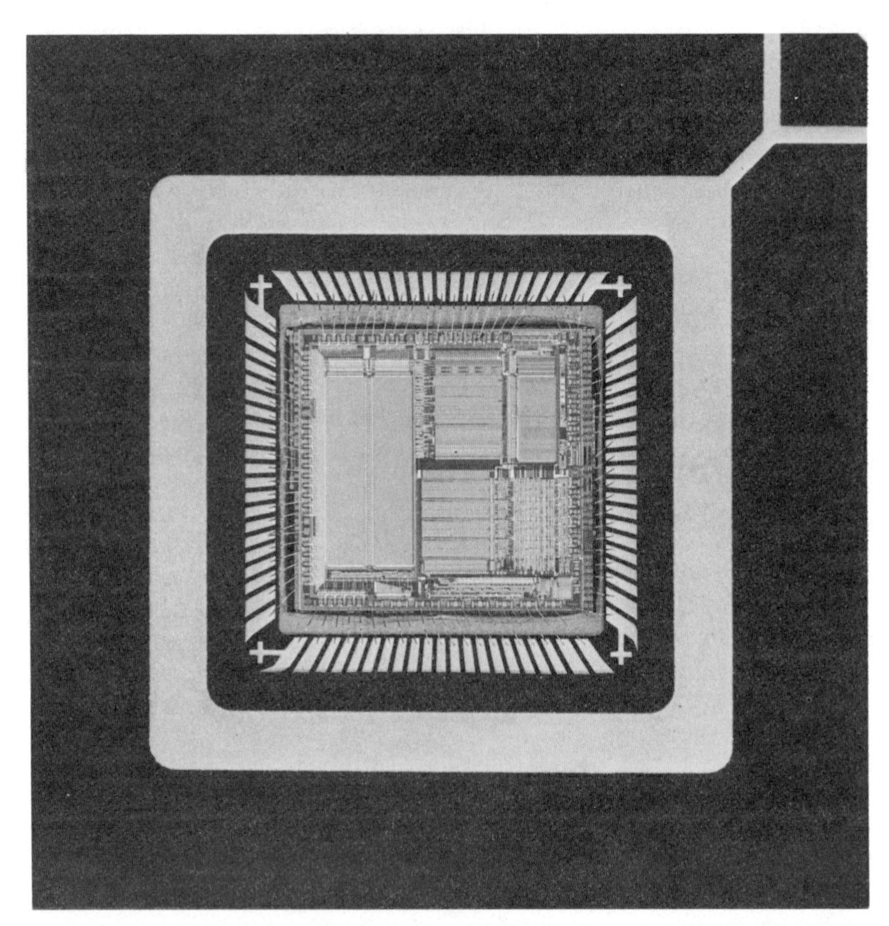

Figure 32.6a *IMS T414 32 bit transputer by INMOS. On a single VLSI chip, smaller than a thumbnail, this transputer combines a 32 bit microprocessor with on-chip memory and four intertransputer links. This makes it possible to carry on high speed communications concurrently with information processing. Designed by the INMOS team, led by Tony Fuge, this product was chosen in the components section of the 1987 British Design Awards programme. Tony Fuge was also chosen to receive the Duke of Edinburgh's Designer's Prize* (Courtesy of INMOS Limited)

leading to a formal press announcement of the award winners in February of the following year. The presentation occasion is in the early spring, and coincides with the announcement of the Duke of Edinburgh's Prize.

Full details of the awards scheme can be obtained from The Manager, British Design Awards, The Design Council, 28 Haymarket, London, SW1Y 4SU.

Figure 32.6b *This photograph gives an indication of the size of the element.*

Award winning examples

The illustrations in Figures 32.1 to 32.6 have been drawn from British Design Award winners for 1987. They are not only examples of good design, but they demonstrate many of the points made in this chapter, including for example the importance of real innovation in the basic concept and the relevance (or otherwise) of good appearance design.

PART EIGHT

Engineering Applications

33

Sales engineering 1

Dennis Lock

All of the chapters in this handbook deal with the activities of companies which are dependent on the profitable sale of engineered products or engineering projects for their existence. Sales engineering is concerned with those who are at the sharp end of the marketing function. It forms the important bridge between the customer and the engineering company. General aspects of sales engineering are described in this chapter, and Chapter 34 deals more specifically with the handling of sales enquiries and preparation of sales proposals and specifications.

Role of sales engineering

The technical aspects

Sales engineering is concerned with the selling of engineering projects, or with products that demand some engineering or technical knowledge on the part of the person making the sale. Thus the selling of general office stationery and supplies, although possibly requiring special selling skills in today's competitive environment, is not a job for a sales engineer. It is not necessary to carry out any engineering or give much in the way of technical advice concerning the sale of, say, a box of paper clips. But move up to the sale of some electronic office equipment, such as a large photocopier or a printing machine that needs special installation arrangements, and the boundary into sales engineering may be crossed. Staying with the office example, the specification and selling of a new office telephone exchange, a computer system, or a proposal for a new air conditioning system would be jobs calling for the special skills of sales engineers.

It is, perhaps, by considering the differences between these simple examples that some essential features of sales engineering emerge. Paper clips can be sold straight from a shelf or ordered from a catalogue, and the only discussion concerns their size, quantity, price and (possibly) colour.

The sale of a printing machine might, again, be possible from stock or from a catalogue. But the person representing the seller must discuss its siting, use, initial stocks of consumables, power supplies, servicing and maintenance requirements, optional extra facilities and many other factors with the customer. The seller must do this sensibly, in the technical language of the potential customer. In other words, the salesman must have a sufficient level of technical training, in some cases up to full professional engineering standards. Selling and engineering can hardly be described as similar subjects, and people who can combine the two successfully are valuable company assets.

Selling a new passenger lift (elevator) is another example of a job that needs the services of a sales engineer. Detailed measurements must be taken of the proposed site, structural calculations are needed, and the customer will want to be shown sketches or artist's impressions of how the finished project will look. Note that the word project has now crept in as we move up the scale of complexity. Catalogues may be available to show car sizes and finishes, but every site is different. If the lift manufacturer and the customer are in different countries, the sales engineer may have to study local special safety regulations or other engineering standards. The seller must investigate these problems thoroughly before deciding the engineering solution. He must then carry out a certain amount of preliminary design, with rough layout drawings and cost estimates, before he can prepare a formal sales proposal that aims to meet the customer's requirements.

The highest involvement of sales engineering is seen in projects which have to be designed from scratch, where there is no reference precedent, which cannot be sold by quoting a set of catalogue options, and which contain few or no standard products or components from the seller's range. These include projects for the provision of special purpose machines, for the design, erection and commissioning of process plants, for construction, mining, and other major works. These involve the sales engineers in considerable engineering design work, often supported by the company's engineering department, during which several quite different solutions might emerge for fulfilling the customer's needs.

As a relatively small scale example, consider a sales team asked to propose the supply, installation and commissioning of a machining facility to a customer. The workpiece to be machined has already been designed by the customer, who has provided drawings. The seller has been asked to prepare a priced proposal (in competition with others). The sales engineers might have to choose between the options of recommending a purpose built machine that could perform with high productivity and accuracy, or of trying to sell a more versatile general purpose machine that could be adapted or reprogrammed in the future to machine other products. Between these two extremes, there might be many more subtle design options that affect the total mix of estimated costs, expected performance and reliability, tilting the customer's choice one way or the other against competing sellers.

Engineering sales proposals can obviously range greatly in complexity, from enormous projects at astronomic costs down to the specification of a single component in answer to a problem, or perhaps simply the recommendation of a particular grade of a consumable such as lubricating oil or paint. In many of these cases the customer must rely entirely upon the word of the sales engineer that the proposed item will, in quality management terms, be fit for the intended purpose. The customer is entitled to expect that the sales engineer is thoroughly conversant with the product, and able to give sound advice as to its use and performance with complete honesty.

So the sales engineer must work with the customer to find out what is wanted, and ensure that the problems are understood, with all relevant data collected. Then one or more solutions have to be developed, evaluated, priced and possibly discussed again with the customer before a formal proposal is prepared and presented. The process of establishing the customer's needs is common to all sales engineering: it is only the scope that varies with the novelty or complexity of the proposed product or project.

The commercial aspects

Expressed in very crude terms, the principle objectives of sales engineering might be seen as being to sell as much as possible, as profitably as possible, and without getting the engineering information or solutions wrong at any stage. Indeed many managers would be very happy if their sales engineering departments managed to achieve just that.

But sales engineering is only part of the wider function of a company's marketing function. In addition to customer orders, sales engineers gather valuable information in the course of their work which can often help the company to formulate plans for product development, marketing strategy and long range plans. More specifically, sales engineers for many industrial products will be expected to play a big part in compiling sales forecasts, especially in the shorter term such as quarterly and annually (there is a brief section on sales forecasting later in this chapter).

It is recognized that sales engineers' contacts with potential customers, even if they do not result in immediate orders, are valuable in establishing goodwill, leading to possible orders in the future. This is known as customer development, and depends largely on courteous, regular contact with the prospective customers in person, by telephone or by letter.

When orders are received, it is useful to find out what prompted the customers to opt for the company's goods or proposals: analysis of the answers collected by sales engineers can give important clues as to customers' perception of product quality and pricing, and of the effectiveness and benefits resulting from expenditure on advertising or other forms of sales promotion. Similar information relating to reasons for not winning orders is usually even more valuable in helping a company to develop its products and marketing policies.

Secondary roles in sales engineering and selling

Although this chapter concentrates on sales engineers as those principally responsible for gaining orders for new work, many other company employees support the sales engineering and general selling process.

Many sales proposals need the professional expertise of the company's engineering department in developing solutions. The estimating, cost and planning engineers, and (where relevant) the production management may all be asked to provide information that will help to establish prices and delivery promises. Any of these professional staff may be asked to assist the selling function by duty on exhibition stands, or by backing up sales engineers on visits to important clients.

These are the obvious secondary roles. But many other members of a company play their parts in influencing the general impression made on potential customers. The attitude of the switchboard operator when answering and connecting calls, the standard of correspondence produced by secretaries, and the level of competence in the printing department for producing proposals free from missing pages, misspelt words and poor layout all play their part.

Potential customers often visit a seller's offices, either by invitation or at their own request. It is taken for granted that they will be shown into a comfortable conference room or office, be given suitable refreshments, and a first class presentation that puts the company in a good light. There will probably be good food and wine, and perhaps some evening entertainment too. The effect of all this will be offset or ruined altogether if, on touring the engineering offices, there is a general attitude of sloppiness, with drawings strewn about haphazardly, and no sense of purpose or order. Above all, everyone should appear to be busy and well motivated, even if this has to be 'stage managed' for a particular occasion.

Adequate reception arrangements must be made when visitors arrive. A courteous and efficient receptionist is a valuable asset in this respect. On no account should important visitors be subjected to the discourtesy of being kept waiting beyond the time fixed for their appointed arrival.

These are not cosmetic requirements. The potential customer will want to see that his project is going to be run by an efficient company, with adequate resources and expertise. He wants to know that the company is going to be a pleasant and courteous organization to deal with. He will identify lax standards in the engineering offices with the risks of late delivery, technical incompetence and a potential waste of project money – his money. It is sometimes very difficult for a potential customer to make an objective judgement of a new contractor, and the subjective impression may be all important where all other factors are equal.

These remarks on company image apply also to existing customers. They are future potential customers, and they too must receive a good impression during visits and all contacts with company staff in general.

The sales engineer

The nature of a sales engineer's duties varies greatly according to the type of project and the degree of design engineering associated with the task. At one extreme nearly all of the engineer's time will be spent in travelling and calling on potential customers. Such engineers often work from home, and are responsible for a particular territory. Right at the other extreme is the sales engineer who is practically office bound, spending long periods in developing possible solutions to complex engineering problems resulting from enquiries received, and only visiting a small number of customers within a year (but some of these may involve considerable travel and periods spent overseas).

The degree to which a sales engineer's activity is devoted to selling and dealing directly with customers depends generally on the value of the orders to be taken.

In general, low cost products sold in high numbers are likely to involve frequent contacts with customers and much time spent on the road. High cost, low volume work, such as the sale of engineering projects, is more likely to require the sales engineer to work as a solution engineer in the sales office, with calls to customers made for investigative or customer development purposes only. For valuable projects, many companies entrust the actual presentation of proposals and subsequent sales negotiations to higher levels of management (so that the sales manager, the managing director, or even both acting together would visit the senior management of potential customers).

The task of the average sales engineer is often arduous, involving travel, periods away from home, exposure to critical or even hostile customers, constantly working to deadlines, long hours inside and outside the office, and yet having to remain reasonably dressed and keeping a cool head.

The work of engineers selling low cost high volume products is easily assessable in quantitative terms, unlike most other engineering jobs. Financial sales targets can be set for each month. The performance of each sales engineer can be measured easily, if not always entirely fairly, by comparing the value of orders actually taken in each period against the set targets. There is also a strong competitive element, because in many companies sales achievements are compared not only with targets, but also between different individuals in the team. The rewards for success, in terms of commission, incentives, special awards and promotion can be great. But failure is soon recognized and, to put it euphemistically, can very quickly jeopardize an individual's long-term future with the company.

Assessing the work of sales engineers working on proposals for high value contracts may not be at all easy in financial terms. It may be that only a few orders a year, taken as a result of work from the whole sales team, will be enough to maintain a satisfactory workload and produce adequate sales revenue and profits. In these cases, bonus payments to the group as a whole may be a fairer way of rewarding achievement. Otherwise, the value of each individual will be seen in terms of the quality of engineering work produced.

Most sales engineers have to combine stamina, pleasant personality, technical competence, the ability to sell, pride or at least a belief in the product, and a taste for a competitive challenge. The sales engineer is usually the vital link between customer and seller during the crucial time when orders are won or lost. The impression created on the customer may be a critical factor in beating the competitors to a sale. Confidence without arrogance, willingness to listen, and persuasive power of argument are all plus factors in a salesman's make up. John Fenton, guru of selling, once listed the following characteristics as those needed by a good sales engineer:

1 Acceptability to people of all levels.
2 Good health.
3 Fluency of communication in speech and in writing.
4 Resilience: the ability to get up and back in fighting.
5 A positive temperament: the ability to make things happen.
6 Sincerity, honesty, integrity and loyalty.
7 Enthusiasm.
8 Empathy: the ability to understand why a person holds a particular opinion, without necessarily agreeing with it.

Recruitment and training

When a vacancy for a sales engineer arises, two options for recruitment must be considered. One is to appoint someone from within, from the existing design engineering staff. The other is to recruit from outside through advertising, an employment consultant or an agency.

Recruiting from within

The biggest advantages of appointing from within are familiarity, in two senses:

1 The engineer should already be familiar with the company's products and product policy.
2 The company should be familiar with the engineer's capabilities and potential for development.

However, the qualities that go to make a good design engineer are not the same as those needed for a successful sales engineer. It could be that the proposed appointee has proved to be a first class, dedicated and reliable technical person, but is lacking in one or more of the essential characteristics of an ideal sales engineer.

A useful approach to this problem is to carry out some advance training and development. It is common for design engineers to have to support the sales engineers technically: this involvement can be extended by arranging for suitable design engineers occasionally to accompany senior sales engineers on visits to clients, or possibly to spend a few weeks working in the sales engineering department. This may help the company to assess whether an individual engineer is likely to develop into an effective sales engineer, and it will give the individual a foretaste from which he can judge whether or not the sales engineer's life is for him (or her).

When a company is short of engineering work, this may be the time when a maximum sales drive is required. The temporary transfer of design engineers into the sales engineering function might be indicated under these circumstances, subject to the availability of sufficient funds and willing cooperation from the staff involved. (The engineering design staff may find the alternative of redundancy less attractive.)

If a mistake is made in an internal appointment, then the engineer may have to be returned to the design engineering department. This can be doubly damaging. The company will have suffered a setback in its attempt to strengthen its selling potential, and may actually have lost business. The individual's previously successful design career will have been interrupted and, upon resumption, will probably bear the scars of the failed experiment for some time to come.

Recruiting externally

If recruiting from within has the advantages of dealing with known quantities, then external recruiting conversely carries the risks associated with the unknown. The most careful assessment and interviewing of candidates can still result in mistakes. In all probability, even a very good recruit will take many months, if not years, of training and experience before he/she becomes fully familiar with the company's products and policies.

There are two principal advantages to be gained from recruiting externally:

1 The catchment area is not limited, and it should be possible to attract one or more candidates who match a recruitment specification that has been based on a careful study of the optimum requirements for a sales engineer for this company and its products (the best available internal candidate may only be a compromise).
2 A newcomer to the company can bring fresh ideas, valuable experience of the market, detailed knowledge of competitors (in many cases from within) and might have many useful contacts and much influence among potential customers.

Recruiting procedures are covered in general terms in Chapter 5. For sales engineering applications, unless the company has the appropriate experience, it will probably be sensible to reduce the risk of a disastrous mistake by recruiting through an agent or consultant who specializes in this type of appointment: one who can attract the best available candidates and assist by carrying out penetrating interviews.

Wilson (1983) gives a remarkably full account of recruiting for sales positions, including an extremely detailed checklist designed to identify a comprehensive range of character traits and motivational factors.

Training

A new recruit to sales engineering, no matter how experienced externally, will need considerable training before he/she becomes fully effective as a company sales engineer. Indeed, a minimum amount of training is essential before the sales engineer can be let loose on unsuspecting customers. Any buyer or manager subjected to the time wasting and embarrassing experience of giving audience to a sales engineer who is inadequate through lack of training or experience is likely to feel insulted, and to be deterred from admitting any other sales person from that company into his office ever again.

In a reputable company which markets complex engineering products or projects, the depth of training needed for each of its sales engineers is unlikely to be provided by a few lectures, or the provision of wadges of reading matter. It is far more likely that the recruit will be asked to spend suitable periods, ranging from a few days to weeks or even months in several departments of the company as part of an in depth familiarization process. Very large companies will have specialized training centres for this purpose.

Essential training can be classified under four headings:

1 The company.
2 The product.
3 The market.
4 Selling and negotiating techniques.

Company training

Company training for any new recruit is often referred to as induction training. The usual aim is to familiarize the newcomer with the company in general, ranging from the location of creature comfort facilities to some training in the use of company procedures and an insight into the organization structure. For new sales engineers, induction training must be a far more intensive exercise, with a very detailed syllabus. The content and purposes of this training (not given in sequence of priority) are likely to be:

1 A detailed briefing of the company's origins and history of development.
2 The current group organization structure, with all active subsidiary companies and their functions described.
3 The head office organization structure.
4 The sales organization structure.
5 Company procedures, especially those relating to design engineering and to sales engineering.
6 Commercial information, such as current sales turnover, targets for profit margins and growth, and the sales engineer's own expected role in the future pattern of objectives.

Product training

The sales engineer may have been recruited to sell a particular product, or to sell a general range of products. If the technical training is to be centred on one product, then the sales engineer should be given an overall briefing about the other products which the company sells or manufactures in order that he can recognize potential sales openings for them and refer prospects to the sales engineers responsible for those products.

If the company does not sell standard products, but is engaged solely in the conduct of engineering projects, then the technical training will be more in the form of teaching company design philosophy, and in the methods needed to work with customers in recognizing their needs and defining possible solutions.

Items in product training might include some or all of the following:

1 In depth learning of all features of a particular product, or of a range of products. This would include, performance characteristics, optional extras, installation methods, recommended operating methods, servicing and maintenance requirements and the spares and consumables that a customer would need to operate and support the product. This training should emphasize the strong selling points, and suggest ways of countering expected criticism of the product or its price.
2 A knowledge of the company's design policy. This especially concerns any factor to which the company attaches great importance or of which it is

especially proud and declares as part of its corporate image (e.g. building for reliability). The company's design policy is also relevant in guiding the sales engineer if he has to propose technical solutions to a customer's perceived problems. In the sale of complex projects, this training might involve the sales engineer spending time with the design engineering department, perhaps actually producing some drawings, and becoming familiar also with internal company engineering standards and drawing practice.

3 A general appreciation of other products that are sold by the company. Although these may not be the subject engineer's responsibility, he may be asked about other products by customers and must know whether or not the company sells such items and, if so, where to direct each sales enquiry.
4 Information on past products or projects which won reputations for being:
   (a) Notable successes – something to build a reputation on.
   (b) Dismal failures – in case a potential customer raises one of these examples, it is well to be prepared and have an answer ready.

The market

The engineer will expect to be given as much information as possible about the market in which he is to operate. This will include such things as:

1 The way in which the company divides its sales territories.
2 Specific information on the sales territory to be covered by the engineer, including details of its existing and potential customers with suggestions for new prospects. This training may involve accompanying an experienced sales engineer on calls.
3 Information on competitors' products. Some companies are able to compile enough data for tables to be drawn up in which individual technical and commercial features of their own products are compared one for one against all known competitors. In due course, the new sales engineer may be expected to play his part in collecting such data from competitors, by whatever means is available (often by posing as a possible customer and asking for catalogues and price lists).
4 General market intelligence, perhaps gained from market research, in order to give the engineer a grasp of the full potential and so that he can see his role and that of his new company as part of the total market picture.

Selling and negotiating techniques

Several educational establishments conduct training courses that concentrate on imparting the skills peculiar to selling. Topics covered should include:

1 Initial preparation before meeting the prospective customer.
2 Methods of securing an appointment.
3 How to make a sales presentation.

4 Negotiating techniques, and ways of handling criticism and sales resistance.
5 Techniques for closing the sale.
6 Ways of planning calls to make the most effective use of costs and time.

The company will probably want to augment any external training with its own detailed instructions on the way in which it expects its products to be presented and the way in which the company itself is to be represented.

The Institute of Marketing is an organization which can give valuable advice on training. In addition to offering a very comprehensive range of seminars on all aspects of marketing, specially designed in-company training programmes can be arranged. The address for enquiries is Director, IM Marketing Training, The Institute of Marketing, Moor Hall, Cookham, Maidenhead, Berkshire SL6 9QH.

Sales forecasting

The market planning framework

The amount of revenue which any engineering company expects to earn from sales is obviously one of the key ingredients in its strategic and short-term planning and budgeting. Sales engineers have an important part to play in compiling the input data for such planning, helping to set sales forecasts for the company which will be reflected in their own personal performance targets.

In Figure 33.1, a project engineering company is shown surrounded by its operating environment, with some of the more important external factors depicted as a kind of wheel of fortune. The company's organization is shown in simple functional terms, but highlighting the position of the sales engineers as lying between the customers and new prospects on the one hand, and the solution engineering and design engineering functions within the company on the other.

In this general context, the sales engineers should be in a strong position to advise their management on the likely market potential for individual products or projects for which they are responsible. Sales engineers are usually expected to make annual forecasts of the value of orders that they will take, and the sum of these forms the basis for annual sales budgets.

Trying to estimate the level of demand from a company's customers for twelve months ahead is an art that is apt to generate wrong answers. The most accurate forecasts are those made where the demand exceeds supply. For example if it is known that a particular motor car cannot be produced in large enough quantities to satisfy customer demands, then the sales forecast for the year ahead will be determined by the constraints of production capacity. 'Everything we can make we can sell'. The company's production plan itself

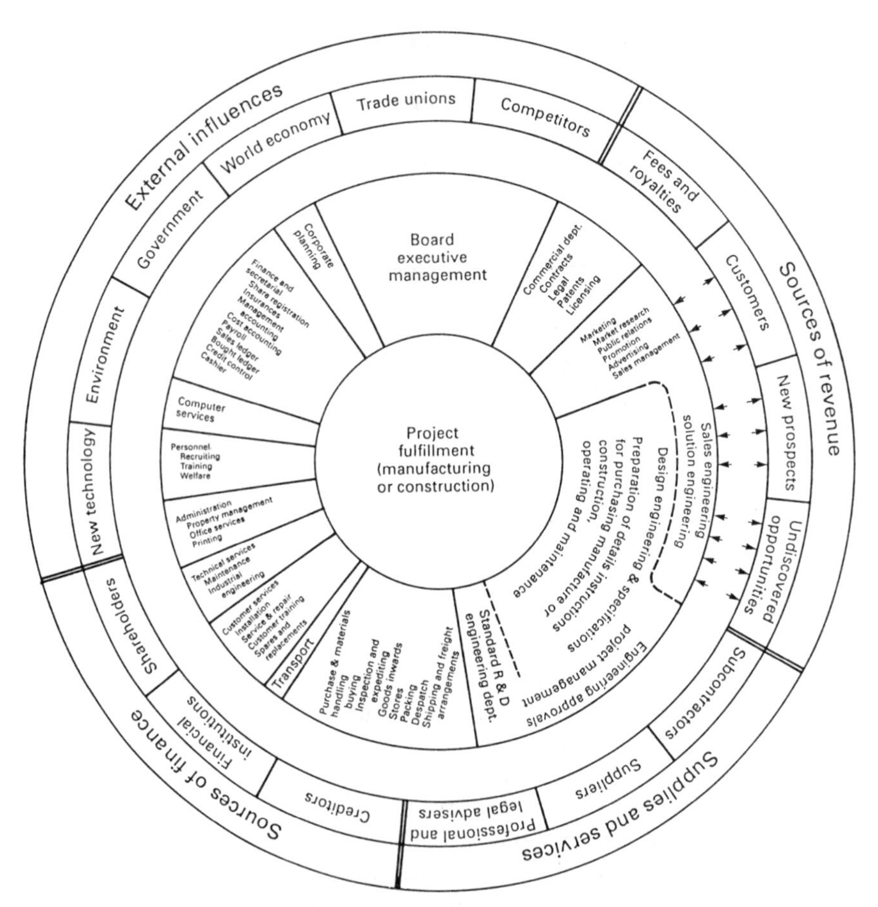

Figure 33.1 *The wheel of fortune. This shows a project engineering company in relation to its operating environment and the sales engineers in their bridging role between the company and its customers*

will lead to a sales forecast that, barring production setbacks, must be achieved.

But, as the diagram in Figure 33.1 illustrates, in project and most product engineering work the sales engineering/customer relationship is only part of a much wider pattern of influences. The actual levels of sales achieved will be dependent on a diverse range of factors, many of which are outside the control of the sales engineers and their managers. Sales engineers' forecasts, therefore, cannot be depended upon for accuracy (although they may be the best forecasts that can be made).

Figure 33.2 sets sales forecasting in its context between overall company strategy and individual performance targets. All engineering companies which aim to perform successfully develop long range strategic plans in order that they shall fulfil their overall corporate objectives. These long range plans may cover periods of five or ten years: in some industries with heavy research or capital investment the plans will extend farther into the future. In the aircraft industry, for example, the point at which sales revenue recovers total investment costs may typically be in excess of fifteen years.

In order to achieve the long range plan, a company will break it down into several more detailed plans, in a hierarchical structure. Important among the detailed operating plans are those for annual production and sales, both contributing to the company's set of annual budgets. The sales engineer's principal contribution to this financial planning process is made at the annual level, through sales forecasts of products or projects.

Once annual plans have been developed and agreed, quarterly and monthly breakdowns are established that provide a basis for monitoring performance. This includes measuring the performance of the sales engineers, collectively and individually.

In a typical manufacturing company there will be one or more identifiable product ranges, each of which must be the subject of a separate sales forecast and subsequent control. Another way of viewing the sales structure is to consider the total sales that will be achieved by each individual sales engineer or group of engineers (according to the organization). A breakdown of sales by geographical region may also be important. Some companies identify particular customers who buy regularly and in large quantities as major accounts, and these, too, may be a separate area for forecasting.

Apart from the purpose of contributing to overall corporate plans, sales forecasts are going to be used as the basis for assessing performance. It follows that for any breakdown chosen in classification of sales by product, sales engineer, customer or region there must be a corresponding arrangement to gather statistics from actual sales under the same headings. Otherwise it will obviously not be possible to weigh performance against targets in every case.

An important aspect of measuring performance is in using the results as feedback with which to modify the plans in upper levels of the hierarchy. This measurement and reporting process must not be too long delayed. In engineering terms, planning, measurement and control are a closed loop system, and phase lag should be kept to a minimum. In the example of Figure 33.2, for example, there is a lag of one month shown at level 5.

Some methods of sales forecasting

Engineered products

A typical method for forecasting the number of items to be sold within each product range, and for establishing individual sales engineer's targets, is to have each sales engineer fill in a questionnaire. This will allow each individual

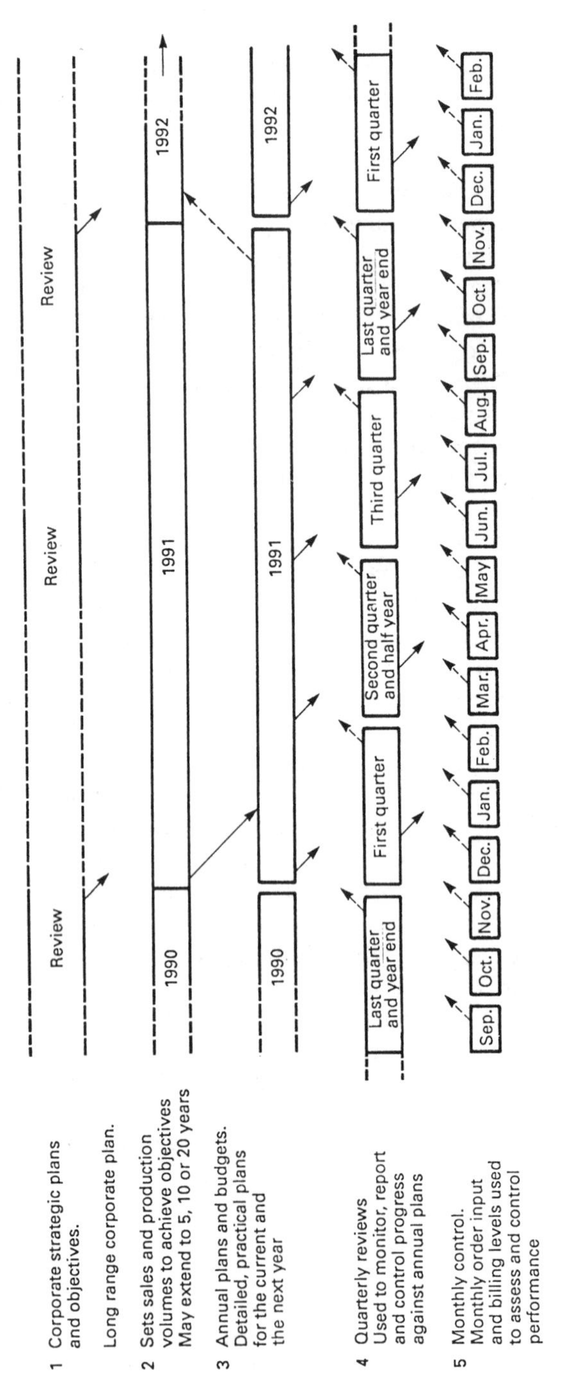

Figure 33.2 *Sales forecasting and targets as part of the planning hierarchy*

to state what sales he thinks can be achieved during the forthcoming year. The sales manager will then review the figures with each engineer, either in a general sales meeting or individually. The figures agreed with each sales engineer will become that engineer's sales target for the following year.

In simple theory, the figures agreed are then entered on a summary form to arrive at the sales forecast for the coming year. Figure 33.3 shows an example for a company with eight sales engineers whose task is to sell three related products (Minilog, Midilog and Maxilog).

Systems division sales forecast for 1991

Sales engineer	Minilog		Midilog		Maxilog		Other £	Personal targets £	Remarks, confidence in forecast, etc.
	Units	£	Units	£	Units	£			
Adams	220	44 000	50	35 000	6	12 000	5000	96 000	
Anderson	400	80 000	50	35 000	—	—	5000	120 000	
Chambers	100	20 000	75	52 500	12	24 000	2000	98 500	
Duffy	500	100 000	10	7 000	—	—	10 000	117 000	
Harris	250	50 000	40	28 000	10	20 000	2000	100 000	
Jones	200	40 000	30	21 000	2	4000	1000	66 000	Trainee
Murgatroyd	200	40 000	50	35 000	2	4000	2000	81 000	
Wilberforce	180	36 000	45	31 500	6	12 000	5000	84 500	
Direct sales	200	40 000	—	—	—	—	20 000	60 000	
Totals	2250	450 000	350	245 000	38	76 000	52 000	823 000	

Figure 33.3 *An annual engineering product sales forecast derived from sales engineer's estimates*

In practice, the sales manager, although leaving the individual sales engineers' targets intact, may wish to adjust the total sales forecast levels before submitting them to the higher levels in the planning hierarchy. Some engineers can be inexpert at forecasting. While, on the other hand, there could be a temptation for engineers to set themselves low, easy-to-attain sales targets, it is not uncommon for individuals to place a value on their selling ability which has no chance of ever being realised in practice. In order that production plans and company budgets are set at the most realistic levels, they must be reviewed and judged by the company's most experienced managers.

In dividing sales forecasts into quarterly and monthly elements, it may be necessary to take account of seasonal or other expected fluctuations. If ice cream or coal were the commodity being sold, an obvious seasonal pattern could be expected. But, seasonal sales patterns are also experienced with some engineering products (heating and ventilating replacement components at the onset of winter, for example) and exceptional sales demands can sometimes be identified with trade exhibitions or other events.

Engineering projects

If there is room for error in sales forecasting for engineered products, there is the possibility for total disaster in estimating new project work. When a company is handling several medium-sized projects simultaneously, then forecasting errors may tend to offset each other. But a company handling only one or two major projects at a time will be seriously disrupted if, as often happens, a hoped-for order is not received or is delayed while the client attempts to obtain project funding. At the very extreme end of uncertainty, the very existence of a company could depend on the winning of an order, where a joint venture partnership has been set up to bid for the work in order to spread the forces, resources and risk.

Most companies handling projects usually have several in hand, one or more about to start for certain, and some for which they are bidding. The gestation time for an order for any sizeable project can easily run to many months or years: the client takes time to consider the feasibility of the project, then there is a call for tenders, time has to be allowed for bids to be prepared, the bids have to be assessed, and there may be more delays while technical anomalies are ironed out and the client assembles the necessary funds. Companies handling fair sized projects, therefore, generally have some idea of the work that will be in prospect for the coming year.

One way in which forecasts are quantified for possible new engineering projects is illustrated in Figure 33.4. All known projects for which the company is bidding, or is expecting to bid are listed. Then the probability of winning an order in each case is assessed subjectively, with a probability factor being declared on a percentage scale. The possible timescale, start date and value of each project is estimated (on a ball-park basis if necessary). From these time and cost estimates, the value of work likely to fall within the sales forecast year can be judged. Finally, the probability factor is applied to each case to obtain a set of weighted values.

This procedure can be very inaccurate. However, provided the company is not actually running out of work, errors in a typical project engineering company should be reduced in significance owing to the presence of a core of known work from actual orders running through the forecast period. This known work can include engineering service agreements or other work renewable with clients on an annual contract basis, plus considerable work in completing existing contracts (some of which could be of several years' duration).

Performance statistics

There are three principal sets of data which, according to circumstances, are used by companies to measure the performance of their sales engineers. Each

Projects division sales forecast for 1991

All figures are in £000s

Project	Total value	Duration (years)	Potential this year	How certain?	Weighted forecast	Remarks
A	4 100	4	800	25%	200	
B	2 450	3·5	700	100%	700	ORDER EXPECTED MARCH LETTER OF INTENT RECEIVED
C	480	2·5	100	50%	50	
D	2 150	3	500	40%	200	
E	270	1·5	100	80%	80	
F	3 500	2·5	1,100	40%	440	
G	800	3	200	10%	20	
H	1 050	3	300	100%	300	CONTRACT WILL BE SIGNED IN JANUARY
Sundries	250	1·5	160	100%	160	
Total new orders			3960	54%	2150	
Service contracts	200	ANNUAL	200	100%	200	
Residue of existing orders	4600	—	1900	100%	1900	
Totals			6060	75%	4250	

Figure 33.4 *A method sometimes attempted for project sales forecasts*

item will be measured and checked against the corresponding target or forecast for each given period (typically calendar months).

Call statistics

Some companies use the number of calls made by their sales engineers as one indication of performance. This approach is somewhat mechanistic, and might be totally irrelevant for a company selling high value capital projects (where the sales effort is concentrated on only a relatively few potential customers). Further, the number of calls by itself takes no account of whether the calls were effective in achieving or developing sales.

However, where the products are suitable, call statistics can be a useful pointer to the amount of time and effort put in by an engineer: how well he/she organizes appointments and route planning to cover a territory effectively. This is particularly relevant to those given free rein within a designated sales territory, especially when operating from home rather than from the head office.

Too many calls on one customer can be counter-productive, and call frequency statistics need to take into account how the calls have been spread over existing customers and newly identified prospects. It is possible for a company to establish guidelines for calling on existing customers, perhaps specifying regular monthly visits to customers with whom there is a good level of business and rapport. Less important customers can be classified as providing average or low levels of business, with call frequencies reduced to, say, three and six months respectively. Each company will learn to develop its own recommendations to its sales engineers for calling, writing and telephoning, depending (among other things) on the nature of the product and the size of territory allocated to each sales engineer.

For every call made, whether or not a sale results, a visit report should be completed by the engineer and filed in the sales office. This is especially important for engineers working from home. Call reports need not be complicated, and standard preprinted pads can be issued for the purpose (see Figure 33.5).

Apart from their use in following up actions arising from visits, call reports have other valuable uses for the sales manager. They provide information relevant to the compilation of lists for such things as general mail shots and the issue of invitations to exhibitions or promotional open days. Above all they safeguard against the perils faced when a sales engineer leaves employment, complete with his personal list of customers ready for his new employer. The place for customer lists is in the general sales office files, as part of company intellectual property, and not secreted away in personal notebooks.

Value of orders taken

The monthly value of orders actually won by a sales engineer, measured against the agreed sales forecasts, is the most telling mark of achievement in the sale of most engineered products. Many incentive schemes are used to induce sales engineers to gain orders. The payment of a monthly commission in addition to salary is appropriate for many industries. Commission is typically credited to the sales engineer when the goods are invoiced (with a corresponding deduction for credit notes which have to be issued when a customer returns goods or otherwise cancels or reduces the value of an order). Monthly competitions for highest individual achievement are common, with special gifts or short holidays (preferably over weekends and not in potential selling time) given as prizes.

For sales engineering work with a high solution engineering content, team effort rather than individual performance will have to be monitored. In such cases, special rewards (if any) will be applicable to the group as a whole, possibly given in the form of an annual bonus. Another form of reward is to arrange a sales conference or seminar in some exotic or otherwise attractive part of the world.

For recording and reporting individual or group sales achievements, it is usual to issue a monthly tabulation arranged along the lines shown in Figure 33.6.

REPORT OF SALES VISIT

<table>
<tr><td colspan="2">Name and address of customer: Date of visit:</td></tr>
<tr><td>Person(s) seen</td><td>Position(s)</td></tr>
<tr><td colspan="2">Reasons for visit:</td></tr>
<tr><td colspan="2">Result/interest shown/products discussed</td></tr>
<tr><td colspan="2">Background information:</td></tr>
<tr><td colspan="2">Follow up action required:</td></tr>
<tr><td>Next visit date planned:</td><td>Sales engineer reporting:</td></tr>
</table>

Figure 33.5 *Format for routine field sales reports*

1991 Sales Performance – Systems Division

All figures in £000s

Product	Jan.		Feb.		Mar.		Apr.		May		Jun.		Jul.		Aug.		Sep.		Oct.		Nov.		Dec.	
	Plan	Act.	Plan	Act.	Plan	Act.	Plan	Act.	Plan	Act.	Plan	Act.	Plan	Act.	Plan	Act.	Plan	Act.	Plan	Act.	Plan	Act.	Plan	Act.
Minilog																								
Month	37		38		37		38		37		38		37		38		37		38		37		38	
Cumulative	37		75		112		150		187		225		262		300		337		375		412		450	
Midilog																								
Month	20		20		21		20		21		21		20		20		21		20		20		21	
Cumulative	20		40		61		81		102		123		143		163		184		204		224		245	
Maxilog																								
Month	7		6		6		7		6		6		7		6		6		7		6		6	
Cumulative	7		13		19		26		32		38		45		51		57		64		70		76	
Sundry items																								
Month	4		5		4		4		5		4		4		5		4		4		5		4	
Cumulative	4		9		13		17		22		26		30		35		39		43		48		52	
Division totals																								
Month	68		69		68		69		69		69		68		69		68		69		68		69	
Cumulative	68		137		205		274		343		412		480		549		617		686		754		823	

Figure 33.6 *A tabulation of monthly divisional sales targets by product. Actual results would be entered as the year proceeds. Similar schedules can be kept to monitor the performance of individual sales engineers against their targets*

Value of billings made

For a company which is able to supply orders from stock, or at least with very little lead time, the value of invoices issued can be a statistic that is almost contemporary with the value of orders taken. This is true where, although sales engineering expertise is necessary to recommend the right product for a particular application, the goods are already available in a stores or warehouse for immediate delivery. This would apply, for example, in the case of many engineering components and manufacturing tooling and consumables. It also applies to the sales of much office machinery.

Where billing figures are used as the control, reports would be similar to the layout of Figure 33.6 except that monthly billing targets and achievement would be used instead of order values.

Billing figures are a more comprehensive form of assessment, since (barring serious debtors) they record the performance of the company as a whole in achieving sales revenue and are directly relevant to the profit and loss account.

When the company has to manufacture products against orders taken, billing targets are less relevant to the performance of individual salesmen owing to the time lag that must elapse following receipt of orders. In these cases, however, monthly billing targets are still a valid control tool, but they will be directed primarily at the manufacturing department.

Order, billing and backlog statistics

Although not used as a direct measurement of sales performance by individual sales engineers, a combination of order value and billing values can be used in a technique of order backlog analysis to provide management with a useful guide to the level of the forward order book. The method is particularly useful for companies which have to manufacture products against orders, and where there is a delay of several months or even years between the receipt of a typical order and its completion.

A simple order backlog analysis statement is shown in Figure 33.7. The table starts by looking at the residual value of orders received in the past – the amounts on these orders that yet have to be invoiced. This total is shown in the bottom left-hand corner of the illustration (£1 950 000).

In January, no new orders were received, and invoices totalling £80 000 were issued, so that the cumulative order backlog remaining was reduced by £80 000 (accountants always put negative amounts in brackets) to £1 870 000. Various small orders are shown as having been received for the months of February, March and April but it was not until May that an order big enough to reverse the reducing backlog was received.

At year end, the cumulative backlog (£1 593 000) was £357 000 less than the year's opening backlog. This means that the company's level of activity is reducing through lack of orders, and increased selling activity is indicated.

Input – output– backlog statement

All figures are in £000s

		Backlog brought forward	Jan.	Feb.	Mar.	Apr.	May	Jun.	Jul.	Aug.	Sept.	Oct.	Nov.	Dec.	Year end
Input (new orders)			—	10	65	20	512	24	—	—	18	-	43	-	692
Output (billings)			80	75	85	92	45	112	82	126	80	91	111	70	1049
Backlog (excess of orders over billings)	Month		(80)	(65)	(20)	(72)	467	(88)	(82)	(126)	(62)	(91)	(68)	(70)	(357)
	Cumu-lative	1950	1870	1805	1785	1713	2180	2092	2010	1884	1822	1731	1663	1593	1593

Figure 33.7 *An input – output table used to show the backlog of orders in hand*

In very rough terms, the average amount billed by the company for the year shown was £87 400 per month. As a crude (but useful) control guide, the company's management can say that the year began with orders equivalent to (1 950 000/87 400) months of working at current capacity (which gives 22.3 months). At the end of the year, given that work would continue into the future at the same rate, the backlog of work had reduced to only 18.2 months. If this trend were allowed to continue, the company would eventually run out of work and have to reduce its capacity for work and for earning revenue and profit.

Note that this interpretation of work in hand must not be confused with work in progress. The backlog of orders or work in hand is based solely upon the unbilled value of orders being handled, and much of the related expenditure has yet to come in the future. Work in progress is the value of labour and materials actually used but not yet billed to customers: it represents money tied up in work for which no return has been earned. In other words, it is good to have plenty of work in hand, but the value of work in progress should be kept to a minumum.

Further reading

Adams, Tony, *The Secrets of Successful Selling*, Heinemann, 1985. An entertaining account of selling products from encyclopedias to commercial refrigeration systems.

Bayliss, J. S., *Marketing for Engineers*, Peter Peregrinus, 1985. Published for the Institution of Electrical Engineers.

Fenton, John, *How to Sell Against Competition*, Heinemann, 1984.

Holmes, George and Smith, Neville J., *Sales Force Incentives*, Heinemann, 1987.

Wilson, M. T., *Managing a Sales Force*, 2nd edn, Gower, 1981.

34

Sales engineering 2

Dennis Lock

This chapter continues the subject of sales engineering by concentrating on the procedures for handling sales enquiries, producing sales proposals and writing sales specifications.

Administration of sales enquiries

It is necessary to establish and manage formal procedures for handling all aspects of enquiries. These arrangements should ensure that:

- Every enquiry is acknowledged on receipt.
- No enquiry is lost or forgotten.
- The potential importance of each enquiry is assessed before deciding upon action and the commitment of resources.
- Actions decided and authorized are actually carried out so that, for example, tenders are presented on or before the specified cut-off date.
- Follow-up takes place with each potential customer after tendering until it is known for certain whether the order has been won or lost (and, if possible, why).

Sales coordinator

The day to day administration of a sales engineering department will be greatly improved if one individual is made responsible for the official receipt and subsequent progressing of all customer sales enquiries. For the sake of simplicity in this text this person will be referred to as the sales coordinator but, in practice and depending on the size of the organization, this person might be a secretary, an experienced sales clerk or some more senior member of the sales engineering staff.

Obviously the asssignment of sales coordination duties should not be given to an individual who spends many days out of the office on visits to customers.

The aim should be to create a focal point of information that can be relied upon to be available when wanted, with a sufficiently detailed knowledge of progress on the administration of all current enquiries and sales activities to be able to give sensible answers to day to day queries. This especially means being able to respond sensibly and helpfully to telephone calls or other messages from potential customers when the relevant sales engineers or managers are unavailable.

If the company has several fields of activity or product ranges, with a busy sales department comprising separate sections for each aspect of the business, then clearly it may be necessary to identify a coordinator within each section.

Receipt of enquiries

Every incoming sales enquiry should be seen by the coordinator. This is preferably arranged by having them all directed to him/her in the first instance. The reasons for this are twofold:

1 To ensure that the coordinator is made aware of every active sales enquiry right from the start.
2 So that the coordinator can enter all but the most general enquiries into a progressing system to ensure that they are acted upon without undue delay and then progressed through all subsequent stages.

This means that even those enquiries addressed to the managing director should be seen first by the coordinator if possible, or at least on the day of receipt.

It is important that every enquiry, whether specific, vague or general, is immediately and courteously acknowledged.

General enquiries

General requests for brochures or catalogues are easily dealt with by return, perhaps needing a standard covering letter from the sales engineer responsible for the particular industry or region. But even these general requests, although requiring no specific tender preparation, may warrant registration by the coordinator or the sales engineer for follow up. If requests arise from an advertising campaign or an exhibition, the company may need to record all resulting requests and follow them up to analyse and assess the benefits obtained from the particular promotional expenditure.

Screening

Engineering companies, especially those working internationally, are accustomed to receiving all manner of requests for work on engineering projects from a wide variety of sources. Some of these may be very attractive, offering a strong chance of lucrative work at low risk. Many will be less attractive, for a number of reasons. The preparation of tenders can involve much time and expense, especially where a considerable amount of investigation and conceptual engineering design is needed. Before committing such expenditure, the company's management must evaluate the potential worth of each enquiry in a screening process. Undesirable enquiries will be politely refused and attractive enquiries will be pursued vigorously. Borderline cases can be actioned in cautious stages, with further screening stages as the risks and likely advantages become clearer.

Screening decisions are often made by a management committee which meets regularly for the purpose. It is important that the decision to reject or proceed with all significant enquiries is made at senior executive level. A decision to bid can be the signal for hard work and considerable expenditure, involving not only the sales engineers but also other departments in the company including engineering, estimating, planning and printing.

Some factors likely to influence screening decisions favourably

The following items are not given in any order of importance. In any case different companies will attach their own values to the various factors, and through their own knowledge and policies will be able to add to the factors listed here.

1 Work is needed to fill capacity and occupy available resources.
2 Familiar technology means low risk.
3 Retained engineering from previous projects allows scope for saving on engineering costs.
4 Previous projects for this customer have been successful.
5 The customer is well known and respected by the industry: therefore the finished project has potential for good publicity and resulting kudos.
6 The customer's financial stability checks out all right (in other words, the project is unlikely to fail through lack of customer funding).
7 The project is either in-house or at a convenient location.
8 If the project is for an overseas customer or site:
   (a) There are no known onerous national technical standards or regulations.
   (b) The climate is tolerable.
   (c) No expected difficulty with local labour relations.
   (d) The region is politically stable.
   (e) Road, rail, air and sea communications are good.
   (f) Information communications are good.
   (g) There are no known political trade embargoes or unfair tariff levies or restrictions.
   (h) Bribes are not expected.

(i) Reasonable port and customs formalities.

9 Management have good grounds, either objectively or intuitively, for believing that a contract would be won.

Factors likely to deter a company from bidding

These are generally the converse of the factors just listed as being favourable to bidding.

1 There is no spare capacity with which to undertake a project of the size envisaged, and no desire to expand the company temporarily for the purpose.
2 The project would use unproven new technology, at high risk, and is unsuitable for fixed price tendering (but cost-plus might be suggested).
3 Although the technology is not new, the company has not previously conducted work of this kind. There is no in-house expertise or retained engineering from previous projects. Although these problems could be overcome, the additional costs would indicate uncompetitive pricing in the tender. The company will only be able to tender with any chance of getting the order if it is willing to absorb these additional costs as internal engineering development.
4 The customer is known to be difficult, with a poor record from past projects. This might include such problems as:
   (a) Failure to pay invoices on time or at all.
   (b) Frequent changes to specification, especially at the last minute.
   (c) Unreasonable refusal or delay in approving drawings and specifications.
   (d) Failure to provide agreed facilities for construction, installation or commissioning.
   (e) Unacceptable standards of plant operation and maintenance, leading to post completion disputes.
   (f) A liking for litigation.

   . . . and many more possibilities that would deter a reasonable engineering company from ever wanting to work for them again.
5 The company is known to have cash flow problems, or to present a financial risk (this sort of financial intelligence can be obtained through organizations such as Dun and Bradstreet).
6 The project location would pose administration difficulties or actual threats. Here are some of the considerations:
   (a) The environment is hostile, through climate, disease, earthquakes, volcanic activity, or other problems.
   (b) The national government involved is unlikely to be cooperative, or will indeed be obstructive (through tariffs, regulations and unfairly tough engineering standards).

(c) The area is liable to erupt in riot, commotion, civil war or any other unpleasant form of strife likely to be injurious and uninsurable.
(d) There is unavoidable corruption and bribery involved in the proposed arrangements.
(e) The local workforce has a bad reputation in one or more respects (left to the reader's imagination).
(f) Communications are difficult or non-existent.

7 The management feels that, no matter how much effort goes into preparing this bid, the company is unlikely to secure a contract and the effort and expense is likely to be wasted.

Action plans

When a screening decision is reached, an effective method of authorizing and implementing any resulting action is for the screening committee to issue an action plan. An example of an action plan pro forma is shown in Figure 34.1. The operation of an action plan is similar to that of a works order, since it sets out the work that is authorized and required and is issued to all who are affected by its instructions.

In practice, the sales coordinator would register each enquiry on receipt, allocate an enquiry number from a register, and fill in the top two sections of a blank action plan form. The form would then be passed for consideration at the next meeting of the screening committee. When the committee have made their decision, the coordinator can distribute copies of each action plan as required, retaining the original or one copy for the purpose of monitoring progress.

It often happens that a company will prepare a proposal without having received a direct enquiry from a potential customer. Perhaps a call for tenders has appeared in a trade journal or in the national press, for example, and the company has decided to respond. Once an attractive prospect has been spotted, it should strictly be subjected to registration, screening and the provision of an action plan in the same way as direct enquiries from customers. Whatever the reason for preparing a sales proposal, expenditure, progress and the quality of the proposal must all be properly controlled.

If clarification has to be obtained before a clear screening decision can be given, the coordinator must issue a new action plan, bearing the same enquiry number but raised to revision 2. This is used to submit the revised enquiry to the committee for reconsideration.

Action plans are the sales engineering equivalent of project authorizations or works orders: all authorize expenditure and announce what has to be done, by whom and when. The main difference with action plans is that they authorize the expenditure of company funds which cannot be charged out directly to a customer, and which may prove to have been wasted if no order

PROJEX LTD CUSTOMER ENQUIRY

Customer Address:	Enquiry number	revision
Enquiry date: Telephone: Contact: Customer's reference: Telex:		

ENQUIRY SUMMARY OR TITLE

SCREENING COMMITTEE

	Comment:
We will not bid. Inform customer ☐	
We will bid. Inform customer ☐	
Clarify with customer and rescreen ☐	
	..............Signed..............

ACTION PLAN

Action	For action by	Authorized costs for this action plan			Wanted by (date)
		Labour	Travel	Other expenses	
Define the task					
Review task definition with customer					
Develop engineering solutions					
Evaluate customer's operating costs					
Review proposed solutions with customer					
Estimate our project costs					
Write proposal and prepare artwork					
Printing and binding					
Transit time					
Deadline for proposal:presentation					
Sales engineer in charge:					

Figure 34.1 *Action plan for a sales enquiry. A form such as this can be used to authorize and progress the work needed to deal with a customer enquiry for an engineering project up to the preparation of a sales proposal or tender*

results. Action plans can provide a simple but effective system for managing expenditure and progress during the preparation of sales proposals for engineering projects.

Preparation of sales proposals

Companies with considerable experience of preparing sales proposals for engineering projects will have developed a logical procedure for dealing with them, based on identifiable stages from the initial customer enquiry until the customer's order has been won or lost. The steps listed on the action plan form in Figure 34.1, for example, are those used by the manufacturer of special purpose manufacturing systems when developing sales proposals. The most important steps are:

1 Screening (as just described).
2 Task definition – document by means of drawings, listed data and descriptive text exactly what it is believed the customer wants.
3 Task definition review – which will probably involve one or more meetings with the customer to ensure that the company's view of what the customer wants agrees with what the customer thinks he wants. This stage can sometimes result in the customer having to consider his needs in greater detail: the sales engineers can sometimes help in this process.
4 Solution engineering – during which the sales engineers, almost certainly in consultation with the company's engineering department, develop one or more design concepts that might be expected to fulfil the customer's needs.
5 Solution review with customer – each solution is discussed with the customer's buyer or management, supported by budget estimates of the likely capital, installation, operating and maintenance costs and performance characteristics. The aim is to identify the solution which appeals most to the customer, and develop that into a formal sales proposal. In some cases a sale could result at this stage: in more complex projects more engineering work may be needed before the company can commit itself to a firm sales proposal.
6 Proposal – comprising a technical specification, with a budget or firm price (as required), a delivery schedule, lists of essential and optional extras and a schedule of commercial terms.
7 Final sales specification – which becomes essential when the customer places an order. This document, based on the sales proposal, incorporates all changes up to the point when the order or contract is signed, and gives definitive instructions to the company's design engineering and production or construction departments.

Some of the more important aspects of these stages will now be described. The significance and amount of work needed for each step will obviously depend on the size, complexity and novelty of each project, and the review stages especially might not be necessary for some straightforward projects.

Task definition

Task definition for the purposes of preparing a sales quotation is an important business. It may be necessary to visit the customer or a proposed project site in order to check on all manner of customer requirements and local conditions. Since the data collected is going to be used as the basis for pricing and a possible contractual commitment, it is essential that no significant item is overlooked. For this reason it is strongly recommended that checklists should be used by sales engineers to support the task definition. Every engineering company with a continuous record of project work should have a fund of experience from which it can develop checklists with specific relevance to their type of work.

If the proposed customer or project site happens to be a few thousand miles away, it is even more important that every possible aspect is considered during the fact-finding visit to the prospective customer. As an example, Figure 34.2 shows a checklist that was developed by a mining engineering company for use by its engineers before, during and after the initial visit to each overseas client. This company also developed a meteorological fact sheet for each region in which it carried out regular work (see the example in Figure 34.3): a sheet such as this, but with the data omitted, would provide another useful checklist for a potential overseas project.

There is, however, a danger in being too ambitious in the compilation of checklists. If every insignificant item that can possibly be dreamed up is included (this author remembers such a case) then the closely worded, multipaged document that results ceases to be a convenient aide-memoire: the engineer, perhaps pressed for time and suffering from jet-lag, is likely to skip sections of it and miss one or more vital items. The most effective checklist keeps to the point, and is well organized and itemized.

At its simplest, task definition consists of the customer pointing to a standard item in a catalogue and saying 'That's what I want' (in which case solution engineering is not required at all). At its most complex, the customer may have a manufacturing, process or construction problem to which there is no clear-cut solution, and for the solution of which he is totally reliant on proposals from competing contractors.

The task definition for a new lift installation might be relatively straightforward, with fairly well defined requirements such as the number of floors to be served, number of passengers to be carried, shaft dimensions available, motor room accommodation, door and car finishes, and so on. But there might be

Project site
Availability of utilities (power, water, sewerage, etc.)
Local taxes
Import restrictions
Political restraints on purchasing sources
Statutory or mandatory local regulations relating to labour and environmental controls
Site accessibility by road, rail, air, sea
Site conditions – seismicity tectonics
Climatic conditions (temperature, humidity, precipitation, wind direction and force, sunshine, dust, barometric pressure)
Site plans and survey
Soil investigation and foundation requirements
Local manufacturing facilities
Sources of materials, manpower and construction equipment
Transportation, insurance, etc.
Access restrictions (e.g. low bridges)

Contractual and commercial
How firm are proposals?
Client's priorities (time/money/quality)
Termination points and responsibility
Client's cost expectations/budget available
Client's programme requirements
General contract matters
Scope of work envisaged:

- Full detailed design or basic design only?
- Procurement responsibility – ourselves, client of other?
- Construction responsibility – ourselves, client or managing contractor?

How accurate are existing estimates (ball park, comparative?)
Check below-the-line estimate items with the checklist in the estimating manual
How is project to be financed?
Any restraints on purchasing owing to financing requirements?

Initial design and technical definition
Flowsheets
Layouts
Urgent specifications for long-delivery purchases
Identify other critical jobs
Are further investigations necessary?
Is further information required from the client?
Process parameters
Design parameters
Design standards, drawing sheets, etc.
Engineering standards
Design parameters
Is information from previous similar projects available for retrieval and use?

Organization

Names, addresses, telephone and telex numbers of participants and nominated project representatives for correspondence
Communication methods available (telephone, telex, mail, couriers, facsimile, etc.)
Work breakdown
Sectional responsibilities for project management, specialist functions, procurement, construction
Manpower resources
Organization chart

Programme

Timescale targets
Are targets realistic?
Have long-delivery items been considered in target dates?
Transport time expected to site
Initial bar chart or network

Control techniques and procedures

Methods and management tools involved
Site capability for handling techniques (e.g. Is computer available?)
Procedures for updating the programme and updating frequency
Methods of expediting and programme control
Methods of cost control
Transportation procedures to be used
Arrangements for project meetings
Arrangements for site visits
Design approvals
Drawing approvals
Purchasing approvals

- Enquiry specifications
- Bid evaluations and choice of vendors
- Authority for placing orders and for committing costs

Purchasing procedures

- Arrangements for paying vendors' invoices
- Inspection arrangements

Numbering systems to be used on this project – as specified by the client, or our usual standards?
Distribution of documents, who gets what and how many copies?
Define all procedures by issuing project standing instructions

Figure 34.2 *Task definition checklist for mining projects*
(Courtesy of Seltrust Engineering Limited)

SPECIFICATION

.0	METEOROLOGICAL & GENERAL DATA FOR ZAMBIA COPPER MINES	
.1	AMBIENT TEMPERATURE IN SHADE	
	Maximum recorded	36 °C
	Minimum recorded	-2 °C
	Average of daily maxima in hottest month (October)	32 °C
	Average of daily minima in coldest month (June)	6 °C
.2	RELATIVE HUMIDITY	
	Maximum recorded	98 %
	Maximum monthly average of daily maxima (January)	96 %
	Minimum monthly average of daily minima (September)	24 %
.3	PRECIPITATION - RAIN AND SNOW	
	Maximum annual recorded rainfall	1 745 mm
	Average annual rainfall	1 220 mm
	Maximum rainfall in 24 hours	124 mm
	Maximum rainfall in 3 minutes	20 mm
	Thunder storm days per annum	130
	Snow	None
.4	WIND	
	Maximum recorded gust	30 m/s
	Prevailing direction - January variable - predominantly N to E - July - E to SSE	
.5	SUNSHINE	
	Hours per day (mean) - April to October	9 h
	- November to March	5.2 h
.6	GENERAL	
	Dry and dusty - April to October	
	Heavy rains - November to March	
	Hottest months - September/October	
	Coldest Months - June/July	
.7	LOCATION	
	Latitude S	* 12° 40' ± 40'
	Longitude E	* 28° 15' ± 15'
	Altitude of surface above mean sea level	* 1 260 m ± 40 m
.8	BAROMETRIC PRESSURE	
	Average in mbar at 20° surface ambient temperature	* 875 ± 10 mbar

* These are approximate and averages for a number of sites.
Do not use for calculations.

TITLE –	SPECIFICATION No. amendment sheet

Form 253 Rev 2

Figure 34.3 *A specification which can double as a checklist for climatic conditions during task definition*
(Courtesy of Seltrust Engineering Limited)

complications such as special local fire authority requirements, possibly demanding steel roller shutters at each landing or certificated fire rated doors. The time to find these requirements out is during task definition, and not after the sales proposal has been presented.

To take another example, the task definition for the proposed design, manufacture, supply of an automated machining system would have to establish the configuration of the workpiece, the machining operations required, the production rate expected, the machining characteristics of the material, the accuracies required, the quality of machined finish needed, and details of the clamping points or locating lugs that the customer can provide (if the workpiece is a moulding, casting or weldment). Some of these parameters might be changed as task definition discussions suggest improvements or the need to modify the requirements.

Solution engineering

Solution engineering is needed in some degree for every proposed engineering project, the only exception being the sale of goods straight from a catalogued source of standard items. The amount of work needed will obviously depend on the scale of the proposed project and the difficulty of the problem. This section summarizes activities that would be needed for a medium- to large-sized project.

Feasibility

Even if an action plan has decreed that the company will bid, initial engineering studies may show that the project as conceived by the customer is not feasible, either technically or economically. This discovery is better faced early, before abortive design effort is committed. The options are to pull out of the proposal (with the agreement of the screening committee), or to go back to the customer and get the task redefined.

On very large capital projects, a preliminary feasibility study may itself involve many months of investigation and engineering. This may be commissioned by the client, or by his financers (who will want to know if their money is going to be invested safely and wisely). For a specially commissioned study of this sort the engineering company becomes a paid consultant. Some such studies can cost millions of pounds, last several years, and require that all the solution engineering for the proposed real project is carried out. In other words, solution engineering becomes a project in its own right, and the resulting study report takes the place of the sales proposal. The engineering company enjoys the luxury of being paid for its solution engineering. Studies of this sort can be required for major natural resource development schemes, or for some very large civil engineering projects. It is no longer sales engineering, but the methods and most of the objectives are similar.

Design

Considerable design work may be necessary to think up and then prove the viability or otherwise of one or more possible solutions. Time is usually short on these occasions, with the customer having set a deadline for receipt of tenders from the various competing companies. Sometimes a company finds out later than its competitors about a new project, which increases the pressure still further.

It is usually necessary to prepare layouts, flowcharts or block operating diagrams and summary lists of major items to be manufactured, purchased or subcontracted. The more that can be defined at this stage, the better can be the cost estimates. For a fixed price contract, this allows keener pricing with less risk.

It is a paradox that while the whole exercise may have to be conducted in a period of a very few weeks, the decisions made at this stage will have far reaching consequences for a project that might last as many years. The earliest decisions in a project are likely to have the most influence on its final cost and profitability.

The sales engineers will need to investigate the existence of previous designs, proven on earlier projects, which can be re-used or adapted for the new project (these earlier designs are sometimes referred to as a company's retained engineering). If the use of retained engineering is possible, this can reduce the estimated engineering costs, speed up the planned timescale and reduce risk. But the sales engineers must record this intention with their solution specification. Otherwise there is a danger that when an order is won the retained engineering will not be re-used, everything will be designed anew, and costs and time will exceed plan.

The solution proposals must have the blessing of the company's engineering management, but it is likely that senior design engineers and the engineering management will have been involved directly or indirectly in the solution engineering. For manufacturing projects it may be necessary to consult the production management before finalizing any solution, in case aspects need changing to suit the manufacturing facilities available, or to take advantage of cost saving techniques.

It may be possible for the sales engineers to remain in touch with the customer during the solution engineering phase, to ensure that work proceeds in line with his requirements. This approach is seen at its best in the simultaneous engineering method described in Chapter 30, where the contractor and the customer develop the solution jointly to achieve the optimum effect for both companies.

Throughout this period of intense activity, the sales engineers must not lose sight of the dates imposed on them by the action plan. If their activity is hemmed in by strict time constraints, so is that of those who must print and prepare the proposal document. As much information must be fed through as possible while the design in progress, not leaving everything to the last impossible moment. The more time allowed for the proposal document, the better its presentation is likely to be. If the proposal reaches the customer too late, the bid will probably be disqualified, with all the cost and effort wasted, and the sale prospect lost. The sales coordinator's role in monitoring progress is vital in this respect.

Cost estimating and pricing

Project cost estimating was discussed briefly, with illustrations, in Chapter 20. Typically, this task will be entrusted to a cost estimating engineer or department, and the estimates will be assembled as soon as design information becomes available from the sales engineers. Although the function of cost estimating may not be the direct responsibility of the sales department, the results will obviously be of vital interest when the sales proposal is assembled. Apart from the magnitude of the figures produced, a guide to their accuracy will be invaluable when gauging prices.

Some companies like to place their cost estimates in various categories, according to the methods used and the degree of confidence embodied in them. One such classification, applicable to major projects, is as follows:

Classification	*Input data conditions*	*Accuracy*
Ball park	Minimal information available, other than the task definition and the benefit of general experience from past projects	± 25%
Comparative	Preliminary layouts and work definition after solution engineering is complete or nearly so. Major elements of the work can be compared with similar jobs and cost records from the past. Budgetary quotations for major items of equipment obtained from suppliers wherever possible.	± 15%
Feasibility	Solution engineering finished, with general layout drawings available, and all major purchases listed and priced.	± 10%
Definitive	Only possible when project work is actually well under way, when design is complete and all major items have been ordered. This estimate is continually updated as part of cost control and progress reporting as work proceeds, until the error tends to zero at project completion.	± 5%

For the purpose of a sales proposal, it is likely that the estimates used will be comparative or, if time allows, feasibility. It should be borne in mind that, although estimating accuracies are discussed in this context, in reality there can be no such thing as a forecast of project costs which has an error of ± zero. No project can be defined with such accuracy because projects, by their nature, are dynamic and subject to all sorts of unpredictable changes. The best that can be done with estimates is to reduce the element of uncertainty: if the final costs happen to equal those originally estimated, then that must be partly a matter of good fortune.

All project estimates contain two major blocks of entries, known as above-the-line costs and below-the-line costs.

Costs written above line are all those attributable to the direct labour costs (salaries and wages), indirect costs (overheads), and the cost of all bought out materials, equipment and subcontracted services. These costs add up to the estimated total cost of the project if it were all to be carried out at the cost rates and prices ruling currently, and with no unforeseen difficulty.

Costs written below the line typically include allowances for contingencies, inflation of wages and materials (always known as escalation costs in this context) and a mark-up for profit at a rate determined or approved by the company's senior management.

Sometimes it happens that the contractor is aware of items that may be required, but which cannot be properly specified and priced at the time of tendering. For example, a building contractor tendering to refurbish an old building may be concerned that stripping the rendering from a wall could reveal weaknesses showing that part of the wall should be rebuilt. It would not be reasonable to include that work in the main body of the estimate, because no one can be certain that it will need to be done until well after the contract has been signed and work is in progress. The method used is to add provisional items to the schedule of work, and to show the costs separately as provisional sums (p.c. sums), to be spent only if they prove to be justified.

Particular care is necessary in cost estimating and pricing when more than one national currency is involved. It is customary to convert all sums of money into one currency, known as the control currency, during estimating and pricing. Safeguards against future adverse movements in exchange rates can be made by a below-the-line allowance or by attempting to include some clause in the conditions of contract.

In most cases the pricing of major projects is seen as a job for a senior management committee or the board of directors. The more confidence that can be expressed in the estimates, the safer the management will feel in reducing the mark-up rate in order to produce a competitive price.

Under some circumstances a negative mark-up (loss) might be allowed, perhaps if the work is very badly needed to fill a gap in the company's overall workload, or because the project is particularly attractive for some strategic purpose.

There is another important aspect to cost estimating that is often ignored.

This is the evaluation of the benefits that the project can be expected to bring the customer or client. This process involves considering, in as great depth as possible, the life cycle and operating costs of the equipment or plant being offered. There are several purposes for making such estimates, the principal two being:

1 To assist in choosing and recommending the optimum engineering solution.
2 To form part of the sales proposal and strengthen the argument in favour of it.

Delivery promise

Planning and scheduling techniques were described in Chapter 22. If the project design is sufficiently well advanced, and if there is enough time available, it might be possible to carry out some detailed planning, using sophisticated techniques, to arrive at a promised delivery, installation and commissioning programme.

The practical case is usually quite different, and the project delivery quoted to a customer is likely to be based on a broad judgement, probably aided and illustrated by a very simple bar chart. This need not be a serious problem: most project engineering companies know how long a major project should take, based on their past experience. It may even be that the customer has imposed his own delivery requirements. This does not mean that modern planning techniques are irrelevant. The reality is that they are often used to ensure that projects are completed within time limits that have already been set at the sales proposal stage.

The proposal documents

The principles of report writing, given in Chapter 9, are particularly relevant to the format and contents of proposals. For large projects, the intended readers are going to be very senior members of the client's management. It is vital that the first one or two pages summarize the proposal by spelling out very concisely:

- An itemized summary of the essential scope of supply, with optional extras shown separately.
- What each item will cost.
- The time schedule.
- The benefits (quantified if possible) for the client if he takes advantage of this proposal.

A table of contents should then follow. The detailed contents will typically include:

1 A detailed statement of the items and services being offered. Apart from the basic parts of the project, schedules should specify items such as the training and instruction of client's staff, list of installation and operating manuals to be supplied and, if the details are known at this stage, lists of consumable materials and recommended replacement maintenance spares.
2 The technical description, defined precisely by a well written text which refers, where appropriate, to engineering specifications and drawings by their unique serial and revision numbers. All performance data must be included, preferably in clear tabular form.
3 The commercial terms. Chapter 21 deals with this subject, but no sales engineering department would commit their company to commercial terms without having them written by or, at least approved by, a qualified commercial manager or legal adviser.

These are the essential contents but most proposals will contain or be accompanied by a considerable amount of other information designed to amplify the project details and to give a general description of the company.

Information specific to the project may include a proposed organization chart, showing the names of managers and senior engineers who would be employed on it. This may be backed up by summaries of qualifications and career experience of the named individuals.

Non specific information will often give details of the company's history, its ownership, resources and skills available, general organization, and illustrations of recent successfully completed projects.

The proposal may be backed up by artist's impressions, models, charts or other visual aids.

The use of a word processor for the text is a great aid to text preparation, allowing last minute changes to be incorporated without having to retype and recheck whole sections. The word processor is also useful in storing frequently used texts, such as company history, biographical details, and commonly used sets of commercial conditions. Preprinted sheets describing and illustrating recent successful projects can be held in stock, for insertion at the rear of proposals as appropriate.

The sales engineers might have established a very good relationship with the customer during solution engineering, and they may have persuaded their contacts that theirs is the proposal that should be chosen. But in many cases, the people seen by the sales engineer will only be able to recommend acceptance to their seniors. The sales engineer, or his company's more senior managers, may not be allowed to present their case in person when the tenders are opened. In such cases the quality of the proposal document assumes great importance.

It may be necessary to have the text translated into the language of the client: the desirability or need for this must be established from the outset. If the text is translated, great care must be taken to ensure that the legal terms are not made ambiguous.

The availability of a competent in-house printing department is practically mandatory in view of the limited time that is usually allowed for printing and binding. It may be desirable to engage professional design help to establish a corporate style for proposal covers in general. The company should take care to research the particular religious and cultural sensibilities of some overseas clients who might be offended by the use of particular colours, symbols or words.

Sales specifications

When the work of the sales engineer has resulted in the customer's signature on a purchase order or a project contract, it is obviously important that the customer and seller share the same interpretation of what has been sold. For this purpose, the sales engineer must provide a sales specification that sets out the scope of supply in definitive terms.

Specifying standard items to be supplied from a catalogue

For very simple sales, such as the supply of components from a catalogue, it is usually sufficient for the customer to quote the descriptions and relevant catalogue or part number of each item on the order form, along with the quantities to be supplied. The purchase order itself then becomes a sales specification. It should contain all the information that the supplier needs for order fulfilment.

The catalogue numbers give positive indentification of each item, and the descriptions back up this information in case of a numerical error. The specification is really only a simple list. Other information on the order will give delivery and invoicing instructions and may confirm agreed prices. There is no need for the supplier to write any additional specification.

Many suppliers reserve the right to modify or substitute catalogue items at their discretion. A statement to this effect is often printed in catalogues, usually with the explanation that this is in the interest of continual product improvement. There are occasions when a purchaser can only accept goods exactly as they are described in the supplier's catalogue. Perhaps there is some feature of an item which would normally be considered secondary to its main function, but which just happens to be important in a particular purchaser's application. It may be that the purchaser wishes to match exactly items obtained previously from the supplier. A change from plastic material to a metal for a simple bracket or support might seem irrelevant, but it would not be so to a purchaser who needed the item to be acid resistant or who had previously relied upon its electrical insulating properties. In such cases the purchaser would need to write a purchase specification to define the important characteristics, so overriding the supplier's declared freedom to make changes. On receipt and acceptance by the supplier, such purchase specifications become part of the purchase order and act as a sales specification for the supplier.

Purchase orders and purchase specifications are dealt with in Chapter 35. The supplier's sales engineers must check orders and customer's specifications to ensure that they are complete, without obvious errors. It may be desirable for the relevant sales engineer to read through the documents with the customer to make certain that there is no mistake or misunderstanding (essential for high value items). But, there is no good reason why the sales engineer needs to write a separate, formal sales specification.

Specifying catalogue items with options

Sometimes equipment is sold from a catalogue but, although the goods are listed, numbered and priced as standard manufactured items, the purchaser (possibly helped by the sales engineer) must specify certain parameters or choose from a range of optional extras. It may be that the type of power supply has to be specified (voltage, frequency, single or three phase). Perhaps the equipment exists in right- and left-handed versions. In fact, the ordering of a basic catalogue item might require the specification of a whole string of operating and performance data.

In order to save the sales engineer's time (and therefore increase his productivity) standard order forms can be designed and printed for these catalogue items with options. The seller knows the range available and all the questions that have to be asked. So, these can be printed on the standard form, with boxes provided for ticks or for the entry of data, as appropriate.

The sales engineer will always carry a pad of these forms, which are customarily set up as duplicate or triplicate sets on no-carbon-required (NCR) paper. When a customer decides to buy, the engineer can ask the relevant questions, place ticks or data on the form as necessary, and then get the customer to sign it. One copy is left with the customer, and the one bearing the customer's signature is taken back to headquarters by the sales engineer, where it becomes the formal sales specification for the supplier's internal use and control.

Although the sales engineer has had to produce a sales specification in order to specify all the possible variables, the use of a pro-forma has acted as a useful checklist and made the task relatively simple. This procedure should always be considered when a standard range of products is being sold.

Specifying catalogue items with modifications

The production management in companies which offer a standard range of products are apt to tear their hair out in large tufts when one of their company's sales engineers agrees to a customer's request for the supply of a standard product with special modifications. Apart from causing disruption in a product line that should be manufactured in standard batches for stock, there is also the longer term complication of having to repair or service non-standard items in the field.

Customers are apt to ask for products which differ in some degree from those listed in the catalogue. 'Wouldn't it be nice if this unit was fitted with circuit breakers instead of fuses.' 'Please could I have one of these pumps with BSF studs and imperial inlet and outlet – we don't want to start introducing metric sizes in our old plant room.' Inexperienced sales engineers are sometimes tempted to make such concessions in order to gain sales. It may be an easy option for them.

If the sales engineers are also given some discretionary range for discounting or price reduction, they might be tempted to offer maximum discounts for low volume orders in order to make the sale. When the sales engineers start selling nothing but specials, and all at discounts, the company would have been better off if no sales had been made at all.

However, some companies are in business to sell specials, or are willing to do so if a price addition can be agreed with the customer to compensate for the design and increased production costs that must result. In these cases, the sales engineer must ensure that the modifications are properly recorded in a sales specification, so that revised drawings and manufacturing instructions can be checked against the specification before manufacturing starts, and to ensure that the customer gets exactly what he ordered.

The sales specification for this type of modification need not be a long, complicated document. A memorandum accompanied by a dimensioned sketch might suffice. But it must satisfy these conditions:

1 It must be a written document – never a verbal instruction.
2 The document must be dated.
3 The document should be given a unique identifying reference number.
4 It should be checked with the customer.
5 The proposed modification must be agreed with the company's engineering and production managements before the specification is released for design and production.

6 The document must accompany the purchase order, which must quote its reference number and date.

Further:

7 The price uplift must result from a proper estimate of the likely true costs.
8 Any increase in delivery time must be assessed and reported to the customer.

Sales specifications for engineering projects

The sales specification for a project starts life as the formal proposal document against which the customer placed his order. The details of that proposal might have changed substantially during its preparation and during a prolonged period of negotiation and discussion with the customer. The specification must be revised to reflect these changes before it is passed over for project execution to start. There may be several such revisions and issues of the specification before a final sales specification is agreed with the customer. The final version is also often known as the project specification.

The specification will have been given a sales enquiry or proposal number when the enquiry was first received and the action plan issued. This number plus the revised issue number should be sufficient to define the project specification absolutely. This is achieved by observing the following steps:

1 A schedule of all documents contained in or associated with the specification must be prepared. This should be bound in with the specification, and it must list all the associated drawings and specifications, together with their serial and revision numbers.
2 If a change is made to the text of the specification, the document must be reissued at the next higher revision number. Deleted passages should be ruled through, with the relevant revision number written alongside (usually in an inverted triangle, as with common drawing practice). Added passages or pages should be highlighted in the same way.
3 The specification must also be revised if any associated drawing or specification is revised. In this case, the specification must be reissued with a revised documents schedule. The revision number is written in the schedule margin against the changed entries.

When the sales engineering work has finished, the project specification, together with all its listed drawings and specifications, becomes a complete project definition package. When project work is authorized to start, the project definition package should formally be handed over to the engineering project manager. Changes after this date must be the subject of formal engineering change or modification procedures, and these are described in Chapter 25.

35

The engineering/purchasing interface

D. W. Diamond

The engineer's involvement throughout all phases of the procurement cycle is a vital factor in ensuring that the end result – providing a finished product which is in all respects acceptable to the end user – is achieved. To be acceptable, the product must meet the end user's specified requirements with respect to quality, delivery time, ultimate price and subsequent performance. The engineer, in collaboration with the buyer, must have this ultimate objective in mind at all times.

The procurement cycle

The term procurement cycle (or process) is preferred to purchasing throughout this chapter as, to fully understand the engineer's role, it is first necessary to appreciate that his input must start well before the actual purchase is effected and frequently continues beyond the period of manufacture. For reference within this text, the procurement cycle is considered to cover:

1 Enquiry package preparation.
2 Bidder selection.
3 Issue of invitations to bid.
4 Bid review and comparison.
5 Vendor selection.
6 Placing the purchase order.
7 Expediting of manufacture or supply.
8 Inspection/testing/quality control.
9 Delivery.

It should also be borne in mind that, following this phase, the engineer can well be involved with factors arising from installation, commissioning and final acceptance of the completed product.

The engineer's role

It is obviously difficult to define the engineer's role in procurement in a detailed form that would apply to every conceivable requirement, irrespective of the nature of the product and its ultimate service. In order, however, that most aspects of the engineer's input can be considered, the following account assumes that the goods to be purchased comprise equipment that has a high engineering content and exacting performance characteristics and that it must be obtained at a competitive price.

To ensure that the basic requirements and agreed standards are met, the following detailed procurement cycle is usually adopted:

1. The engineer prepares a technical package, consisting of:
   (a) Specification defining technical parameters and scope.
   (b) List of applicable standards.
   (c) Definition of all necessary quality assurance/control, cost and scheduling, inspection and installation interfaces.
2. The engineer issues the technical package to the buyer.
3. The engineer and buyer prepare, jointly, a list of acceptable bidders, including any bidder prequalification or screening considered necessary.
4. The buyer prepares a request for quotation, adding the commercial and contractual requirements to the technical package provided by the engineer. This forms the complete enquiry package.
5. The enquiry package is reviewed by the engineer and, where applicable, by other interested parties such as construction, installation and shipping specialists.
6. The enquiry package is issued to the selected bidders by the buyer.
7. The bids are received, recorded and opened by the buyer in conjunction with the engineer.
8. The buyer makes a first review of the bids, listing them in order of competitiveness by the first-sight price identification.
9. The engineer carries out a technical review of the bids in order of price precedence as listed by the buyer.
10. The buyer carries out a commercial review of the bids which have been nominated by the engineer as being technically compatible.
11. The engineer and buyer make a joint comparison of the technical and commercial reviews.
12. Selected bidders may be invited to meetings with the engineer and buyer to ensure total commercial and technical compatibility with the requirements of the enquiry package.
13. Final selection of a recommended bidder. This choice is made jointly by the engineer and buyer after consultation with other interested parties such as construction, installation, quality assurance, shipping, etc.

14 Approval for the purchase is obtained from the designated signatory (which might include the client's approval for some capital projects).
15 Preparation and issue of a purchase order is carried out by the buyer.
16 The successful bidder (the vendor) acknowledges formal acceptance of the order.
17 The engineer reviews and approves technical submissions made by the vendor.

A flow diagram of this cycle is shown in Figure 35.1.

From this point on in the procurement cycle, the engineer's input is likely to consist of the following, on an as-needed basis:

1 Responding to technical questions raised by the vendor during manufacture.
2 Arbitrating with respect to any technical difference which may arise between the purchaser's inspection/quality control staff and the vendor.
3 Attending and approving the results of performance or acceptance tests.
4 Responding to post-delivery technical questions during erection, installation, construction, commissioning or setting to work.
5 Providing engineering input with respect to the approval of vendor's invoices or other claims for payment.
6 Contributing to obtaining the final acceptance of the ultimate user.

The cycle so far outlined must be regarded as indicative rather than definitive. For example, the engineer's continuous interface with the purchaser's own client (where applicable) is recognized implicitly, and the need for his possible involvement with the vendor's subsuppliers or subcontractors also has to be accepted.

The technical package

Effective procurement must always rely largely upon the clarity of the material/equipment requisitions and specifications prepared by the engineer. In a completely unambiguous form, these base documents must clearly define standards, specify quality and performance and ensure that no misunderstanding of technical interpretation can arise. The importance of this phase of the engineer's input cannot be over emphasized. Errors, omissions and indifferent drafting at this stage can result in lengthened deliveries, escalated costs and a possible reduction in quality standards at a later date.

The temptation to overelaborate the technical package contents is ever present. The inclination to adopt a 'when in doubt, include it' philosophy leads to enquiry packages that are so unwieldy that they often defeat the intended purpose of protecting the purchaser: many bidders rarely have time

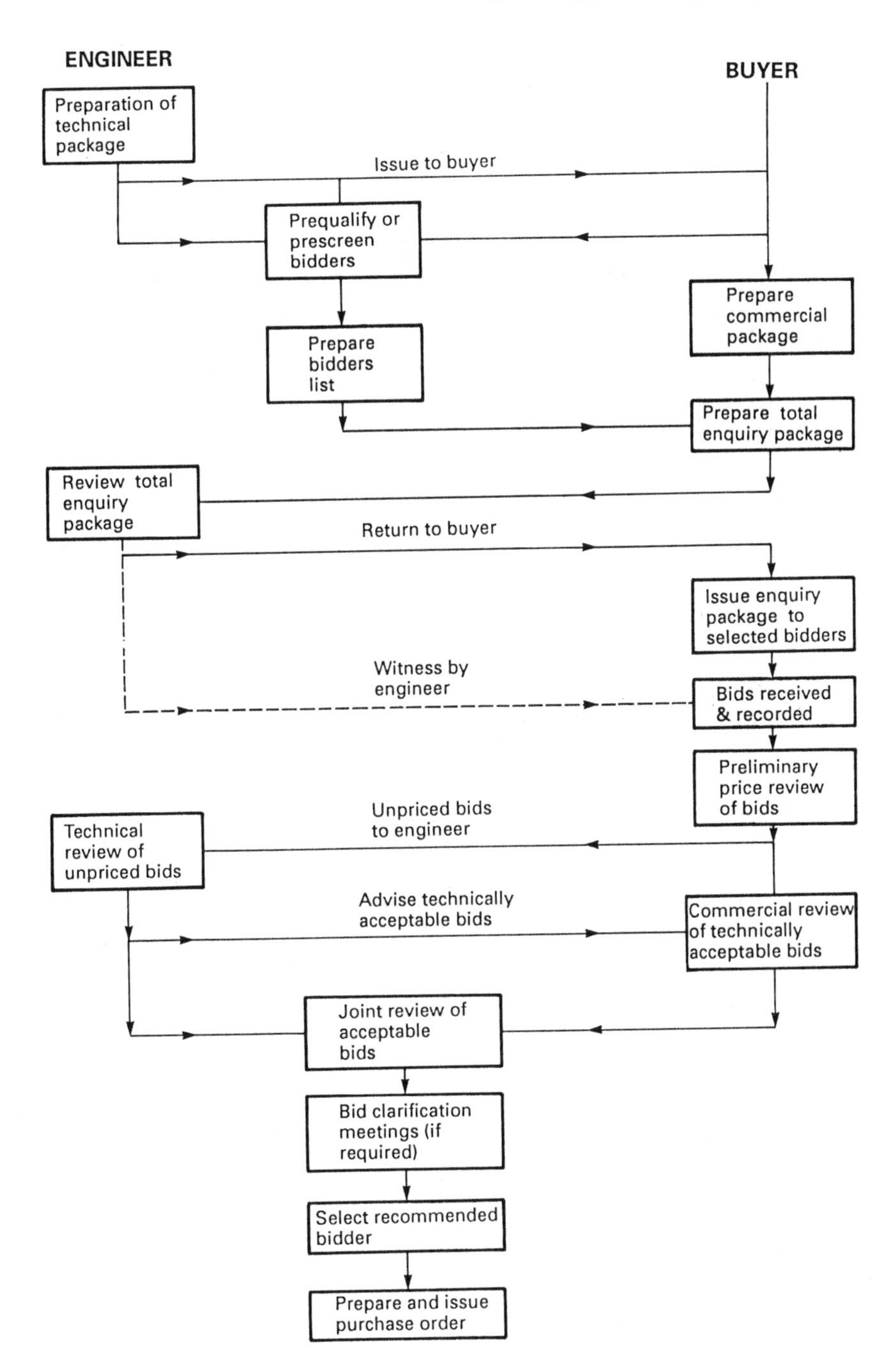

Figure 35.1 *The procurement cycle*

to review adequately an overwhelming number of documents that apparently attempt to cover every contingency known to man with respect to ultimate performance.

Recognizing that brevity must always be the sensible aim, material requisitions and their associated documents must however be comprehensive and complete to the degree needed to ensure that bids will require the minimum amount of post-tender questioning and conditioning. As a minimum, requisitions should contain:

1 Detailed scope of supply (including required ultimate performance).
2 List of technical attachments (including all necessary standards).
3 Drawings and data required to be submitted with bid.
4 Drawings and data required to be provided by the vendor in the event of an order.
5 Quality assurance requirements.
6 Inspection/testing requirements.
7 Details of spare parts required (commissioning and operational).
8 Vendor service representative requirements with respect to commissioning/ installation – where applicable.

In many companies, specialist aspects such as quality assurance are managed by disciplines separate from the design engineering function. The need for such arrangements is generally accepted but experience will confirm the recommendation that the engineer be responsible for coordinating all such inputs to the technical package. The responsibility for ensuring that no one particular facet of the overall engineering requirements is overemphasized to the detriment of the others (with possible risk to the ultimate performance of the product) must always reside with the engineer.

Figure 35.2 illustrates the front sheet of an engineering specification used by a mining engineering company for major and minor purchases. Note that this sheet is introductory, and lays special emphasis on defining the content of the complete specification, including all amendments. If it becomes necessary to amend any sheet of the specification, either before the purchase order is issued or afterwards, the front sheet will always be amended itself, and its amendment number will always be referred to as the amendment number for the complete specification. Using this arrangement, it is only necessary to replace those individual specification pages which have been amended, and the correct constitution of the complete specification is always positively identified by the schedule of sheets, with their individual amendment numbers, at the top of this front page.

Figure 35.3 shows the second sheet to that shown in Figure 35.2. This page, because it is preprinted, acts as a checklist for the engineer and ensures that proper consideration is given by the engineer in specifying the nature,

SPECIFICATION

SHEETS

At the latest amendment as shown below, this specification consists of all the sheets listed in the following table, assembled in ascending numeric - alphabetic order, and with each sheet carrying the amendment number stated. When this specification is amended normally only revised or additional sheets will be issued with this front sheet.

Sheet	amd't	Sheet	amd't	Sheet	amd't	Sheet	amd't	Sheet	amd't	Sheet	amd't	Sheet	amd't	Sheet	amd't
1															

ATTACHMENTS

The following attachments form a part of this specification

AMENDMENTS

No.	Date	Purpose of issue (e.g. enquiry, purchase - requisition no.) and technical details

APPROVALS

Originated by	Checked by	Senior engineer	Project engineer

Client –
Project –
Plant –
Discipline –

Project No –

TITLE	Specification number	Sheet number	Amendment number

Figure 35.2 *Standard form used for the front sheet of an engineering purchase specification* (Courtesy of Seltrust Engineering Limited)

quantities and timing of technical and other documents required from the vendor.

Although this chapter concentrates on the purchase of complex, high value items, it is relevant to stress here that the correct specification of many low cost items is equally important and that the procurement of these, too, may need to be supported by the preparation of a detailed specification. Even where the vendor provides his own catalogue description, it may be necessary to write a specification that safeguards the purchaser against the claim made by most vendors that 'we reserve the right to modify the design of any item in this catalogue . . .'. The vendor might well make a design change that would not affect technical performance but renders his product unsuitable for interchangeability for example.

Bid list preparation

A suitable bid list should be prepared jointly by the engineer and the buyer. This should include possible vendors who have previously prequalified to provide the product required. A review of such prequalification records should satisfy both the engineer and the buyer that each bidder will have:

1 The financial strength to sustain the cash flows likely to arise during the project.
2 The experience of the product, competency and plant capacity to complete the project within the constraints of the likely contract.
3 Technical capability (including human resources) sufficient to satisfy the requirements of the contract.
4 A complete understanding of similar project scopes and the ability to absorb subsequent changes.
5 The experience to understand the commercial requirements of the contract and associated implications.
6 The facilities (testing, quality control, etc.) necessary to endorse assurance of quality and product performance.

It is appreciated that in many cases the engineer and buyer will not have access to suitable prequalification records. It is also accepted that even where the records do exist they cannot guarantee the unquestioned suitability of a given bidder. If correctly prepared and interpreted, however, prequalification records can at least diminish the element of risk.

Where such records are not available and there is insufficient time or resources to establish them, it is often useful to prescreen potential bidders to ensure that they are capable of offering a competitive bid. Such screening can be useful in ensuring that time is not lost by bid invitations being returned by

SPECIFICATION

DRAWINGS AND DOCUMENTS

1. SCOPE OF SUPPLY. Complete engineering drawings, installation, operating and maintenance manuals, parts lists, recommended spares lists and other documents for all equipment/services covered by this specification, all in accordance with the following requirements.
2. QUANTITIES.
   Drawings for approval — 3 prints
   Final drawings -- 1 transparency
   Other final documents — 6 copies
3. LANGUAGE. English
4. QUALITY. High quality suitable for microfilming. Final certified drawings full-size on good quality plastic film and not folded.
5. IDENTIFICATION. Drawings and documents shall bear the purchase order number under which this specification is issued, appropriate equipment and tag numbers as given, and purchaser's drawing numbers when supplied.
6. REVISIONS. Drawings which are revised after initial submission shall be resubmitted immediately showing details of the changes and the new revision number.
7. CERTIFICATION. Final drawings shall be certified as accurate.
8. "AS BUILT" DRAWINGS. Where the specification includes installation and erection, "as built" drawings reflecting any site changes shall be supplied as soon as possible after completion of the work.
9. DATE OF SUBMISSION. Shall be within the number of weeks from order placement specified in the tabulation.
10. APPROVAL. All drawings for equipment specially designed to meet the specification shall be submitted for approval prior to manufacture, unless otherwise agreed.
11. TEST CERTIFICATES. Shall indicate the British or other national standard or code in accordance with which tests were performed and be identified as paragraph 5. The original and five copies, all signed, shall be supplied.
12. LUBRICATION REQUIREMENTS. Shall specify recommended lubricants and quantities required for one year's operation.
13. RECOMMENDED SPARES LISTS. Shall be based on (one)(.......) years operation under specified conditions and shall include for each item:
   - Identification, part or serial number
   - Maker's name and reference
   - Quantity recommended
   - Unit price and delivery

SPECIAL REQUIREMENTS

SCHEDULE OF SUBMISSION TIMES

Drawing/document	Shall be submitted		
	If ✓ with bid	WEEKS (see para 9)	
		for approval	final certified
Arrangement drawings			
Foundation outline drawings			
Foundation loading details			
Engineering flowsheets P & ID's			
Electrical schematics			
Electrical termination and connection diagrams			
Shop details			
Calculations			
Data sheets			
Performance data			
Drawing schedule			
Test certificates			
Parts and materials lists			
Recommended spares list			
Lubricant requirements			
Erection instructions & drawings			
Operating & maintenance instructions manuals			

EQUIPMENT MARKING AND IDENTIFICATION

Each item shall be identified by its mark, equipment or tag number as given in the specification by the method indicated thus ✓

☐ Type stamping & painting
☐ Stamped nameplate
☐ Wired on metal tags
☐ As specification section ....

INSPECTION AND TESTING

Documented tests and inspection as indicated thus ✓ below are required

Inspection by purchaser	Required
During manufacture	
Final Inspection	
Of packing	
As specification section..................	

Testing	Witnessed	Not witnessed
Standard works		
No load running		
Full load performance		
As specification section............		

NOTES

TITLE	Specification number	Sheet number	Amendment number

Figure 35.3 *Standard specification sheet for vendor documentation and inspection/testing requirements*
(Courtesy of Seltrust Engineering Limited)

uninterested suppliers or incompetent bids being offered by unsuitable bidders. Screening enquiries should elicit:

1 Interest in submitting a bid.
2 Technical capability.
3 Estimated delivery period.
4 Likely shop loading during projected manufacturing period.
5 Recent experience in producing the product required.
6 Ability to fulfil any statutory requirements.
7 Adequate quality assurance systems.
8 Financial viability.

After joint evaluation of the screening results, the engineer and buyer should be able to produce a bidders list which recognizes:

1 Satisfactory response to screening.
2 Relevant experience.
3 Satisfactory prior performance.
4 Competitive potential.
5 A realistic chance of success.
6 Financial viability.

It has to be accepted that certain constraints may limit the employment of prescreening techniques. Time available to obtain bids; the ultimate financial value of the likely order; individual resources of the possible bidders – all may be restricting factors in employing the prescreening telex. In such cases, the engineer and buyer may be required to fall back on respective judgements based upon their individual experience of the likely bidders in question. At the end of the day, it can only be emphasized that prequalification or prescreening reduces but does not eliminate the element of risk which is always latent in bidder selection – particularly where international competition is sought.

The enquiry package

An enquiry package prepared by the buyer normally comprises the technical package outlined above plus a commercial package which will identify:

1 Purchaser's project or job reference.
2 Issue date.
3 Bid due date.
4 Required delivery point and duration.
5 Delivery terms (FOB, FOR etc.).
6 Packing, marking instructions.

7 Suborder requirements.
8 Bid bond details (where required).
9 Performance bond details (where required).
10 Retention details (where required).
11 Liquidated damages (where required).
12 General conditions of contract.

The need for clarity and absence of ambiguity noted earlier with respect to the engineering input applies in equal measure to the commercial aspects of the enquiry package. Lack of clear definition in either of these elements at this stage is a guarantee of future problems, some of which could escalate into litigation. As, in the course of preparation, various changes may have been made in the requirements, it is advisable for the buyer to undertake a joint check with the engineer to ensure that the resulting combined package agrees in all respects with the final requirements of the purchaser.

Some buying departments send a covering letter with enquiries using a standard form, while others make use of a standard letter. An example of a standard enquiry letter is shown in Figure 35.4.

Receipt of bids

The need for ensuring full confidentiality of all bids is paramount in the operation of all bid opening procedures. In order that this basic precept is assured, the following procedure is recommended:

1 The incoming sealed quotations should be delivered unopened to a designated recipient (usually the buyer) and placed in secure storage.
2 The bids should be prepared in both priced and unpriced forms, with each noted clearly as such.
3 Bid openings should be made all together at the designated time of the bid-due date.
4 Bid openings should be undertaken by more than one person, and they should be witnessed by the engineer.
5 The receipt of all bids should be recorded formally, each bid being date stamped. For complex bids of high value, each page should be perforated by a hard seal.
6 Bid receipt records should be endorsed by witnesses, and receipt of bid copies by individuals within the purchaser's organization should be acknowledged by signature.

Figure 35.5 shows a form suitable for recording the opening of sealed bids.

REQUEST FOR QUOTATION

To:

Date:

Our ref:

Dear Sirs

Project:

Equipment:

Quotation due date:

AABBEC (UK) Ltd invite you to submit a quotation for the supply of goods/material/equipment described in the attached material requisition number:

Your quotation should include the following:

A Confirmation that the goods/material/equipment will meet the quality and performance requirements detailed in the specifications/drawings/standards nominated in the material requisition reference: and attached hereto. These are incorporated in this Request by reference.

B Where your quotation is based upon alternatives to the specifications/ drawing/standards nominated in the aforesaid material requisition your alternatives are to be fully detailed in your quotation in a separate, clearly designated section, set aside for that purpose only.

C Confirmation of your acceptance of the **terms** and conditions of purchase attached, which must be included as a specific statement with your quotation.

D The quotation must reach AABBEC (UK) Ltd not later than: and the envelope/package must be clearly marked:

Tender reference:

For the attention of:

E Bid validity. State the latest date until which your bid is open for acceptance.

F Delivery. The delivery period from receipt of purchase order shall be clearly stated.

G Contract period. The contract period shall be the bid validity period (E) plus the quoted (or achieved) delivery period to the selected delivery point.

H Basis of price. All prices are to be quoted in £ sterling, and are to be fixed for the contract period (G) and are to be offered against the following delivery alternatives:

(i) Ex works (FOT)

(ii) Delivered to our works at:

(iii) Delivered FOB to nominated UK port.

I Any order placed with respect to your quotation will include appropriate clauses to cover the following:

(i) Liquidated damages at one per cent per week up to a maximum of ten weeks of default with respect to the delivery period quoted (F).

Figure 35.4 *An example of a standard enquiry letter*

(ii) A retention of five per cent of the finally agreed contract price will be retained as a performance bond for the agreed warranty period. Payment of the agreed retention will be effected upon vendor's presentation of an acceptable bank guarantee to the full value of the release.

(iii) Payment by AABBEC (UK) Ltd will be effected in sterling within 45 days of receipt of an acceptable invoice, (accompanied by full shipping documentation), receipt of all associated technical information, or upon completion of installation (whichever is the longer).

J AABBEC (UK) Ltd reserve the right to accept all or part of your quotation and are not committed necessarily to accepting the lowest bid against any particular item.

K The Request for Quotation comprises, in addition to this document:

(i) The material requisition listed in (A) including associated technical scope and notes.

(ii) Technical attachments (listed)

(iii) Commercial attachments (listed)

(iv) Commercial notes

L Your specific attention is drawn to the following:

(i) Inspection instructions (Appendix)

(ii) Requirements for vendor drawings and technical data (Appendix)

(iii) Packing, marking, invoicing and shipping instructions (Appendix)

(iv) Sub order requirements

(v) Request for spare part recommendations

For AABBEC (UK) Ltd

Purchasing manager

Figure 35.4 continued

SEALED BID OPENING RECORD

RECORD NO ____________

EQUIPMENT ____________

BUDGET PRICE ____________ CLOSING DATE ____________

BID REF	BIDDERS NAME	BIDS RECEIVED		BID OPENING RECORD	
		DATE	BY	DATE OF QUOTATION	UNEVALUATED TOTAL PRICE
A					
B					
C					
D					
E					
F					

REMARKS ____________

PREPARED BY ____________ DATE ____________

WITNESSED BY ____________

Figure 35.5 *Example of an opening record for sealed bids*

Bid evaluation

The time taken in evaluating bids can, if not strictly controlled, result in an abnormal expenditure of manhours. The procedure necessary to ensure a viable economic approach has to be chosen carefully. It is not possible to give guidelines for every purchasing situation, but the following procedures are recommended as being appropriate in most cases:

1 The buyer will release to the engineer unpriced copies of those bids which have been commercially shortlisted by first-sight price comparison.
2 The engineer undertakes a preliminary technical review of the released bids. In the event of the released bids not being technically acceptable, the

buyer will release the remainder of those received. The engineer will advise the buyer of those bids which are technically acceptable.

3 After the buyer has confirmed that the technically acceptable bids are also commercially viable, the engineer undertakes a detailed technical examination of the selected bids.

4 On an agreed form of technical bid evaluation sheet, the engineer should record comparisons based upon the following:
   (a) Acceptability of the product quality and performance.
   (b) The bidder's compliance with the relevant technical specifications and standards.
   (c) The bidder's compliance with quality assurance/control requirements.
   (d) The bidder's ability to provide the required technical information within the specified period.
   (e) The bidder's ability to offer adequate manufacturing facilities.
   (f) Likelihood of the promised delivery date being achieved

   and, where applicable:

5 Previous performance for similar products.
6 Maintenance requirements.
7 Availability of spares.
8 Servicing facilities and support.
9 Installation or erection characteristics.
10 The quality of subvendors (subcontractors to the bidder).
11 Adequacy of testing facilities.
12 Operating costs.

It is appreciated that, unlike a straightforward commercial/pricing analysis, a technical comparison must always in some degree be difficult to quantify and it is sometimes suggested that many of the necessary judgements delivered during technical comparisons are therefore subjective. However, providing the engineer is fully aware of the ultimate requirements of his bid comparison and he has a thorough understanding of the equipment/commodity under review, he should be able to produce an objective analysis of the comparative bids. It should be remembered that the engineer produced the technical scope and engineering specifications in the first instance and he should therefore be able to undertake an evaluation by adopting a form of rating (suitably weighted) dependent upon such factors as final use, ultimate location, criteria of performance, importance of maintenance, availability of spares and after sales service.

An example of a weighted factor table of comparisons used in a technical evaluation is shown in Figure 35.6. This is not intended to cover all possible cases, but it has been compiled to include typical factors encountered in the provisioning of complex equipment with a high engineering content. The actual weighting factors chosen in this example are also not intended to be definitive: the choice of factors is dictated by variables such as criticality, location, complexity of maintenance, installation and so forth.

The selection of appropriate factor values is critical, and a matter for very careful judgement on the part of the engineer. The examples illustrated in Figure 35.6 are intended to convey indications of a poor bidder (bidder 1), average bidder (2) and a good bidder (3). It should be emphasized, however, that such tables are intended to be used in conjunction with the normal style of technical and commercial evaluation. Their employment assists in the completeness of overall assessments, but they are not offered as a substitute for any other essential input.

It may prove necessary to hold clarification meetings with bidders in those instances where concern exists with respect to the bidders' interpretation of the purchaser's requirements. In these cases it is strongly recommended that both the buyer and the engineer attend the meetings in order that the commercial implications of possible technical amendments can be recognized immediately and, conversely, that the impact of any corrective commercial measures on engineering performance can be assessed.

Selection and recommendation of a vendor

After completing their respective individual reviews, buyer and engineer should be jointly capable of producing an overall analysis which will set out the technical and commercial comparisons in a form that will lead to final vendor selection. The minimum requirements of this comparison should (in addition to all usual job or contract references) include:

1 Bid date.
2 Bid validity period.
3 Full description of equipment, commodity, material, service, etc.
4 Resumé of the technical evaluation including performance compliance.
5 Operating costs (where applicable).
6 Itemized and total price on a compatible delivery basis.
7 Price basis (firm or otherwise – price escalation formulae where applicable).
8 Terms of payment.
9 Where applicable – transport/freight costs, import duties and taxes.
10 Rates of exchange used to convert currencies where international bidding has been sought.
11 Comparison with any preset budget prices.
12 Acceptance of warranties.

EVALUATION DESCRIPTION	WEIGHTING FACTOR (A)	BIDDER 1		BIDDER 2		BIDDER 3		BIDDER 4	
		RATING (B)	EVALU-ATION FACTOR (AXB)	RATING	EVALU-ATION FACTOR	RATING	EVALU-ATION FACTOR	RATING	EVALU-ATION FACTOR
1.0 SIMILAR EXPERIENCE OF									
1.1 Equipment Manufacture	10	4	40	6	60	8	80		
1.2 Purchaser's Requirements	2	0	0	6	12	10	20		
2.0 EXECUTION									
2.1 Capacity Shop Fabrication	13	5	65	7	91	8	104		
2.2 Execution Plan	8	4	32	8	64	7	56		
2.3 Key People	5	10	50	10	50	10	50		
3.0 GENERAL									
3.1 Planning/Controls Procedures	12	5	60	5	60	7	84		
3.2 Materials Management	10	3	30	5	50	8	70		
3.3 QA/QC	10	3	30	5	50	8	80		
4.0 SPECIFIC									
4..1 Engineering Resources	12	8	96	9	108	9	108		
4.2 Procurement Capability	3	1	3	2	6	3	9		
4.3 Installation Capacity	5	1	5	2	2	4	20		
5.0 GENERAL									
5.1 Understanding /Commitment	5	4	20	5	25	5	25		
5.2 Overall Schedule Confidence	5	2	10	3	15	4	20		
RATINGS	100	52	461	74	603	90	726		

Figure 35.6 *Table for comparing vendors' execution ability using weighted factors*

13 Details of spare parts offered.
14 Confirmation of adequate servicing facilities.
15 Assessment of bidders' financial stability.
16 Rates for specialist commissioning/installation engineers.
17 Assessment of after sales services.
18 Interchangeability with similar equipment/parts.
19 An appraisal of the effect of promised deliveries (based on past delivery performances) as related to the purchaser's scheduled requirements.
20 Where a bid suggests that higher initial capital investment would ensure lower operating costs, the comparison should estimate the payback period with respect to incremental investment and offer a conclusion.

A typical bid comparison schedule is illustrated in Figure 35.7. This, when coupled with the technical analysis schedule and weight factor table (Figure 35.6) can form the basis of any report prepared to support the nomination of a recommended bidder.

It is recognized that the final recommendation of a vendor must result from an integrated process which takes all engineering and commercial factors into account. It follows, therefore, that the selection and recommendation must be achieved by a joint approach between the respective disciplines in order to achieve agreement in all respects. In some cases such agreement is not readily achieved.

Difficulties in this respect often centre on the compatibility of quality with price. The engineer may be led instinctively to choose the solution that offers the best available quality or performance, while the buyer may be motivated towards opting for the lowest acceptable price. Obviously a compromise has to be reached, and the level at which such a compromise is reached is often dictated by a balance of the purchaser's ultimate requirements.

Where the purchaser is himself in a competitive situation, and providing that the lowest bid will ensure equipment of an acceptable (if not the best) overall quality and performance, the low bid has to be accepted. If the purchaser is not constrained by competitive and budgetary considerations, the choice might be made on factors such as ease of maintenance and superior overall performance irrespective of the quoted price. Ultimately, such a decision is a management judgement which can only be made on the basis of an objective assessment made jointly by the engineer and the buyer. Chapter 18 takes the process of financial evaluation farther.

It is essential for the engineer and buyer to operate as a fully integrated team, with a concerted approach and single objective throughout the purchasing cycle. Should either discipline neglect the requirements and views of the other, the end result can only be to the detriment of the purchaser's total requirements.

BID SUMMARY

SELLERS							
Country of origin							
Bid reference							
Bid date							
Period of validity							
Bid currency							
Project exchange rate							
Item	Qty	Description	Price	Price	Price	Price	Price
Quoted price ex works							
Discounts							
Inspection-testing							
Packing-export prep.							
Fob charges							
Fob cost							
Est transport to site							
Customs duty tax etc.							
Est.total cost on site							
Quoted delivery ex works							
Est. transport time							
Est. total delivery to site							
Conditions of payment and notes-							

RECOMMENDED BY PURCHASING AGENT—

Reasons: lowest price ☐ acceptable delivery ☐ (tick for yes)

For purchasing agent

RECOMMENDED BY ENGINEER —

Reasons: meet technical requirements ☐ (tick for yes)

Project/senior engineer

RECOMMENDATION TO CLIENT —

Delivery required	Delivery quoted	Planning Engr	Reason:
Budgeted cost	Price quoted	Cost Engr	Project manager

Project title ____________ Project No. ____________

Specification title ____________ Specification No. ____________

Figure 35.7 *Example of a bid summary or bid analysis form*
(Courtesy of Seltrust Engineering Limited)

Approval

It is often necessary to insert approval steps at various stages throughout the procurement cycle. The level and frequency of such steps are dictated by the purchaser's organization structure, the nature of his relationship with his client (where applicable), the value of the bids under consideration and the basis of the eventual contract with the selected vendor. The time spent in obtaining or correcting information needed for an accurate, objective assessment of expenditure in a competitive bidding situation can be out of all proportion to the value of the goods being purchased.

The inclination of the approving authority to refer back recommendations is understandably strong where these lack clear, positive engineering and commercial endorsements. In some organizations this inclination may be paramount or an even mandatory approach. In conjunction with the buyer, therefore, it is the engineer's responsibility to ensure that the necessary approvals are not delayed, wasting valuable time, simply because further information is requested by the approval authority which should have been provided or clearly defined in the first place.

Purchase order issue

Pre-award meetings are sometimes necessary with vendors before a purchase order can be issued. The need for such meetings frequently arises where the equipment to be purchased is particularly critical or expensive. These meetings need the presence of both the engineer and the buyer, since their main purpose will be to:

1. Ensure that the vendor has completely understood the technical scope and commercial requirements of the purchaser.
2. Revalidate the vendor's promised delivery, and re-affirm the required schedule for submitting drawings and technical information.
3. Check the vendor's current manufacturing capacity and facilities (these may have changed in the time interval between bid preparation and final appraisal).

When these matters have been resolved, together with any other outstanding technical questions, the engineer will review the original enquiry package and update or change it as necessary to provide the technical package that will accompany the purchase order. He will also check the engineering content of the purchase order itself, before it is issued. The engineer needs to recognize that this is the last stage at which any amendments, corrections or scope changes can be finalized without running the risk of contractual differences.

Once the contract has been accepted by the vendor, any such changes could increase the price or qualify for extended timescale.

Purchase order documents vary greatly from one company to another, but the following examples represent a fairly typical set. Figure 35.8 shows a purchase requisition, which is issued by the engineer as the final instruction to issue a purchase order once the appraisal and approval phases have been concluded. Figure 35.9 illustrates the purchase order issued by the buyer, and Figure 35.10 is a set of standard conditions of purchase used by a mining engineering company, and printed on the reverse of its purchase order form.

Purchase order amendments

Any change in scope or technical requirements which arises after the issue of a purchase order must result in the preparation, approval and issue of a purchase order amendment. The commercial position when an amendment is made is quite different from that which obtained during the initial enquiry phase, since there are now no competing bidders and the vendor has a monopoly of supply. Once again, the need for effective cooperation between the engineer and the buyer is paramount.

Once the terms of a proposed amendment have been discussed and agreed with the vendor, the engineer will issue a requisition amendment to the buyer who, in turn, issues a corresponding purchase order amendment to the vendor. The purchase order amendment will be accompanied by the necessary amendments to the technical package. The requisition amendment stage may be omitted when the purchase order amendments are for purely commercial reasons, for example when an amendment is issued to confirm some minor commercial point agreed between the vendor and the supplier.

The approval level for purchase order amendments is often set at a lower level than that for the original purchase order and is usually dependent upon the amount of any additional costs involved.

Distribution for amendments should be identical to that given to the original order.

Progress, control and expediting/inspection

The purchasing organization will typically be responsible for following up the progress of vendors, arranging visits aimed at verifying claims for progress payments, carrying out stage inspections or witnessing tests, or simply checking on the current state of progress. In all of these actions, even where the purchasing organization employs its own team of expeditors and inspectors, the engineer will be expected to collaborate. This often means accompanying the inspectors/expeditors on visits to vendors' works.

REQUISITION

To.

Date:
Our ref:
Your ref:

☐ Please issue a purchase order and arrange transport etc as appropriate
☐ Please amend the purchase order
for the equipment materials or services detailed in the attached specification (at the amendment number given below) in accordance with the following

VENDOR (name and address)

QUOTATION (reference and date)

CONSIGNEE (name and address)

Please instruct the vendor to confirm his arrangements regarding sub-suppliers or sub-contractors

Inspect in accordance with – specification ☐ notes below ☐
The recommended degree of expediting is – intense ☐ normal ☐ none ☐
The recommended standard packing category is
The order is required on site by (date) –
The recommended method of transport is –
Client's charge account –
Client's authority reference –
Your technical contact is telephone ext.

Notes

PRICE DETAILS	
Basis	–
Original quote	–
Total previous amendments	–
This amendment	–
LATEST TOTAL PRICE	–
Budget allowance	–

DELIVERY DETAILS
Original delivery –
Current delivery –
Required delivery –

APPROVALS			
Originated by	Senior engineer	Project engineer	Project manager

Client – Project – Plant – Discipline –	Project number:	
	Specification number	Amendment
TITLE	Requisition number	Amendment

Figure 35.8 *Purchase requisition. Example of a requisition prepared by an engineer as the instruction to the buyer to issue a purchase order*
(Courtesy of Seltrust Engineering Limited)

PURCHASE ORDER

AABEC (UK) Ltd
Randolph House
Beckenham Square
London
HT19 4LX

TELEPHONE: 10-3676-47
TELEX : 87191636
CABLE : AABEC (LON)

468205

ALL CORRESPONDENCE,SHIPPING DOCUMENTS AND INVOICES MUST SHOW THE ABOVE ORDER NUMBER

TO:

PAGE 1 OF PAGES

DELIVERY TERMS

FACTORY LOCATION

PAYMENT TERMS

DATE

Please supply the Goods, Materials, Equipment and Associated Services included in this Purchase Order in accordance with all terms, conditions and instructions set forth herein; and in the attached Terms and Conditions of Purchase, and any document specified herein

ITEM NO	QUANTITY	DESCRIPTION	VALUE/ ITEM
TOTAL VALUE OF THIS ORDER:			

For information on this order contact:

Delivery date ex works:

VENDORS ACCEPTANCE
Signature

Name of signatory
Title

AABEC (UK) LTD

AUTHORIZED SIGNATURE

Please sign and return the acceptance copy within 14 calendar days after receipt of this order.

Acceptance of this implies acceptance of all conditions stated herein as well attached hereto.

Figure 35.9 *Typical purchase order form*

1 DEFINITIONS:

Company	–	means Seltrust Engineering Limited.
Seller	–	means the person, firm or company to whom the company's order is addressed.
Goods	–	means the supply and delivery of the goods, materials or equipment in accordance with the company's order together with any subsequent modifications specified by the company.
Contract	–	means the agreement between the company and the seller for the supply of goods.

2 PAYMENT: Net cash against shipping documents or other proof of delivery unless otherwise agreed (subject to any deductions and retentions authorized in the terms of the order, and subject to the seller carrying out all his obligations).

3 PRICES: All prices are fixed for the duration of the contract and, unless otherwise agreed, are not subject to escalation charges of any description.

4 QUALITY AND DESCRIPTION: The goods and all materials used are to conform to description, be of sound materials and quality and be equal in all respects to any specification given by the company or seller.

5 INDEMNITY: The seller shall at his own expense make good by repair or replacement all defects attributable to faulty design and/or workmanship which appear in the goods within the period of 12 months from date of delivery. The seller shall also indemnify the company in respect of all damage or injury occurring before the abovementioned period expires to any person or to any property and against all actions, suits, claims, demands, costs, charges or expenses arising in connection therewith to the extent that the same have been occasioned by the negligence of the seller, his servants or agents during such time as he or they were on, entering onto or departing from the company's premises for any purpose connected with the contract.

6 PATENTS: The seller will also indemnify the company against any claim for infringement of letters patent, trademark, registered design or copyright arising out of the use or sale of the goods and against all costs, charges and expenses occasioned thereby except is so far as such infringement is due to the seller having followed the design supplied by the company.

7 LOSS OR DAMAGE: All responsibility for any loss or damage, whether total or partial, direct or indirect from whatsoever cause shall lie with the seller until full and complete delivery in terms of the order shall have been made by the seller. But it is agreed that the company will take all necessary steps to ensure that it does not in any way invalidate any claim which the seller may have against any carrier.

8 CHANGES IN THE WORK: No variations of, or extras, to, the order shall be carried out by the seller unless specifically authorized by the company on its official amendment form.

9 SUBSUPPLIERS: The seller shall provide a list of all subcontractors or subsuppliers when requested by the company.

10 EXPEDITING: The company's expediting staff shall be given access at all reasonable times to the seller's works or offices or those of any subcontractor in order to view or discuss work in progress.

11 REJECTION: The company may at any time, whether before or after delivery, reject (giving the reasons therefor) any goods found to be inferior, damaged or if the seller commits any breach of the terms of the order. This condition shall apply notwithstanding that the goods may have been inspected or tested by the company.

12 ARBITRATION: Any dispute or difference arising from the contract shall on the application of either seller or the company, be submitted to arbitration of a single arbitrator who shall be agreed between the parties of who failing such agreement shall be appointed at the request of either party by the President for the time being of the Law Society. The arbitration shall be in accordance with the Arbitration Act 1950.

13 TIME FOR COMPLETION: The seller's promised delivery dates must be firm, but if delivery is delayed through any cause beyond the control of the seller and immediately such cause arises the seller notifies the company in writing giving full particulars then a reasonable extension of time shall be granted. If delivery is not made within the time stipulated or within any extension of time the company shall be at liberty to cancel the contract without prejudice to any right or remedy which shall have accrued or shall thereafter accrue to the company.

14 TITLE OF GOODS: Title of the goods passes to the company on delivery to the specified place of delivery as requested by the company.

15 LAW OF THE CONTRACT: Unless otherwise agreed the contract shall be subject to the laws of England.

Figure 35.10 *Standard conditions of purchase. This example shows how a mining engineering company includes its standard conditions of purchase on the rear of each purchase order form*
(Courtesy of Seltrust Engineering Limited)

A typical inspection/expediting report form, used to record the results of a visit in a succinct manner, is shown in Figure 35.11.

The buyer will keep a schedule of progress covering the original issue of each order and its subsequent amendments. A suitable form is illustrated in Figure 35.12.

Other aspects of purchasing progress were detailed in Chapter 23. It remains to be said here, however, that the purchasing organization often finds itself at the end of a tight project schedule, expected to obtain complex equipment in a period sandwiched between the issue of engineering information and the required delivery date. The engineer has a prime responsibility to ensure speedy release of specifications and enquiries, using judgement and common sense where necessary to give the buyer advance warning of purchases that are to come where delivery is expected to be a problem. Many companies arrange that the engineering department gives priority to the release of information for such items as weldments, castings, bulk steelwork, bearings and the like in order that the buyer can make preliminary enquiries and, in really urgent cases, reserve capacity with suppliers on the basis of a letter of intent.

As each procurement cycle nears its end, another important duty of the engineer is likely to be the approval of vendor's invoices. Good relations, based on sound business ethics, are essential in ensuring the necessary degree of cooperation between purchaser and vendor, if both are to succeed in meeting the demands of increasingly competitive markets. One of the more important aspects of such a successful relationship is the purchaser's willingness and ability to make prompt payments when due to the vendors, in order that vendors are able to operate from a sound commercial platform which is not subject to erosion by erratic cash flow. In extreme cases, vendors have been known to withhold or refuse further supplies when their bills have remained unpaid for long periods. The engineer must play his part in these matters, and prompt and fair attention to invoices is vital.

Conclusion

Any summary of the engineer's role in the procurement cycle must recognize the experience of those unfortunately oft-repeated cases in which the lead in the procurement team is contested between engineers and buyers functioning as rivals. This divisive approach can only operate to the detriment of the interests of all parties including the vendor's performance and the ultimate quality of the product. Successful procurement can only be achieved by the effective cooperation of those responsible for all engineering and commercial requirements of a given project, whether it be the batch production of safety pins or the provision of steam turbines.

INSPECTION / EXPEDITING REPORT

This report No	Sheet 1 of	Date of this visit	Last report No	Date of last visit	Inspector / Expediter

MAIN SUPPLIER DETAILS

Supplier

Location

Supplier's reference

Personnel contacted

Telephone No Telex No

Equipment

Agreed delivery
Week number
Date

Latest delivery
Week number
Date

Next visit, week
To expedite
To continue inspection
For final inspection
To inspect packing

SUB-SUPPLIER DETAILS

Supplier's order No Date

Sub-supplier

Location

Sub-supplier's reference

Personnel contacted

Telephone No Telex No

Equipment

Agreed delivery to main supplier
Week number
Date

Latest Delivery
Week number
Date

Next visit, week
To expedite
To continue inspection
For final inspection
To inspect packing

ORDER STATUS SUMMARY (For details see attached sheets)

Progress ahead of target		Slippage	Tests witnessed as specified		Complies with specification		Released for packing		Released for shipping	
Yes	___ Weeks	___ Weeks	Yes		Yes		Yes		Yes	
No			No		No		No		No	

Action by	Action required

Client	Project number	
Purchasing agent's order number Date of order	Specification(No's)	amendment
TITLE	REQUISITION No	amendment

Figure 35.11 *An inspection/expediting report form*
(Courtesy of Seltrust Engineering Limited)

PURCHASE ORDER NUMBER	VENDOR		LIQUIDATED DAMAGES	INCENTIVE BONUS	RETENTION	BANK GUARANTEE	BUYER	MATERIAL/EQUIPMENT		
ORIGINAL PO/SUPPLEMENT	ORIGINAL	1	2	3	4	5	6	7	8	9
Approval to commit										
Commitment Telex Sent										
Commitment Telex Acknowledged										
Commercial Attachments Received										
Engineering Attachments Received										
Purchase Order Draft										
Purchase Order Final										
Purchase Order Issued to Vendor										
Purchase Order Acknowledgement Copy Received										
VALUE										

DATE	COMMENT:

Figure 35.12 *A buyer's purchase order log*

From the engineer's point of view, the required degree of cooperation can only be effected if:

1 He is fully consulted at the start of the procurement cycle.
2 His views and requirements are respected by all other parties and he has sufficient command of the skills necessary for him to present them with complete clarity.
3 He understands the requirements of others and accepts the advantages inherent in his ability to respond from a basis of a multidiscipline approach.
4 He is kept informed of all developments (including commercial) throughout the cycle and, in turn, ensures that all engineering developments are made known to others.
5 He fully appreciates the ultimate needs and goals of the project.
6 He is willing and able to accept that his role is team-orientated and that the ultimate success of the procurement process is in direct proportion to the effectiveness of a team in which he is a single (albeit important) member.

Providing the above basics are realised and realistically applied, the engineer's contribution will be effective throughout the procurement cycle.

36

Industrial engineering

Dennis Whitmore

The transforming of designs and prototypes into production quantities is one of the most difficult transitions in the engineering process. It is not as simple as just reproducing the prototype in large quantities. Many problems and 'bugs' must be ironed out before it is ready to be mass-produced. This chapter looks at some of the stages which link the design function to manufacture.

From design to production

A good designer will incorporate the requirements of production into his design, but not all of the problems can be anticipated. Prototypes will be made as 'one-off' models, possibly by hand, and as such could be more reliable than their mass-produced descendants.

Some companies ensure that the transition is carried out more smoothly by inserting a *preproduction* phase between the design and manufacture. This is a limited run for which prototype jigs, fixtures, tools, and equipment are made, to simulate production conditions. Many of the potential problems are identified during this dry-run period, and it is the function of *industrial engineering* to supervise this linking activity. Industrial engineering is a part of *management services*, and calls upon the techniques and disciplines of this function.

The design of the product shows its finished appearance and the components which form its constituent parts. However, there are other aspects which must be built into the production. Included among these are:

1 The conditions under which the operatives will manufacture the product, and these are considered by the industrial engineer who uses the techniques of *ergonomics*.
2 The methods by which the operations on the product are carried out (*method study*).
3 The time it takes to perform these operations (*work measurement*).

Ergonomics

While the designer might be engrossed in designing the most effective and aesthetically acceptable product at the cheapest possible cost, he may not be concerned too much with the conditions under which his brainchild will be manufactured. This is the province of a discipline called 'ergonomics', from the Greek ergon (work) and nomos (laws). The subject of ergonomics has been given the title 'fitting the job to the worker' which, it must be agreed, is much simpler than the other way round!

Unsatisfactory working conditions can be the cause of low morale, high and costly labour turnover, and increased stress on workers which can lead to psychosomatic illnesses, heart disease, and strokes. This makes the work of the ergonomist most urgent.

Larger organizations employ ergonomists specially for the purpose of applying the discipline, but there is no reason why industrial engineers and designers should not take into consideration the principles of ergonomics. These principles can be applied to various areas of work, such as:

1 The design of the furniture and equipment used in the work.
2 The layout of the workplace itself including the decor.
3 The design of the job.
4 The environmental conditions.
5 The arrangements for working (e.g. shift work).

Stresses to which people are subjected

One of the major areas of concern for the ergonomically minded engineer or manager is (or at least should be) the stresses which may be imposed upon people by the working conditions under which they are operating.

Stress is the cause, and strain is what people experience. This is very important in job design. The thing which causes illnesses is not the stress directly, but how it affects a person, and everyone reacts differently to the same stress. It is a case of 'One man's meat is another man's poison': some thrive under stress because it gives them a sense of purpose to overcome the problems, while others just worry about it, and it is these people who are at risk. The risks are psychosomatic illnesses (such as ulcers, asthma, some arthritic complaints) and coronary heart disease.

When a person is under stress the body gears itself up for a fight by secreting chemicals into the blood stream. The body does not actually physically fight to burn up these chemicals so they remain to do harm, such as the dreaded cholesterol which attacks the arteries and can cause blockages (thromboses). The job of the ergonomist is to design jobs so that if a person cannot avoid stress, he can work off the somatic changes by taking part in some physical effort. In any case, a person must learn to live with it, because a certain amount of strain is unavoidable.

The foregoing is necessarily a simplified account but it does highlight the importance of job design in avoiding illness. The following sections illustrate some of the areas of potential stress in the working environment.

The workplace and its equipment

One of the first priorities for the comfort of the employee is the workstation. Often the seating is whatever happens to be at hand, or has been ordered in bulk for the average workers. Unfortunately workers do not come in average sizes (it would be very boring if they did). In fact heights of 95 per cent of men vary between about 182 and 206 cm, a difference of 24 cm and furniture must accommodate these variations if people are to be comfortable. The study of the dimensions of people is called anthropometry.

Seat heights for example depend on the popliteal height, or length of the lower leg to the back of the knee. For women wearing shoes with low heels this height can vary, for 95 per cent of women, from 38.5 to 44.5 cm. A chair made for the average woman will be up to 3 cm too high for some, and up to 3 cm too low for others. From this it can be seen that chairs should be adjustable to accommodate these different heights. Besides the heights of chairs, the shape is also an important consideration, so that the body is held correctly without any undue pressures on the nerves of the legs. Working heights of benches, tables, and desks must also be taken into account but are more difficult, of course, to make adjustable.

The layout of the workplace itself is important if people are not going to waste effort in stretching for tools, controls, and materials. Data are available which show the normal working area for the average person, which allows work to be done with bent elbows. These data also give the maximum working area in which the person can work with outstretched arms. Controls and tools which are used infrequently may be put in the area between the maximum and normal working distances. Dials, monitor screens, and VDU screens should be at eye level to avoid backaches and neck complaints.

The actual design of dials, and displays is another area for the attention of the ergonomically minded designer who will ensure that these are easy to read to avoid errors in interpreting the values. Controls should be simple to operate and compatible with the direction of movement required.

Much research has been carried out into the effects of visual display units as used with computers and video machines. These can cause both imaginary and real problems to viewers. It is a fact that a high electromagnetic field exists between the screen and the face of the viewer, and this causes the face to be bombarded by particles producing eye troubles. The screen also emits X-rays which are potentially dangerous. The bombarding of the VDU screen by electrons also causes a high voltage potential to build up on the face of the screen, causing a lack of negative ions in the air, and this is said to cause health problems. Most of these fields and potentials diminish as the viewer moves away, so the distance between viewer and screen is of great importance.

While on the subject of VDU screens it is important to remember that the colour of the screen also has an effect on the viewer. It is not possible to say which is the most restful colour because individual preferences come into it. Popular opinion seems to point to brown screens as most favourable, with green in second place. But the main thing to be borne in mind is the colour which seems most acceptable to the person using the screen.

Something which is often overlooked is the decor, such as the colours used on the walls, ceilings, and soft furnishings. It has been found that certain colours are conducive to high productivity and the psychological well being of workers, while others are disturbing and tiring. Colours can give a feeling of coolness in hot areas and warmth in colder ones, besides enhancing the effects of height and width of a room. Women know too well the effects on their figures of vertical and horizontal stripes in their clothes.

The design of the job

The actual design of the workplace is a very wide-ranging subject, and usually is treated as a separate discipline. In this present text it is dealt with adequately in the section dealing with *method study*. However, many of the principles of ergonomics are incorporated in method study investigations and projects, so the subject rightly belongs here.

Environmental conditions – heat and humidity

This area of study covers all the aspects of the environment which could be potential stressors. These include temperature and humidity, noise, vibration, lighting of the workplace, contrast, and glare.

Again with the comfort of the worker at heart, the ergonomically minded industrial engineer will ensure that both the ambient temperature and humidity are set at levels which feel right for the type of work being undertaken. Although legislation proposes certain maxima and minima these are not always ideal for all circumstances; they must be adjusted to meet individual needs. People can work in relatively high temperatures of, for example, around 30°C provided that their bodies can sweat and get rid of the excess heat. However if the air is very humid this will restrict the evaporation of the sweat and they will become very uncomfortable.

Noise

Another stressful aspect of the environment is distracting noise. Once again this is a personal thing which affects individuals differently. A motorcycle can be music to some, but a horrendous nuisance to others. The intensity of the sound is obviously an important factor which must be considered.

The ear hears noise through the compression waves which the sound creates. A pop group's sound will be many times louder than the noise in a workshop. So it would be very difficult to measure such a large scale of intensity. This is overcome by using a logarithmic scale which compresses the range as the intensity increases, and is known as the decibel (dB) scale. These changes are measured using a special sound-level meter. High intensity noise over a long period can make people deaf.

The pitch (or frequency) of the sound is a second factor which can affect a person's health. Frequency is measured in hertz, or cycles per second. The ear is most sensitive around the middle frequencies of (2500 to 4500 Hz), but can be damaged by ultrasonic sound which may be high in intensity but cannot be heard. There are many examples in industry of the use of ultrasonics for cleaning, testing for cracks in metal, and moving small components about. Machines should be designed with adequate shielding against emission of unwanted noise, or the operators protected in other ways.

A noise may be at an acceptable level and if it is continuous it will not distract people once they get used to it. However, noise which occurs at irregular intervals, and comparatively infrequently is very off-putting and has been the cause of loss of productivity.

Lighting

A poor level of lighting intensity understandably will cause eyestrain and a reduction of working efficiency, but this is not the only consideration of the ergonomist, and these will be considered later in this section.

Various Acts of Parliament in the UK which concern offices and factories lay down minima for lighting levels in different areas of a building. Often these minima are insufficient for the workplace and usually it is found that organizations provide more lighting than the recommended intensities. Illumination is measured in lux, 1 lux being equivalent to 1 lumen distributed over 1 square metre (previously, the intensity of illumination was measured in lumens per square foot, so it is seen that 1 lux is approximately one tenth as bright as 1 lumen per square foot). Unfortunately there are very large discrepancies between the recommendations of different countries. For example, the typical illumination of a drawing board in the UK might be 1000 lx, but 3000 lx is not unknown in US offices. Fine bench work is often quoted as needing 1500 lx, but some researchers suggest that six times this level is necessary. Typical levels for general office work are in the region of 500 lx.

The iris of the eye opens and closes to increase or reduce the amount of light it receives. Contrast is the difference between the light reflected by an object compared with its surroundings. Thus a high contrast might cause the object being worked upon to become dim as the eye accommodates the brightness of the background. On the other hand, with printed materials or drawings a high contrast is an advantage. The contrast must be adjusted consistent with good visual acuity.

Equipment is often made for its aesthetic esteem rather than its functional ability. Consequently its lovely shiny surface may reflect light, as glare, into the eyes of the operator. Besides the design of the machine the positioning of the lights is important, and much glare and unwanted reflections can be avoided.

The workplace designer will choose the best lighting for the circumstances, probably settling upon fluorescent lights for general lighting. For fine work it may be necessary to provide individual, adjustable, lights. These will also help to avoid glare as the operator moves them into the best position. While selecting the lighting the workplace designer will also pay attention to the colour of the illumination. Certain colours are restful, while others are irritating or uncomfortable. Many people prefer the warmer glow of the slightly orange fluorescent tubes to the harsher white tubes.

Arrangements for working

Murrell *et al.* have identified optimum ages for some jobs. Unfortunately a person cannot freeze his age, nor can one keep changing jobs as age creeps on. However, the effects of age are mitigated by experience. As a person ages and personal faculties deteriorate, the accumulated experience makes up for some of the loss. There is little the ergonomist can do about ageing anyway! But he can do something about shiftwork.

The body's functions are controlled by a regular cyclic routine called the circadian rhythm (or diurnal cycle). It is said that the natural cycle of body functions is 25 hours but there are only 24 hours in a day. The body has to operate during waking hours, and shut down during sleep. The circadian rhythm arranges that functions such as heartrate, digestion, arousal, concentration, secretion of chemicals into the blood, and others are at a peak around 2 or 3 o'clock in the afternoon, and fairly dormant at about the same time in the early morning. This is important to someone who is on shiftwork because when that person changes from a day shift to a night shift his body is not attuned to this new pattern. Similar situations arise after a journey to a far destination by air, the so-called 'jet-lag'. Luckily the body's clock can eventually catch up with the time lag and the cycle 'inverts' to accommodate it. This might take three or four days, two weeks, or longer.

So shiftwork patterns must be carefully thought out, and workers should be specially picked for the work. But how many *are* selected for their ability to invert? From this it will be seen that a week on days, followed by a week on nights is not a healthy pattern to adopt. Most of the time the juices are gnawing away at the vitals while the person is tucking in to a good meal and the body is dormant. Longer periods are necessary before shift-changes are made. Against this there is the social aspect too. The person is sleeping while everyone else is enjoying life. Clearly the ergonomist and personnel officer should work closely together.

Ergonomics – concluded

This section has shown how important it is to consider the wellbeing and comfort of the employees, be they managers or operatives. Only by looking after their bodies and their minds can they give their best. Most probably there is a lot more that could be done by organizations to implement the principles of ergonomics. So much time is spent designing machines to be efficient, and as much again on the maintenance of them, and yet relatively little is done by many organizations on maintaining the health of its workers.

How the job should be done

There are two aspects to finding the best way to do a given task. One is planning a job from scratch, where it does not yet exist, and the other is improving a job which has already been in operation. The main difference is that in the latter case one has the choice of whether to improve on what is already available, or to forget the existing job, and start again as though it did not exist.

The name given to the discipline of work improvement is *method study*. Although it follows a traditional pattern, method study is an example of problem solving, and as such any of the usual problem solving techniques may be used. There are many of these, and some may appear as being a little 'gimmicky'. The two most popular are the so-called traditional approach, and the less structured *brainstorming*.

The traditional approach

The traditional approach is a very logical one, and is based on step-by-step progression to the final solution. The steps involved are:

1 Define the problem – what is the *real* problem to be solved (and not that which is apparently the problem).
2 Collect the facts about the job and what is to be achieved.
3 Examine these facts to identify the choices available.

4 Select the most effective of the choices, and build this into a final development of the job.
5 Try out the new method in a dry run, and iron out the problems which must inevitably arise.
6 Monitor the progress of the newly introduced job on a continuous basis, making changes whenever necessary.

When one examines the foregoing it is apparent that this is common sense. Before a scheme can be developed it is necessary to know what needs to be done, and all about the possibilities and choices of action.

The traditional approach has its advantages. It makes its ponderous way through the existing situation, looking at every fact in turn, seeing if an improvement can be effected. Although this is a very logical approach it has been criticized for being ponderous, and likely to come up with something which is merely an improvement on the method which was already being used. One could finish with a stereotyped method when perhaps what is needed is a radically new idea.

Furthermore, many practitioners find the long-winded examination procedure outlined above too time-consuming, and try to circumvent the complete process by leaving out the examination stage (step 3) and go straight for their proposals. This, of course, has the danger that not everything may be taken into account, and relies on the preconceived ideas of the originator. An answer may be found in brainstorming.

An illogical approach – brainstorming

There is a well-known saying 'If you can't beat them, join them', and this could be applied to brainstorming, because it allows the problem solver to jump straight into suggesting an alternative or choices. The technique was first introduced in the US, and credited to A. F. Osborn. It has been widely used in problem solving.

In method study brainstorming may be used in the following steps.

1 Define the problem – 'What are we really trying to achieve? Never mind how we are doing it now!'
2 Speculate on possible solutions – throw ideas into a pool, regardless of their feasibility and without inhibition, in other words, brainstorm the ideas.
3 Consider the ideas in the pool very carefully, not dismissing the 'way out' ones out of hand. Those which show promise are offered to specialists for opinions and even trial runs.
4 Assemble those ideas which are selected, and from these form the proposed method.
5, 6 These steps are similar to those in the traditional approach.

The success of brainstorming relies heavily on step 1, the skill of the participants in suggesting innovative ideas rather than the weary old stereotyped ones and on step 2, the degree of freedom given to participants to throw in ideas without fear of criticism or ridicule. *All* ideas are put forward for consideration because some of the seemingly ridiculous notions may be the very ones which score.

Other problem-solving techniques relevant to method study

There are other problem-solving techniques, but these are mostly used in very specialized cases. Two are outlined below. Although their use is quite restricted, they are worth considering by the dedicated problem solver, who can consult the books listed at the end of this chapter.

Morphological analysis

Morphological analysis is a systematic approach used mainly when trying to think up a new product from existing methods and technology. The variables are listed, and all the possible variations of the variables determined. The most likely variations of each variable are selected, and combined into the final idea. For example, if packaging were the product, one of the variables might be labelling, and another, the container itself. Variations for labelling could be paper printed label, print straight on to the container, impress the words into the container surface, a printed tag attached to the container, and so on. Variables for the container itself might be: box, cylinder, bottle, can, bag, tube, etc. In just the ones quoted there are no less than $4 \times 6 = 24$ different combinations. The best of each variable is chosen as the final packaging format.

Analogies

Analogies are things (or situations) which, while different, bear a resemblance to each other, and one can be used to describe the other. When trying to think up new methods, create new products and designs, and generally solve problems, some people look for similar things or situations which already exist in production or even in nature, and which bear a resemblance to the problem with which they are confronted to see how others, or nature, arrived at a solution. They then adapt the solution to fit their own problem.

A method study example

A simple case will now be considered to illustrate the procedure used in method study. This is concerned with the manufacture, in large quantities, of a computer controlled valve unit.

The description is, necessarily, a simplified account of what actually went on, but it will give a good idea of the way in which the method study phases are used. The method of examination used is *function analysis*, which is based on brainstorming.

Phase 1 Define the problem

The operation is to assemble the control valve units. The present unit is too expensive and the price is, therefore, not competitive. There is a danger of losing the market for this product. Part of the cost is attributable to the relatively high reject rate of partly finished units.

Phase 2 Collect the facts

Present method. The present method of producing the components is summarized in an outline process chart (OPC), which is reproduced in Figure 36.1. This chart indicates all the jobs making up the complete operation. The OPC uses two main symbols: a circle denotes a job or operation; and a square is used for each inspection or test.

Constraints. The unit must essentially conform to the design and perform the specified functions.

Other charts. It may be necessary to draw additional charts to define individual methods. One example, for the inspection of components after machining, is displayed as a *two-handed process chart* (Figure 36.2). The layout of equipment, and the flow of people and traffic around the area may be shown as a *flow diagram*.

Statistics. The function performed by the control unit must be specified. The costs of the jobs at various stages of production, and the costs of components also need to be known. The volume of production and the reject rates at each inspection stage are other essential data.

Phase 3 Speculation

The functional analysis approach is brainstorming, in which members of the team throw in ideas as they occur. Here are some examples for this case study:

1. Buy in the castings ready finished.
2. Set up a multispindle machine to tap all the holes in one operation.
3. Buy the printed circuit boards (PCB) already assembled.
4. Abandon all inspection and take the risk.
5. Reduce the amount of inspection.
6. Change the inspection sequence.
7. Get the supplier to inspect his cases and compensate for rejects and loss of production.
8. Change the supplier in order to get cases of better quality.

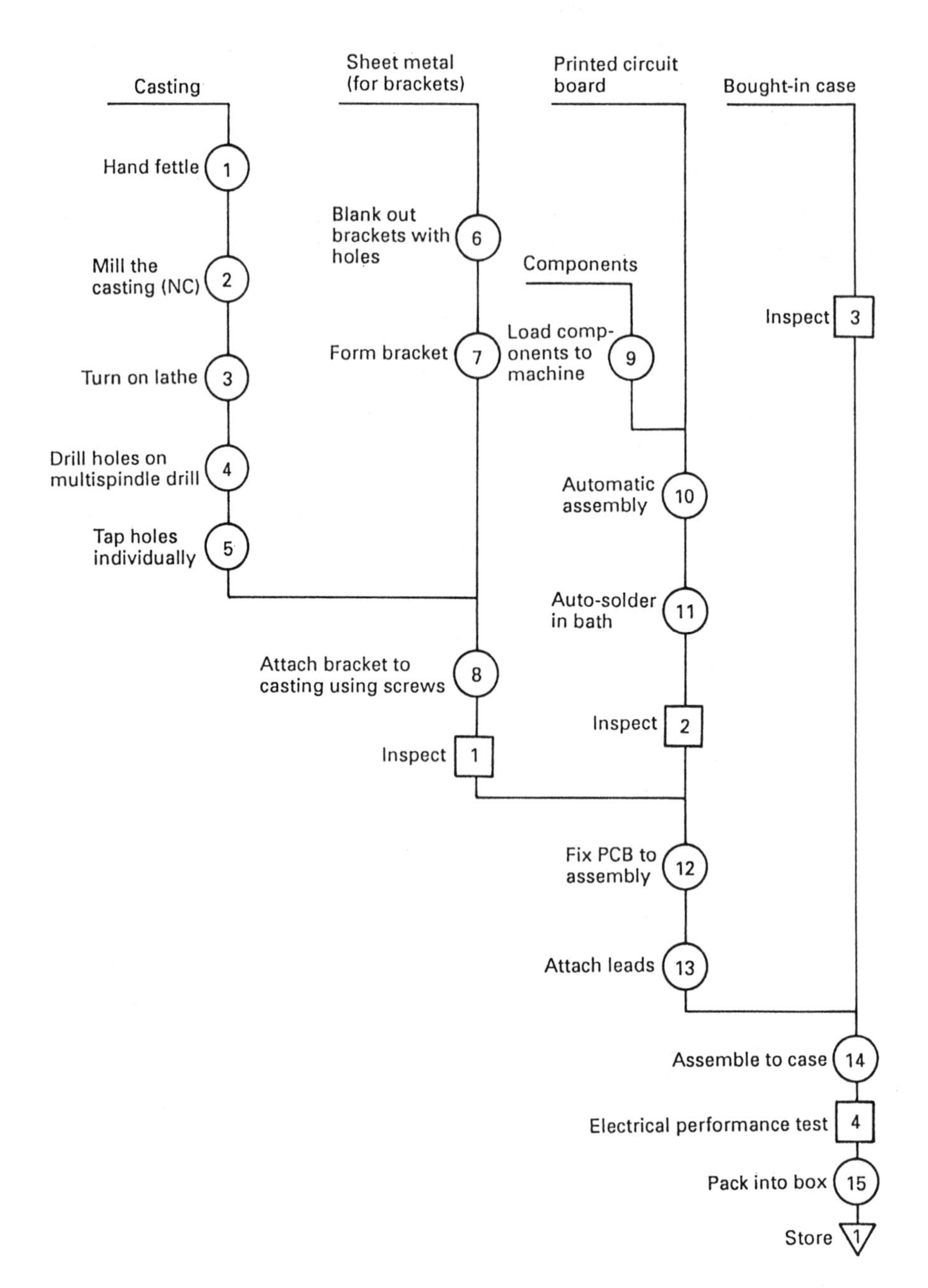

Figure 36.1 *An outline process chart for the manufacture of the computer controlled valve unit of the method study case history (present method)*

Two-handed flow process chart
(operator type)

Sheet no. 1 of 1

Department: Machine shop   Charted by: EFJ   Date: 2 April

Brief job description: Visually inspect for blow-holes etc.,and gauge dimensions
Chart begins (LH): Wait
Chart ends (LH): Place aside component to reject box.

Tools/equipment: Standard gauge   Part no. QCT 23

Left hand				Right hand
No.	Description			Description
1	Wait	D	⇨	Pick up component
2	Hold while making visual inspect	□	▽	Hold while making visual inspect
3	Turn component to new face	⇨	⇨	Turn component
4	Hold component	▽	⇨	Pick up gauge
5	Hold	▽	□	Check major dimensions with gauge
6	Turn component 90°	⇨	D	Wait
7	Hold	▽	□	Check minor dimension with gauge
8	Hold	▽	⇨	Place aside gauge to bench
9	Turn component to reverse side	⇨	⇨	Turn component to reverse side
10	Move component about while inspecting	□	□	Turn component about
11	If accepted: Place aside to box	⇨	⇨	If accepted: Reach for next one
12	If rejected: Place to bench	⇨	⇨	If rejected: Pick up pen
13	Hold pad	▽	○	Write out reject note
14	Hold pad	▽	○	Tear off, tie to component
15	Place aside component to reject box	⇨	⇨	Reach for next component
16				
17				
18				

Figure 36.2 *A two-handed flow process chart for the inspection of components after machining (see Figure 36.1)*

9 Use a cheaper packing medium.
10 Redesign the casting to incorporate brackets as lugs.
11 Attach parts by a different method to eliminate tapped holes.

Note that other suggestions were made, too numerous to list here.

Phase 4 Analysis

The team is now asked to investigate each of the items in the speculation list, and report back. Here is the result.

1 The cost would be higher.
2 Superseded by 11, below.
3 No supplier can be found to take this on, but enquiries are continuing.
4 The reject rates at the inspection stages are currently:

Stage 1	Blow holes exposed by milling	= 24 per cent
	Bad tapping	= 0.5 per cent
	Bad milling	= 0.1 per cent
Stage 2	Bad soldering	= 0.4 per cent
Stage 3	Damage (various problems)	= 1.2 per cent
Stage 4	Electrical performance	= 3.1 per cent

Because of the high reject rates on stages 1, 3 and 4, inspection cannot be eliminated.

5 Stage 2 inspection could be abandoned, because the faults would be found at stage 4 testing.
6 Most faults are due to blow-holes revealed by the milling. At the first inspection stage, the *added value* of the component is far higher than it was at the milling stage (because of all the work done on it). Therefore, move the inspection stage to follow immediately after milling, in order to catch the rejects early.
7 This is recommended. The supplier has agreed to improve.
8 This is superseded by item 7.
9 A cheaper box has been found, to be used in future.
10 The prototype looks promising and the design is to be changed.
11 Trials with rivets have proved successful. Rivets are to be incorporated.

All other suggestions are treated in the same way.

Phase 5 Recommendations

Having investigated all the suggestions made during phase 4, the most promising can be accommodated in a new design. The proposed new OPC is shown in Figure 36.3.

Results

The biggest savings achieved were on added value. Moving the inspection (a change which cost nothing to effect) reduced the factory cost by 14 per cent. The other improvements added a further saving of 9 per cent, giving a net total reduction of 23 per cent, after initial costs for tooling and others had been taken into account.

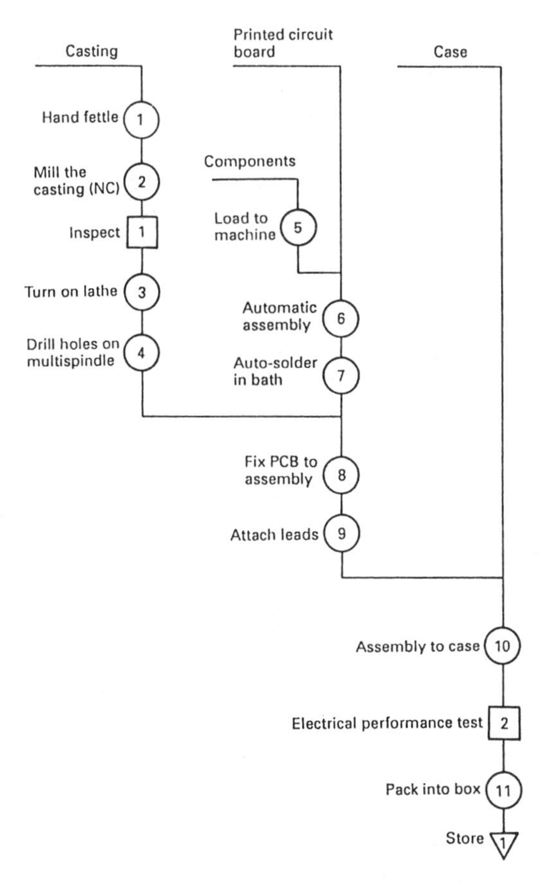

Figure 36.3 *An outline process chart for the manufacture of the computer controlled valve unit (proposed method)*

Measuring the work

The labour content of the production of designs needs to be costed. As workers are paid for their time it is time which must be measured. Industrial engineering has devised several techniques which will measure the work content of jobs under different circumstances.

In the case of jobs in progress it is possible to time the work using a stopwatch, but in the case of a design which is not yet in production there is nothing to time. It is necessary, therefore, to have a set of work measurement techniques which measures work *before* it is actually done. Thus the techniques may be identified according to whether they are used for assessing work content of future jobs (prospective work measurement) or jobs which already exist (retrospective work measurement).

There are only three ways of establishing times for jobs, or indeed for any situations. These are shown in Figure 36.4. Jobs can be measured:

1 By timing, using some sort of chronometer such as stopwatch, clock, or computer.
2 By estimating the time, based on experience of the work.
3 By extracting times for the various subdivisions of the job from tables of either predetermined motion time systems or synthetic data.

Timing techniques can only be used retrospectively, but estimating, and the use of tables can be used either retrospectively or prospectively. In using estimating or tables it is necessary to analyse the job into its elements. If the job does not yet exist this means defining the job and *visualizing* the way it ought to be done.

The theory of work measurement

Work measurement is based on the setting of a time it should take an experienced, trained worker to do a job under defined conditions and method of working. The principle depends upon the interpretation of the words 'should take'. The only way to compare work contents of different jobs is to standardize on the actual doing of the work. Clearly the time taken to do a job will depend upon the speed at which a person works. This is how long it *does* take. Work measurement is interested in how long it *should* take. But people do not all work at this hypothetical 'standard rate of working', so it is necessary to time the rate at which they are working and relate this to how long they *would* have taken had they been working at the standard rate. Thus if a person took 2 minutes to do a certain job, but was only working at half what is considered to be the standard rate, it is assumed that the person would take only half this time (i.e., 1 minute) at the standard rate. This adjusted time is known as the *basic time*, and it would be, in this case, one basic minute.

Virtually all work measurement systems use this standard rate of working, which the British Standards Institution has called 100 rating. In techniques using actual timings the observer is required not only to time the work, but also to estimate the rate of working. For example, if a certain job was timed at 3.5 minutes, but the observer assessed the rate at which the job was being done as only 80 per cent of the standard pace (or 80 rating), the basic time would be:

3.5 minutes $\times$ 80 rating $\div$ 100, or 2.8 basic minutes

In other techniques which do not rely on the direct timing of the work, the rating is already built into the times, as shown in later sections of this chapter.

The techniques provide only the basic time for doing the job, assuming that

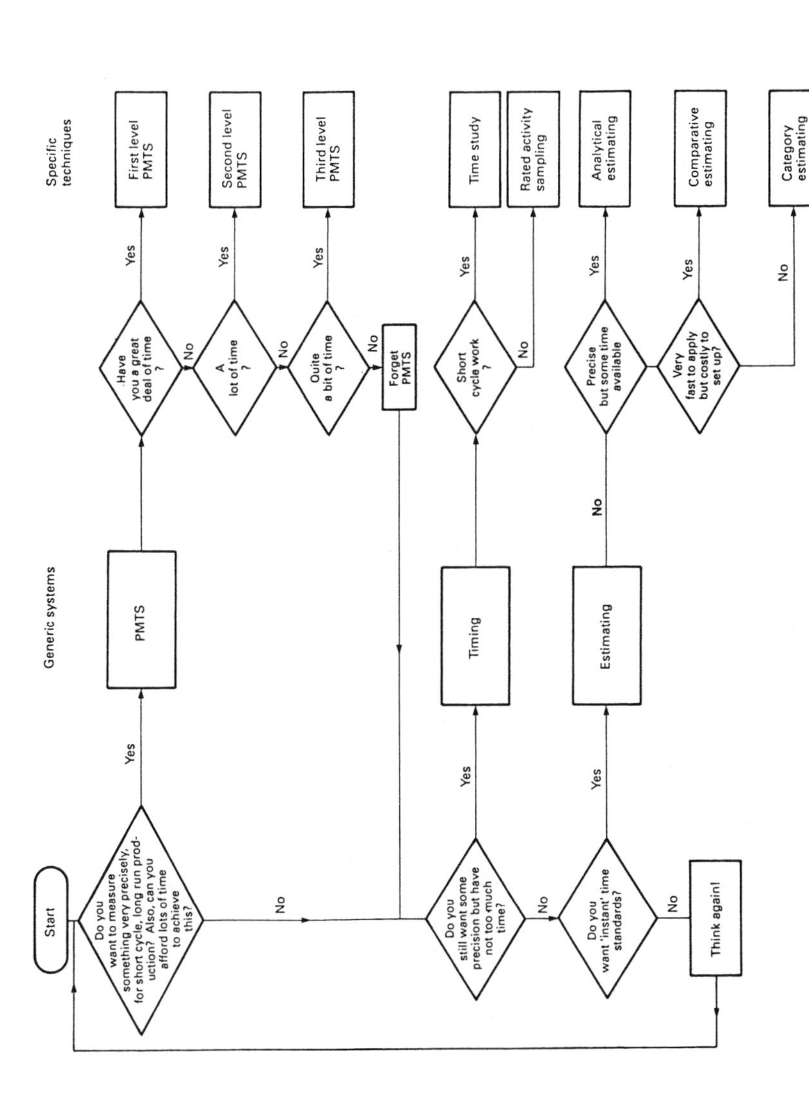

Figure 36.4 ***The three fundamental methods for deriving basic times for work: predetermined motion time systems (PMTS), timing techniques, and estimating methods***

the operator does not break off for rest. This is considered to be unreasonable, so allowances are made for rest, personal needs, and for adverse conditions in the job which are likely to affect the performance of the worker. These allowances are obtained either by estimating the extent of the interference, or extracted from tables of allowances, many of which are published in textbooks (see the bibliography).

Using predetermined time tables

There are two main categories of predetermined tables for measuring work, and these are:

1 Predetermined motion time systems (PMTS).
2 Synthetic data (also known as standard data).

Predetermined motion time systems differ from synthetic data in two ways. The first is that PMTS data are universally applicable (almost worldwide) whereas synthetic data usually are generated for use by the particular organization for which they are intended.

PMTS are based on the idea that all jobs can be analysed into fundamental human motions with times established for these, the sum of the individual times being the total time for the complete job.

At its lowest (or first) level PMTS uses elements which reflect these motions, such as finger, arm, leg, and trunk movements. Times for grasping objects are listed, as are walking, bending, stooping, kneeling, and foot motions. A hypothetical, but typical set of elements is shown in Figure 36.5. An accompanying table of basic times is illustrated in Figure 36.6.

Some commercial systems of PMTS at this first level are Detailed Work-Factor (Science Management Corporation), Methods-Time Measurement (MTM), Simplified PMTS, and Primary Standard Data (PSD).

These first-level systems often are considered to be too fiddly, and time consuming to apply, and in an attempt to overcome this disadvantage second-level systems were developed. These are the result of combining first-level elements, averaging the times on the tables, and other methods of simplification. For example, to produce MTM-2 from MTM-1 the elements *reach*, *grasp*, and *release* were combined into the MTM-2 element of GET, while *move* and *position* were made into the element PUT. Similar combinations reduced the number of times in the tables from about 260 to a mere thirty-nine. The simplification cut the application time drastically.

Further simplifying of the systems produced third-level PMTS. In the case of MTM-3 there are only ten basic times. Figure 36.7 summarizes these developments.

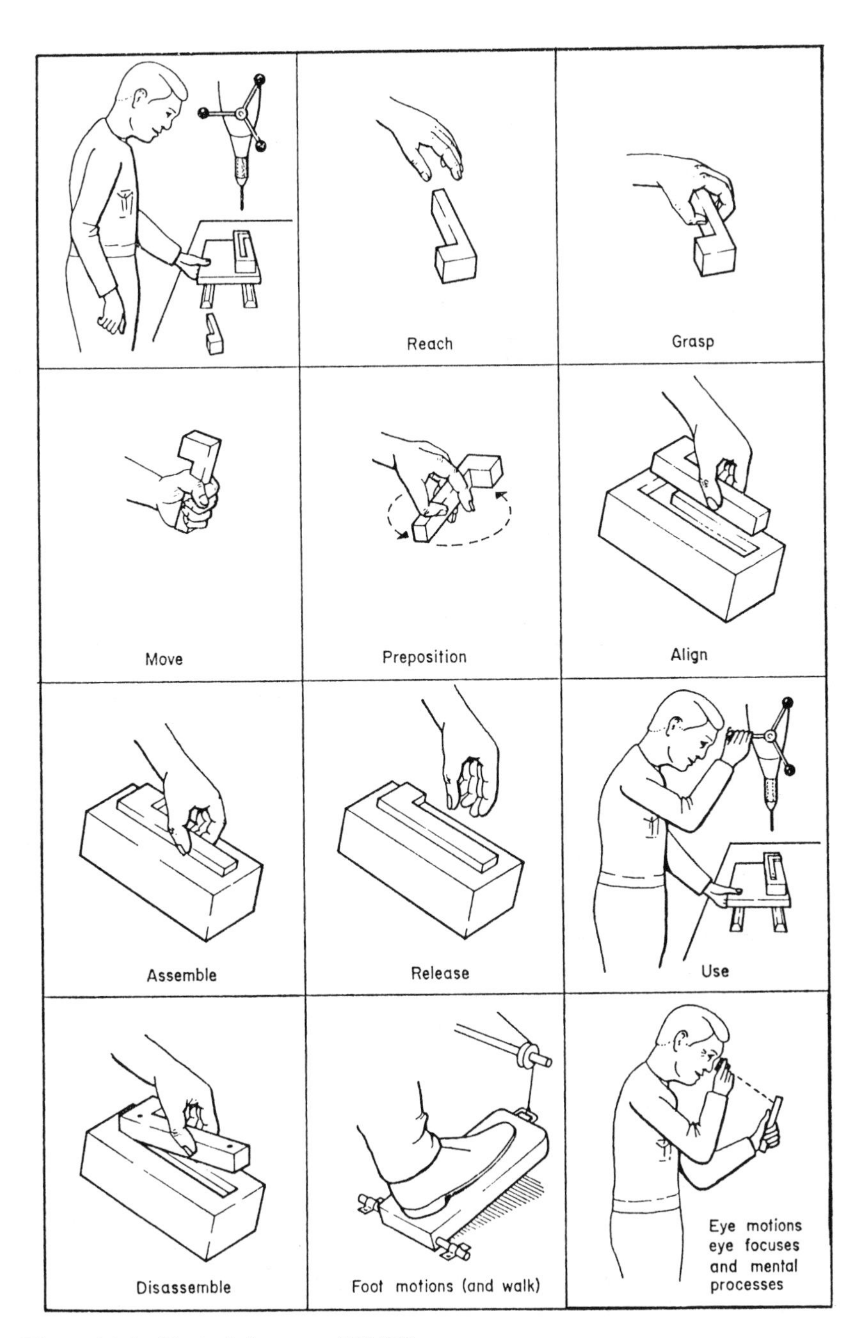

Figure 36.5 *Typical elements of PMTS*
Reproduced from Whitmore, D. A., *Work Measurement*, Heinemann, 2nd edn, 1987.

RATED ACTIVITY SAMPLING

ELEMENT DESCRIPTION

Job: Folding, decorating, and boxing '21st Birthday' cards, Type B21F

File No.: EX 327

Unit of production: per card.

El. No. ELEMENT DESCRIPTION

1. Take up card with L.H. and move to work area, simultaneously taking up bone with R.H. Make first fold with both hands, carefully aligning bottom edge to top edge. Make crease with bone held in R.H. Turn card through 90° and make second fold, aligning bottom edge to top edge. Make crease with bone in R.H. Place aside bone.
   Break-point: As bone touches bench (audible break-point). Freq: 1/1

2. Take up gold key from tote-bin, picking up adhesive with R.H. Apply spot of adhesive to key back and replace tube of adhesive. Transfer key to R.H. and holding card with L.H. position key to front of card. Press.
   Break-point: On release of pressure (visual break-point). Freq: 1/1

3. Lift corner of card with L.H., taking up adhesive with R.H. Squeeze tube and apply a spot of adhesive to each corner of the top sheet of the pile of coloured foils. Replace adhesive, take up foil and position under top fold of card. Place aside adhesive.
   Break-point: As tube touches the bench (audible break-point). Freq: 1/1

4. Press card down to stick foil to card. Place aside card to bench.
   Break-point: As card touches bench (visual break-point). Freq: 1/1

5. When four cards have accumulated on the bench move four boxes from pile to work area. Insert the four cards one at a time to the boxes. Take up two lids, one in each hand from their respective piles, and place these on boxes. Repeat for the other two boxes.
   Collect boxes up into a pile of four and place to pile of finished boxes.
   Break-point: As hands release boxes (visual break-point). Freq: 1/4

NOTE: Materials are replenished, and full boxes removed periodically by a separate servicing operative, so these elements are not included.

Figure 36.6 *An example of the application of Work-Factor Moving Time Tables.* Reproduced from Whitmore, D. A., *Measurement and Control of Indirect Work*, Heinemann, 1980.

Procedure

1 The job is defined and the method charted, or if still in the design stage it is visualized and charted.
2 Next the job is analysed into its constituent elements which are consistent with those defined by the system to be used.
3 The corresponding basic times for the elements are extracted from the appropriate tables and added to the analysis sheet.

Level	*Elements*	*Characteristics*	*Examples*
First level PMTS (microPMTS)	Very detailed elements such as *reach*, *grasp*, *release*, *move*, *regrasp*, *step*, *bend*, *arise*, etc. Very small time unit, e.g. one Detailed Work-Factor Time Unit = 0.0001 minute one Time Measurement Unit (MTM) = 0.00001 hour	Very precise, but very long time needed to derive a time standard	Detailed Work-Factor Methods-Time Measurement (MTM) Primary Standard Data
Second level PMTS (microPMTS)	Achieved by any of (i) combined elements, (ii) combining moving distances, (iii) the averaging of times in the tables to reduce the number of individual times, (iv) the simplifying of elements, and (v) the eliminating of certain elements. In MTM *reach*, *grasp*, and *release* combined together into one new element *get*. *Bend*, and *arise* became just *bend*. *Stoop*, *kneel*, and *arise* from these, are analysed under *bend*, and *sit*, and *stand*, and also analysed as *bend*.	Not quite as precise but quicker to derive the time standards	Ready Work-Factor MTM-2
Third level PMTS (microPMTS)	Achieved by further simplification of elements, and combining of element times.	Less precise, but much faster to apply.	Abbreviated Work-Factor MTM-3
Higher level systems (macroPMTS)	Achieved by building the basic elements into fundamental subtasks, e.g. Open filing cabinet drawer Use stapler Insert floppy disk into disk drive Find tool in tool box	Relatively fast to apply (in PMTS terms)	Basic Work Data Clerical Work Data Clerical Work Improvement Programme
Standard data (synthetics)	Even larger basic jobs, using tabulated data, formulae, graphs, or computer programs, e.g. Run off x copies on Xerox copier Machine components of different sizes Wallpaper rooms of different sizes	Fast to apply	Individually derived for each organization

Figure 36.7 *The various levels and development of predetermined motion time systems*

4 The total of these times is the basic time for the job.
5 Because this is the basic time, an allowance for relaxation is added to produce the standard time.

An example of an MTM-1 analysis for assembling a transformer can to a printed circuit board is shown in Figure 36.8. This analysis shows the basic motions of *reach*, *grasp* and others, with their elemental times in time measurement units (TMU). As each TMU is worth 0.00001 hours (0.0006 minutes), the basic time of 208 TMU, when converted from the 82 rating of MTM to the 100 rating of BSI is equal to 0.102 basic minutes ($0.0006 \times 208 \times 82 \div 100$).

It will be seen that it would not even be necessary to see the job being done. A designer could mime the task, pretending to assemble an imaginary can to a circuit board so that the necessary movements could be written down in elemental form. This could not be done if a timing method were used instead.

Synthesis and synthetic data

Rather than waste effort setting basic times for work which may be repeated in slightly different form over and over again, it makes sense to store time-data previously obtained and retrieve these when required. A computer is ideal for such files.

Synthetic data are similar to PMTS but the times are for much larger elements (see Figure 36.7). Thus the system is reasonably quick to apply. An example of a table applicable to manufacturing custom-built sheet metal forming tools is shown in Figure 36.9. So from this, a former 180 cm long and 12 cm wide would take 25.2 hours to produce.

The synthetic times for such work may be obtained (a) by using the basic techniques of timing, estimating, or PMTS, or (b) by purchasing readymade data where these are available. A well-known example is found in maintenance work on cars, where garages are provided with standard time data by the manufacturers.

Estimating

Estimating is by far the quickest technique of work measurement to apply. One could estimate the time it would take to service a car (say, 2.5 hours) in a few seconds whereas the PMTS calculations would require a staggering 100 hours or more. However the speed of application is no guarantee of reliability. All estimating should be for times for work done at standard performance.

To cope with the variety of work, several forms of estimating have evolved.

Overall estimating

Overall estimating is a quick assessment of a total time for a job: 'I reckon that will take about a day to do'.

MTM ANALYSIS

MTM-1 ✓
MTM-2
MTM-3

Job description:
Assemble transformer to plate
Method "B"

Sheet 1 of 1
Analyst P. G.
Date 2 March

El.		LH	tmu	RH	
	To spring retainer	R8C	11.5	R-E	Toward transformer
	Grasp retainer	G4B	9.1		
	Lift clear of container	M.B	6.4	R4B	To transformer
	and regrasp if required		2.0	G1A	Grasp transformer
			16.9	M14C	Transformer to plate
			19.7	P28SE	Locate 1st. stud
			5.6	P18E	Locate 2nd. stud
		M2C	5.2		Hold transformer
		P28D	21.8		"
		APB	16.2		"
		APA	10·6		"
	Contact release	RL2	—		"
	To spring retainer	R8C	11.5		"
	Grasp retainer	G4B	9.1		"
	Retainer to stud	M8C	11.8		"
	Locate retainer to stud	P28D	21.8		"
	Press home retainer	APB	16.2		"
		APA	10.6		"
	Contact release	RL2	2.0	RL1	Release transformer
			208.0		

Figure 36.8 *An example of an analysis of a job using Methods-Time Measurement*
Reproduced from Whitmore, D. A., *Work Measurement*, Heinemann, 2nd edn, 1987.

Basic times for various forming tools (in basic minutes)

	Widths of forming tools (cm)		
Lengths (cm)	9	12	15
80	18.2	20.9	23.7
100	19.0	22.0	25.0
120	19.7	22.9	26.1
140	20.2	23.7	27.2
160	20.9	24.5	28.1
180	21.3	25.2	29.0
200	21.8	25.8	29.8
220	22.3	26.2	30.6

Formula: basic time (basic minutes) $= 10 + (\text{width}) \times (0.01 \times \text{length})^{0.4}$

Figure 36.9 *A table of synthetic (or standard) data produced for the manufacture of custom-built sheet-metal forming tools of different lengths and widths*

Analytical estimating

Analytical estimating is more precise than overall estimating, analysing a job into fairly large elements. The stages are:

1 Define the work, and analyse it into elements of appropriate size for estimating.
2 Search the synthetic data bank for any data which apply to any of the elements of the job.
3 Those elements for which no synthetic data are available are estimated from experience.
4 Sum the elements to obtain the complete job basic time.
5 Add any allowances necessary to obtain the standard time.

Comparative estimating

It has been found simpler to estimate into ranges of times rather than attempt to estimate the job time exactly. This concept is embodied in the technique called comparative estimating. When using this system it is necessary to first set up the categories called 'time slots' and then add descriptions of typical jobs to these slots as guides. The typical jobs (known as bench marks) can be purchased from consultants, or can be generated by the organization itself.

Figure 36.10 shows a typical spreadsheet of benchmarks. It will be seen that the slots are not regularly spaced, but are in geometric or logarithmic progression; that is to say, the width of each slot is some multiple of the previous one.

Procedure for comparative estimating

1 The job to be measured is compared with the benchmarks in the first slot. If

BENCH-MARK SUMMARY SHEET

Department: Electrical | *Trade*: Electrician II | *Sheet*: 1 of 4

Group: A *Range*: up to 20 minutes *Slot basic time*: 12 minutes	*Group*: B *Range*: 20 to 45 minutes *Slot basic time*: 30 minutes	*Group*: C *Range*: 45 minutes to $1\frac{3}{4}$ hours *Slot basic time*: 70 minutes	*Group*: D *Range*: $1\frac{3}{4}$ hours to 4 hours *Slot basic time*: $2\frac{3}{4}$ hours
Wire up 3-pin plug	Mount and install 24-hr timer into lead between switch and immersion heater	Change standard fitting for fluorescent fitting	Cut to size metal conduit, fix to floor, and wire up four workstations from point in room
Change damaged lighting bayonet socket for new	Prepare and renew brushes on type 87 generator	Chase out hole in wall, fit pattress and wire in outlet to existing ring main	Cut into wall, mount three separate flush pattresses into holes, mount and connect new outlets to ring main
Mount junction-box on wall	Mount junction-box, prepare and connect 25 to 30 wires	Do complete standard overhaul on motor/ generator	
Identify, skin, and connect 10 wires to terminals		Mount junction-box, prepare 30 to 40 wires, and connect	

Figure 36.10 *A spreadsheet for electrical jobs, as used in comparative estimating. This is the first sheet; there would be others for Groups* E, F, G *etc. for jobs of longer duration*

the work content of the job being measured is similar to that of the benchmarks the job is allocated the basic time for that slot. If not, the job is compared with the benchmarks of the next slot, and so on until the most appropriate one is found. This is usually done by exception, by working from both ends of the ranges, eliminating each slot in turn until only one is left; the correct one.

2 When the most appropriate slot is located, the basic time for this slot is allocated to the job. The basic time is the geometric mean of the slot boundaries.

3 Once again, it is necessary to add allowances to the basic time.

When using this system, if the estimator thinks that the most appropriate time for the job is not the geometric mean, but needs slightly more or less time than this, he does not add more time on to the mean: he uses it as it stands. The system works on the statistical equivalent to 'swings-and-roundabouts'. Those jobs which take longer than the mean basic time counteract those which take a shorter time. Thus on average the estimating is reliable over a period of time, although not for any one particular occasion.

Category estimating

A variation on the theme of comparative estimating is a sister technique called category estimating. This is very similar to comparative estimating, with its logarithmic slots but without the benefit of the guiding benchmarks. The most appropriate slot is found purely from the experience of the estimator who argues on the lines, 'I think the job would take longer than one hour, longer than one to two hours, longer than two to four hours, but could be done within four to eight hours.' In this case the 'four to eight hours' slot is chosen, and the geometric mean basic time of 5.66 hours is given to the job ($\sqrt{[4 \times 8]}$).

The main advantage of category over comparative estimating is that it is very easy to set up. No time is taken up deriving the benchmarks, which can take six months or more to collect.

Timing techniques

As previously pointed out, timing techniques are only useful when setting times for existing jobs or deriving synthetic data.

There are two major techniques which require some form of timing: time study; and rated activity sampling. Both use the concept of rating, previously described. They are very similar in application and in the statistical basis on which they are founded. The main difference is that the elements of time study are convenient parts of the job whereas the elements of rated activity sampling are periods of fixed intervals.

Time study

In time study the following procedure is adopted. It has been abridged for clarity, but a full account is given in the author's official text on the subject for the Institute of Management Services. The following assumes that an electronic study board is used with internal timing.

1 The job is first analysed into elements of a convenient size for timing and rating (usually about 20 to 30 seconds).
2 The operative doing the job is observed, and the rating of the pace of working of the operative assessed and entered into the handheld computer study board. At the end of the element the pressing of a button causes the board to record the elapsed time for the elements.
3 The procedure is continued until all elements have been timed. The observer thanks the operative, and proceeds to print out the study.
4 The computerized board is plugged into a computer and the data down-loaded, calculations made, and the study sheets printed out. Figure 36.11 shows a typical printout for a study.

Although the timing by the computer is extremely accurate the rating is still only as good as the observer can make it. There is no way to mechanize the rating function. As with the other methods, allowances must be added to obtain the standard time for the job.

Rated activity sampling

Rated activity sampling is more suited to long cycle jobs, and is carried out by making snap ratings of the working pace, not at the ends of elements, but at predetermined time intervals, which may be every quarter, half, or one minute, or longer than this in some cases. The basic time is found by summing the ratings, multiplying by the time interval, and dividing by 100. For example, if a time interval of half a minute is chosen, and one cycle of the job is studied, and the following ratings are taken, the calculation is:

Time intervals:	1	2	3	4	5	6	7	8	9
Ratings:	90	80	80	100	90	95	85	90	95

Total ratings: 805, multiplied by 0.5 minute = 402.5 minutes divided by 100 = basic time = 4.025 minutes.

Applications of work measurement standards

Once the standard time for a job has been determined it may be used for various purposes, such as costing, planning, manning, incentive schemes, and others.

If, for example, a standard time were 2.5 minutes per cycle of the work, and

OBSERVATION STUDY SHEET

OPERATION: FIRST INSPECTION OF COMPUTER CONTROLLED VALVE UNIT TYPE QCT23

OPERATOR: J C KELLEY | OBSERVER: K WILLIAMS

START TIME: 10:57.24
FINISH TIME: 11:01.26
ELAPSED TIME: 00:04.02

STUDY NUMBER: M1351

SHEET NO. 1 OF 2

ELT	CODE	DESCRIPTION	START	FINISH	OBS.TIME	RATING	BASIC TIME
01	IV50	PU COMP., VISUALLY INSPECT	10:57.24	10:57.36	0.12	100	0.1200
02	IM36	PU GAUGE, CHECK MAJOR DIM.	10:57.36	10:57.47	0.11	90	0.0990
03	IM42	TURN COMP., CHECK MINOR DIM.	10:57.47	10:57.57	0.10	90	0.0900
04	IV28	VISUALLY INSPECT	10:57.57	10:57.75	0.18	100	0.1800
01	IV50	PU COMP., VISUALLY INSPECT	10:57.75	10:57.88	0.13	95	0.1235
02	IM36	PU GAUGE, CHECK MAJOR DIM.	10:57.88	10:58.00	0.12	85	0.1020
03	IM42	TURN COMP., CHECK MINOR DIM.	10:58.00	10:58.10	0.10	90	0.0900
04	IV28	VISUALLY INSPECT	10:58.10	10:58.27	0.17	105	0.1785
01	IV50	PU COMP., VISUALLY INSPECT	10:58.27	10:58.41	0.14	90	0.1260
02	IM36	PU GAUGE, CHECK MAJOR DIM.	10:58.41	10:58.52	0.11	90	0.0990
03	IM42	TURN COMP., CHECK MINOR DIM.	10:58.52	10:58.63	0.11	90	0.0990
04	IV28	VISUALLY INSPECT	10:58.63	10:58.81	0.18	100	0.1800
05	CL63	WRITE OUT REJECTION NOTE	10:58.81	10:59.07	0.26	80	0.2080
01	IV50	PU COMP., VISUALLY INSPECT	10:59.07	10:59.20	0.13	90	0.1170
02	IM36	PU GAUGE, CHECK MAJOR DIM.	10:59.20	10:59.32	0.12	85	0.1020
03	IM42	TURN COMP., CHECK MINOR DIM.	10:59.32	10:59 42	0.10	90	0.0900
04	IV28	VISUALLY INSPECT	10:59.42	10:59.60	0.18	95	0.1710
01	IV50	PU COMP., VISUALLY INSPECT	10:59.60	10:59.73	0.13	95	0.1235
02	IM36	PU GAUGE, CHECK MAJOR DIM.	10:59.73	10:59.84	0.11	95	0.1045
03	IM42	TURN COMP., CHECK MINOR DIM.	10:59.84	10:59.94	0.10	90	0.0900
04	IV28	VISUALLY INSPECT	10:59:94	11:00.13	0.19	85	0.1615
01	IV50	PU COMP., VISUALLY INSPECT	11:00.13	11:00.25	0.12	100	0.1200
02	IM36	PU GAUGE, CHECK MAJOR DIM.	11:00.25	11:00.36	0.11	90	0.0990
03	IM42	TURN COMP., CHECK MINOR DIM.	11:00.36	11:00.46	0.10	100	0.1000
04	IV28	VISUALLY INSPECT	11:00.46	11:00.64	0.18	95	0.1710
		INTERRUPTION BY SUPERVISOR	11:00.64	11:01.05	0.41		
01	IV50	PU COMP., VISUALLY INSPECT	11:01.05	11:01.16	0.11	105	0.1155
02	IM36	PU GAUGE, CHECK MAJOR DIM.	11:01.16	11:01.26	0.10	100	0.1000

Figure 36.11 *Printout of data collected during a time study using a computerized study board. The figure shows the observation study sheet. Analysis, and summary sheets would also be produced by the computer*

operatives worked for 7.5 hours at a rate of £6.45 per hour (including overheads), for five days each week, the following calculations may be made:

Costing This job has a standard time of 2.5 standard minutes for each component made, thus each one has a labour cost of $2.5 \div 60 \times 6.45$, or 26.9p.

Planning How long will it take to complete a contract for 2300 components?

At 2.5 standard minutes per component the total standard time for 2300 components is 5750 minutes or just over 5 man-days.

Manning Suppose a day's work on maintenance tasks consists of Job A with a standard time of 1.6 hours, Job B with a standard time of 2.6 hours, Job C at 1.2 hours, but to be done on 10 vehicles and Job D at 0.8 hours, but repeated 35 times. The total standard time is $1.7 + 3.6 + 12 + 28 = 45.3$ hours, which requires 6 people to complete in a day of 7.5 hours, and if they work at standard performance.

If they do not work at standard performance, but their average is 85 performance, then $100/85 \times 6$, or 7 people will be necessary.

Further reading

Allen, M. S., *Morphological Creation*, Prentice Hall, 1961.
Gordon, W. J. J., *Synectics*, Harper & Row, 1961.
Osborn, A. F., *Applied Imagination*, Charles Scribner, 1963.
Grandjean, E., *Fitting the Task to the Man*, Taylor & Francis, 1985.
Murrell, K. H. F., *Ergonomics*, Chapman & Hall, 1965.
Whitmore, D. A., *Work Measurement*, Heinemann, 2nd edn, 1987.
Whitmore, D. A., *Work Study and Related Management Services*, Heinemann, 1980. (Obtainable from Ainsworth-Maine, Whitehill Farm, Sewell, Dunstable, LU6 1RP.)

Acknowledgement

Extracts from Work-Factor Moving Time Tables used in this chapter are reproduced by kind permission of the copyright holders:

J. H. Quick, J. H. Duncan, and J. A. Malcolm, *Work-Factor Time Standards*, McGraw-Hill Book Co. Inc, 1962 Wofac Company (now Science Management Corp.) H. B. Maynard (ed.), *Industrial Engineering Handbook*, McGraw-Hill Book Co. Inc 1956, 1963.

37

Post design services

David Warby

The most obvious reason for providing design information is to enable the production of goods or the construction of a project that meets the purchaser's requirements. But design information must also be provided for:

1 Technical communication with the customer and suppliers.
2 Installation manuals.
3 Training for the customer's staff.
4 Commissioning manuals.
5 Operating and maintenance manuals.
6 The technical definition for any warranties.
7 The reference point for incorporating modifications during service.

These additional applications of design information may be described as post design services. Such post design uses will be at least as significant as the primary use for manufacture. Recognition of the importance of these phases of the engineering task, and attention to their completion, are essential for the engineering company and the customer alike.

Design and post design responsibilities

The range of products covered in engineering is substantial, as is the related design and post design activity. Nevertheless, there are fundamental steps which have to be taken, regardless of the product, through the sequence from the original specification of performance to operation by the customer (the purchaser). Similarly, there are fundamental relationships between the parties involved. The location of the responsibility for specification and design should be very clearly identified and understood. Figure 37.1 illustrates these sequences, relationships and responsibilities for an engineering system.

Four design levels have been identified to clarify the question of responsibi-

Design level	Design content	Post design content	Responsible party	Defect correction
1A	Specify overall performance – inputs, outputs, life. Specify environmental inputs Specify standards Specify dimensional limits Specify purchaser's extent of supply	Taken over plant after commissioning Operate Maintain Report defects	Purchaser	
Design interface 1B	Specify overall process Specify overall equipment parameters		Purchaser or consultant	Report / correct
Design interface 2	Design process Specify systems Specify equipment in details Specify dimensions – layouts, GAS Specify materials of construction	Write overall manual for commissioning Assemble overall operating and maintenance manual Commission plant/performance test Arrange correction of defects	Main contractor	Report
Design interface 3	Design systems Design equipment	Write equipment manuals for – installation – commissioning – operation and maintenance Manufacture, install and commission Correct defects	Sub-contractors	Report / correct
Design interface 4	Design components	Manufacture components	Component manufac-turers	

Figure 37.1 *Responsibilities and relationships in the design and post design activities for an engineering system*

lity for specification and design. Any problems in operation are most likely to have their roots in either specification or design.

Design level one: the purchaser's requirements

1 Specification of the overall performance which the purchaser requires, together with the constraints which will be imposed on the suppliers. The performance will have been determined by market assessment, and achieving or bettering it will be the primary interest of the purchaser.
2 Specification of the process and main equipment types by which the overall performance is to be achieved.

Design level two: overall engineering design

Design of the specified process, and the detailed specification of the systems and equipment required to achieve the process to meet the overall performance.

Design level three: systems and equipment design

Design of the systems and equipment which have been specified. At this level there will be a large number of equipment items which are well proven, standardized or proprietary. The level of design skill and knowledge in each item will be high, but the degree of uniqueness and innovation low.

Design level four: component design

Design of components which will be selected by the systems and equipment suppliers. These will largely be standardized catalogue items. The design skills required will vary, but their serviceability will probably be well proven.

The roles of the purchaser, consultant, main contractor and subcontractors in the design and post design activity may vary considerably. This variation will reflect differences in approach by nation, industry, and company. With the largest undertakings the respective roles set out in Figure 37.1 will probably apply. However, with smaller and particularly specialized processes, the purchaser may carry the responsibilities down to design level 2 or even 3. This will require the purchaser to employ an experienced and comprehensive design team. The compensating benefits might be improved design from specialized operating experience. At the other extreme, the 'design and build' approach would lead to the main contractor picking up responsibility at design level 1/1 and continuing through level 2 to level 3.

Within these broad extremes there will tend to be many variations, depending upon the purchaser and his advisers. The 'design interfaces' identified between the different design levels each define a clear transfer of a level of information and responsibility. Any blurring of these design interfaces is likely to give rise to confusion if more than one party is involved, with consequent loss of clarity of responsibilities.

These issues have been dealt with at some length because of the difficulties which frequently arise in establishing responsibilities for the execution of post design work, its coordination and, in the event of problems, liability for correction and the consequences of failure.

The link with the customer

The supplier's links with the customer are important for the execution of work in hand and also for the future. That future may entail support of the product which has been supplied, and also possible new business. The nature of the link will vary with the service or product supplied, and whether or not the relationship is bound by contract.

Contractual links

If a supplier contracts to provide goods or services he will only have a contractual relationship with his immediate customer, even though ownership passes eventually to another purchaser. This limit on the contractual relationship between the two parties is the principle of the 'privity of contract'. The way in which this operates in the case of the organizations previously cited is shown in Figure 37.2. There is no legal connection between contract A, contract B or contract C. Thus the purchaser cannot instruct or otherwise require the subcontractor or the component manufacturer to do anything under the terms of their contracts. Similarly the main contractor cannot have any direct dealings with the component manufacturer, or vice versa.

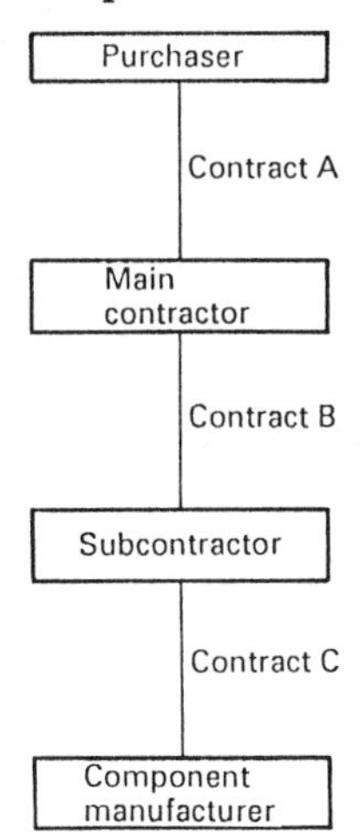

Figure 37.2 *Contractual relationships*

In practice, the normal contact between parties in an engineering project means that there is considerable informal communication between all concerned. This is probably necessary and desirable, but it should not be confused with the formal contractual links. The purchaser, for instance, might quite reasonably suggest some change to the subcontractor that he believes will benefit the works. However, the main contractor must receive any requirements for change from the purchaser. When these have been fully defined, the implications of performance, cost and time for completion should be established with the subcontractor. If these implications are acceptable to the purchaser, the instructions for change can then be given, with all relationships having been correctly maintained. To avoid confusion, it is good practice to hold meetings with only the two parties to a particular contract present.

The formal links and constraints apply through the contract period. The 'contract period' commences at a date designated in the documents, usually the date of signing. It continues through the time the equipment is 'taken over' for use by the customer and terminates when any defects which have become apparent during an initial period of use have been corrected, this being the 'defects liability period' (commonly twelve months). Ideally each subcontract should have the same 'take over' date and 'defects liability period' as the main contract. This will result in a tidy coincidence of obligations for correcting any defects revealed by the initial operations. However, in large capital projects, the overall contract period can be many years, and suppliers undertaking a small part of the overall works in a short period may be unwilling to have extended obligations hanging over them long after their work has been completed. With the situation being repeated and multiplied down the pyramid of sub-subcontracts its resolution can be considerable. The answer to this problem can only be reached through the balance of supply and demand in the letting of subcontracts.

In Figure 37.3 these points are illustrated. The different contract periods for main contractor, subcontractors and component manufacturers can be seen clearly. The timescale is illustrative only and may be months or years. However, the considerable delays that occur between a subcontractor or component manufacturer completing their supply and the eventual commissioning are apparent, as are the delays to the defects liability period. For clarity only one subcontract and one component manufacture contract are shown.

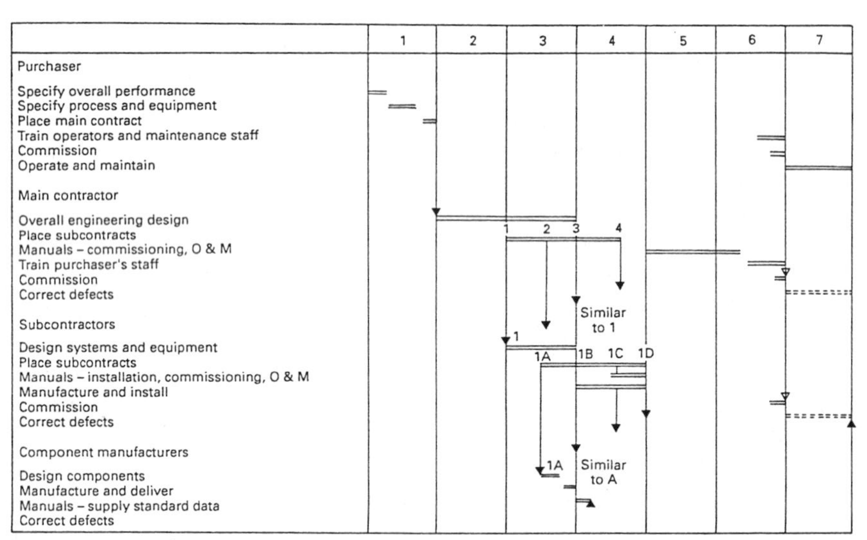

Figure 37.3 *Programme relationships*

Warranties

In engineering contracts, the term warranty is most commonly used to describe the obligations of a supplier to rectify any defects that might develop in the product he has supplied. There is however a wider meaning of the word in contract law. All contracts will have obligations other than the correction of defects, and some of these will be classed as 'conditions', and others as 'warranties'. In this context a condition is fundamental to the contract and, if not observed, will give the injured party the right to regard the contract as discharged (i.e., ended), and accordingly sue for damages. A warranty is less important, and if breached gives grounds for a claim for damages, but not discharge of the contract.

In addition, there are common law obligations which apply to any contract which, at least in part, involves the supply of work and materials. These require the goods to meet two conditions.

Merchantable quality

This can be interpreted as that quality which could customarily be expected for the particular article, having due regard to price. The article need not be perfect, and the purchaser cannot expect to have anything superior to that which he has asked for and paid for. If there are specific imperfections in the article, the supplier should point these out. For instance, a manufacturer may offer a standard range of electric motors which are of low cost and suitable for general service within certain limits of temperature, humidity, duty cycle and life. Those limits will be determined by imperfections in design, materials and manufacture, but are clearly identified by the expressed and implied terms of the manufacturer's product specification. These motors will be of merchantable quality if they comply with that product specification.

Fitness for purpose

Goods must be suitable for the specific purpose for which they are intended. To continue the example of the electric motor, the full operational requirements should be stated by the purchaser and understood by the supplier.

If the motor fails in service, say, because the duty cycle had proved too severe, it would not be fit for purpose, although it was of merchantable quality in that it conformed to the supplier's specification. In these circumstances the supplier would be liable if the purchaser had fully defined the required duty, or had asked the supplier to ascertain the duty and supply a suitable motor.

Services

In contracts where there is no supply of work and materials, the Supply of Goods and Services Act 1982 will require any professional services, particularly design and advice, to be carried out with an exercise of reasonable skill and care. Most offers of such services imply a warranty of such skill and care.

Limitation

Contracts may be either 'under hand' (that is signed only) or 'under seal', in which case the company seals of the two parties will be impressed on the contract documents, normally with the signature of two directors. The Limitation Act 1980 provides that debts or claims under a contract cannot be pursued after six years for a contract under hand, or twelve years for a contract under seal. The periods run from the date of the cause for action.

An important consideration for both purchaser and supplier will be the performance of the goods after they have been taken into use, and the resolution of any problems that might arise. Unless there is a clause to amend the general law of contract, there may be a liability on a supplier through his common law obligations for six or twelve years beyond the end of his supply. A claim from the purchaser under his common law rights will be complicated because it will have to demonstrate that the goods were either not of merchantable quality or not fit for purpose. After a significant lapse of time, such a claim might be difficult to prove, and be contested with counter claims. For the supplier, a long period of liability will be difficult to price and service.

For these reasons it is customary for engineering contracts to define a 'defects liability period' during which it is clearly understood that the supplier will either repair or replace the goods at his cost in the event of a failure. Some forms of contract clarify the situation further by stating that such a clause defines the supplier's liability 'in lieu of any condition or warranty implied by law as to the quality of fitness for any particular purpose . . .'. Such a defects liability period is most commonly twelve months, but may be shorter or longer at the discretion and agreement of the two parties. It must be emphasized that the terms of a contract cannot set aside statutory obligations.

The terms warranty and guarantee are interchangeable in the context of defects liability, the choice varying with different contract conditions. The same words are used to cover the matter of product performance when the supplier gives undertakings, say, of the inputs and outputs of all items. Performance warranties are also commonly restricted to a demonstration of specified performance during the defects liability period.

Supplementary warranties

In the typical contractual 'pyramid' used in the examples, there may be a situation where the purchaser finds it inconvenient or impossible to pursue a claim under the contract via the normal route of the main contractor. This will tend to be the case particularly after the end of the contract period. The main contractor may have disbanded his team, and in practical terms find it very difficult to deal in detail with some claim from the purchaser which relates to a subcontractor's work. This difficulty can be provided for by incorporating into the subcontract a supplementary warranty agreement from the subcontractor to the purchaser. If this is done, the purchaser will have the option of dealing with the main contractor or directly with the subcontractor. The arrangement is illustrated schematically in Figure 37.4.

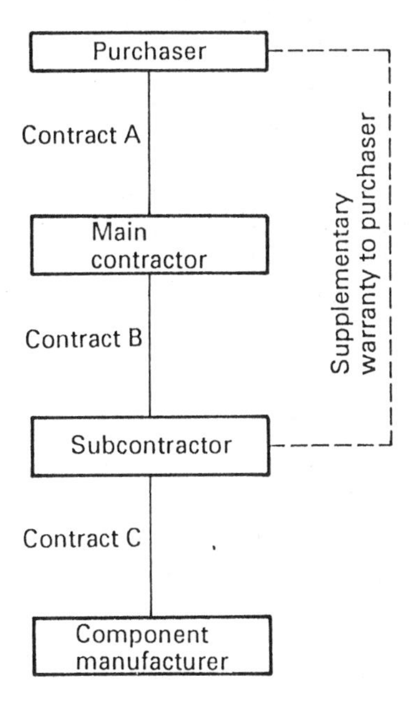

Figure 37.4 *Supplementary warranty*

Correct use

For any warranty claim against a supplier to be enforceable, the purchaser must be able to demonstrate that the goods have been correctly installed and used. In this context use will cover commissioning, operation and maintenance. It will be normal good practice to arrange all contracts to allow the supplier to (as far as possible) install and commission his equipment, and accept full responsibility for these activities. Thereafter, clear and comprehensive operating and maintenance manuals should be provided to the purchaser, supported by adequate training of the purchaser's staff.

Installation

Installation is an integral part of the production process, and as such needs to be treated with the same formality as any activities in the supplier's factory. In addition, the installation phase normally brings the supplier into a location which is not under his control, and where there are many coordination interfaces with other organizations.

The difficulties of installation are frequently not appreciated by designers. A sound practical discipline is to consider the whole production process in reverse, commencing with commissioning and working backwards through installation, delivery and production in order to arrive at a practical method before the design approach is decided.

After due consideration, a method statement covering the installation should be produced. Ideally this will be an integral part of the supplier's offer, clearly defining how the work is to be done. It will deal with the associated matters of coordination with others, the services which are required, safety and the arrangements for client inspection and acceptance. It is probable that the method statement will need to be refined and developed during the course of the contract, but it will nevertheless be a solid reference point for everyone concerned.

Operating and maintenance manuals.

The production of operating and maintenance manuals is a crucial component of the post design activity. Some purchasers will have very specific requirements for the structure, format and quality of operating and maintenance manuals but, whether or not this is the case, the fundamental requirements will not differ. A factor which should be considered in assessing the task is the considerable time and cost which may be involved, frequently representing as much as 10 per cent of the overall design cost. The technical content of the operating and maintenance manual will clearly be a matter for the designers concerned with each particular part of the total system. Their style in both text and diagrams will differ and, at the highest standard of manual, the purchaser may require these differing styles to be uniformly presented. In such cases, the editing, artwork and printing would customarily be undertaken by a specialist organization.

In dealing with the basic principles of the production of operating and maintenance manuals, it will be convenient to refer back to Figure 37.1 and 37.3. Using the examples of structure of subcontracts and suborders to component manufacturers, the process of creating an operating and maintenance manual is set out diagramatically in Figure 37.5. This will begin with the component manufacturers supplying product leaflets with their equipment which has been ordered by various subcontractors. Each subcontractor will write a set of operating instructions and a set of maintenance instructions relating to the subsystem for which they are responsible. Those operating and maintenance instructions will then be combined into a manual together with the component manufacturer's product leaflets.

There will therefore be as many subsystem operating and maintenance

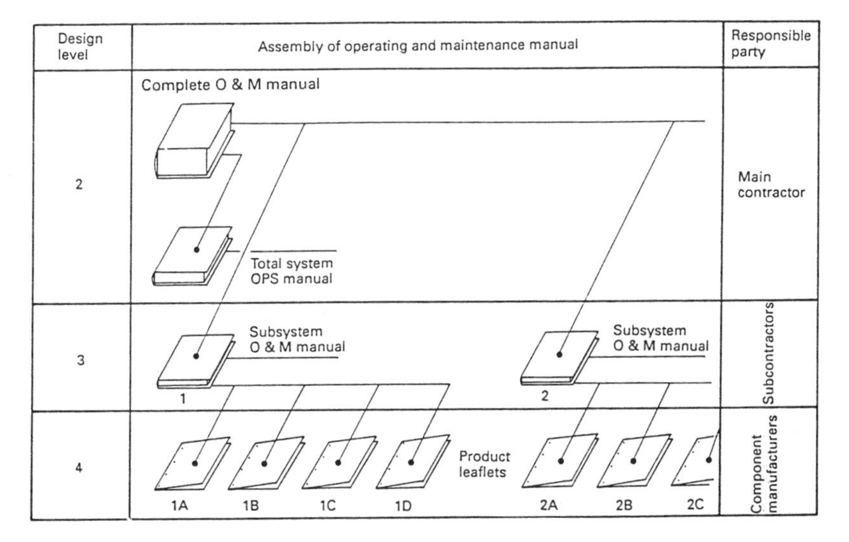

Figure 37.5 *Creation of operating and maintenance manuals*

manuals as there are subcontracts. The main contractor will have the task of writing the operating manual for the total system, and this can only be finalized when all the subsystem operating manuals are available. To finalize the documentation process, all the various manuals will then be incorporated into a complete operating and maintenance manual, suitably indexed. With large systems, it may be impractical for all the information to be in one volume.

Training the customer's staff

Once the operating and maintenance manuals are available, the customer's staff can begin to study the new equipment for which they will be responsible. In some cases they will be very familiar with the new equipment, perhaps because it is an addition to existing plant. In other cases, it may be completely new to the people involved, and in these circumstances an early start on familiarization will be essential. This might start with visits to the manufacturers to see the equipment being produced. This is of particular value for maintenance people.

For the training of operating personnel, the ideal circumstance is the opportunity to carry out the task on similar equipment. In many cases this will be possible, but if this is not so and prior experience is of vital importance, a simulator may be used. In order to be sure of a smooth commissioning programme and subsequent operations, the recruitment and training of the customer's staff should be an important feature of the contract programme.

Commissioning

The commissioning of the plant will take place at the end of a considerable period of design, production and installation. The new equipment will either be brought into service as an additional system which is parallel to other facilities, or in series with existing equipment as a modification. The criticality of the commissioning programme will depend on which of these circumstances applies.

If the equipment is in parallel, a delay to the commissioning will be unfortunate and result in a loss of the new output. However, if a modification is being incorporated in series with other plant, a delay in commissioning will result in the total output being lost, normally with disastrous consequences. In either case, the full specified output for the new equipment will almost certainly not be reached immediately. After start-up there will be a period of caution while the equipment is subjected to operating conditions for the first time, and there may be difficulties revealed at this stage which require the system to be shut down for modification. The extent to which the attainment of specified output is delayed will depend very much on the degree of innovation involved in the system. These points are illustrated in Figures 37.6 and 37.7.

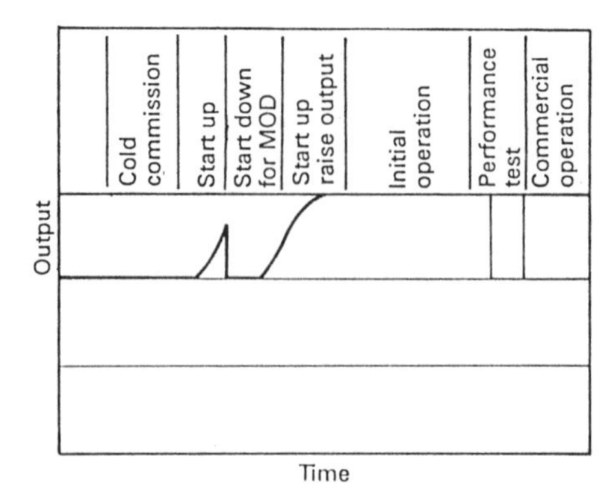

Figure 37.6 *New equipment commissioned in parallel with existing*

The issue of training the customer's staff is a critical factor in bringing new equipment successfully on stream. Apart from the practical training which might be necessary, there are human problems to consider. The customer's staff may feel uneasy about the new equipment being introduced, and in spite of reasonable consultation may be resistant to the change. Where new equipment is being introduced in series with operating plant, the task of training and familiarization is particularly difficult because the staff concerned will be fully committed on their routine duties during the manufacturing and installation period. Frequently there will be a short planned shutdown to tie in

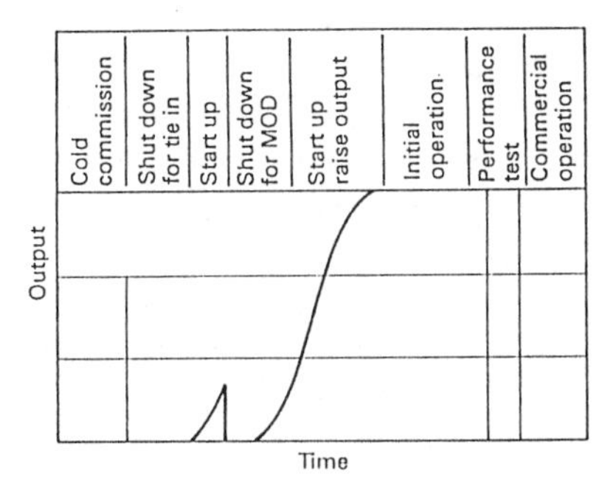

Figure 37.7 *New equipment commissioned in series with existing*

the new equipment, and then the modified system will be expected to go into operation.

Whatever can be done to involve the people who are going to operate and maintain the plant should be put in hand soon enough to generate confidence and a sense of ownership. The main contractor would be well advised to agree with the purchaser a process of consultation and information to take place on a regular basis in the run up to commissioning. Depending upon the scale of the new equipment the programme might include meetings to report progress, photographs of construction, a news sheet and, most importantly, arrangements for feedback from the customer's staff.

Commissioning is the final stage of the process of production, and requires a comparable systematic approach. All steps should be planned carefully, clear instructions written and record sheets designed to coincide with those instructions. Work will commence with the final inspection of equipment items, followed by pre-start checks in accordance with the manufacturer's product data.

When all equipment items in the subsystem have completed their pre-start checks, the subsystem will be ready for cold commissioning. In this sense cold will describe the operation of the system literally cold, or with a passive operation medium, or in some other way functioning in a safe and stable mode. When the subsystems have been cold commissioned, cold commissioning of the total system can be undertaken. The client's staff will become involved at this stage since they will be operating the systems themselves for the first time.

At the completion of cold commissioning of the total system, the new plant will be ready to commence operation. From this point, the process of hot commissioning can commence. In this sense hot will mean literally hot, or with the designed working medium and other operating conditions. The systems will be gradually brought into service and the output raised to the specified level. As soon as the plant has operated at the specified output for a reasonable period, the main contractor's commissioning team will withdraw and leave the purchaser to operate the plant on a routine basis.

At a point as soon after hot commissioning as possible, it will be customary to undertake a performance test to demonstrate that the inputs and outputs of the plant conform with the specification. The main contractor will define and conduct the test, with the results being witnessed by the purchaser. The satisfactory completion of the performance test will mark a vital point in the contract at which the new plant is taken over by the purchaser, and the defects liability period commences. These steps in the commissioning sequence, together with the appropriate documents, are set out diagrammatically in Figure 37.8.

Design level	Equipment items	Subsystems	Total system				Responsible party
	Final inspect. pre-start check	Cold commission	Cold commission	Hot commission	Operate	Performance test	
Action 1 Document			Operate Witness	Operate	Operate O & M manual	Operate Witness	Purchaser
Action 2 Document	Witness	Witness	Commission Cold commission Procedure and record	Raise output to specification Hot commission Procedure and record		Conduct test Performance test Procedure and record	Main contractor
Action 3 Document	Inspect Check Inspect and pre-start record	Cold commission Procedure and record					Subcontractor
Action 4 Document	Nil Product data						Component manufacturers

Figure 37.8 *Commissioning sequence and documents*

Informal links

During the course of the formal contract period, the various parties may well have concurrent informal links with each other, resulting from contracts in the past and the search for future business. The formal relationships dictated by privity of contract will not apply in these circumstances. Consequently a subcontractor or component manufacturer might be dealing directly with the purchaser in supporting equipment previously supplied at the same time as being distanced contractually on the current work. Such informal links may have an effect on the formal actions of all parties when they consider their need for continuing business dealings.

Customer support

The decision to purchase an item of equipment will certainly have been influenced by the support that is offered by the supplier. This support might be essential or merely desirable, depending upon the purchaser's skills, resources and policy. A carefully evaluated maintenance programme will weigh the alternatives of in-house and supplier support.

Spares
The ready availability of spares is the first element of customer support, and it will be customary for the purchaser to ask the supplier to recommend the spares required for a given period of operation (say, two years).

Spares may be manufactured by the supplier or may be standard items manufactured by others and incorporated in the supplier's equipment. Some economies can be achieved with this category of spares by the purchaser dictating the manufacturer of a class of components (say electrical relays) in his initial specification. The ordering of such spares by the purchaser can either be through the original supplier, or directly from the component manufacturer. The simplicity of re-ordering against the supplier's spares list with his handling charge must be balanced against the possible extra work but cheaper direct order. Spares business can be an important element of a supplier's sales, and purchasers can underestimate the true cost of making the purchases direct from manufacturers.

Service
The regular maintenance of equipment can be carried out either by the purchaser's staff, or possibly by the supplier. Properly executed supplier's maintenance contracts can offer significant economies for a purchaser through reduced staff and spares inventory costs. Furthermore, the supplier's specialized skills should ensure proper attention to the equipment, and may be supported by a warranty.

Apart from regular support of this sort, short-term help may be requested by a purchaser for major equipment overhauls or modifications to reduce down time.

Feedback
Operation of equipment by the purchaser provides the supplier with the most valuable data on its suitability. It is therefore sensible for purchasers and suppliers to maintain a close relationship to secure this valuable information. For the supplier the feedback is vital to eliminate any defects that might emerge in his product. The purchaser will benefit by advance warning of type faults, or by advice on modifications or improvements that may have been incorporated elsewhere. Because of their familiarity with the equipment that

they operate and maintain, purchasers frequently become more knowledgeable in some aspects of the product than the suppliers. Nevertheless a little knowledge can be a dangerous thing, and a purchaser will be well advised to consult the supplier before incorporating modifications. Any modifications should be fully recorded by revising the as built drawings and specifications.

Index